FORD

SHOP MANUAL FO-201

Models ■ Fordson Dexta ■ Fordson Super Dexta
■ 2000 Super Dexta ■ New Performance Super Dexta

Models ■ Fordson Major Diesel (FMD) ■ Fordson Power Major (FPM)
■ Fordson Super Major (FSM)
■ New Performance Super Major (New FSM) ■ 5000 Super Major

Models ■ 6000 ■ Commander 6000

Models ■ 1000 ■ 1600

Models ■ 8000 ■ 8600 ■ 8700 ■ 9000 ■ 9600 ■ 9700
■ TW-10 ■ TW-20 ■ TW-30

I&T
SHOP MANUALS

Information and Instructions

This shop manual contains several sections each covering a specific group of wheel type tractors. The Tab Index on the preceding page can be used to locate the section pertaining to each group of tractors. Each section contains the necessary specifications and the brief but terse procedural data needed by a mechanic when repairing a tractor on which he has had no previous actual experience.

Within each section, the material is arranged in a systematic order beginning with an index which is followed immediately by a Table of Condensed Service Specifications. These specifications include dimensions, fits, clearances and timing instructions. Next in order of arrangement is the procedures paragraphs.

In the procedures paragraphs, the order of presentation starts with the front axle system and steering and proceeding toward the rear axle. The last paragraphs are devoted to the power take-off and power lift systems. Interspersed where needed are additional tabular specifications pertaining to wear limits, torquing, etc.

HOW TO USE THE INDEX

Suppose you want to know the procedure for R&R (remove and reinstall) of the engine camshaft. Your first step is to look in the index under the main heading of ENGINE until you find the entry "Camshaft." Now read to the right where under the column covering the tractor you are repairing, you will find a number which indicates the beginning paragraph pertaining to the camshaft. To locate this wanted paragraph in the manual, turn the pages until the running index appearing on the top outside corner of each page contains the number you are seeking. In this paragraph you will find the information concerning the removal of the camshaft.

More information available at haynes.com
Phone: 805-498-6703

Haynes Group Limited
Haynes North America, Inc.

ISBN-10: 0-87288-367-1
ISBN-13: 978-0-87288-367-3

© Haynes North America, Inc. 1990
With permission from Haynes Group Limited

Clymer is a registered trademark of Haynes North America, Inc.

Cover art by Sean Keenan

Disclaimer

There are risks associated with automotive repairs. The ability to make repairs depends on the individual's skill, experience and proper tools. Individuals should act with due care and acknowledge and assume the risk of performing automotive repairs.

The purpose of this manual is to provide comprehensive, useful and accessible automotive repair information, to help you get the best value from your vehicle. However, this manual is not a substitute for a professional certified technician or mechanic.

This repair manual is produced by a third party and is not associated with an individual vehicle manufacturer. If there is any doubt or discrepancy between this manual and the owner's manual or the factory service manual, please refer to the factory service manual or seek assistance from a professional certified technician or mechanic.

Even though we have prepared this manual with extreme care and every attempt is made to ensure that the information in this manual is correct, neither the publisher nor the author can accept responsibility for loss, damage or injury caused by any errors in, or omissions from, the information given.

FORD

Models ■ Fordson Dexta ■ Fordson Super Dexta
■ 2000 Super Dexta ■ New Performance Super Dexta

Previously contained in I&T Shop Service Manual No. FO-18

SHOP MANUAL

FORD

FORDSON DEXTA, FORDSON SUPER DEXTA
FORD 2000 SUPER DEXTA
NEW PERFORMANCE SUPER DEXTA

Tractor serial number is stamped on left side of clutch housing flange and prefixed by model number. Engine serial number is stamped on left hand side of cylinder block.

INDEX (By Starting Paragraph)

CONDENSED SERVICE DATA

GENERAL

Torque Recommendations . . . . See End of Shop Manual
Engine Make . . . . Perkins
Cylinders. 3
Bore—Inches, Fordson Dexta. 3.5
Bore—Inches, Fordson Super Dexta, Ford 2000 Super Dexta,
   New Performance Super Dexta 3.6
Stroke—Inches 5
Displacement—Cubic Inches, Fordson Dexta 144
Displacement—Cubic Inches, Fordson Super Dexta,
   Ford 2000 Super Dexta, New Performance Super Dexta 152.7
Compression Ratio (144 cu. in.) 16.5:1
   (152.7 cu. in.) 17.4:1
Pistons Removed From: . . . . Above
Main & Rod Bearings Adjustable? . . . . No
Cylinder Sleeves—Type . . . . Dry
Generator & Starter Make . . . . Lucas

TUNE-UP

Firing Order 1-2-3
Valve Tappet Gap—Intake & Exhaust . . . . 0.010 H
Valve Face Angle—Degrees 44
Valve Seat Angle—Degrees 45
Engine Low Idle—RPM. 550
Engine High Idle—RPM (New Performance Super Dexta 2450

Engine High Idle—RPM (All Other Models) 2200
PTO High Idle—RPM . . . . See Paragraph 47 or 48
Battery Terminal Grounded . . . . Positive

SIZES—CAPACITIES—CLEARANCES

Crankshaft Journal Diameter 2.749
Crankpin Diameter 2.249
Camshaft Journals Diameter (Front) 1.87
   (Center) 1.86
   (Rear) 1.84
Piston Pin Diameter 1.25
Valve Stem Diameter, Intake. 0.3115
Valve Stem Diameter, Exhaust 0.3115
Main Bearing Diametral Clearance. 0.0025-0.0045
Rod Bearings Diametral Clearance 0.002-0.0035
Piston Skirt Clearance (144 cu. in. engine) 0.0035-0.0055
Piston Skirt Clearance (152 cu. in. engine) 0.0045-0.0065
Crankshaft End Play 0.002-0.010
Camshaft Bearing Diametral Clearance. 0.004-0.008
Cooling System—Quarts 9
Crankcase—Quarts (with Filter) 8
Transmission—Quarts 14
Differential, Final Drive & Hydraulic Reservoir—Quarts 20.4
Steering Gear Housing . . . . 1 Pint

FRONT SYSTEM AND STEERING

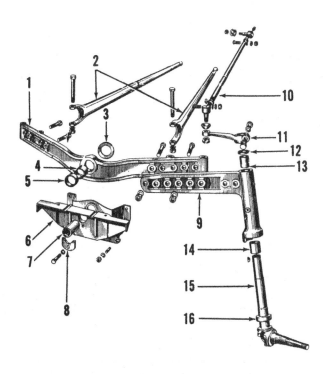

Fig. FO500—Exploded view of front axle and related parts used on the Fordson Dexta. Similarity to American produced Ford tractors will be noted.

1. Axle center member	5. Front spacer	9. Axle extension	13. Upper bushing
2. Radius rod	6. Front support	10. Drag link	14. Lower bushing
3. Rear spacer	7. Pivot pin	11. Steering arm	15. Spindle
4. Pivot pin bushing	8. Pivot pin retainer	12. Dust seal	16. Thrust bearing

SPINDLE BUSHINGS

1. To renew the spindle bushings, support front of tractor and disconnect steering arms from the wheel spindles. Slide spindle and wheel assemblies from axle extensions and remove old bushings using a cape chisel. Install new bushings using a piloted drift of the appropriate size. Internal diameter of new bushings are 1.2495-1.2515 for the upper bushing (13—Fig. FO500) and 1.3425-1.3445 for the lower bushing (14). Diameter of a new spindle (15) is 1.245-1.246 at the upper bearing surface and 1.338-1.339 for the lower.

AXLE CENTER MEMBER AND PIVOT PIN BUSHING

2. To remove the axle center member (1—Fig. FO500), support front of tractor and unbolt radius rods and axle extensions from the axle center member. Remove the axle pivot pin clamping bolt and retainer (8) and remove the pivot pin using a pilot bearing puller and slide hammer. Slide axle center member sideways out of front support (6).

Axle pivot pin bushing (4) can be renewed at this time and should be installed with a piloted drift. New

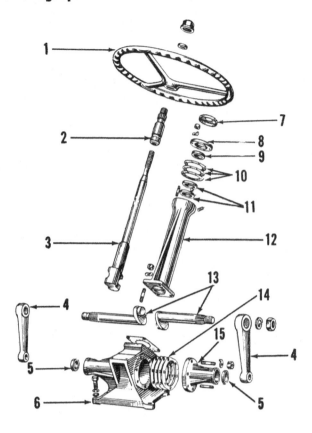

Fig. FO501 — Exploded view of worm and nut type steering gear. Rotary movement of the main nut (2) moves the worm shaft (3) vertically, rotating the two rocker shafts (13) in opposite directions.

1. Steering wheel
2. Steering gear main nut
3. Steering worm shaft
4. Steering arm
5. Oil seal
6. Main housing
7. Dust cap
8. Top cover plate
9. Oil seal
10. Shim pack
11. Ball bearings
12. Steering coulmn
13. Rocker shafts
14. Shim pack
15. Rocker shaft housing

bushing will require no final sizing if not distorted during installation. Make certain, however, that pivot pin has a free fit in the bushing before reinstalling the axle center member.

Tighten the pivot pin retaining cap screw to a torque of 75-85 ft.-lbs. and the front axle extension bolts to a torque of 100-110 ft.-lbs.

FRONT SUPPORT

3. To remove the front support (6—Fig. FO500), remove front axle as outlined in paragraph 2, drain cooling system and remove the hinged hood, grille assembly and radiator. Unbolt and remove the front support from the engine mounting bolts.

DRAG LINKS AND TOE-IN

4. Drag link ends are of the non-adjustable automotive type. The procedure for renewing the drag link ends is evident. Correct toe-in is ¼ to ½-inch. During original factory assembly, toe-in is correctly set and chisel marks are made on the spindle steering arms and axle extensions to mark the setting. In servicing the front end, make sure that each drag link is varied an equal amount to obtain the correct toe-in.

STEERING GEAR

5. The worm and nut type steering gear used on the Dexta tractor is of unique design. An examination of the exploded view shown in Fig. FO501 will assist in understanding the steering gear operation.

The steering gear main nut (2) is secured to the upper end of steering column (12) by the loose ball bearing (11) which controls both end and side thrust. The main nut ball bearing is adjusted by means of the cover plate (8) and adjusting shims (10). Rotation of the steering wheel acts on the worm shaft (3) to raise or lower the shaft in the steering column. The two rocker shafts (13) act directly in machined slides in the lower end of the worm shaft to rotate the steering arms (4) in the proper direction to perform the steering action. End float of the rocker shafts is controlled by shims (14) between the left shaft housing (15) and the main housing (6).

6. **ADJUSTMENT.** Both the rocker shaft end float and the main nut bearing should be adjusted to eliminate slack without applying preload. The steering shaft main nut bearing controls side play as well as end float. To check the main nut bearing adjustment, grasp the rim of the steering

wheel and check for excessive rocking motion. If excessive motion is found, rig a dial indicator to contact the top of the steering wheel nut and measure the end float while moving the main nut back and forth with the steering wheel. To correct the bearing adjustment, first remove steering wheel nut and steering wheel, drive pin from the throttle lever and remove lever. Remove the four screws retaining upper instrument panel and move panel to the side out of the way. If necessary, the warning light bulb holders can be pulled out of their sockets. Bend back the locking tabs on the six cover plate retaining nuts and remove the nuts and cover plate. The cover plate oil seal can be renewed at this time. Remove shims corresponding in thickness to the measured end play and reinstall the cover plate and retaining nuts. Shims are available in thicknesses of 0.002, 0.004 and 0.010. Disconnect the steering drag links at the rear and check steering gear for free rotation before reassembling. Tighten the cover plate nuts to a torque of 10-15 ft.-lbs. and lock in place. Reassemble by reversing the disassembly procedure.

To check the end float of the rocker shafts, rig a dial indicator to bear on the end of the left rocker shaft and check for end float in the shaft by moving it back and forth with the steering arm. If end float is excessive, the steering gear should be removed and overhauled as outlined in paragraphs 7 and 8.

Fig. FO502—To remove the steering worm shaft from main housing, remove lower stud and withdraw shaft through side opening as shown.

7. REMOVE AND REINSTALL. To remove the steering gear assembly, first remove the hood, battery and steering housing lower side plates. Drain approximately ½ gallon of coolant from the radiator and remove the temperature indicator sending unit from the engine block. Remove the steering wheel, drive the pin from the throttle lever and remove lever. Remove the four screws retaining the upper instrument panel plate and remove the plate. Disconnect the fuel, primer and bleed back lines from the fuel tank and unbolt and remove the tank complete with upper instrument panel and temperature gage. Disconnect and remove the vertical throttle rod. Disconnect the drag links from the steering arms and unbolt and remove the steering gear assembly.

To reinstall, reverse the removal procedure. Secure the warning light bulb holders to the upper steering column with a piece of string before installing the fuel tank. Make sure the three rubber mounting pads are in place before fastening the fuel tank in place.

8. OVERHAUL. First remove the unit from the tractor as outlined in paragraph 7, then unstake and remove the six nuts retaining the steering gear top cover and remove the cover

(8—Fig. FO501). Remove the main nut upper race and the fifteen loose balls and unscrew the main nut (2) from the steering worm shaft (3). Invert the steering assembly over a drain pan and drain the oil. Unstake and remove the four nuts retaining the steering column (12) to the main housing and remove the column. The main nut bearing lower race can be removed from the upper end of the steering column at this time by drifting it out from below with a suitable drift.

Unstake and remove the six nuts retaining the left rocker shaft housing (15) to the main housing and remove the housing and left rocker shaft (13). Keep the adjusting shims (14) together in a safe place to avoid damage to the shims. Remove the lower stud from the main housing and remove the steering worm shaft from the side opening as shown in Fig. FO502.

Clean the parts in a suitable solvent and examine. Renew those that are scored, worn or otherwise damaged. Always renew the oil seals when the steering gear is disassembled. The outer and inner rocker shaft bushings are serviced and should be sized after installation if necessary.

Assemble the steering gear by reversing the disassembly procedure. When installing the left rocker shaft and housing, omit the shim pack and tighten the six retaining nuts evenly finger tight. Measure the gap in several places as shown in Fig. FO503 and equalize the gap by adjusting the nuts. When the gap has been equalized, use a feeler gage to determine the thickness of shim pack necessary to remove all end float without binding. Steel shims are available in thicknesses of 0.005 and 0.030 and paper gaskets in thicknesses of 0.002 and 0.010. The 0.010 gasket will compress to approximately 0.007 when nuts are

properly torqued. A paper gasket should be fitted on each side of the shim pack. Tighten the six retaining nuts to a torque of 55-65 ft.-lbs. and stake in place.

Reinstall the steering column and tighten the retaining nuts to a torque of 55-65 ft.-lbs.; then, install the steering nut, the fifteen bearing balls and upper race. Remove the shim pack (10—Fig. FO501) from the cover plate and install the plate and retaining nuts finger tight. Determine the thickness of the shim pack as outlined for the rocker shaft adjustment, remove the cover and install a shim pack of the proper thickness to just eliminate end float without causing any binding tendency. Tighten the six retaining nuts to a torque of 10-15 ft.-lbs. and stake in place. Center the steering worm shaft between the stops by counting the revolutions of the main nut and fill the steering gear with 90W lubricant as follows:

Remove the bleed plug in the upper end of the steering column and pour the lubricant down the center of the main nut until the upper end of the worm shaft is just covered. Reinstall the bleed plug and reinstall the steering gear on the tractor.

There are no master splines for the location of the steering arms on the rocker shafts. When correctly installed, both arms will incline to the rear approximately 7 degrees with the worm shaft half way between the stops. To aid in installing the arms, one chisel mark has been made on the outer end of the rocker shafts and two chisel marks are located on the outer face of the large boss on the steering arm. The chisel mark on the shaft should index with one of the marks on the steering arm, depending on which side of the tractor it is installed. If the chisel marks are in alignment and both arms incline to the rear an equal amount, the steering arms are properly installed.

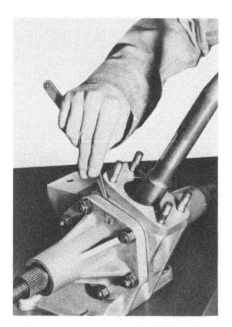

Fig. FO503—Rocker shaft and main nut bearing are adjusted to zero clearance by means of shims. Method of determining thickness of shim pack is shown, see text for details.

ENGINE AND COMPONENTS

The Fordson Dexta tractor is equipped with a three-cylinder diesel engine having a bore of 3.5 inches, a stroke of 5 inches and a piston displacement of 144 cubic inches.

The Fordson Super Dexta, Ford 2000 Super Dexta and New Performance Super Dexta are equipped with a three-cylinder diesel engine having a bore of 3.6 inches, a stroke of 5.0 inches and a piston displacement of 152 cubic inches.

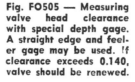

Fig. FO504 — Cylinder head nuts should be tightened to a torque 55-60 Ft.-Lbs. in the sequence shown. Cylinder head bolts on left side of engine extend to retain injectors; a deep socket is therefore required.

shown in Fig. FO504 to a torque of 55-60 ft.-lbs. Make certain that the bores and seats for the injectors are clean and free from dirt and carbon. Install the injectors using a new copper sealing washer and tighten the injector holding nuts evenly to avoid cocking the injector. After restarting the engine, check the injectors for compression blow-by. Adjust the valve tappet clearance to 0.010 for each valve after the engine has been brought up to operating temperature.

Both the 144 and 152 cubic inch engines are similarly constructed. Differences in the 144 cubic inch engine due to production changes and between the 144 and 152 cubic inch engines are noted in the text where service procedures and/or specifications are affected.

R&R ENGINE ASSEMBLY

9. To remove the engine and clutch assembly first drain the cooling system, remove hood and if the engine is to be disassembled, drain oil pan. Disconnect the radiator hoses, headlight wire and the grille braces from top water outlet. Disconnect the radius rods and steering drag links at rear, support the tractor under transmission housing and unbolt the front support assembly from engine. Roll the front axle, front support and radiator as an assembly away from tractor and block same in an upright position.

Shut off the fuel and disconnect the fuel primer and bleed back lines. Remove the battery, battery support and fire wall. Disconnect wires from the oil pressure warning light sender, generator, starter, cold starting unit and lights. Disconnect the tachometer cable, heat indicator sending unit, starter cable and starter control rod. Remove the air cleaner, throttle link or governor control rod, swing the engine from the two engine mounting hooks and unbolt engine from the transmission case.

CYLINDER HEAD

10. To remove the cylinder head, first drain cooling system and remove the hood. Loosen the injector lines at the pump and disconnect lines from injectors. Remove the bleed-back line. Immediately cap all exposed fuel line connections to protect the system from dirt and remove injectors. Re-

move the rocker arm cover and rocker arms assembly. Disconnect battery cables and remove battery. Remove the heat indicator sending unit and disconnect the water outlet casting from cylinder head. Disconnect the fire wall from the two brackets located on cylinder head, then disconnect the air cleaner mounting bolts and upper hose. Unbolt and remove the right fire wall bracket and rear engine lifting plate from the cylinder head. Disconnect the external oil feed line from the right rear corner of the cylinder head. Loosen the two bolts securing the fire wall to the clutch housing and rock the fire wall back enough to remove the clip attaching the governor vacuum line to the cylinder head. Disconnect the primer fuel line, governor vacuum line, heater unit electrical lead and throttle link from the intake manifold. Unbolt and remove the cylinder head.

The head gasket is marked "Top—Front." Coat the head gasket on both sides with an approved sealing compound before installing. Tighten the cylinder head nuts in the sequence

VALVES AND SEATS

11. Exhaust and intake valves seat directly in the cylinder head. Valve heads and seat area of the cylinder head are numbered consecutively 1 to 6 from front to rear of engine. Valve seats for both the intake and exhaust valves should be ground to an angle of 45 degrees. A seat width of $\frac{1}{16}$-$\frac{3}{32}$-inch should be maintained, using appropriate narrowing stones. Valves should be ground to a face angle of 44 degrees. No more material than is necessary should be removed when regrinding valves and seats. After reinstalling the valves in their appropriate seats, the maximum clearance between the top of the valve head and the face of the cylinder head should be measured using a straight edge and feeler gage or special depth gage, as shown in Fig. FO505. If the clearance for any valve exceeds 0.140, the valve must be renewed. If clearance is less than 0.059, reface valve so that clearance is 0.059-0.087. Valve stem diameter is 0.311-0.312.

Fig. FO505 — Measuring valve head clearance with special depth gage. A straight edge and feeler gage may be used. If clearance exceeds 0.140, valve should be renewed.

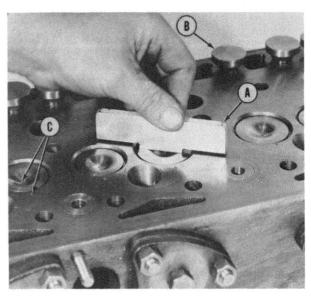

VALVE GUIDES AND SPRINGS

12. Intake and exhaust valve guides are interchangeable and should be pressed from the cylinder head if renewal is required. New valve guides have an inside diameter of 0.314-0.3155 and a stem to guide clearance of 0.002-0.0045. Renew the guide and/or valve if the clearance is excessive. Valve guides on early Fordson Dexta models were pressed into cylinder head until machined shoulder on guides contacted top surface of head. Later production guides have no shoulder and must be installed so that they protrude 0.584-0.594 from top surface of cylinder head. It is recommended that a collar which is 0.584-0.594 wide be fitted around the guide and used as a stop when pressing new guides into cylinder head. Install guides with counterbored end down. New valve spring seat washers with smaller I.D. must be used with new type guides.

Springs, retainers and locks are interchangeable for the intake and exhaust valves. The valves are fitted with an inner and outer spring which may be installed on the valve with either end up. The inner spring has a free length of 1.365-1.405 and should test 21-25 lbs. when compressed to a length of 0.838. The outer spring has a free length of 1.783-1.803 and should test 48-52 lbs. at 1.151 inches compressed length. Valve springs should be renewed if they are discolored, distorted or fail to meet the test specifications given.

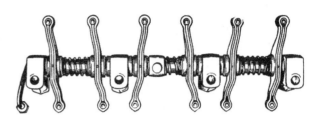

Fig. FO507 — Top view of rocker shaft assembly, showing proper positioning of rocker arms.

Fig. FO508 — Exploded view of rocker shaft assembly showing proper location of the various parts.

VALVE TAPPETS

13. The mushroom type valve tappets operate directly in machined bores in the cylinder head and are retained in the head by the adjusting screw lock nuts. The diameter of the tappet stem is 0.62225-0.62375, with a recommended clearance of 0.00075-0.0035 in the cylinder head bore. To remove the tappets, first remove the cylinder head; then, remove the adjusting screw and locknut and slide the tappets from the cylinder head bores.

Tappet gap should be set to 0.010 for both the intake and exhaust valves. The gap should be set with the engine at operating temperature. To adjust the tappet gap, open the timing cover on the left side of the flywheel housing and turn the engine until the TDC mark on the flywheel is aligned with the mark on the housing. If No. 1 cylinder is on the compression stroke (both valves closed), adjust the gap on valve Nos. 1, 2, 3 and 5 to the recommended 0.010 clearance. If both valves on No. 1 cylinder are partially open, adjust valve Nos. 4 and 6. Turn the engine one full revolution until the TDC mark is again aligned and adjust the remainder of the valves.

ROCKER ARMS AND SHAFT

14. To remove the rocker arms assembly, raise hood and remove the cover. Disconnect oil feed line from the cylinder head at rear of rocker arms shaft and unscrew the four retainer nuts evenly.

To disassemble, remove the snap ring at rear of shaft and withdraw parts from the shaft. The rocker arms are right hand and left hand assemblies and should be installed as shown in Fig. FO507. The rocker shaft has a diameter of 0.62225-0.62375 with a recommended diametral clearance of 0.00075-0.0035 between the rocker arm and shaft. If the clearance is excessive, renew the rocker arm or shaft as required. When reassembling the rocker shaft, refer to Fig. FO508 for the correct location of the parts.

Note: If sufficient lubricant does not reach the rocker shaft, check oil lines shown in Fig. FO512 for obstruction.

TIMING GEAR COVER AND GEARS

15. **GEAR COVER.** To remove the timing gear cover, first drain radiator, remove hood and disconnect the headlight wires. Disconnect the grille braces from water outlet on cylinder head and disconnect the radiator hoses. Disconnect radius rods and drag links at rear end, support tractor under flywheel housing, unbolt front support from engine and roll the assembly away from tractor.

Unbolt top water outlet from cylinder head, and water pump from timing gear cover; then, remove both units from the tractor as an assembly.

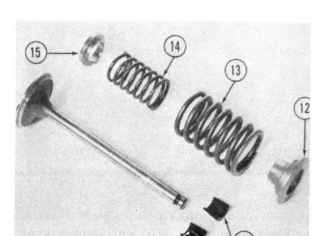

Fig. FO506 — Exploded view of valves & springs.

11. Spring keepers
12. Spring retainer
13. Outer spring
14. Inner spring
15. Spring seat

Remove the starting jaw and crankshaft pulley, then unbolt and remove the timing gear cover. The crankshaft front oil seal can be renewed at this time and should be installed with a suitable driver so that lip faces rear of engine.

16. **TIMING GEARS.** Before removing any gears in the timing gear train, first remove rocker arm cover and rocker arms assembly to avoid the possibility of damage to the pistons or valve train if either the camshaft or crankshaft should be turned independently of the other.

The timing gear train consists of the crankshaft gear, camshaft gear, pump drive gear and an idler gear connecting the other three gears of the train.

Timing gear backlash should be 0.003-0.006 between the idler gear and any of the other timing gears. Replacement gears are available in standard size only. If backlash is not within recommended limits, renew the idler gear, idler gear shaft and/or any other gears concerned.

Unstake and remove the idler gear retaining bolt and slip the gear off the idler shaft. The shaft has a light press fit in the engine block and is further positioned by the pin shown in Fig. FO509. Pry shaft from its place in the block if renewal is indicated.

The crankshaft gear is keyed in place and fits the shaft with 0.001 press fit to 0.001 clearance. If the old gear is a loose fit, it may be possible

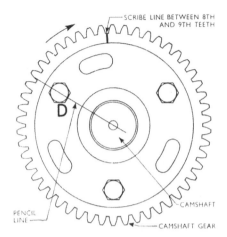

Fig. FO510—Suggested method for marking camshaft timing gear if replacement gear is unmarked. See text.

to pry it off the shaft with a heavy screwdriver or pry bar. If a puller is needed, it will first be necessary to remove the oil pan and lower timing gear housing.

The camshaft gear and pump drive gear are identical and replacement gears may not be marked When renewing camshaft or pump drive gear, proceed as outlined in the appropriate following paragraph.

17. CAMSHAFT TIMING GEAR. The gear is attached to the camshaft by cap screws through the three round holes located in the gear face. Hole spacing is equal and the gear can be attached to the camshaft in any one of three positions; however, only one of these positions will per-

mit correct timing of the engine. One of the round attaching holes is marked with a stamped letter "D" on the front face of the gear as shown in Fig. FO510. On later production tractors, the camshaft is marked with a stamped letter "D" on the front face of the hub center, slightly to one side. Install the gear on the camshaft so that the letter "D" on the camshaft is nearest the attaching hole marked "D" on the camshaft gear. Using a straight edge, draw a line from the center of the camshaft, through the center of the cap screw in the marked hole, to the outer rim of the gear as shown in Fig. FO510. Count 8 gear teeth clockwise from this marked line and scribe a timing mark between the eighth and ninth teeth as shown.

On older production tractors, where the stamped letter "D" is on **gear flange camshaft, install the camshaft** in its journals and rotate the shaft while applying pressure to the third tappet from the front of the engine. When the third tappet is raised to the top of its stroke, the uppermost hole in the camshaft flange should be marked to align with the stamped letter "D" on the camshaft gear. From this point, proceed as directed above for the timing mark location.

18. INJECTION PUMP DRIVE GEAR. The gear is attached to the adapter by means of cap screws through the three slotted holes in the face of the gear. The holes are not equally spaced and the gear can be

Fig. FO509—Idler gear shaft is a light press fit in block and is retained in place by idler gear bolt. Note timing gear oil feed hole in block and corresponding hole in shaft. Locating pin keeps oil feed holes aligned.

Fig. FO511—View of engine timing gears with marks properly aligned showing (A) camshaft and gear locating marks and (B) timing marks.

attached to the adapter in only one position. Before installing a new un-marked injection pump drive gear, first install and time the remainder of the gears as outlined in paragraph 19, and remove the timing window cover plate from the timing gear housing flange. Slip a socket wrench with a short extension through the hub of the gear and fit the socket over the nut securing the drive gear adapter to the injection pump cam-shaft. Rotate the injection pump clockwise until the timing line marked "S" on the gear adapter flange of pumps with vacuum governor; or timing line marked TC on gear adapter flange of pumps with mechan-ical governor, aligns with the timing mark on the housing. Slight pressure will be required on the wrench han-dle to keep the two marks in line. Install the gear on the adapter so that the three attaching cap screws can be installed approximately midway in the three slots. Install the retaining washer and three cap screws in the gear adapter and tighten securely. After the gear has been installed and tightened, affix a timing mark on the gear tooth which aligns with the mark on the idler gear.

19. TIMING THE GEARS. Due to the odd size of the idler gear, the tim-ing marks will all be in alignment only once in 18 crankshaft revolutions. To time the engine, unstake and re-move the idler gear attaching bolt and remove the idler gear. The rocker arm cover and rocker arms assembly should previously have been removed.

Rotate the crankshaft until the key-way and timing mark are in a vertical position, rotate the camshaft until the timing mark aligns as nearly as pos-sible with the center of the idler gear shaft, rotate the injection pump until the timing mark aligns with the cen-ter of the idler gear shaft. Pressure will need to be applied to hold the injection pump timing mark in align-ment. Install the idler gear so that the three timing marks on the gear align with the timing marks on the crankshaft, camshaft and injection pump drive gears. The engine will then be in proper time. Reinstall the idler gear retaining bolt and washer and stake in place. Fig. FO511 shows the engine with the timing marks in proper alignment.

CAMSHAFT

20. To remove the camshaft, first remove the timing gear cover as out-lined in paragraph 15, remove the rocker arm cover and rocker arms assembly, lift the tappets and with-draw the camshaft and gear.

Camshaft end thrust is controlled by means of a leaf type spring at-tached to the timing gear cover. Be-cause of the spur type gears used in the timing gear train, camshaft end play does not present a problem.

The camshaft journals ride in three machined bores in the engine block. The center journal is pressure lubri-cated by means of an external oil line. Oil for lubrication of the rocker arms and valve train is metered by the cen-ter camshaft journal and fed to the rocker shaft by means of the second external oil line. Fig. FO512 shows the

location of these two external lines. The front and rear camshaft bearing journals are gravity lubricated by the return oil from the rocker arms as-sembly.

Camshaft journal diametral clear-ance is 0.004-0.008 for all three jour-nals. Journal diameter is as follows:

Front journal diameter....1.869-1.870
Center journal diameter...1.859-1.860
Rear journal diameter.....1.839-1.840

CONNECTING ROD AND PISTON UNITS

21. Connecting rod and piston units are removed from above after the cylinder head and oil pan have been removed. To remove the oil pan it is first necessary to remove the front end assembly as outlined in paragraph 15. Connecting rod and bearing caps are numbered to correspond to their respective bores. When installing the rod and piston units, make certain that the correlation numbers are in register and face away from the cam-shaft side of engine.

The connecting rod caps are secured with self locking nuts which should be renewed and tightened to a torque of 50-55 ft.-lbs. with cadmium plated nuts or 65-70 ft.-lbs. with unplated nuts.

PISTONS, RINGS AND SLEEVES

22. Three compression rings and two oil control rings are fitted to each piston. Fig. FO513 shows the arrange-ment of the rings on the piston. The top compression ring is chrome plated.

Fig. FO512 — Camshaft and valve train are lu-bricated by means of two external oil lines shown. Lower line feeds cam-shaft center bearing. Oil metered by camshaft center bearing is fed to rocker shaft through up-per line. Front and rear camshaft journals are gravity lubricated by re-turn oil from rocker shaft.

Fig. FO513—Engine piston, showing ar-rangement of the three solid and two seg-mented piston rings.

The third compression ring and the top oil control ring are of laminated steel construction. End gap of these rings should be as follows:

Third compression ring...0.008-0.010
Top oil control ring......0.018-0.037

Installation of the first and second compression rings and the lower oil control ring is conventional. These rings should have 0.002-0.004 side clearance in the ring grooves and end gaps as follows:

First compression ring.....0.010-0.015
Second compression ring...0.009-0.013
Lower oil control ring.....0.009-0.013

The solid rings are non-directional and may be installed either side up.

The third compression ring consists of four spring steel segments which should be installed as illustrated in Fig. FO514. When the ring segments are installed, the end gaps should be positioned 180 degrees apart, and gaps should be above each end of piston pin.

To install the segmented oil control ring, first place the expander in the back of the groove and spiral in two of the segments. Next fit the center spring, then the other two segments. The last segment installed will require a slight pressure to overcome the action of the center spring. The gaps of the four segments should be positioned at equal distances around the piston with none of the gaps in line with the piston pin. The high silicon aluminum alloy piston is available in standard and 0.030 oversize. for 144 cubic inch engine and in standard size only for the 152 cubic inch engine. New piston should be fit with 0.0035-0.0055 clearance in the sleeve in 144 cubic inch engine and 0.0045-0.0065 clearance in 152 cubic inch engine when measured at the piston skirt.

The dry type cast iron cylinder sleeve is furnished as an unfinished sleeve only and must be bored to size after installation. The sleeve should have 0.002-0.004 press fit in the cylinder block in 144 cubic inch engines and 0.002-0.005 press fit in 152 cubic inch engines and should be pressed in until the top of the liner is flush with the top surface of the engine block.

Finish bore the cylinder sleeve to 3.501-3.502 for a standard piston or 3.531-3.532 for the oversize in 144 cubic inch engine and 3.600-3.601 for the standard piston in 152 cubic inch engine.

PISTON PINS

23. The floating type piston pin has an outside diameter of 1.24975-1.2500 and is retained in the piston by snap rings. Piston pin should have a clear-

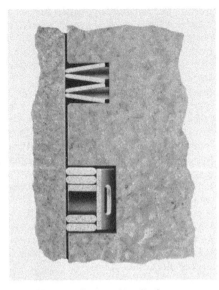

Fig. FO514 — Proper installation arrangement of segmented third compression and upper oil control rings. See text for details.

ance of 0.00025-0.0005 in the piston bosses and a clearance of 0.0005-0.00175 in the connecting rod bushing. These clearances will give a light thumb press fit in the piston and slightly looser fit in the connecting rod bushing.

The connecting rod bushings are renewable and should be final sized after installation.

CONNECTING RODS AND BEARINGS

24. Connecting rod bearings are of the non-adjustable, precision type, renewable from below after the oil pan and connecting rod bearing caps have been removed. When installing new

bearing shells, make certain that the projections engage the milled slot in the connecting rod and bearing cap and that the correlation numbers on the rod and cap face away from camshaft side of engine. Bearing inserts are available in 0.010, 0.020 and 0.030 undersize as well as standard.

Check the crankshaft rod journals and the rod liners against the values which follow:

Crankpin journal
diameter 2.2485-2.249
Bearing diametral
clearance 0.002-0.0035
Rod side play on
crankshaft 0.0095-0.0133
Connecting rod center to center length is 8.999-9.001.

When reassembling, install new self-locking connecting rod nuts and tighten to a torque of 50-55 ft.-lbs. if nuts are cadmium plated or 65-70 ft.-lbs. if nuts are unplated.

CRANKSHAFT AND MAIN BEARINGS

25. The crankshaft is supported in four main bearings of the non-adjustable, precision type. Bearing inserts are available in 0.010, 0.020 and 0.030 undersizes as well as standard. Normal crankshaft end play of 0.00225-0.01025 is controlled by renewable flange type thrust washers located in the rear main bearing cap. Washers are available in standard and 0.0075 oversize.

NOTE: Several different types of main bearing inserts have been used and it is important that the type number (not Ford part number) stamped into the back of inserts be the same on both the top and bottom

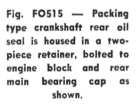

Fig. FO515 — Packing type crankshaft rear oil seal is housed in a two-piece retainer, bolted to engine block and rear main bearing cap as shown.

insert of any one bearing journal. It is recommended that the main bearing inserts be all of one type number. Also, the locking tab on the bottom (bearing cap) thrust washer has been changed in production at engine Serial No. 1449364 and the correct thrust washer must be used.

To remove the front main bearing cap, it is first necessary to remove the timing gear cover, lower timing gear housing and oil pump. To remove the rear main bearing cap, it is first necessary to remove the engine as outlined in paragraph 9 and remove the clutch, flywheel, adapter plate and rear oil seal retainer.

Check the crankshaft and main bearing inserts against the values which follow:

Crankpin diameter 2.2485-2.249
Main journal diameter. . .2.7485-2.749
Main bearing diametral
  clearance 0.0025-0.0045
Main bearing bolt
  torque . . . . .90-95 ft.-lbs.

CRANKSHAFT OIL SEALS

26. The crankshaft front oil seal is mounted in the timing gear cover and may be renewed after the front cover is removed. Removal procedure for the timing gear cover is outlined in paragraph 15. Press the new seal into the cover with the lip to the inside.

The crankshaft rear oil seal can be renewed after detaching engine from clutch housing as outlined in paragraph 56 and removing the clutch, flywheel and engine adapter plate. The rear oil seal retainer is secured to the rear of the engine block and rear main bearing cap by six cap screws and the two halves of the retainer are held together by two long bolts and self-locking nuts as shown

Fig. FO516—After installation of flywheel, runout should be checked as shown. Runout should not exceed 0.004.

in Fig. FO515. The packing type seal halves should be soaked in clean engine oil for a period of one hour before installing in the retainer halves. Press the packings firmly in their retainers, leaving 0.010-0.020 of the packing ends protruding above the face of the retainer. Using new gaskets, fit the two halves of the retainer around the crankshaft and install and tighten the long bolts and self-locking nuts holding the retainer together; then, install and tighten the six cap screws securing the oil seal assembly to engine block and main bearing cap.

FLYWHEEL

NOTE: The same flywheel part number is listed for all models. However, after introduction of the Simms "Minimec" fuel injection pump with a mechanical governor, two "SPILL" timing marks were placed on the flywheel. One mark at 26° BTDC is for use with fuel injection pumps having a vacuum governor and the second mark at 20° BTDC is for use with the mechanical governor type fuel injection pump. If a service flywheel having only one "SPILL" timing mark at 26° BTDC is being installed on a tractor equipped with a mechanical governor type fuel injection pump, a second "SPILL" timing mark at 20° BTDC must be located and stamped into the flywheel so that the pump may be timed. Refer to Fig. FO516A and proceed as follows: Scribe a mark on the flywheel in relation to established "TDC" and "SPILL" marks as shown in Fig. FO516A. Permanently mark flywheel along scribed mark with a sharp cold chisel and, using metal letter dies, stamp "SPILL" and "20° BTDC" on flywheel. Also stamp "26° BTDC" on flywheel at original "SPILL" mark.

27. The flywheel can be removed after detaching the engine from the transmission case and removing the clutch assembly.

Caution: The flywheel is secured to the crankshaft flange by six cap screws which are safety wired. The flywheel or crankshaft flange are not fitted with locating dowels. Use care after removing the retaining screws to prevent the flywheel from falling from its place on the crankshaft and inflicting possible personal injury.

The starter ring gear is secured to the front of the flywheel by six cap screws and lock washers. Install a new ring gear with the rounded ends of the gear teeth to the front of the engine.

When reinstalling the flywheel, carefully clean the mating surfaces of the flywheel and crankshaft flange.

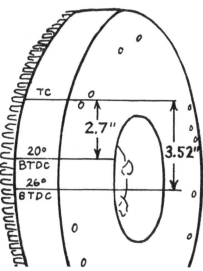

Fig. FO516A—If installing a flywheel having only the 26° BTDC timing mark in engine equipped with mechanical governor type fuel injection pump, locate and stamp the 20° BTDC timing mark on flywheel according to dimensions shown above. Measurements are made around circumference of flywheel.

Tighten the retaining cap screws to a torque of 75 ft.-lbs., affix a dial indicator to the adapter plate and check the flywheel runout as shown in Fig. FO516. Runout should not exceed 0.004 when measured 3¾ inches from the outside edge of the flywheel. A greater runout would indicate the presence of dirt or foreign matter between the flywheel and crankshaft flange, a sprung crankshaft or an improperly ground flywheel face.

NOTE: Prior to Fordson Dexta tractor Serial No. 957E-33407, models equipped with double clutch have spacer washers located between the clutch adapter plate and engine flywheel at each adapter plate retaining bolt. Spacer washers are not required with double clutch adapter plate used on later production tractors.

OIL PUMP AND RELIEF VALVE

28. The oil pump assembly is mounted to the front face of the front main bearing cap and secured to the cap by three screws. The pump is driven from the crankshaft timing gear through an idler gear mounted on the pump flange.

To remove the oil pump, first remove the timing gear cover as outlined in paragraph 15 then unbolt and remove the oil pan. Remove the pump suction and pressure lines and unbolt and remove the lower section of the timing gear housing. Remove the snap ring retaining the oil pump idler gear and lift off the gear. Remove the three cap screws retaining the oil pump to the main bearing cap and remove the pump from its doweled position on the cap.

Disassemble the pump, clean the parts in a suitable solvent and examine for wear or broken parts. Reassemble the rotors in the pump body. CAUTION: When pressing drive gear on rotor shaft, support shaft, **do not** apply pressure against pump rotor. Clearance should not exceed 0.006. Check the clearance between the rotor and pump body as shown in Fig. FO519. Clearance should not exceed 0.010. Check the clearance between the top of the rotors and the surface of the pump body as shown in Fig. FO520. Clearance should not exceed 0.003. The drive and driven rotors are serviced as a matched assembly. Renew the front cover plate if the machined surface is worn or scored. The oil pump idler gear is fitted with a non-renewable bushing; idler gear should be renewed if shaft clearance is excessive. Late production pump bodies are fitted with a sealing ring at front cover plate.

The opening pressure of the relief valve is pre-set at 60 psi and the normal operating pressure of the lubrication system is 40 psi. Relief valve is

Fig. FO518 — **Check clearance between rotors with a feeler gage as shown. Clearance should not exceed 0.006.**

Fig. FO519 — **Check clearance between outer rotor and pump body as shown. Clearance should not exceed 0.010.**

not adjustable. On later tractors the relief valve ball was replaced by a solid plunger type valve which was subsequently replaced by a hollow plunger type valve and longer spring. This latest type valve and spring is used to service all previous types. With the later plunger type valves, spring is retained by circular spring seat and cotter pin. The pressure switch connected to the lubrication warning light is designed to break connection when the engine oil pressure reaches 7-9 psi. The warning light should not come on at any time when the engine is running.

Fig. FO520 — **Check clearance between top of rotors and body gasket surface as shown. Clearance should not exceed 0.003.**

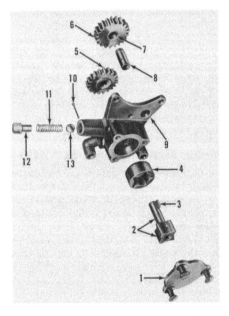

Fig. FO517—**Exploded view of engine oil pump showing component parts.**

1. Cover	8. Idler shaft
2. Inner rotor	9. Pump body
4. Outer rotor	10. Cotter pin
5. Drive gear	11. Spring
6. Idler gear	12. Plunger
7. Bushing	13. Relief valve

DIESEL FUEL SYSTEM

The diesel fuel system consists of three basic components: the fuel filters, injection pump and injection nozzles. Refer to Fig. FO521A. When servicing any unit associated with the fuel system, the maintenance of absolute cleanliness is of utmost importance. Of equal importance is the avoidance of nicks or burrs on any of the working parts.

Probably the most important precaution that service personnel can impart to owners

of diesel powered tractors, is to urge them to use an approved fuel that is absolutely clean and free of foreign material. Extra precaution should be taken to make certain that no water enters the fuel storage tanks. This last precaution is based on the fact that all diesel fuels contain some sulphur. When water is mixed with sulphur, sulphuric acid is formed and the acid will quickly erode the closely fitting parts of the injection pump and nozzles.

Fig. FO521 — Lubricating oil must be maintained at proper level and be changed at regular intervals on mechanical governor type fuel injection pumps. Drain, fill and oil level plug locations are shown. Refer to paragraph 29.

ROUTINE SERVICING

29. Due to the extremely delicate machining operations made necessary by the high injection pressures in a diesel engine, expense and delay associated with emergency repairs to the system increases the importance of routine preventive maintenance to operators of diesel equipment. If the tractor is supplied with clean fuel at all times, the injection pump should not require servicing between the periods of major engine overhaul.

In addition to regular maintenance of the fuel system and air cleaner normally performed by the operator, the manufacturer recommends that after every 200 hours of operation (more often under dirty conditions), the air filter (See Fig. FO-534) in the governor of vacuum governor type fuel injection pumps be removed, cleaned and coated with a light coating of engine oil.

Every 200 operating hours, on fuel injection pumps equipped with a mechanical governor, drain oil from the pump cambox and refill to the oil level plug with new SAE 30 motor oil in temperatures above 90° F., SAE 20 in temperatures between 20° F. and 90° F. or SAE 10 in temperatures below 20° F. Capacity is approximately ½-pint. Refer to Fig. FO521 for location of drain, oil level and filler plugs.

Every 600 operating hours, the fuel filter cartridge should be renewed, the injector assemblies removed, cleaned and adjusted, and the engine air cleaner removed from the tractor and completely cleaned.

Because of the fact that a diesel engine inducts a full charge of air at all times, regardless of the throttle setting or load imposed, the work load on the air cleaner is increased and greater attention in maintenance is required. A partially plugged air cleaner will immediately result in a loss of engine power and speed, increased fuel consumption and excessive smoking. This is especially true of the Fordson Dexta because the injection pump governor is operated from manifold vacuum. For this reason, the air cleaner screen, oil cup and center tube should be the first parts examined in complaints of loss of power or engine speed.

TROUBLE SHOOTING

30. By following a logical sequence, trouble shooting on a diesel engine is more simple than for a gasoline or LP-Gas engine. The following trouble shooting paragraphs follow a logical pattern for locating the cause of the more common troubles encountered. The method of eliminating the cause, in most cases, is obvious.

31. **ENGINE WILL NOT START.** The most common causes are lack of fuel, insufficient compression pressure or improper timing. The starting motor is designed to turn the engine over at 200 rpm when the engine is warm; a cranking speed substantially lower, especially in cool weather, might contribute to difficult starting. If the recommended cranking speed is not approached, check for improperly charged battery, dragging starter motor or too heavy engine oil.

If starter action is lively and even, loosen the bleed plug at the top front of the injection pump body and activate the hand primer lever on the fuel lift pump. Fuel should flow freely from the injection pump bleed screw. If it does not, bleed the system

Fig. FO521A—Cutaway view of the Simms multiple plunger injection pump and injectors used on the Fordson Dexta. Injection pump used on Fordson Super Dexta and Ford 2000 Super Dexta is similar, but is equipped with a mechanical governor mounted on the pump camshaft.

as outlined in paragraph 36. If proper fuel flow cannot be obtained by bleeding the system, check for plugged lines or filter or improperly operating lift pump. If adequate fuel flow was obtained from the injection pump bleed screw, tighten the screw and loosen the pressure lines at the three injector assemblies. Open the throttle for full fuel delivery and check the flow from the injector connections while turning the engine over with the starter. If fuel flow is adequate, the trouble is with the injectors, timing or compression. Check the timing as outlined in paragraph 45 and the injectors as outlined in paragraph 38. Check the compression while the injectors are out.

If no fuel flow was obtained through the loosened injector lines, but adequate flow was available at the pump, the trouble is in the injection pump. Remove the injection pump governor housing and diaphragm as outlined in paragraph 47A and check to see that the governor link and control rod moves freely back and forth in the injection pump body. If no binding is found, or if binding exists which cannot be corrected without further disassembly, the injection pump will need to be removed and returned to a qualified diesel service shop for servicing.

32. **LACK OF POWER.** First check the air cleaner thoroughly for plugging or restriction; then, check in the following order: full fuel delivery by loosening the injection pump bleed screw and operating the hand primer lever, injection pump timing (paragraph 45), injector condition (paragraph 38). Check the engine compression while the injectors are out. If all

of the above checks have not eliminated the trouble, remove the injection pump and return it to a qualified diesel service station.

33. **ROUGH OR UNEVEN OPERATION.** Start the engine and operate at idle speed. Loosen each injection line connection in turn at the injector, and note the engine operation after loosening each connection. The one least affecting the engine operation, indicates a missing or partially missing cylinder. Remove the injector and check as outlined in paragraph 39, or install a spare injector. Check the engine compression while the injector is out. Uneven operation can also be caused by a sticking governor, or worn plunger or delivery valve in the injection pump.

34. **ENGINE KNOCKS.** To determine whether the knock is of fuel system or mechanical origin, speed the engine up to top idle speed and pull the stop control. If the knock immediately stops, it is due to the fuel system; if still audible, the cause of the knock is mechanical. Fuel knock is caused by improperly operating injectors, incorrect engine timing or a poor grade of diesel fuel.

FILTERS AND BLEEDING

35. To renew the fuel filter element, unscrew the filter body retaining bolt at the bottom of the fuel filter and remove the body, bolt and element. Discard the element, clean interior of filter body with a brush and clean fuel. Install the body with a new element and sealing ring in place and tighten

the retaining bolt to a torque of 10 ft.-lbs. Bleed the fuel system as outlined in the following paragraph.

36. **BLEEDING.** To bleed the system, first make sure there is an adequate amount of fuel in the tank and that the fuel tank shut-off is turned on. Remove the two bleed screws (1 and 2—Fig. FO522) from top of filter body and operate the hand priming lever until air-free fuel flows from the bleed screw holes. Install and tighten first the inlet (1), then the outlet (2) bleed screw while continuing to actuate the hand priming lever. Loosen the bleed screw (3) on the fuel injection pump approximately three turns and operate the lift pump as before. Tighten the bleed screw when air-free fuel flows from the pump.

If the tractor does not start readily, loosen the injector lines at the injectors and turn the engine over with the starter until fuel flows from the connections. Retighten the connections and start engine.

INJECTOR ASSEMBLIES

37. **REMOVE AND REINSTALL.** To remove one or all of the injector assemblies, raise the hinged hood and remove the leak-off pipe from the three injectors and the rear connection. Disconnect the pressure lines at the injectors and loosen the connections on the injector pump.

Caution: On vacuum governor type pumps, because of the lighter torque on the injection pump delivery valve holders it is good practice to hold the delivery valves from turning in the pump body while loosening the lower ends of the lines.

Immediately cap all broken connections to avoid dirt entry into the fuel system and unbolt and remove the injector assemblies.

Fig. FO522 — Fuel system showing the location of the three bleed screws in the order of tightening. (1) Filter inlet bleed screws. (2) Filter outlet bleed screw. (3) Injection pump bleed screw. Vacuum governor pump is shown; refer to Fig. FO534A for location of bleed screw on mechanical governor type pump.

Fig. FO523—To completely test an injector nozzle requires the use of a tester as shown.

Make sure that the injector and the seating surface in the cylinder head are absolutely clean and always use new injector seat washers when reinstalling the injectors. Tighten the holding nuts down evenly to avoid cocking the injector assembly. Before releasing the tractor for service, check to make sure that the injectors are making a gas-tight seal in the cylinder head by running the engine for a short time. Stop any combustion leaks by slightly loosening one holding nut and tightening the other while the engine is running.

38. TESTING AND LOCATING A FAULTY NOZZLE. If the engine does not run properly and a faulty injector nozzle is indicated, such a nozzle can be located by loosening each pressure connection in turn at the injector. As in checking a spark plug in a spark ignition engine, the faulty unit is the one which, when its line is loosened, least affects the running of the engine.

WARNING: Fuel leaves the injection nozzle tips with sufficient force to penetrate the skin. When testing, keep your person clear of the nozzle spray.

The spray pattern of the suspected injector can be observed after removal, by reconnecting the injector to its pressure line with nozzle directed away from the tractor, loosening the other two lines so that the engine will not start, and observing the spray pattern while turning the engine over with the starter.

39. NOZZLE TESTER. A complete job of testing and adjusting the injector requires the use of a special tester such as shown in Fig. FO523. The nozzle should be tested for opening pressure, spray pattern, seat leakage and leak back.

WARNING: Fuel leaves the injection nozzle tips with sufficient force to penetrate the skin. When testing, keep your person clear of the nozzle spray.

To test a nozzle, operate the tester until oil flows from the connector pipe and attach the injector assembly to the connector. Close the tester valve and operate the tester handle a few quick strokes. If undue pressure is required to operate the lever, the nozzle valve is plugged and the injector should be disassembled and serviced as outlined in paragraph 44.

40. OPENING PRESSURE. After making sure that the nozzle valve is not plugged, open the tester valve slightly and observe the gage pressure at which the spray occurs. This pressure should be 2200-2350 psi. If the pressure is not as indicated and the spray pattern and leakage tests prove satisfactory, reset the opening pressure by loosening the cap nut, inserting a small screw driver through the leak-off pipe hole, and turning the adjusting nut until the correct pressure is obtained.

Note: After continued usage, injector valve spring will wear in at approximately the lower value of 2200 psi in an injector in perfect condition. When a new injector or spring is installed in a tractor with two used injectors, use the higher setting for the new injector and the lower figure for the used, serviceable injectors. After the new injector has been in use for a short period of time the pressures will be approximately equal. If three serviceable used injectors are installed set to an equal pressure within the range given.

If the correct opening pressure cannot be obtained, the injector must be overhauled as outlined in paragraph 44.

41. SPRAY PATTERN. Two spray holes are located in the nozzle tip. Both spray patterns should be symmetrical, well atomized and spread to a width of about two inches before breaking up into a very fine mist. If spray pattern is streaky, distorted, or not finely atomized the injector must be overhauled as outlined in paragraph 44.

42. SEAT LEAKAGE. Wipe the injector dry with clean blotting paper and bring the injector pressure up to 2200 psi. Hold the pressure at this level for one minute and apply a clean piece of blotting paper to the nozzle tip. The fuel oil stain should not exceed ½-inch in diameter. If the above conditions are not met, overhaul the injector as outlined in paragraph 44.

43. NOZZLE LEAK BACK. Operate the tester handle and bring the gage pressure back to 1850 psi. Note the length of time required for the pressure to drop from 1850 psi to 1350 psi. This time should be between six and 45 seconds.

If the elapsed time is not as specified, dirt, wear or scoring is indicated between the nozzle body and needle valve, and the nozzle should be disassembled and overhauled as indicated in paragraph 44.

Note: Leakage in the injector tester at the connections, gage or check valve will show

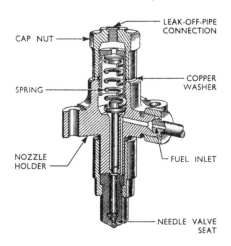

Fig. FO525—Cross sectional view of Simms injector used in all models. Two spray holes are located in the nozzle tip. Holes are 0.32 mm. diameter on New Performance Super Dexta and 0.35 mm. diameter on all other models.

up as a fast leak-back in this test. If all of the injectors tested show consistently faulty when this test is made, a faulty tester should be suspected.

44. OVERHAUL. Unless a suitably clean area and proper injector cleaning equipment is available, do not attempt to overhaul diesel nozzles. The tool set should include injector holding fixture, valve seat scraper, pressure chamber reamer and reverse flush adapter for Simms injectors as well as 0.012 spray hole cleaning wire for both 0.32 and 0.35 mm. spray holes.

Fig. FO525 shows a cross section of the Simms injector used in all models. Refer to this illustration if necessary, for assembly and disassembly sequence. The 0.32 spray hole nozzles for New Performance Super Dexta can be identified by number NH-389 stamped on nozzle.

Secure injector holding fixture in a vise and mount injector in the fixture. Never clamp the injector body in the vise. Remove the cap nut and back off the adjusting nut; then, lift off the spring seat, spring and spindle. Remove the nozzle retaining nut and withdraw the nozzle and valve. Nozzle bodies and valves are a lapped fit and must never be interchanged. Place all parts in clean diesel fuel or calibrating fluid as they are disassembled. It is good shop practice to keep the parts from each injector separate in a compartmented pan.

Immerse those parts of the injector assembly containing hard carbon deposits in a suitable solvent for a short period. When the carbon has softened remove with a brass wire brush. Ace-

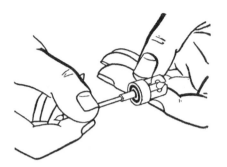

Fig. FO526—Clean the pressure chamber of the nozzle with special tool as shown.

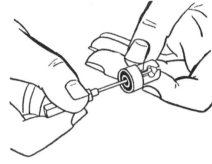

Fig. FO529—Pressure chamber scraper is used to clean lower pressure chamber and annular groove on top of nozzle.

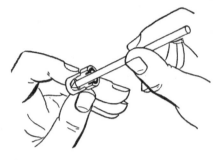

Fig. FO530—Use polishing stick and tallow to polish nozzle seat.

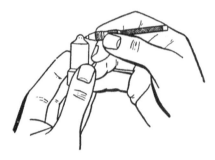

Fig. FO527—Cleaning orifices with 0.012 wire probe held in pin vise. Refer to text.

tone, or a commercial carbon solvent designed for this purpose should be used. **It must be remembered that acetone is a highly inflammable liquid and proper precautions must be taken.** Follow the manufacturers directions and do not leave the parts in the solvent long enough to corrode the metal surfaces. Rinse the parts in clean diesel fuel or light oil immediately after cleaning and place in a clean container of diesel fuel. Clean the pressure chamber of the nozzle body with the reamer as shown in Fig. FO526. Clean the spray holes of the nozzle body with the 0.012 probe held in a pin vise as shown in Fig. FO527. To prevent breakage, the wire probe should protrude from the pin vise only far enough to pass through the spray holes. Rotate the pin vise without applying undue pressure.

Clean the valve seats by inserting the small end of the valve seat scraper into the nozzle and rotating while applying light pressure. Reverse the scraper and clean the upper chamfer with the large end. Clean the annular groove in the top of the nozzle and the lower pressure chamber with the pressure chamber scraper as shown in Fig. FO529.

Thoroughly rinse all parts in clean diesel fuel to remove any dislodged carbon particles and polish the valve body seat with a wood polishing stick and a small amount of tallow.

If scratches are visible on the plunger piston, or if the plunger tends to stick in the nozzle body, the assembly can be remated by using a very small amount of special lapping compound mixed with tallow. Clamp the small upper end of the needle valve in a slow-speed electric drill and polish the large piston area of the valve with a small amount of the lapping compound on a piece of felt. An alternate method is to clamp the tip end of the nozzle body in the drill, apply a small amount of the lapping compound to the needle valve and insert in the nozzle body. Be sure to hold the needle valve up off the valve seat during the lapping operation to avoid damage to the seating surfaces.

Examine the mating surfaces of the nozzle body and injector body to make sure that they are clean and free from nicks or scratches which would prevent a perfect metal to metal seal.

Clean the nozzle body and needle valve thoroughly in clean fuel, insert the valve in the nozzle body, install the nozzle assembly in the back flush attachment and back flush thoroughly with the tester pump. Without removing the needle valve from the nozzle body, remove the assembly from the back flush attachment and position the nozzle assembly carefully on the injector body. Install the nozzle retaining nut and tighten to a torque of 60-75 ft.-lbs.

Note: Place injectors in holding fixture to tighten nut.

Install the spindle, spring, spring seat and spring adjusting nut. Tighten the adjusting nut until pressure from the spring is felt. Connect the injector to the nozzle tester and adjust the opening pressure to 2350 psi, use a new copper gasket and install the cap nut.

Retest the injector as outlined in paragraphs 40 through 43, and if injector fails to pass these tests renew the injector nozzle assembly.

Note: If injectors are to be stored, it is recommended that they be flushed with calibrating oil or other preservative oil similar in viscosity to diesel fuel—prior to storage. Diesel fuel in small amounts, such as that present in an injector assembly, tends to break down in storage. The resulting varnish deposits may cause the units to stick, making it necessary to reclean the injectors prior to use.

INJECTION PUMP

NOTE: Fordson Dexta tractors produced prior to April, 1962, were equipped with a Simms fuel injection pump, Part No. 957E-993101-A, which incorporated a vacuum (pneumatic) governor. Later production Fordson Dexta tractors, Fordson Super Dexta and Ford 2000 Super Dexta tractors are equipped with a Simms "Minimec" fuel injection pump, Part No. 957E-993101-D, which incorporates a mechanical governor. CAUTION: The "Minimec" fuel injection pump cannot be interchanged between a Fordson Dexta tractor (144 cubic inch engine) and a Fordson Super Dexta or Ford 2000 Super Dexta tractor (152 cubic inch engine) without first having the pump recalibrated for proper fuel delivery at an authorized service station. The New Performance Super Dexta is equipped with a Simms "Minemec" fuel injection pump, part No. 960E-993101.

The fuel injection pump is a self-contained unit which includes the engine speed governor, and pump components capable of delivering fuel in properly metered amounts. The pump is horizontally mounted on left side of engine and is driven by the timing gear train. The Simms injection pump proper is a sealed unit and service other than timing to the engine or limited governor service as outlined in paragraph 47A should not be attempted. Factory approved service on Simms injection pumps in this country is exclusively conducted by the Ford Tractor Operations, Ford Motor Co., and their dealers.

45. TIMING. Open the timing window on left side of the flywheel housing and remove the injection pump timing plate from the timing gear housing as shown in Fig. FO531. Turn crankshaft in the normal direction of rotation (clockwise) until number one piston is coming up on compression stroke and continue turning until the proper "SPILL" mark on the flywheel rim is in alignment with the notch on the flywheel housing. At this time, the lower of the two marks on the timing gear adapter, when viewed through the timing gear housing window, should be in alignment with the timing pointer cast on the injection pump mounting flange.

NOTE: "SPILL" mark on flywheel is at 26° BTDC on engines equipped with vacuum governor type fuel injection pump. Flywheels produced after introduction of "Minimec" mechanical governor type fuel injection pump have two "SPILL" timing marks; one mark is at 26° BTDC and a second mark is located at 20° BTDC. If a flywheel with two "SPILL" timing marks is encountered when timing a vacuum governor type fuel injection pump, be sure to use mark located at 26° BTDC on flywheel. When timing "Minimec" mechanical governor type fuel injection pump, be sure to use "SPILL" timing mark located at 20° BTDC. If a flywheel with only the 26° BTDC "SPILL' timing mark

has been installed in a tractor that is equipped with a mechanical governor type fuel injection pump, the 20° timing mark must be located and stamped into the flywheel as outlined in note preceding paragraph 27 so that the pump may be accurately timed.

If the pump timing marks are not aligned, remove the inspection plate on the timing gear cover immediately in front of the pump drive gear and loosen the three cap screws securing the injection pump drive gear to the adapter. Fit a socket with a short extension over the nut attaching the adapter to the pump camshaft and turn the pump and adapter, within the drive gear, until the timing marks are in proper alignment. While applying slight pressure on the socket wrench to hold the marks in alignment tighten the three cap screws securing the drive gear to the adapter. Reinstall the inspection cover and the two timing plates.

46. INJECTION PUMP R&R. To remove the complete injection pump, first thoroughly clean all dirt from the pump, injector lines and supply lines. Remove the pump timing window and front inspection plate and loosen the timing plate on the flywheel housing. Turn the engine crankshaft clockwise until the proper "SPILL" mark (see

note in paragraph 45) is aligned with the timing mark on the flywheel housing and the injection pump timing marks are in alignment.

Disconnect the proofmeter drive shaft, governor vacuum line on tractors with vacuum governor or throttle link on tractors with mechanical governor, fuel stop cable and fuel stop housing bracket from the pump. Shut off the fuel at the tank and disconnect the fuel supply lines from the primary pump and injection pump. While holding the delivery valve holders from turning, disconnect the injector lines from the pump and injectors. Immediately cap all open fuel line connections to avoid dirt entry into the system. Unbolt and remove the injection pump from the engine timing gear housing.

To reinstall the injection pump make sure the flywheel timing marks are properly aligned and install the pump so that lower timing mark on the adapter is in alignment with the mounting flange timing mark to the nearest tooth. If the alignment is exact, no further adjustment is necessary. If the alignment is not exact, retime the pump as outlined in paragraph 45. Reinstall the lines, shut-off stop, governor vacuum line or throttle link and proof meter cable and bleed the system as outlined in paragraph 36.

47. VACUUM GOVERNOR ADJUST. Before attempting to adjust the governed speed, first thoroughly inspect and clean the air cleaner and fill to the proper level with the correct grade of engine oil. Because the governor is operated by manifold vacuum, the condition of the air cleaner is of utmost importance in governor operation. After the air cleaner has been inspected and cleaned, reinstall the primary cleaner stack and check and retighten the air cleaner hose connections. To adjust the governor, start the tractor and bring engine to operating temperature. Position hand throttle lever for low idle speed which should be 550 rpm. If low idle speed is not as specified, loosen lock nut on idle adjusting screw (1—Fig. FO533) and reposition the stop screw until the correct speed is obtained.

Move hand throttle to wide open position and check high idle speed. If the high idle speed is not the recommended 2200 rpm, readjust the maximum speed stop screw (2) until the proper speed is obtained.

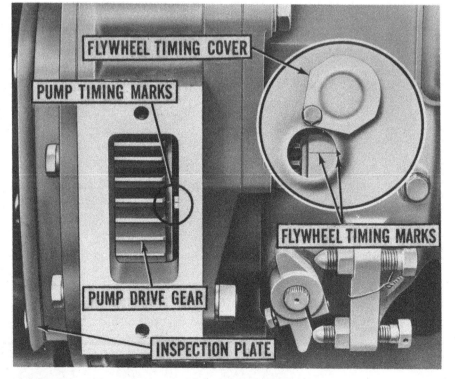

Fig. FO531 — View showing fuel injection pump timing marks on mechanical governor type pump. Inset shows flywheel timing window open and flywheel timing mark aligned with pointer. Timing marks on vacuum governor type fuel injection pumps are similar. Refer to paragraph 45 for timing procedures.

Engine speeds can be noted on the tractor proof meter or checked with a tachometer at the PTO shaft. On Fordson Dexta tractors prior to Serial No. 957E-63953, PTO RPM is 760 at 2200 engine RPM and 190 at 550 engine RPM. After this Serial Number and on prior units with new ratio PTO gears service installed, PTO RPM is 660 at 2200 engine RPM and 165 at 550 engine RPM.

47A. VACUUM GOVERNOR OPERATION AND SERVICING. The pneumatic governor used on the Fordson Dexta consists of a throttle unit in the intake manifold connected by a suction pipe to the spring loaded diaphragm mounted on the rear of the injection pump. Fig. FO534 shows a schematic view of governor assembly.

When the engine is stopped, the governor spring pushes the diaghragm inward, moving the control rod to the position of maximum fuel delivery. When the engine is started and the manifold throttle unit moved to the closed position, a high engine vacuum is created at the manifold end of the suction line which, acting through the diaphragm, moves the control rod in the direction of minimum fuel delivery. The balance of governor spring pressure and manifold vacuum tends to maintain a constant engine speed over the full governed speed range regardless of variations in engine load.

If there is reason to suspect the governor diaphragm of leakage, it can be tested as follows:

Disconnect the governor vacuum line from inner side of governor cover, and the stop cable from governor stop arm. While observing the usual standards of cleanliness, unbolt and remove the inspection cover from outside of injection pump body. Pull the stop lever

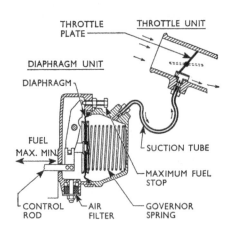

Fig. FO534 — Schematic view of vacuum governor and throttle mechanism showing component parts.

rearward to move the pump control rod to the shut-off position and place one finger over the vacuum connection in the governor cover. Release the stop lever and observe the pump control rod for movement towards the open position. If undue movement is noted, it will indicate a leak in the governor diaphragm which will interfere with proper governor operation.

To renew the diaphragm, unbolt and remove the governor cover and governor spring. Pull the diaphragm assembly outward from the governor body and remove the cotter pin securing the diaphragm to the control rod. When installing a new diaphragm, make sure that the slight protrusion on the diaphragm rim is located in the small recess in the governor body. Reinstall the spring and cover and test as before to make sure the new diaphragm is properly seated in the governor body.

NOTE: Diaphragm is made of leather which can become porous when dry and result in air leak through diaphragm. A diaphragm not otherwise defective can be reclaimed by coating the leather with a light grease such as vaseline and rubbing the grease into the diaphragm. A new diaphragm should also be treated in this manner before installation as it is likely that the leather may have dried out while in storage. Caution must be taken not to subject the diaphragm to solvent or oil that would shrink the leather material as this would limit the stroke of the governor and result in improper governor operation.

The governor spring should exert 3.81-4.06 lbs. when compressed to a length of 1.969 inches.

Reinstall the pump inspection cover and reconnect the stop cable and vacuum line. Caution: Under no circumstances should an attempt be made to move the control forks on the control rod while the inspection cover is off. Each pump plunger is calibrated individually and proper location of the forks is impossible without the use of a calibrating stand.

Normally, no adjustment of the throttle linkage is necessary. In case of damage to the linkage or renewal of parts, remove the battery and adjust the length of the rear throttle link until the throttle arm on the intake manifold moves fully to contact both stop screws.

48. "MINIMEC" MECHANICAL GOVERNOR, ADJUST. Refer to Fig. FO534A for location of speed adjustment screws. With hand throttle in slow idle position, adjust idle (lower) screw to obtain 550 engine RPM. Then, with hand throttle in wide open

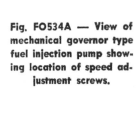

Fig. FO534A — View of mechanical governor type fuel injection pump showing location of speed adjustment screws.

Fig. FO533—Intake manifold showing (1) idle speed stop screw, (2) high speed stop screw and (3) governor vacuum line connection.

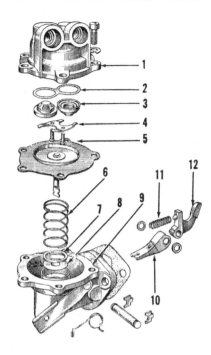

Fig. FO535—Exploded view of primary fuel pump. Pump is mounted on injection pump and driven by injection pump camshaft.

1. Upper body	7. Oil seal washer
2. Gasket	8. Oil seal
3. Fuel pump valve	9. Lower body
4. Retainer	10. Diaphragm link
5. Diaphragm	11. Return spring
6. Spring	12. Activating arm

screw from inlet side of fuel filter and operate the hand priming lever on the fuel pump. There should be a well defined surge of fuel through the filter bleed hole at each working stroke of the pump. Tap a pressure gage into the pump outlet line, bleed the system and check the outlet pressure by operating the engine through the normal engine speed range. Pressure reading should be 6-10 psi throughout the entire speed range.

Refer to Fig. FO535 if disassembly

of the primary fuel pump is indicated. The outlet and inlet valves in the upper body are identical but reversed in their position in the body. The valves should be installed as shown. If renewal of the diaphragm is indicated, push down on the diaphragm and rotate 90 degrees to disconnect pull rod from the operating link. Procedure for further disassembly is evident. When reinstalling diaphragm in lower body, rotate diaphragm in the proper direction so that flange holes will align with holes in the diaphragm.

position, adjust high speed (upper) screw to obtain 2200 engine RPM on Super Dexta models or 2425-2450 RPM on New Performance Super Dexta (tractor serial No. 09C-913383 and later). After proper speed adjustment has been obtained, be sure to tighten lock nuts on adjustment screws.

Engine speed can be observed on proof meter or checked at PTO shaft with tachometer. Power take-off RPM is 660 at 2200 engine RPM, 735 at 2450 engine RPM and 165 at 550 engine RPM. (Multiply power take-off RPM by 3.33 to obtain engine speed.)

As the mechanical governor is an integral part of the fuel injection pump, it is not recommended that governor repairs be attempted outside of an authorized injector pump repair station.

PRIMARY FUEL PUMP

49. The primary fuel pump is mounted on the outside of the injection pump and driven from a cam on the injection pump camshaft. The fuel pump is of the diaphragm type, the component parts of which are available for service.

To test the fuel pump, remove bleed

COOLING SYSTEM

The radiator is fitted with a pressure type cap which raises the boiling point of the coolant. The by-pass type thermostat starts to open at 156-165 degrees F. and is fully open at 185 degrees F.

RADIATOR

50. To remove the radiator, first remove hood and grille. Drain cooling system and disconnect radiator hoses; then, unbolt and lift radiator assembly from front support.

WATER PUMP

51. To remove the water pump, first remove the radiator as outlined in paragraph 50. Remove the fan blades and cap screws securing the top water outlet housing to the cylinder head. Loosen the by-pass hose at water pump and lift off the top water outlet and by-pass hose. Remove the nuts retaining water pump to timing gear front cover and lift off the pump.

To disassemble the pump, remove pulley from shaft using a split type bearing puller, then press shaft and bearing, seal and impeller from the housing as an assembly.

When reassembling, press the shaft and bearing unit in the housing from the front until front end of bearing race is flush with edge of housing. Use a 1½-inch length of 1-inch ID pipe,

so that the pressure is applied to the outer race of the bearing. Reverse the housing and bearing assembly and re-fit the slinger on rear of shaft using a ⅛-inch deep socket as a driver. Reinstall the shaft seal in housing and press the impeller on the shaft using a press and suitable adapter until 0.005-0.020 clearance is obtained between the front edge of impeller blades and the housing. This clearance can be measured with a feeler gage, working through the pump outlet neck. Support rear end of the shaft on a suitable adapter and press the pulley on the shaft until pulley hub is flush with front end of shaft. Check the water pump for binding and correct as necessary before reinstalling on engine.

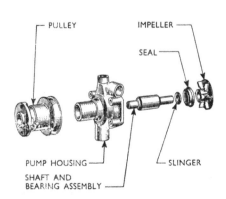

Fig. FO536—Exploded view of water pump used on Fordson Dexta tractor.

ELECTRICAL SYSTEM

GENERATOR AND REGULATOR

52. A Lucas two-pole shunt wound generator and Lucas regulator is used in the Fordson Dexta electrical system. British terminology is used on the generator and regulator markings A comparison of the symbols with those currently in use in this country is as follows: The five terminal posts on the regulator are marked E, D, F, A and A₁. The terminal E (earth) corresponds to G (ground); D (dynamo) to A (armature); F stands for field, as in American terminology; A corresponds to the "hot" wire B, on most American wiring diagrams. A₁ is a special current regulating terminal. The current from A, passes through the light switch giving a higher generator output when the tractor lights are used.

Service specifications of the generator and regulator are as follows:

Generator:
Renew brushes if shorter than . . 0.35
Field Draw
Volts 13.5
Amperes 2.2
Output (Cold)
Maximum amperes 14.0
Volts 16.0
RPM 1500

Regulator:
Cut-Out Relay
Air gap 0.020
Point gap 0.020
Closing range 12.7-13.3
Voltage Regulator
Air gap 0.015
Voltage range 15.6-16.0
Ground polarity . . . . Positive

STARTING MOTOR

53. The 12-volt starting motor is equipped with a manually engaged starter drive incorporating an overrunning clutch. The drive mechanism must be adjusted so that the relay switch, mounted on starter housing, is activated at the time the starter pinion is almost fully engaged in the flywheel ring gear. Refer to Fig. FO537 and adjust as follows: Disconnect the starter leads and remove starter from tractor. Attach wires connected to a battery and light bulb to the two connections on the relay assembly so that bulb will

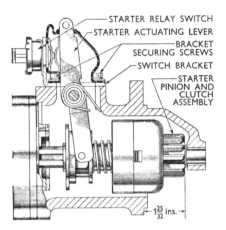

STARTER RELAY SWITCH
STARTER ACTUATING LEVER
BRACKET SECURING SCREWS
SWITCH BRACKET
STARTER PINION AND CLUTCH ASSEMBLY

1 25/32 ins.

Fig. FO537 — Adjust starter actuating mechanism so that front face of pinion is advanced to 1 25/32 inches from starter mounting flange when electrical contact is made. Refer to text.

light when the relay switch is closed. Depress the starter actuating lever until rear face of the starter pinion is 1 17/32 inches from the mounting flange as shown. At this point, the test light bulb should just flash on. Readjust as necessary by loosening the four screws securing the switch bracket to starter housing and shifting the switch and bracket as required.

Service specifications for the starter motor are as follows:

Lock Test
Amperes 450
Torque, ft.-lbs. 28
Normal cranking speed
(engine), rpm 200

CLUTCH

Tractors may be equipped with either a single clutch, for models without live power take-off; or a dual clutch, for models with live power take-off. On models equipped with live power take-off, a single clutch pedal controls both tractor operation and power take-off operation. Depressing the clutch pedal through the first stage disengages the transmission clutch and stops the forward motion of the tractor. Depressing the pedal through the complete range of its travel disengages both the transmission and power take-off clutches; stopping the transmission of all power from the flywheel rearward.

CLUTCH PEDAL ADJUSTMENT
Single Clutch Models

54. Clutch pedal free play is adjusted by means of the adjusting clevis at front end of clutch pedal link. The adjustment is correct when 1¼ inches free play is obtained when measured at center of pedal as shown in Fig. FO538.

Double Clutch Models

55. On models equipped with the double clutch, pedal free play is adjusted by means of the stop screw (A —Fig. FO539). The adjustment is correct when 1¼ inches free play of the clutch pedal is obtained. Clutch pedal height on tractors equipped with the double clutch is adjusted by means of

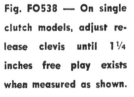

Fig. FO538 — On single clutch models, adjust release clevis until 1¼ inches free play exists when measured as shown.

the adjustable clevis at the front end of the pedal link. To make this adjustment, proceed as follows:

Attach the clutch release arm in the rear hole of the clevis on pedal link, engage the power take-off and adjust pedal height so that the clutch pedal contacts the step plate the moment the power take-off clutch is completely released.

Caution: If pedal linkage is too short, release bearing will bottom in the clutch assembly before the pedal contacts the step plate, resulting in possible damage to the clutch release bearing and/or clutch fingers.

After adjustment has been made as indicated in the preceding paragraph, move the clutch release arm to the forward hole in the clevis and check to be sure that the transmission clutch is fully released when clutch pedal is in contact with the step plate.

If both adjustments cannot be obtained, it will be necessary to readjust or overhaul the double clutch assembly as outlined in paragraph 58.

TRACTOR SPLIT

56. To obtain access to the clutch and flywheel, it will first be necessary to detach (split) engine from clutch housing as follows:

Unbolt and remove the hinged hood and disconnect and remove the tractor battery. Disconnect starter cable from starter and detach wiring harness from lights, generator, starter relay and cold starting unit. Disconnect throttle linkage and primer line from the intake manifold, disconnect upper air cleaner hose clamp and remove the bolts connecting air cleaner and firewall to bracket on engine block. Disconnect the steering drag links and radius rods at rear end. Disconnect the fuel and bleed back lines, proof meter cable, heat indicator sending unit and wire from the oil pressure sending unit. Remove the battery support and fire wall, support both halves of tractor independently, then unbolt and separate the units.

R&R AND OVERHAUL CLUTCH

Single Plate Models

57. Clutch disc and pressure plate are available only as assemblies. Service of the clutch will be confined to renewal of the units as indicated. The adjusting screws in the clutch release fingers are set and staked at the factory and adjustment in the field is not recommended.

Check the clutch release and pilot bearings while the clutch is out and renew as required. The clutch release shaft rotates in renewable bushings in the clutch housing and may be serviced at this time. During reassembly, pack the recess in inside of the clutch release bearing hub with a good grade of high melting point grease. The release bearing is of the pre-lubricated type.

Double Clutch Models

NOTE: Use of a special fixture is required when reassembling the double clutch unit so that both clutch disc hubs will be correctly aligned with each other and with flywheel pilot. Also, fixture is necessary for proper adjustment of clutch fingers.

58. The dual clutch on tractor equipped with live pto is attached to the engine flywheel by means of an adapter plate which is doweled in place and secured with six cap screws. Before removal of the clutch, it is good shop practice to mark the relative position of the clutch on the flywheel so that it can be installed in the same position. Because of the heavier weight of the double clutch assembly, proper correlation of the component parts is more important to maintain proper balance. For most service operations, it is not necessary to remove the adapter plate from the flywheel. Remove the six cap screws attaching the clutch center drive plate to the adapter plate and remove the clutch assembly.

Before disassembling the clutch, punch-mark the center drive plate and the two pressure plates so that the unit can be reassembled in the same relative position. To disassemble the clutch unit, use a valve spring compressor as shown in Fig. FO540, compress the springs and remove the keepers, spring seats and springs. Remove the cap screws retaining cover to center drive plate and the three pins retaining the struts to the transmission pressure plate.

Separate and examine the assemblies. The two clutch discs can be renewed at this time. Renew any of the other parts which show signs of wear or damage. The spring retainers should be discarded and new ones installed during reassembly. Renew any spring that does not test 98-108 lbs. when compressed to a length of 1.670 inches.

NOTE: Although the Dexta series double clutch unit is similar to those produced in the U. S., the regular Nuday NDA-7502 clutch fixture used for U. S. produced Ford tractor clutch units must be modified slightly for use with the English Dexta series clutch units. Also, special Dexta spacers and gage block, obtainable from the Nuday Company, must be used. To adapt the Nuday NDA-7502 fixture for assembly and adjustment of Dexta clutches, proceed as follows: Cut down the transmission clutch disc pilot on the fixture spindle to a diameter of 0.862-0.864. Remove material from the pto disc pilot as shown in Fig. FO542A. Obtain new gage block, NDA-7502-5A, and three new spacers, NDA-7502-8A, from the Nuday Company.

The balance of individual clutch parts are checked at the factory and a spot of yellow paint is placed on the edge of the clutch center plate and pressure plates to indicate the

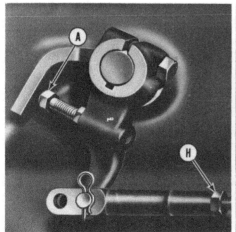

Fig. FO539—Adjust double clutch stop screw (A) until 1¼ inches pedal free play exists. Pedal height is adjusted by means of linkage adjustment (H), see text for details.

heavy point of each assembly. When reassembling the clutch, the paint marks should be placed as near to 120 degrees apart as possible to provide a balanced unit. If the paint marks are no longer visible, assemble the unit by aligning the correlation marks that were placed on the clutch prior to disassembly.

To reassemble the double clutch unit, proceed as follows: Place the transmission clutch pressure plate on the fixture and install the transmission clutch friction disc over the pilot diameter of the center spindle and with long hub of clutch disc down. Position the center driving plate on the fixture with yellow paint mark as near as possible to 120 degrees away from paint mark on the transmission pressure plate. Turn the assembly on the fixture until the pins on the fixture will enter the untapped holes in the center driving plate. Place the pto friction disc (with long hub up) on the fixture and insert the pto disc pilot with tapered end down in hub of disc as shown in Fig. FO542. Then install the pto pressure plate and clutch cover assembly with the three connecting links passing through the holes in the center driving plate and with the yellow paint mark as near as possible to 120 degrees away from the paint marks on the other two plates. NOTE: If paint marks are no longer visible, reassemble clutch by aligning correlation marks made prior to disassembly of clutch unit. Reinstall the cap screws securing cover to center drive plate. NOTE: If any of these cap screws are renewed, be sure that threaded end of cap screw does not protrude from front face of center drive plate.

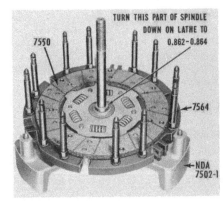

Fig. FO541 — Use the fixture tool to properly align and adjust the double clutch during reassembly. Refer to text.

Using a valve spring compressor, install the springs, spring retainers and new spring keepers. Close the open end of keepers. Pin and key the transmission pressure plate connecting links.

Install adjusting tool (NDA 7502-5A) over the fixture center spindle with large end down as shown in Fig. FO543. Install the flat washer and nut and tighten nut until the three adjusting spacers can be positioned against the lined discs between the pressure plates as shown. Make sure that spacers are positioned 120 degrees apart and installed with blocks marked "TOP" toward the clutch cover. Make sure also, that top portion of spacer is in completely against the pto clutch disc and between the **machined** portion of the pto pressure plate and the center drive plate. Invert tool No. NDA 7502-5A and reinstall it as shown in Fig. FO544. The highest machined surface of the tool is used

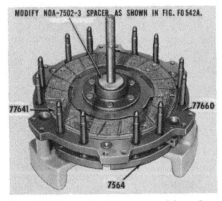

Fig. FO542 — Position center drive plate and pto clutch disc over fixture as shown. Be sure to align correlation marks or properly space balance marks as outlined in text.

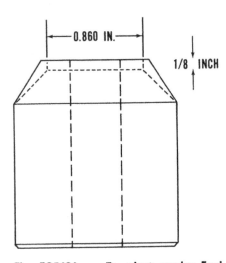

Fig. FO542A — To adapt regular Ford clutch pilot for use in assembling and adjusting Dexta double clutch units, remove material from pilot to dimensions shown by dotted lines. Do not remove material from end of pilot; length must remain the same.

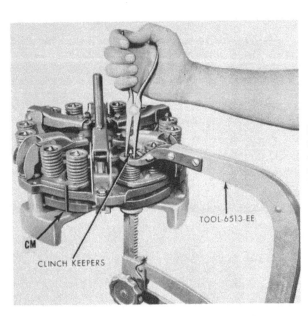

Fig. FO540 — Use valve spring compressor to disassemble and assemble double clutch as shown. (CM) Correlation marks.

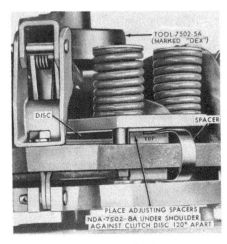

Fig. FO543 — Using adjusting spacers to adjust clutch fingers. Make certain the three adjusting spacers are top side up and are in contact with both clutch discs.

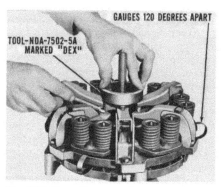

Fig. FO544 — Use gage block as shown to adjust each finger until 0.005 clearance exists between screw head and block.

NOTE: Be sure that pto disc pilot butts against shoulder of spindle and gage block is resting in pto disc pilot when making adjustments. Spindle must be tight in fixture.

After adjustment is completed and checked, invert the adjusting tool, install the flat washer and nut and tighten the nut until the three adjusting spacers can be removed. Lift the ad-

justed clutch assembly from the fixture and reinstall on tractor flywheel.

Before reconnecting the tractor, check the clutch pilot and release bearings and renew as required. Pack the recess in the center of the clutch release bearing hub with a good grade of high melting point grease before reassembly.

to check and adjust the transmission release finger screws and the lowest machined surface to check and adjust the pto release finger screws.

While holding the adjusting tool firmly against the aligning spacer, adjust each finger in turn, until there is a clearance of 0.005 between head of adjusting screw and machined surface of adjusting tool.

TRANSMISSION

NOTE: At tractor Serial No. 957E-63953, changes were made in the gear ratios of the Fordson Dexta transmission. Some of the old ratio gears are no longer available. If necessary to substitute a new ratio gear in a transmission prior to the above serial number, the mating gear or gears must also be renewed. The Ford Tractor Parts Catalog

contains a chart which will identify the different gears and indicate which other gears are required when a substitution is made.

It would be advisable to check any gears being renewed in a transmission to be sure that the new and old gears are alike or that a mating gear set is being installed.

Fig. FO545—Cross-sectional view of transmission assembly with live pto.

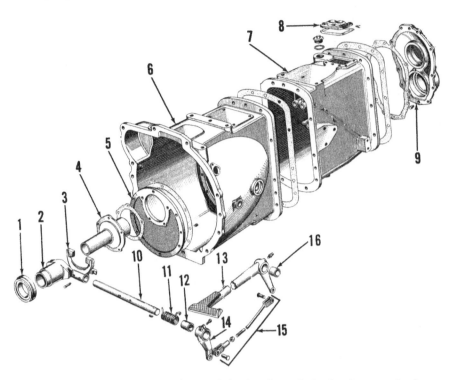

Fig. FO545A—Exploded view of transmission housing and clutch release mechanism.

1. Clutch bearing	6. Transmission	11. Clutch return
2. Clutch release	front housing	spring
hub	7. Transmission	12. Bushing
3. Clutch release	housing	13. Clutch pedal
fork	8. Shifter housing	14. Clutch release
4. Input shaft	9. Rear cover plate	arm
retainer	10. Clutch release	15. Clutch release
5. Adapter plate	shaft	rod
		16. Bushing

INPUT SHAFT SEAL RENEWAL

59. The transmission input (clutch) shaft seal can be renewed after first detaching (splitting) the engine from the cluch housing as outlined in paragraph 56. To renew the seals, proceed as follows:

Disconnect the clutch release rod (15 —Fig. FO545A) from release shaft arm (14) and unhook return spring (11) from release fork. Rotate release shaft forward and withdraw clutch release bearing (1) and hub (2). Remove the two pins securing release fork (3) to shaft (10) and withdraw shaft, fork and return spring from housing.

Remove safety wire or straighten the locking tabs, remove the five cap screws securing input shaft retainer (4) to transmission housing and withdraw the retainer. On live pto models, the pto input shaft will be withdrawn with the retainer. On all models, the transmission input shaft will remain in the transmission.

On live pto models, remove the snap ring (7—Fig. FO547A) at rear of pto input shaft bearing and bump shaft (8) and bearing (6) rearward out of retainer. The input shaft retainer

seal (4), and the seal (2) and needle bearing (3) in the forward end of pto input shaft can be renewed at this time. To install new needle bearing, press on lettered end of cage only. CAUTION: Do not press bearing cage in against shoulder in input shaft. When properly installed, front end of bearing cage should be 1.01 inches below flush with front end of input shaft.

On all models, install seal or seals with lip to rear and reverse the disassembly procedure, being careful not to damage lip of seal during installation. Note: The upper of the five securing cap screws passes through the adapter plate and threads into transmission housing. This screw is longer than the remaining four. Tighten the five cap screws to a torque of 40 ft.-lbs. and secure with safety wire or bend locking tabs against heads of cap screws.

REMOVE AND REINSTALL TRANSMISSION

60. To remove transmission assembly from tractor, first drain the transmission and hydraulic system and detach engine from clutch housing as outlined in paragraph 56. Remove

closure panels from each side of steering gear housing. Remove the four cap screws securing steering gear housing to transmission case and the three bolts securing fuel tank rear support to transmission case and lift fuel tank, steering gear and instrument panel from tractor as a unit. Disconnect brake and clutch linkage and unbolt and remove step plates. Attach a hoist to transmission housing, support rear axle center housing separately, unbolt transmission from rear axle center housing and remove transmission assembly.

OVERHAUL

61. **FRONT HOUSING AND INPUT SHAFT.** To remove the transmission front housing (6—Fig. FO545A) and input (clutch) shaft, first detach (split) the engine from clutch housing as outlined in paragraph 56, and remove fuel tank, steering gear and instrument panel as outlined in paragraph 60. Disconnect clutch release rod (15—Fig. FO545A), unhook return spring (11) and rotate release shaft forward to remove release bearing (1) and hub (2). Remove the two pins securing release fork (3) to shaft (10) and withdraw fork, release shaft and return spring from the transmission housing. Remove the cap screws securing adapter plate (5) to the transmission front housing and withdraw adapter plate and input shaft retainer as a unit from the transmission. Note: Pto input shaft will be removed with adapter plate and retainer on live pto models.

On live pto models, remove snap ring (13—Fig. FO547A) from front of pto countershaft and remove bearing (14) using a suitable puller and knife edge adapter. On all models, support the transmission front housing with a suitable hoist and unbolt and remove the housing and transmission input shaft. On live pto models, slip pto countershaft drive gear (15) from pto countershaft (16) as the two housings are separated. During reassembly, make sure gear (15) is in place in transmission front housing, and that the splines of gear (15) and shaft (16) are engaged as the two housings are fitted together.

To remove the transmission input shaft from transmission front housing, remove snap ring (12) from rear of bearing and bump shaft (9) and bearing (11) rearward out of housing. The needle pilot bearing in the rear bore of transmission input shaft of Fordson Dexta models may be removed at this

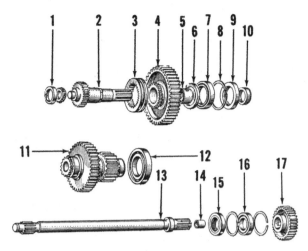

Fig. FO546—Exploded view of secondary transmission and pto counter shaft assembly.

1. Pilot bearing
2. Output shaft
3. High-low coupling
4. Gear
5. Bushing
6. Washer
7. Bearing
8. Snap ring
9. Seal
10. Seal sleeve
11. Secondary countershaft
12. Bearing
13. Pto countershaft
14. Bushing
15. Seal
16. Bearing
17. Pump drive gear

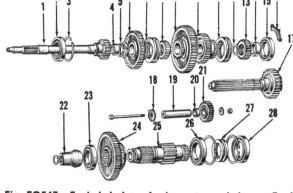

Fig. FO547—Exploded view of primary transmission on Fordson Dexta single clutch models. Fordson Super Dexta and Ford 2000 Super Dexta transmissions have a floating type roller bearing instead of needle bearing (4). Also, all models produced after Fordson Dexta Serial No. 957E-63953 have cap screw threaded into reverse idler shaft (19) to retain idler instead of long bolt, washer and nut shown.

1. Input shaft
2. Input bearing
3. Snap ring
4. Needle bearing
5. Washer
6. 2nd-5th gear
7. Coupling
8. Connector
9. 1st-3rd gear
10. Reverse gear
11. Coupling
12. Snap ring
13. Connector
14. Spacer
15. Bearing
16. Retainer
17. Main shaft
18. Thrust washer
19. Shaft
20. Bushing
21. Reverse idler
22. Pto coupler
23. Bearing
24. Countershaft gear
25. Countershaft
26. Bearing
27. Coupling
28. Bearing

time if renewal is indicated. When re-installing, press the bearing into the bore until rear face of bearing is 0.090 beyond rear face of shaft. Fordson Super Dexta and Ford 2000 Super Dexta transmissions have a floating type roller bearing in the rear bore of the input shaft.

62. PTO COUNTERSHAFT AND REAR COVER. To remove the pto countershaft and transmission rear cover, first remove the transmission as outlined in paragraph 60 and the transmission front housing as outlined in paragraph 61. On models without live pto, remove the snap ring from the front of the pto countershaft and with-draw pto coupling (22—Fig. FO547). If the pto countershaft (13—Fig. FO-546), pto countershaft rear bearing (16), seal (15) or hydraulic pump drive gear (17) are to be renewed, re-move the gear shroud, the snap ring from rear of pto countershaft and withdraw the hydraulic pump drive gear. Unbolt and remove transmission rear cover plate (9—Fig. FO545A) with pto countershaft attached. If necessary, tap the transmission output shaft (2—Fig. FO546) forward as the plate is removed, to free bearing (7) from the rear cover. The output shaft and transmission countershaft, with rear carrier bearings attached, will re-main in the transmission housing as the rear cover is removed.

The output shaft rear seal (9) can be renewed at this time. To renew the pto countershaft rear oil seal, first re-move the pto countershaft from the rear cover plate by removing the snap ring at the rear of bearing (16) and bumping the shaft and bearing rear-ward out of cover. When installing seals (9 and 15) in the rear cover plate, make sure the main sealing lip is fac-ing the transmission (front) side of plate. Using suitable adapters, drive output shaft seal (9) into rear of cover until seal seats against the snap ring in the cover. Pto rear oil seal is in-stalled from front of cover and bot-tomed against snap ring.

The bushing in the bore at rear of pto countershaft is renewable. A new bushing has an inside diameter of

0.502-0.503 and should be installed with a suitable driver until rear face of bushing is 0.010 beyond rear face of pto countershaft.

During installation, tighten the rear cover plate retaining cap screws to a torque of 40 ft.-lbs.

63. SECONDARY TRANSMISSION. The secondary transmission can be dis-assembled after removing transmission as outlined in paragraph 60, and rear cover as outlined in paragraph 62. To disassemble, withdraw the secondary countershaft (11 — Fig. FO546) far enough to free the front carrier bear-ing from its location in the housing and lower the countershaft to the bot-tom of transmission housing. With-draw the output shaft assembly (2)

Fig. FO547A—Input shaft assembly used on trac-tors equipped with live pto. Input shaft (9) and pto countershaft (16) are identical on tractors with or without live pto.

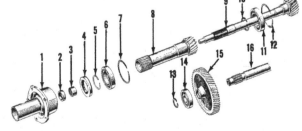

1. Input shaft retainer
2. Oil seal
3. Needle bearing
4. Oil seal
5. Snap ring
6. Bearing
7. Snap ring
8. Pto input shaft
9. Transmission input shaft
10. Snap ring
11. Bearing
12. Snap ring
13. Snap ring
14. Bearing
15. Pto countershaft gear
16. Pto countershaft

from rear of transmission while retaining the high-low sliding coupling (3) in position in shifter fork. Remove the sliding coupling, then withdraw the secondary countershaft.

To disassemble the output shaft assembly, remove gear (4), bearing (7) and seal sleeve (10) at the same time, using a suitable puller. The output shaft gear is fitted with a 1.875-1.876 renewable bushing (5), which should have 0.001-0.003 diametral clearance on the shaft. During reassembly, note that thrust washer (6) has a flat machined into its inside diameter which must match a corresponding flat in the output shaft.

The output shaft (2) and high-low coupling (3) are serviced only as a matched assembly and must be installed with the locating marks in register. The self-aligning roller bearing (1), at the front end of output shaft, is matched with its outer race which is installed in the rear bore of the primary transmission main shaft, and the two must be renewed as an assembly. Renew the bearings if rough or excessively loose, and seal sleeve (10) if nicked or grooved at the oil seal contact area.

During reassembly, place the secondary countershaft in the housing and check locating marks on high-low coupling and output shaft. Reinstall coupling in shifter fork, move fork and coupling forward into the high position, and insert output shaft into coupling with locating marks in register. A coating of heavy grease on the rollers of the output shaft front bearing will hold them in place and assist in the entry of the bearing in its race. Reinstall the countershaft front bearing in its bore in housing and seat the bearing by bumping rear of shaft with a soft hammer; then, install the transmission rear cover plate.

64. **MAIN TRANSMISSION.** The main transmission can be disassembled after removing the transmission as outlined in paragraph 60, the transmission front housing as outlined in paragraph 61 and the secondary transmission as outlined in paragraph 63. To disassemble, insert a step plate of the proper size in the front end of the transmission countershaft and, using a suitable puller, remove the countershaft front bearing (23—Fig. FO547) and the countershaft cluster gear (24). Remove the snap ring from the front of transmission main shaft and withdraw the thrust washer (5) and second gear (6). Loosen the set screw in the shifter fork and remove fork, sliding coupling (7) and connector (8) from

the main shaft. Rotate the countershaft until milled flat (see Fig. FO-547B) is at the top of the shaft and remove first gear (9—Fig. FO547) from the main shaft, followed by the reverse driven gear (10).

Loosen the set screw retaining the high-low shifter fork (2—Fig. FO548) to its rail, remove shifter housing (4) and withdraw the high-low shifter fork and rail. Unbolt and remove the main shaft rear bearing retainer (16—Fig. FO547) and remove main shaft (17), bearing (15) and reverse connector (13) by tapping shaft rearward with a soft hammer. Remove the retaining screw, withdraw reverse idler gear (21) from housing and remove countershaft (25) by bumping it forward out of the housing.

NOTE: Reverse idler on all models produced after Fordson Dexta tractor serial No. 957E-63953 is retained by cap screw threaded into front end of reverse idler shaft. On Fordson Dexta models prior to this Serial Number, reverse idler was retained by a bolt passing through hollow idler shaft as shown in Fig. FO547. Bolt was secured by self-locking nut and flat washer in rear transmission compartment.

To remove main shaft rear bearing (15) from the shaft, first remove snap ring (12), reverse coupler (13) and spacer (14). Remove bearing by using a split type puller and suitable press. To renew the output shaft front bearing race located in the rear bore of main shaft (17), first remove bearing (15) from shaft as indicated above and drive out the race with a suitable punch, through the holes provided.

The three idler gears (6, 9 and 10), on the transmission main shaft have an inside diameter of 1.801-1.802 and a diametral clearance of 0.0011-0.0026 on the main shaft splines. The reverse idler gear is fitted with a renewable bushing having an inside diameter of 1.1245-1.1255 and a diametral clearance of 0.0015-0.0035 on its shaft. When installed, the reverse idler gear should have an end play of 0.010-0.025. The reverse idler shaft is a press fit in the

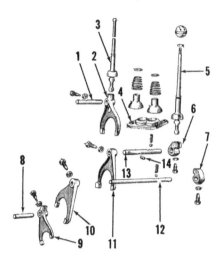

Fig. FO548—Exploded view of shifter rails and forks.

1. High-low rail	8. 4th-6th rail
2. High-low fork	9. 4th-6th fork
3. High-low lever	10. 1st-3rd fork
4. Lever housing	11. Reverse fork
5. Main shift lever	12. Rail, 1st-3rd
6. Selector	13. Rail, reverse
7. Selector	14. Interlock pin

transmission housing and may be drifted out if renewal is indicated. Newer type threaded reverse idler shaft and cap screw may be used to renew hollow shaft, bolt and self-locking nut in early Fordson Dexta Models. Install the replacement shaft so that it extends 1.30-1.31 from the machined surface into the front transmission compartment. The reverse connector and coupler are serviced only as a matched assembly and should be installed on the main shaft so that the locating marks are in register.

NOTE: If less than 2500 lbs. force is required to press reverse idler shaft into housing, it is suggested that the shaft be removed and the following Fordson Major parts be installed: 1—E27N-7140 Shaft, 2—DKN-7717 Washers, 1—BB-5721-B Bolt and 1—34420-ES2C Nut. Install as follows: Press shaft into housing until it protrudes 1.30-1.31. Grind head of bolt to 3/16-inch thick. Grind ¼-inch from side of one washer. Place other washer on bolt with flat side against bolt head and install bolt and washer in shaft and gear from front. Install modified washer at rear with protrusion to front. Apply Loctite to bolt threads and tighten nut only enough to prevent front washer from turning. Check idler gear for 0.010-0.025 end play and be sure no interference exists when reassembling transmission.

65. **SHIFTER RAILS AND FORKS.** Renewal of the shifter rails and forks requires the disassembly of the complete transmission with the following exceptions:

Fig. FO547B — Transmission countershaft showing milled flat which provides clearance for removing first gear from main shaft.

The high-low shifter rail (1—Fig. FO548) and fork (2) may be removed after removal of the secondary transmission assemblies as outlined in paragraph 63, and the shifter levers and housing.

The first-third shifter fork (10) may be removed after removing the primary countershaft cluster gear and the transmission main shaft second gear as outlined in paragraph 64.

The fourth-sixth shifter fork (9) interlocks with the right hand finger of the reverse shifter fork as shown in Fig. FO549, and the two forks are shifted together by the same shifter rail. To remove the reverse shifter fork (11—Fig. FO548) and rail (13) after the remainder of the transmission has been disassembled, remove the set screws securing the fork and selector (6) to the rail, then withdraw the rail from the housing. Remove the fourth-sixth fork, loosen the retaining set screw and slide the rail out of fork and housing. Be sure to interlock the two forks during assembly. Remove the remaining shifter rail by removing the set screw in selector (7) and withdrawing the rail. An interlock pin (14) is installed in the housing between the

Fig. FO549—Front view of rear transmission housing with gears removed, showing interlocked reverse and 4th-6th shifter forks. These forks must be in place before any of the transmission gears can be installed.

two shifter rails at the front end. To remove the interlock pin, remove the rails as outlined above, remove the expansion plug on the left side of transmission housing and withdraw the interlock pin.

thickness of 0.0065 (uncompressed thickness 0.009-0.012). After determining the correct number of gaskets to install, reassemble tractor in the normal manner.

An alternate method is to detach the rear axle center housing from the transmission housing, remove the drive pinion as outlined in paragraph 70, and remove the rear pto shaft from the center housing. Wrap a piece of heavy string around the differential case so that a pull rearward on the string will rotate the differential assembly in a forward direction. Install the differential assembly in the tractor and install the left axle housing with the number of gaskets removed prior to overhaul. Tighten the retaining nuts to a torque of 45-50 ft.-lbs. and, working through the pto opening in rear of center housing, attach a pull scale to the string. The preload is correct if a pull of 4½ lbs. is required on the pull scale to rotate the differential assembly.

67. **OVERHAUL.** After the differential unit has been removed from the tractor as outlined in paragraph 66, examine the two halves of the case and place correlation marks to aid in reassembly if they are not already present. If equipped with differential lock, remove the snap ring (1—Fig. FO550B), washer (2), sliding coupling (3), return spring (4) and coupling adapter (5). Remove the safety wire

DIFFERENTIAL, BEVEL GEARS AND REAR AXLE

DIFFERENTIAL

66. **REMOVE AND REINSTALL.** To remove the differential unit, first drain rear axle center housing and hydraulic reservoir and block up under the center housing to raise the rear wheels. Remove the left rear wheel and disconnect left brake linkage. Disconnect left hydraulic lift link from pivot stud. Unbolt left axle housing from center housing and remove assembly from tractor. Carefully remove and count the number of gaskets installed at the axle housing flange so that the same number can be installed during reassembly. Withdraw the differential unit from center housing.

Differential carrier bearing preload is established by the number of gaskets (A—Fig. FO550) installed between left axle housing and rear axle center housing during assembly. The maximum recommended preload on the differential carrier bearings is 30 in.-lbs. rolling torque. The allowable preload in terms of pinch on housings is 0.003 clearance to 0.003 preload.

The manufacturer recommends that whenever major repairs have been made to the differential assembly, bearings or axle housings, a check and adjustment of the preload be made as follows:

Detach rear axle center housing from transmission case and block the center housing up so that right axle housing is toward bottom and left hand side of center housing is uppermost. Locate the differential assembly in the cup of the right axle housing. Remove axle shaft from left hand axle housing and position the housing over the differential assembly with no gaskets installed. Rotate the differential assembly by hand to make sure that carrier bearings seat properly in their cups and position four retaining nuts equally around the left hand axle housing flange. Tighten the nuts finger tight and, using a feeler gage, adjust the nuts until the gap is equal at all points around the housing. Measure the gap and install the number of gaskets which, when compressed, will most nearly equal the measured gap. The gaskets supplied have a compressed

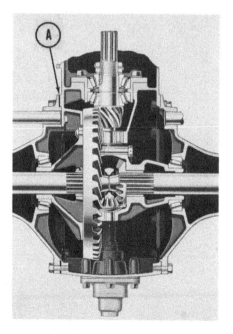

Fig. FO550—Cutaway view of final drive used on Dexta tractor without differential lock. Carrier bearing preload is adjusted by means of gasket pack (A).

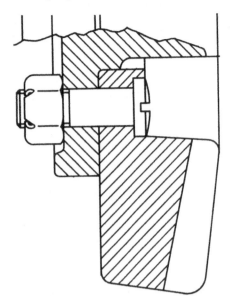

Fig. FO550A — Cross-sectional view of differential case and ring gear showing proper installation of slotted head bolts and self-locking nuts when renewing ring gear.

and remove the cap screws retaining the right half of the differential case. On models equipped with differential lock, the cap screws must be loosened equally as the right half of the differential case is removed due to interference with the large carrier bearing cone.

The bevel ring gear and bevel drive pinion are available only as a matched set, either with or without the differential case. To renew ring gear on differential case, proceed as follows: Center punch the upset (ring gear) end of each rivet in exact center of counterbore in ring gear. Then, using a $\frac{9}{16}$-inch drill, drill into each rivet until the upset end is cut off. Use suitable drift punch to remove rivets from assembly. Special bolts and self-locking nuts are available to attach new ring gear to differential case. Assemble with slotted heads of bolts to ring gear side of assembly and tighten the self-locking nuts to a torque of 50-60 ft.-lbs.

Examine the axle shaft splines, pinion gears, differential spider, thrust washers and the axle gear bushings and renew if necessary. When renewing the axle gear bushings, press the new bushings 0.600 below flush with inside face of differential case halves. The differential carrier bearings should be checked and renewed if necessary. If renewal is indicated for the carrier bearing cups, the axles must be removed from the axle housings and the right hand axle housing from the center housing to provide access for removal.

After reassembly, tighten the cap screws retaining the right half of the differential case to a torque of 70 ft.-lbs. On differential lock equipped models, the right hand carrier bearing cone must be removed to tighten the cap screws with a torque wrench if a special torque wrench adapter is not available. After tightening the cap screws, check the axle side gears to be sure there is no binding and in-

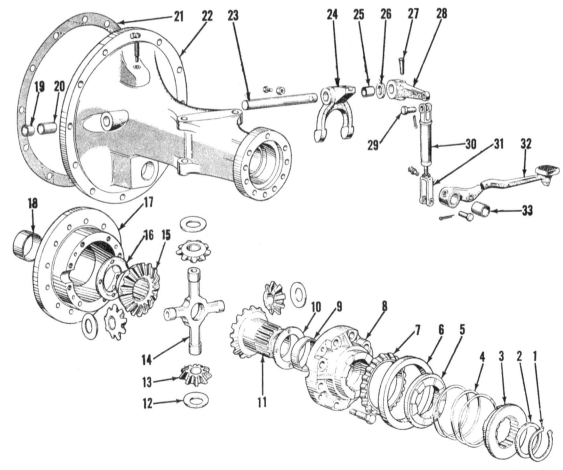

Fig. FO550B — Exploded view of differential lock and related parts.

1. Snap ring
2. Washer
3. Sliding coupling
4. Coil spring
5. Coupling adapter
6. Bearing cup
7. Cone & roller assy.
8. Differential case (R.H. half)
9. Bushing
10. Thrust washer
11. R. H. axle gear
12. Pinion thrust washer (4)
13. Pinion gears (4)
14. Differential spider
15. L. H. axle gear
16. Thrust washer
17. Differential case (L.H. half)
18. Bushing
19. Plug
20. Bushing
21. Gasket
22. R. H. axle housing
23. Shaft
24. Fork
25. Bushing
26. Seal
27. Pin
28. Lever
29. Pin
30. Operating link
31. Yoke
32. Foot pedal
33. Bushing

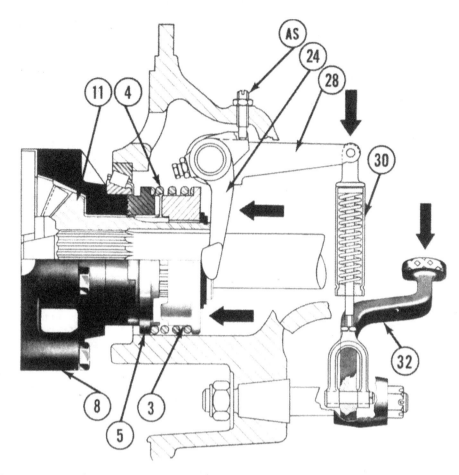

Fig. FO550C — Schematic view showing operating principles of differential lock. To engage lock, pressure is appled to sliding coupling (3) from foot pedal (32) through spring loaded link (30), lever (28) and fork (24). Spring loaded link will snap coupling to engaged position at instant dogs on sliding coupling and coupling adapter (5) are aligned. Lock will be disengaged by coil spring (4) whenever traction is equalized on rear wheels.

AS. Adjusting screw
3. Sliding coupling
4. Coil spring
5. Coupling adapter

8. Differential case
11. R. H. axle side gear
24. Operating fork

28. Operating lever
30. Spring loaded link
32. Foot pedal

stall new safety wire through drilled heads of cap screws. Reinstall the differential unit and check carrier bearing preload as outlined in paragraph 66.

DIFFERENTIAL LOCK

68. DIFFERENTIAL LOCK OPERATION AND ADJUSTMENT. The differential lock consists of a dog type coupling which can be engaged to lock the right hand axle gear (11—Fig. FO550B) to the differential case. This results in both rear wheels being turned at the same speed, regardless of any difference in traction of the rear wheels.

In operation, when one rear wheel starts to spin, the foot pedal (See Fig. FO550C) is depressed which applies spring pressure to the sliding coupling. When the dogs on the coupling are aligned with the notches in the coupling adapter, the spring pressure applied through the operating fork will snap the coupling into engaged

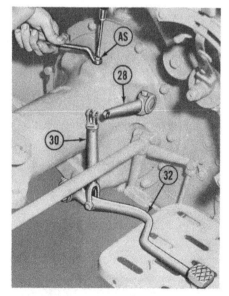

Fig. FO550D — Adjusting the differential lock. Refer to text for adjustment procedures.

AS. Adjusting screw
28. Operating lever

30. Operating link
32. Foot pedal

position. The foot pedal can then be released and the differential lock will remain engaged until the traction on the rear wheels becomes equalized. As there will then be no side pressure on the coupling dogs, the coil spring between the sliding coupling and the coupling adapter will push the sliding coupling to disengaged position. If necessary to make a turn before the differential lock is automatically disengaged, the lock can be manually disengaged by momentarily depressing the transmission clutch pedal or applying the brakes on the wheel with least traction.

Proper adjustment of the differential lock mechanism requires minimum clearance between the operating fork and face of sliding coupling when the lock is in disengaged position, and that the lock be fully engaged when the foot pedal is depressed until it strikes the right hand foot rest. Before attempting to adjust the differential lock, be sure that the foot rest is not bent out of position and there is nothing on the foot rest to prevent full travel of the foot pedal. To make adjustment, proceed as follows: Disconnect the spring loaded operating rod from operating lever as shown in Fig. FO550D. Back-off the locknut on adjusting screw in axle housing. Position operating lever so that operating fork just contacts face of sliding coupling. Turn adjusting screw in until contact is made with operating fork; then, back adjusting screw out ¼-turn and tighten locknut.

Block up right rear wheel and turn wheel while pushing down on operating lever to fully engage the differential lock. Hold operating lever in this position and with foot pedal against foot rest, adjust length of spring loaded operating rod so that pin can be inserted through operating rod yoke and operating lever. Then, shorten rod two turns, tighten locknut and reinstall pin connecting operating rod to operating lever.

69. OVERHAUL DIFFERENTIAL LOCK. The differential lock foot pedal pivots on an extended hydraulic lift lower link shaft. Pedal has renewable bushing; be sure to align grease hole in bushing with hole in pedal when renewing the bushing. The spring-loaded operating rod is renewable only as an assembly. The operating lever and fork shaft pivots in renewable bushings in the right axle housing. To renew the operating shaft, bushings, fork and/or pedal pivot shaft, remove right axle housing as follows: Drain oil from differential and hydraulic lift compartments. Disconnect differential lock operating rod from operating lever and remove foot pedal and hydraulic lift lower link from pivot shaft. Block up under center housing and unbolt and remove right rear wheel, fender and axle assembly from axle housing. Be careful not to lose or damage shims located between axle housing and axle bearing support. Then, unbolt and remove axle housing from center housing. The differential lock sliding coupling, spring and coupling adapter can also be inspected and renewed at this time. See Fig. FO550E. If service of differential unit is indicated, remove left axle housing and differential as outlined in paragraph 66.

When reassembling tractor, install only one gasket between right axle housing and center housing and install the same number of shims between axle housing and axle bearing support as were removed during disassembly. Tighten axle housing to center housing retaining nuts to a torque of 50 ft.-lbs. and axle housing to axle support retaining nuts to a torque of 40-45 ft.-lbs. If adjustment of differential carrier bearing pre-load is indicated, refer to paragraph 66. To readjust axle end play, refer to paragraph 74.

MAIN DRIVE BEVEL PINION

70. To remove the main drive bevel pinion, first detach the rear axle center housing from the transmission housing and remove the hydraulic pump as outlined in paragraph 102. Remove the hydraulic lift cover as outlined in paragraph 91 and remove the left

axle housing and differential assembly as outlined in paragraph 66. Remove the six mounting cap screws in the pinion carrier and remove the carrier and pinion assembly by using two $\frac{9}{16}$-NC jack screws as shown in Fig. FO-551. Disassemble pinion and bearings and renew worn or damaged parts.

When reassembling the bevel drive pinion, adjust the bearings until a rolling torque of 12-16 in.-lbs. is established. When checked with a spring scale as shown in Fig. FO553, the scale reading should be 16-21 lbs. When adjustment is completed, secure the adjusting nuts by bending the tabs on the lock washers.

NOTE: The heavier duty tapered roller bearings and bearing cups cataloged for Super Dexta bevel pinion shaft may be used to renew bearings and cups on Dexta pinion. Do not use the narrow Dexta pinion shaft tapered bearings in Super Dexta models.

The main drive bevel pinion is available for service only in a matched set with bevel ring gear, either with or without differential case assembly.

REAR AXLE SHAFT, BEARINGS AND SEALS

74. **BEARING ADJUSTMENT.** The rear axle shafts are carried on one tapered roller bearing at the outer end of each shaft. The bearings are retained in their cups by contact of the inner ends of the two axle shafts in the differential assembly. The recommended end play of 0.004-0.012 is adjusted by means of shims placed between the axle side housings and the outer bearing retainer on each side of the tractor as shown in Fig. FO553A.

A quick check of the bearing adjustment can be made by supporting rear of tractor and removing the wheel and tire assemblies. To make the check, shift the transmission into neutral and rotate either axle shaft. If the opposite shaft rotates in the same direction, the bearing adjustment is too tight.

To accurately measure the bearing clearance, remove both brake drums and make sure one axle shaft bearing is fully seated in its cup by driving a wedge between the axle flange and bearing retainer as shown in Fig. FO-554. Push inward on the opposite axle flange and thread a long $\frac{3}{8}$-16 bolt in the brake drum retaining hole in the axle flange as shown in Fig. FO555. A lock nut installed on the bolt will assist in holding it in place. Thread the bolt into the axle flange until contact is just made with the bearing retainer flange while pressure is being applied inwardly on the axle. When the bolt has been adjusted as outlined above, wedge the axle outward and measure the clearance from the same position on the bearing retainer, as shown. If the measured clearance is not within the recommended range of 0.004-0.012, correct the adjustment by adding or removing shims (2—Fig. FO553A) between the bearing retainer and axle housing. Adjustment may be made on either axle shaft bearing but an effort

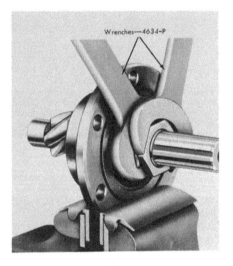

Fig. FO552—Adjusting main drive bevel pinion preload before final assembly.

Fig. FO553—Suggested method of checking bearing pre-load with a pull scale. Scale should read 16-21 pounds.

Fig. FO550E — Differential lock coupling can be inspected or renewed with R. H. axle housing removed as shown.

Fig. FO551 — View of rear axle center housing with lift cover removed, showing suggested method of pulling drive pinion. Differential unit must be removed to provide clearance for pinion pilot bearing.

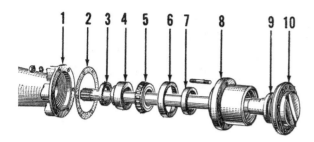

1 2 3 4 5 6 7 8 9 10

Fig. FO553A — Exploded view of rear axle assembly showing bearing, seals and shims. Later production uses self-locking nut instead of collar (7).

1. Axle housing
2. Shim pack
3. Inner seal
4. Collar
5. Bearing
6. Bearing cup
7. Spacer
8. Bearing retainer
9. Outer seal
10. Axle shaft

should be made to keep the total shim pack equally divided between both axle shafts. To add or remove shims it will be necessary to remove the axle shaft as outlined in the following paragraph:

75. R&R AXLE ASSEMBLY. To remove either rear axle shaft, support rear of tractor, disconnect the brake linkage and remove the wheel and tire assembly and brake drum. Unbolt the bearing retainer from the axle housing and withdraw the axle shaft, brake assembly and bearing retainer from the tractor. The axle bearing inner oil seal (3—Fig. FO553A) can be renewed at this time and should be installed with the lip toward the differential assembly. The shim pack (2) is located between the brake backing plate and axle side housing. Five thicknesses of shims are available ranging from 0.015 to 0.058. The minimum number of shims to obtain the proper bearing adjustment should be used.

When reassembling, adjust the end play as outlined in paragraph 74 and tighten the bearing retainer bolts to a torque of 40-45 ft.-lbs.

75A. OVERHAUL AXLE ASSEMBLY (BEARING RETAINED BY STEEL COLLAR). To renew the bearing, bearing cup, retainer, outer seal or axle shaft, drill through the steel collar as shown in Fig. FO556 and split the collar with a suitable chisel. Pull the bearing and bearing retainer from the axle shaft with a suitable puller or large press.

To reassemble, stand the axle, outer flange downward, on a wooden block to protect the wheel stud threads and install the bearing spacer with the tapered inside edge downwards. Fit the retainer, with bearing cup and outer seal in place, over the axle shaft and pack with wheel bearing grease. Install the bearing cone over the shaft and make sure the bearing is seated firmly against the spacer. Heat the

collar evenly with a suitable torch until the color changes to a dark blue and immediately drop the collar over the axle shaft and seat firmly against the bearing using a piece of heavy pipe as a driver.

75B. OVERHAUL AXLE ASSEMBLY (BEARING RETAINED BY SELF-LOCKING NUT). Unscrew the self-locking nut from axle shaft. Pull the bearing and bearing retainer from axle shaft with suitable puller or large press. The axle outer oil seal may be renewed in bearing retainer without removing the bearing cup. If renewing bearing cup, be sure that new cup is driven into retainer until firmly seated against shoulder. Install new seal with lip of seal towards bearing cup.

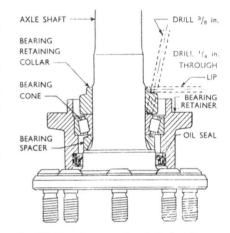

Fig. FO556—Remove shrunk fit bearing retaining collar from axle shaft by drilling through collar as shown, then cracking collar with chisel.

Fig. FO554—To accurately measure axle shaft end play, remove both wheels and brake drums, wedge one axle shaft outward as shown, and measure end float of opposite axle as shown in Fig. FO555.

Fig. FO555—Using a long bolt and feeler gage as shown, measure the axle shaft end play as outlined in text.

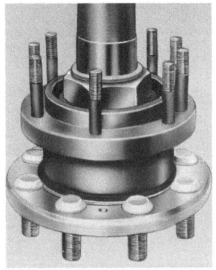

Fig. FO556A — View showing self-locking nut that is used to retain rear axle bearing on late production tractors. Always renew self-locking nut when reassembling axle unit.

To reassemble, stand axle on wood block to prevent possible damage to threads on wheel studs and install bearing spacer with tapered inside edge against shoulder on axle. Position bearing retainer over axle shaft and pack retainer with wheel bearing grease. Pack bearing cone and roller assembly with wheel bearing grease and drive cone into place against bearing spacer. Install **new** self-locking nut on axle shaft and tighten nut to a torque of 230-250 ft.-lbs. Install axle assembly in axle housing and adjust bearing end play as outlined in paragraphs 74 and 75.

NOTE: Set axle in wheel disc to hold axle while removing or installing nut. Use of Nuday Tool No. 4235-D is recommended. This special tool multiplies torque wrench reading by 2; therefore, 115 - 125 ft.-lbs. torque reading is required.

76. **AXLE HOUSINGS.** Procedure for renewal of the rear axle side housings is evident. When renewing housings, check and adjust the differential carrier bearings as outlined in paragraph 66 and the axle bearings as outlined in paragraph 74.

BRAKE SYSTEM

ADJUSTMENT

77. When disengaged, the brake shoes are held in proper alignment with the brake drum by means of adjustable steady posts (A—Fig. FO-556A), to prevent dragging. To adjust, loosen the lock nut on the upper and lower adjustable steady posts and back the posts out until they are free of the

shoes. Firmly apply the brake and lock in the applied position with the parking lock. Screw the adjustable steady posts into the back plate until the end of the post just contacts the brake shoe and lock in place by tightening the lock nut.

To adjust the brake shoes, jack up rear wheels, open the adjusting hole cover on backing plate at rear of axle housing as shown in Fig. FO557 and turn the star wheel with a screw driver. Push forward on the screw driver handle to tighten the adjustment. Continue tightening until a slight brake drag is observed when turning the wheel, then back off until the shoe just clears the drum. Adjust the left brake clevis to equalize the pedals when the brakes are applied.

R&R BRAKE SHOES

78. Jack up rear end of tractor and remove the rear wheels and brake drums. Disconnect the two secondary springs from the brake shoe anchor pins and detach and remove the anchor plate. Force rear end of the two shoes apart and remove the adjuster unit and rear retracting spring. Remove the four hold down pins, springs and cups by compressing the spring and turning the outer cup 90 degrees in either direction to disconnect it from the pin. Lift off the shoes, together with the front retracting spring. If anchor pins are worn, they may be renewed by loosening the securing nut and driving the pin from backing plate. Brake shoes are interchangeable but tend to wear in to the drum when used. If the shoes are not to be renewed it is good shop practice to mark the shoes so that they may be reinstalled in the same location.

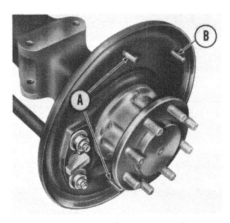

Fig. FO556A — Adjustable steady posts (A), located in brake back plate, are used to align the shoes with the brake drum. Stationary post (B) is built into plate and is not adjustable.

Fig. FO557 — Adjust each brake shoe as shown until shoes just clear drums.

POWER TAKE-OFF

Aside from the input shaft and gear at the front of the transmission assembly, the power take-off train on all models is identical whether or not the tractor is equipped with a live power take-off. Service procedures covering the input shaft and pto countershaft are included in the transmission section of this manual. The following paragraphs will cover the power take-off train from the shifter unit to the rear of the tractor. Fig. FO558 shows a sectioned view of the rear power take-off train.

OUTPUT SHAFT

79. To remove and/or overhaul the pto output shaft (5—Fig. FO558) first drain rear axle center housing and hydraulic system and remove the four cap screws retaining rear pto support cover to rear axle center housing. The output shaft and rear support assembly can now be withdrawn from the tractor.

To disassemble the unit, remove the snap ring retaining rear bearing to the support housing (10) and bump the shaft and bearing forward out of housing. The rear seal can be renewed at this time by driving it forward out of support housing. Using a suitable driver, install the new seal from the front of the support housing, lip up, until front edge of seal is 1/8-inch below the bearing locating shoulder.

To renew the pto output shaft, rear bearing or rear seal sleeve, first press the front collar forward off the shaft. This collar has a 0.0005-0.002 press fit

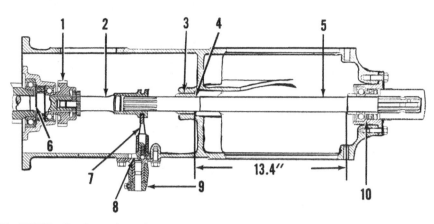

Fig. FO558—Sectional view of rear pto train. Sliding connector pilots in pto countershaft and meshes in internal splines in hydraulic pump drive gear when pto is engaged.

1. Pump drive gear	4. Front collar	6. Oil seal	8. Oil seal
2. Sliding connector	5. Output shaft	7. Shifter fork	9. Shifter lever
3. Bushing			10. Oil seal

on the shaft. The front bearing retaining collar is a shrunk fit on the output shaft and must be cracked with a suitable chisel before removing. After removing the front bearing retaining collar the rear bearing can be moved forward off the shaft. If renewal of the seal sleeve is indicated, press this sleeve forward off the shaft. The seal sleeve has a 0.0003-0.0018 press fit on the output shaft.

To reassemble the unit, press the seal sleeve, then the bearing on the shaft until the two units are firmly seated on the shoulder at the rear splines. Heat the bearing retaining collar evenly with a suitable torch until it reaches a dark blue color and immediately seat it over the shaft against the bearing. Press the front collar on the shaft with the chamfered edge forward until the rear edge of the front collar is 13.4 inches from the front edge of the bearing retaining collar as shown in Fig. FO558.

SHIFTER UNIT

80. To remove the pto shifter unit, first drain the rear axle center housing, remove the left hand step plate and disconnect the clutch rod from the clutch pedal. Unbolt and remove the shifter unit as shown in Fig. FO559. Slip shifter fork from inner end of shifter crank, and remove the nut from the tapered lever retaining pin. Drive the pin from lever, remove the lever and slide the shifter crank from cover plate. When the shifter crank is removed from the cover plate, the detent ball and spring will be free to fall from the cover. Care should be used not to lose the ball or spring. The shifter lever oil seal may now be renewed. Install the new seal with lip towards inside of cover plate.

PTO SLIDING CONNECTOR

81. The front end of the pto sliding connector (2—Fig. FO558) pilots into a bushing fitted into a bore in rear end of the pto countershaft. The rear end of the sliding coupling splines over the front splines of the pto output shaft. When the pto coupling is engaged, external splines at the front end of the coupling engage in the internal splines machined into the rear face of the hydraulic pump drive gear. To remove the pto sliding connector, detach the rear axle center housing from the transmission housing, remove the shifter unit and withdraw the sliding coupling from the front.

The bronze bushing (3) supporting the center of the pto shaft may be renewed from the front after removing the pto output shaft. Press new bushing into its bore until front edge of the bushing is 0.022 to the rear of the front face of the housing boss.

BELT PULLEY

The belt pulley is supplied as extra equipment and may be mounted and operated in right or left horizontal position or in down vertical position.

82. Overhaul procedure is as follows:

Drain lubricant and remove housing cover plate (12—Fig. FO560). Straighten locking tabs and remove pulley shaft nuts (10). Remove the shaft (9), outer bearing (7) and seal (8) from the housing. Withdraw the

Fig. FO559—Shifter unit can be unbolted and removed from the outside if service is indicated.

Fig. FO560 — Cut-away view of belt pulley unit showing gear arrangement.

1. Oil seal
2. Bearing
3. Drive gear
4. Shaft pinion
5. Housing
6. Inner bearing
7. Outer bearing
8. Oil seal
9. Pulley shaft
10. Adjusting nut
11. Bearing
12. Housing cover

pulley shaft pinion (4) and inner bearing (6). The drive gear (3) may now be lifted from the housing. To renew the drive gear shaft oil seal (1) it is first necessary to remove the large bearing cup from the housing

and drive the seal to the inside. To renew the pulley shaft oil seal (8) pull the outer bearing from the shaft, fit a new seal over the shaft and re-install it in the housing as the unit is reassembled. Adjust the pulley shaft

end float to 0.002 by means of the adjusting nut (10) and lock in place with the locknut and tab washer. Adjust the drive gear shaft bearings to 0.002 end float by means of the aluminum shims under the cover plate (12).

HYDRAULIC LIFT SYSTEM

The hydraulic lift system incorporates automatic draft control and automatic implement position control. Fluid for the system is common with the rear axle final drive, but separated from the transmission by oil seals. Hydraulic power is supplied by a gear type hydraulic pump mounted in the rear axle center housing and driven by the power take-off shaft. The system is protected by a gauze type strainer on the intake side of the pump and a partial flow renewable element type filter on the system return line. Both filters are located in the oil reservoir in front part of the rear axle center housing.

NOTE: Production changes have been made from time to time in the Dexta series hydraulic system which will affect both parts procurement and service procedures. These changes will be noted in the text as well as information on identification and interchangeability of the different parts.

TROUBLE-SHOOTING

83. Trouble in the hydraulic lift system or malfunction of any of its parts will usually show up in: (a) failure to lift, (b) inability to hold implement in raised position without up and down bobbing motion, (c) over correction in draft control, (d) erratic action or overtravel or (e) a noisy pump. The probable causes of trouble and methods of checking to locate the source are outlined in the following paragraphs.

84. **WILL NOT LIFT.** First make sure that the system contains the proper amount of oil. An oil level plug is located on the left hand side of the rear axle center housing. Check the mechanical power train by engaging the pto shifter and checking for correct rotation of the pto output shaft. Move the touch control lever to the top of its quadrant and run the engine with the selector lever in both the down, and forward position. If the lift still fails to operate, move the auxiliary service knob on the acces-

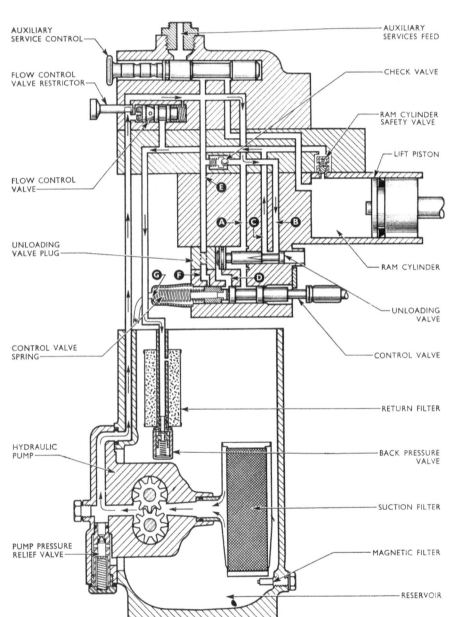

Fig. FO561 — Schematic view of oil flow with control valve in neutral position on late production hydraulic system. Oil flow in early production Fordson Dexta is similar except that rear land of control valve holds oil in lift cylnder instead of front land as shown in diagram. Moving the control valve to rear (right) will open passage (F) and allow cylinder to discharge oil to sump. Moving the control valve forward (left) will allow low pressure oil in passage (A) to both sides of unloading valve piston. Due to larger area of valve piston head, unloading valve will be forced to rear (right) blocking passage (C). Pump will then build up pressure in passage forcing check valve open and oil will flow to tractor lift cylinder or to auxiliary service port depending upon position of auxiliary service control valve. Rate of oil flow from pump may be varied by adjusting the flow control valve restrictor. The flow control valve will then by-pass more or less oil to sump depending upon position of restrictor.

sory plate or flow control valve to the out position and remove the plug from the top of the plate or flow control valve. Turn the engine over with the starter and observe if a flow of oil is pumped through the plug opening. If no flow is present, remove the lift cover and turn engine over with starter again. If oil flows from pressure tube in side of center housing, service the lift cover assembly as outlined in paragraphs 91 through 99. If no oil flows from pressure tube when turning engine over with starter, remove and overhaul the hydraulic pump as outlined in paragraphs 102 and 103.

Possible causes of failure to lift within the cover assembly would be improper adjustment of the control linkage, sticking or binding of the control valve, sticking safety valve, sticking back pressure valve, sticking unloading valve or broken ram cylinder, piston or seals.

Possible causes of failure to lift within the pump assembly would be shearing of the drive key or shaft, broken or extremly worn pump body or gears, plugged intake filter, ruptured seals in pump or sticking relief valve.

85. BOBBING (HICCUPS). Although usually caused by an internal leak in the lift cylinder hydraulic circuit, this condition may result from a springy implement causing a rebound in the control linkage. With hydraulic oil at operating temperature and a dead weight of approximately 1250 pounds on the lift arms, three or less corrections in a period of two minutes should be considered normal. If corrections, or bobbing, occurs at shorter intervals of time, service of the hydraulic lift is indicated.

To determine the cause of leakage, mount a heavy implement and raise the three-point linkage. Remove the hydraulic filler cap and visually check rear end of cylinder for leakage around piston. If a leak is noted, renew the piston seal.

If piston was not leaking, shut off engine with implement raised. If implement falls about 6-12 inches, then stops; or if rate of fall decreases noticeably after falling about one foot, renew the control valve and bushing as outlined in paragraph 95 or 96.

If implement falls all the way to ground at a steady rate and piston is not leaking, a leaking check valve or safety valve is usually indicated.

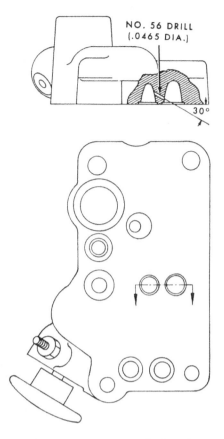

NO. 56 DRILL (.0465 DIA.)

30°

Fig. FO561A — On early Fordson Dexta tractors not equipped with flow control valve, pump flow may be reduced by carefully drilling hole in accessory plate as shown above.

Remove the accessory plate or flow control valve and renew the safety valve as outlined in paragraph 93; then repeat the test. If leak still occurs, renew the check valve and seat as outlined in paragraph 92.

Additional points to check are the "O" rings located between lift cylinder and cover and the fit of the unloading valve bore plug in the lift cylinder. While lift cover is off, make sure that valve linkage operates without binding.

85A. OVER-CORRECTION IN DRAFT CONTROL. Under some field conditions or implement applications, uneven depth control may result from excessive oil flow from the hydraulic pump which will cause the system to over correct for a change in draft. Later production Fordson Dexta, Fordson Super Dexta and Ford 2000 Super Dexta tractors are equipped with an adjustable flow control valve which is used to regulate the hydraulic pump output to meet different field conditions and implement requirements.

Where excessive pump output is causing over correction in draft control on early Fordson Dexta tractors, a by-pass hole may be drilled in the hydraulic accessory plate to reduce pump output as shown in Fig. FO561A. An alternate and preferable method of correcting this difficulty would be to install a flow control valve and associated linkage.

If adjusting the flow control valve on units so equipped does not change the rate of hydraulic lift, check the flow control valve plunger (52—Fig. FO564) to be sure it is not sticking or that the plunger spring (53) is not damaged or broken.

86. ERRATIC ACTION. Usually caused by binding of the control valve, back pressure valve, unloading valve or linkage. Before removing top cover, check to see that the rockshaft is properly adjusted so that the lift arms will drop of their own weight.

87. NOISY PUMP. Usually caused by a worn pump or plugged intake filter. May be caused by use of oil which foams excessively and allows air to enter pump.

CONTROL ADJUSTMENTS

88. MAIN CONTROL SPRING. This adjustment should be made before making adjustments on the internal valve linkage. To make the adjustment, rotate the control spring yoke (63 — Fig. FO563 or FO564) until a slight pre-load is placed on the spring. Preload is correct for normal operation when the control spring (62) can just be rotated when grasped by the finger and thumb of one hand. When heavy draft loads are encountered, it may be necessary to tighten the main control spring ½-turn over normal setting to obtain desired implement depth. Main control spring should be readjusted when returning to normal draft loads.

89. DRAFT CONTROL. Accurate and easy adjustment of the hydraulic linkage is not practical without the use of special Ford adjusting gages. If such gages are available, use the rockshaft locating arm designed for the model NAA Ford tractor (N—503), modified as shown in Fig. FO561B to clear the Dexta linkage. Use the adjusting gage NCA 502 designed for the model 600 and 800 Ford tractors

Fig. FO561B—Modify the Ford NAA locating arm by removing metal where indicated, until arm will clear Dexta linkage.

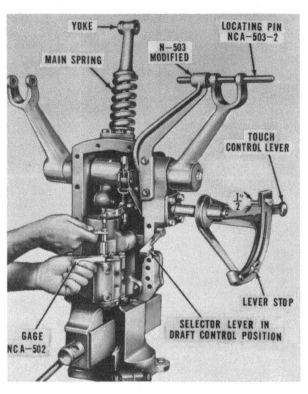

Fig. FO562 — Method of properly adjusting control valve for constant draft position, using Ford adjusting tools.

and proceed as follows: Clamp the hydraulic lift cover in a vise and affix the locating arm to the housing flange as shown in Fig. FO562. Move selector lever to the down, or "Draft" position and move the touch control lever until it is located exactly ½-inch from upper stop on quadrant. Use adjusting gage NCA 502, and with the small end, measure the clearance between the machined edge of the control valve land and the machined surface of the valve housing as shown. The gage should just enter this gap without any binding or side clearance. If adjustment is incorrect, make the proper correction by lengthening or shortening the valve linkage as shown. If the recommended adjusting gages are not available, clamp the cover assembly in a vise and locate the selector and touch control levers as recommended and proceed as follows:

Move the rockshaft in the lowering direction until front end of ram piston contacts the closed end of the cylinder. Measuring at the hole in the end of the lift arm, move the rockshaft back ½-inch and lock in position by tightening the cap screw on one end of the rockshaft (45—Fig. FO563 or FO564) until the shaft will maintain its position. (The rockshaft will be maintained in this position also, for the position control adjustment outlined in paragraph 90.) With the rockshaft, selector lever and touch control levers in the positions indicated, the clearance between the machined land on the control valve and the machined surface on the cylinder should be 0.396. The plate at rear of housing which retains the control valve is cut away to provide clearance for measurement.

After completing the position control adjustment as outlined in paragraph 90, and installing the lift cover on the tractor, readjust the rockshaft retaining cap screw until the lower links will just drop of their own weight.

90. **IMPLEMENT POSITION.** To make this adjustment, first adjust the main control spring as outlined in paragraph 88 and the draft control linkage as outlined in paragraph 89. Both of these adjustments must be made before the position control linkage can be adjusted. With cover mounted in vise and rockshaft correctly positioned as outlined in paragraph 89, move selector lever to position control (parallel to gasket surface of cover) and touch control lever against lower stop of quadrant as shown in Fig. FO562A.

90A. If adjusting gage NCA 502 is available, use thicker (Position) end of gage and measure clearance between control valve land and machined surface of cylinder housing. The gage should just enter the gap with no binding or side clearance. If clearance is not correct, loosen lock nut at upper end of position control rod (23 — Fig. FO563 or FO564) and adjust rod length by turning the hexagon head of the rod until the correct clearance is obtained. Note: When loosening the lock nut, hold stamped nut (32) with a wrench to avoid shearing the locating pin in the position control arm (31). When correct clearance has been obtained, tighten lock nut and recheck the adjustment.

90B. If adjusting gage is not available, adjust the linkage as outlined in

paragraph 90A until the clearance between the control valve land and the machined surface of the housing is 0.449.

NOTE: If in attempting to make position control adjustment, proper adjustment cannot be obtained as outlined in paragraph 90A or 90B, the hydraulic lift system linkage is either improperly assembled, damaged or worn, and the linkage must be serviced as outlined in paragraph 99.

LIFT COVER AND CYLINDER, OVERHAUL

The lift cover assembly includes the rock (lift) shaft, control quadrant, lift cylinder, main control valve, unloading valve, safety valve and check valve. The system relief valve is located in the hydraulic pump, and the back pressure valve is located at the lower end of system return line.

91. **REMOVE AND REINSTALL.** To remove the lift cover, disconnect lift arms (65—Fig. FO563 or FO564) from lift links and remove clevis pin from main control spring yoke (63). Remove tractor seat, push selector valve knob in, place touch control lever (27) in lowering position and place selector lever in draft control position; then, force the oil from the lift cylinder by pushing the lift arms to their lowermost position. Remove the cover re-

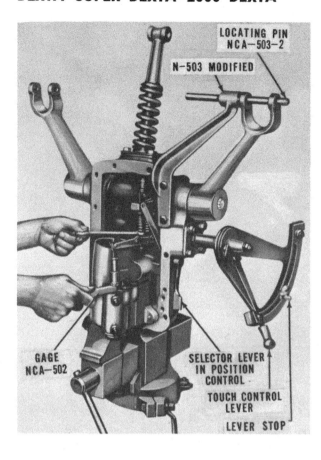

LOCATING PIN NCA-503-2

N-503 MODIFIED

GAGE NCA-502

SELECTOR LEVER IN POSITION CONTROL

TOUCH CONTROL LEVER

LEVER STOP

Fig. FO562A—Method of setting and adjusting the linkage for implement position control.

taining cap screws, attach a suitable hoist to cover and lift the assembly off the center housing.

When reinstalling the cover, use a new gasket and "O" rings and tighten the retaining cap screws to a torque of 30-35 ft.-lbs.

92. **CHECK VALVE AND SEAT.** To remove the check valve (41—Fig. FO-563 or FO564) first remove the plug (36) located in front flange of cover (44). With a pair of needle nose pliers, grasp the protruding end of pilot (37) and pull same from cover. The spring (39), spring seat (40) and ball (41) can then be withdrawn from cover. Examine the check valve seat (43), ball (41), spring (39) and "O" rings (38 and 42) and renew if they are chipped, worn or scored. The check valve seat should always have a sharp edge. To remove the valve seat (43) use Ford puller NCA997A, or a suitably threaded rod, nut and large washer as a puller. Oil passage in front of seat is threaded to receive the puller. Caution: When pulling the check valve seat, be extremely careful that the pulling screw remains centered in the bore. The hardened seat is extremely brittle, and misalignment may break the necked portion at the location of the hole.

93. **CYLINDER SAFETY VALVE.** If the cylinder safety valve (16—Fig. FO563 or FO564) only is to be removed, this can be accomplished without removing the lift cover from the tractor. To remove the valve, unbolt and remove the accessory plate (56— Fig. FO563) or flow control valve (54 —Fig. FO564) exposing the safety valve. The valve can then be unscrewed from the cylinder using the correct size deep socket. Early production safety valve was factory preset at 2400 psi. The safety valve in all models after Fordson Dexta Serial No. 957E-59444 is factory pre-set at 2750-2850 psi. If service is indicated renew the valve assembly, preferably with the later 2750-2850 psi valve assembly.

93A. **ACCESSORY PLATE.** The accessory plate (56—Fig. FO563) furnished as standard equipment on Fordson Dexta models prior to tractor Serial No. 957E-68355 is equipped with a plunger type selector valve (48) and jack tapping threaded for ½-inch tapered pipe threads. To remove the selector valve or accessory plate, first place selector lever in draft control, push remote cylinder selector knob in and move hydraulic control lever to bottom of quadrant. After exhausting

all oil from lift cylinder by pushing lift arms to bottom, unbolt and remove plate from lift cover. To remove valve, loosen locknut on locking plunger (49) in front of housing next to the selector knob. Back out the plunger assembly and withdraw selector valve (48). The selector valve is a selective fit in the accessory plate bore. Replacement plungers are 0.7482-0.7497 in diameter and are color coded green, white, blue, yellow, and orange; orange being the largest size. In fitting a new valve, use the largest size that will operate without binding. Screw the locking plunger (49) into the housing until detent can be felt, but so that the valve will still operate freely. Lock in place with the locknut.

93B. **FLOW CONTROL VALVE.** Fordson Dexta models after tractor Serial No. 957E-68355, Fordson Super Dexta and Ford 2000 Super Dexta tractors are equipped with a flow control valve to regulate hydraulic pump output instead of the accessory plate described in paragraph 93A. Although the flow control valve also incorporates a remote cylinder selector valve, the selector valve spool is not interchangeable with that used in the accessory plate. An exploded view of the flow control valve unit is shown in Fig. FO564. Turning the knob (51) in or out moves a restrictor valve (51B) between maximum and minimum flow positions. Restricting the flow of oil from the hydraulic pump causes the shuttle valve (52) to move against pressure of the spring (53) and also against hydraulic back pressure at the spring end of the valve plunger. Movement of the valve, which is related to the position of the restrictor valve, by-passes a varying amount of oil back to the sump and thereby regulates the amount of oil flowing to the tractor lift cylinder or remote cylinder. Marks "F" (fast) and "S" (slow) cast into the valve housing indicate maximum and minimum flow positions.

The main hydraulic control lever is equipped with a moveable spacer which, when moved to a position between the lever and flow control valve linkage, contacts the linkage and moves the valve to "F" (maximum flow) position whenever the control lever is moved to full raise position.

To remove the flow control valve assembly, first push the remote cylin-

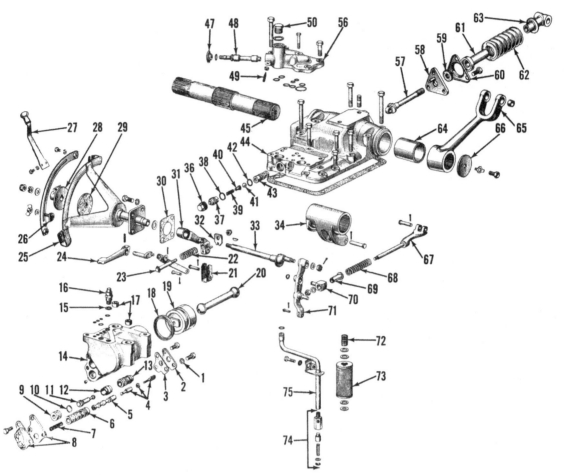

Fig. FO563 — Exploded view of early Fordson Dexta hydraulic lift cover, cylinder and linkage. Accessory plate (56) can be replaced with flow control valve (54—Fig. FO564) if complete valve and linkage are used. Although component parts are different, complete lift cylinder assembly (14) is interchangeable with later type complete lift cylinder assembly (14A — Fig. FO564).

1. Sealing washer	18. Piston seal	36. Check valve plug	59. Felt seal
2. Rear cover	19. Piston	37. Check valve pilot	60. Seat support
3. Gasket	20. Piston rod	38. "O" ring	61. Spring seat
4. Control valve link	21. Control cam	39. Check valve spring	62. Main control spring
5. Control valve	22. Position control spring	40. Check valve spring guide	63. Control spring yoke
6. Control valve bushing	23. Position control rod	41. Check valve	64. Bushings (2)
7. Control valve spring	24. Position control selector lever	42. "O" ring	65. Lift arm
8. Baffle plate	25. Quadrant	43. Check valve seat	66. Retaining washer
9. Unload valve plug	26. Lever stop	44. Lift cover	67. Draft control link
10. Unload valve "O" ring	27. Control lever	45. Lift arm cross shaft	68. Over-ride spring
11. Unload valve	28. Friction plate	47. Remote cylinder selector knob	69. Bushing
12. Unload valve bushing (front)	29. Friction disc	48. Selector valve spool	70. Draft control swivel
13. Unload valve bushing (rear)	30. Gasket	49. Detent assembly	71. Valve control lever
14. Lift cylinder	31. Position control arm	50. Jack tapping plug	72. Spring
15. Copper gasket	32. Stamped adjusting nut	56. Accessory plate	73. Oil filter element
16. Safety valve	33. Control lever shaft	57. Control spring plunger	74. Back pressure valve
17. Dowel pins	34. Ram lift arm	58. Retaining plate	75. Return tube

der selector knob (47) in, move the hydraulic control selector lever to draft control position and push the hydraulic control lever to bottom of quadrant. Exhaust all oil from tractor lift cylinder by pushing lift arms to bottom; then, unbolt and remove the flow control valve assembly from hydraulic lift cover.

Disassembly procedure for the flow control valve is evident after inspection of unit and reference to Fig. FO565. The flow control valve plunger (52) and remote cylinder selector spool (48A) are selective fit. When renewing a plunger or spool, select the largest size that will fit in the bore without binding. Sizes are color marked as follows:

	Flow Control Valve Color	Selector Valve Spool Color
Smallest Dia.	Red	Green
. . . .	Yellow	White
. . . .	Blue	Blue
. . . .	Green	Yellow
Largest Dia.	White	Orange

Lubricate all valve parts and reassemble using new "O" rings and gasket.

LIFT CYLINDER

94. **CYLINDER R&R AND INSPECT.** To remove work cylinder (14 —Fig. FO563 or 14A—FO564) from lift cover, disconnect the control valve linkage pin and remove the link (4— Fig. FO563 or FO564). Remove the

four cap screws retaining cylinder to lift cover and remove the cylinder and control valve housing. The piston (19) can be removed from cylinder with compressed air. Examine the piston and cylinder for wear or scoring and renew the piston seal (18) when reinstalling.

NOTE: The hydraulic lift cylinder was changed in production at Fordson Dexta tractor Serial No. 957E-68355 and, although the cylinder is interchangeable as a unit with prior production cylinders, the control valve, control valve bushing and certain other cylinder parts are not interchangeable. It is important therefore that the cylinders can be correctly identified when obtaining cylinder service parts. The early production cylinders can be identified by the valve retaining plate (2 and 8—Fig. FO563) at each end of

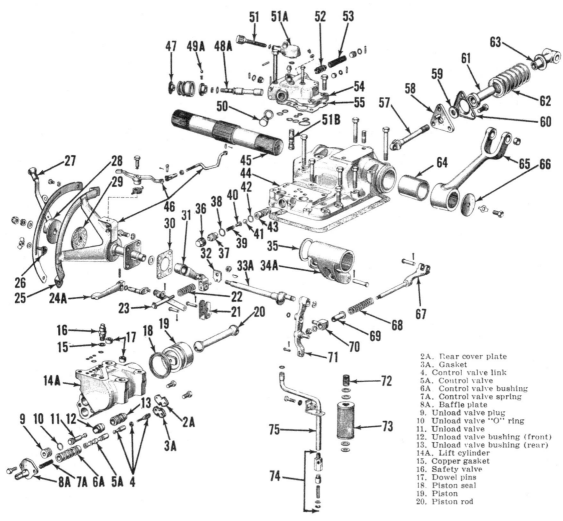

Fig. FO564 — Exploded view of late hydraulic lift cover, cylinder and linkage. Ram lift arm (34A) and spacer (35) may be used to replace early production lift arm (34—Fig. FO563). Control lever shaft (33A) and snap ring may be used to replace early production shaft (33—Fig. FO564), washer and nut.

2A. Rear cover plate			
3A. Gasket			
4. Control valve link			
5A. Control valve			
6A. Control valve bushing			
7A. Control valve spring			
8A. Baffle plate			
9. Unload valve plug			
10. Unload valve "O" ring			
11. Unload valve			
12. Unload valve bushing (front)			
13. Unload valve bushing (rear)			
14A. Lift cylinder			
15. Copper gasket			
16. Safety valve			
17. Dowel pins			
18. Piston seal			
19. Piston			
20. Piston rod			

21. Control cam	35. Spacer washer	49A. Detent assembly	62. Main control spring
22. Position control spring	36. Check valve plug	50. Jack tapping plug	63. Control spring yoke
23. Position control rod	37. Check valve pilot	51. Restrictor adjusting knob	64. Bushing (2)
24A. Position control selector lever	38. "O" ring	51A. Restrictor control lever	65. Lift arm
25. Quadrant	39. Check valve spring	51B. Restrictor valve	66. Retaining washer
26. Lever stop	40. Check valve spring guide	52. Flow control valve spool	67. Draft control link
27. Control lever	41. Check valve	53. Flow control valve spring	68. Over-ride spring
28. Friction plate	42. "O" ring	54. Flow control valve housing	69. Bushing
29. Friction disc	43. Check valve seat	55. Gasket	70. Draft control swivel
30. Gasket	44. Lift cover	57. Control spring plunger	71. Valve control lever
31. Position control arm	45. Lift arm cross shaft	58. Retaining plate	72. Spring
32. Stamped adjusting nut	46. Flow control valve linkage	59. Felt seal	73. Oil filter element
33A. Control lever shaft	47. Remote cylinder selector knob	60. Seat Support	74. Back pressure valve
34A. Ram lift arm	48A. Selector valve spool	61. Spring seat	75. Return tube

the cylinder being secured with three cap screws; whereas, the valve retaining plates (2A and 8A—Fig. FO564) on late production cylinders are secured with only two cap screws. Thus, if a cylinder has three valve retaining plate cap screws, order parts identified in Ford Tractor Parts Catalog as being used in year range 11/57/60/10; if two valve retaining plate cap screws are used, order parts identified for year range 10/60.

95. R&R CONTROL VALVE AND BUSHING. (USED 11/57/60/10.) To remove the control valve (5 — Fig. FO563) from the housing, remove the baffle plate (8) and spring (7) from front (closed) end of cylinder then remove the plate (2) and valve (5)

from the open end. Note that the retaining cap screw directly above the control valve is fitted with a copper sealing washer. The hole into which this cap screw is threaded serves as a lateral oil drilling in the control valve body and must always be sealed with the sealing washer.

The control valve bushing (6) may be renewed by pressing it from the housing. Select a new bushing of the correct color code and press it into the housing in the same relative position. The end of the bushing containing the exhaust (small) hole is located at the rear (open) end of the cylinder and the front end of the bushing must be

flush with the machined face of the front (closed) end of housing. Fig. FO566 shows the special Ford designed tool recommended by the manufacturer for removing and reinstalling the valve bushings.

The land at the rear end (end towards linkage) of the control valve covers the cylinder exhaust (lowering) port in the control valve bushing; wear on the valve and bushing at this location subjects the cylinder circuit to leakage which could cause bobbing (hiccups).

96. R&R CONTROL VALVE AND BUSHING (USED 10/60/-.) To remove the control valve from the

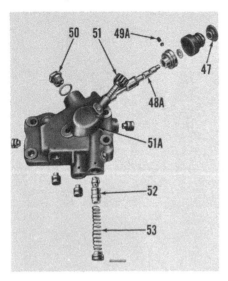

Fig. FO565 — Exploded view of flow control valve. Refer to Fig. FO564 for legend. Restrictor valve (51B) is not shown in above view.

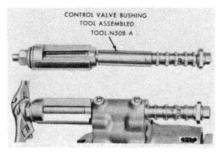

Fig. FO566—Threaded rod type push puller designed by Ford, for removing and installing valve bushings.

housing, remove the baffle plate (8A —Fig. FO564) and spring (7A) from closed (front) end of cylinder and remove the plate (2A) and valve from rear of cylinder. A gasket is used between the rear plate and cylinder similar to prior production but the cap screw that was threaded into an oil passageway has been eliminated.

To renew the control valve bushing, press bushing out towards rear (open) end of cylinder. Select a new bushing of correct color code and press bushing into front (closed) end of cylinder. When in position, bushing must be flush with machined surface at front (closed) end of cylinder. End of bushing with deep counterbore must be to rear (open) end of cylinder.

The land on the front end of the control valve covers the cylinder exhaust (lowering) port in the control valve bushing; wear on the valve and bushing at this location subjects the cylinder circuit to leakage which could cause bobbing (hiccups).

97. SELECTING PROPER SIZE OF CONTROL VALVE AND BUSHING. When renewing the control valve and bushing, care must be taken that the proper size valve and bushing are selected due to the exceptionally close fit required of these parts. Good service procedure requires that both the valve and bushing be renewed whenever renewal of either is indicated.

Control valve bushings are sized in steps of 0.0002 from an outside diameter of 1.0000-1.0002 to 1.0014-1.0016. Color codes for each size range are as follows:

Blue/White..........Smallest dia.
White
Blue
Yellow
Green
Orange
Green/White
Red/White............Largest dia.

The control valve bushing bore in each lift cylinder is measured at the factory and a spot (or spots) of paint is applied on the outside of the cylinder adjacent to the control valve bore to indicate the bore size. When renewing the control valve bushing, select a bushing having the same color code as appears on the lift cylinder. NOTE: Do not confuse the streak of paint near the control valve bore with the color code paint spot(s). The paint streak is marking applied for factory assembly only. Also, color code is applied on the cylinder near the unloading valve bore to indicate the size of the unloading valve bushing to use. The control valve bore and the unloading valve bore may or may not be of the same color code in a cylinder, so care must be taken that the correct color code mark is observed. The size code color mark will be a spot of paint as with blue color code, or two spots as with a blue/white color code.

Control valve spools are sized in steps of 0.0002 from a diameter of 0.5917-0.5919 to 0.5927-0.5928. Color codes for each size range are as follows:

White...............Smallest dia.
Blue
Yellow
Green
Orange...............Largest dia.

The correct size control valve spool can be selected only after the bushing is pressed into the lift cylinder. By trial and error, select a valve spool that will have a slight drag when moved back and forth in the bushing in its normal position, but without any binding or sticking tendency. Valve and bushing must be absolutely clean and lubricated with motor oil or hy-

draulic oil when checking fit of valve to bushing. As the color code indicates a size range only, it may occur that a valve will appear too tight or too loose when another valve of the same color code will fit properly. Also, if all valves on hand have been tried without finding one that fits and another bushing of the correct color code is available, install another bushing in the housing and try the fit of the control valves again. The inside diameter of the bushing is not related to the bushing color code.

98. UNLOADING VALVE AND BUSHING. The unloading valve (11—Fig. FO563 or FO564) can be removed after extracting the pressed in plug (9) at closed end of cylinder. Sealing between the unloading valve and the bushing is by means of an "O"-ring. If valve or bushings are eroded or scored, they should be renewed. The bushings (12 and 13) for the unloading valve can be removed using the same Ford threaded rod-type puller N508A used to extract and install the control valve bushing. The unloading valve front bushing (12) should be fitted with the two large notches against the rear bushing. The unload valve plug (9) seals the front end of the bushing except for the area of the small notch. Press unloading valve bushings out towards front (closed) end of cylinder. To install bushings press thin front bushing in with rear bushing. Shoulder on rear bushing must be flush with machined face of lift cylinder.

98A. UNLOADING VALVE BUSHINGS, PLUG, VALVE AND "O" RING SIZE SELECTION. The unloading valve bushings and the bore sealing plug are sized in steps of 0.0002 from an outside diameter of 1.0000-1.0002 to 1.0014-1.0016. Color codes for each size range are the same as outlined for the control valve bushing in paragraph 97. When renewing the unloading valve bushings or plug, select new parts of the same color code as appears on the lift cylinder adjacent to the unloading valve bore.

As the unloading valve functions as a flow director only, a close fit is not required between the unloading valve and bushings. Therefore, there is only one size unloading valve available and it should fit all bushings. However, it is very important that no binding condition is present when the valve is installed. After the two unloading valve bushings have been installed as outlined in paragraph 98, lubricate and install unloading valve **without** the sealing "O" ring. The valve should slide back and forth freely. Correct any binding condition before proceed-further. Remove the unloading valve and install the sealing "O" ring; then, lubricate valve and reinstall it. A slight drag from the "O" ring should be noted. If not, or if valve then binds in bushings, select another "O" ring that will effectively seal valve to bushing without binding. CAUTION: Do not attempt to install an "O" ring of unknown quality at this location. Some materials used in "O" ring manufacture may shrink or swell when subjected to hydraulic oil and cause malfunction of the hydraulic system.

NOTE: It is possible that misalignment of the two unloading valve bushings may cause the unloading valve to bind. If binding condition cannot be eliminated by changing "O" rings or installing another valve, renew the unloading valve bushings as outlined in paragraph 98.

ROCK SHAFT AND CONTROL LINKAGE

99. The rock shaft (45 — Fig. FO-563 or FO564) and lift arms (65) can be removed from the lift cover by removing the retaining cap screw and washer (66) from one end of the rock shaft and sliding the shaft in the oposite direction out of lift cover. The loose rock shaft bushing (64) in one side of the cover will be removed with the shaft. The ram arm (34 or 34A) and two lift arms (65) on the rock shaft are fitted with a master spline and cannot be assembled in the wrong position. Ram arm (34) is interchangeable with late ram arm (34A) and washer (35). Unscrew the control spring yoke (63) from the rear of the control spring (62). Remove the three cap screws retaining the control spring seat (58) and remove the seat and retainer (60). Remove the nut or snap ring from the inner end of control lever shaft (33 or 33A) and remove the draft control linkage. If difficulty has been experienced getting or keeping the correct draft control adjustment, disassemble and remove the draft control link (67) and check the link against a new one. This link will sometimes become bent, and due to its shape, is very difficult to detect. Examine the remainder of the parts for breakage, bent parts or wear and renew if necessary. The self-locking nut at forward end of draft control link (67) on early models should be tightened until it bottoms on the link shoulder. Adjust the castellated nut at inner end of control lever shaft until the parts

are free but no end play exists and install the cotter key. Late control lever shaft (33A) and snap ring are interchangeable with early control lever shaft (33), washer and self-locking nut. Readjust the hydraulic linkage as outlined in paragraphs 88, 89 and 90 before installing lift cover on the tractor.

RETURN LINE, FILTER AND BACK PRESSURE VALVE

100. The return line (75—Fig. FO563 or Fig. FO564), complete with filter (73) and back pressure valve (74), can be removed after lift cover is off, by removing the screw retaining return pipe to rear axle center housing. Lower the return line to clear the cover flange and withdraw the assembly from the housing.

The manufacturer states that the return line filter (73) will not need to be renewed except at times of major overhaul or when the oil has become contaminated. To renew the filter, unscrew the back pressure valve housing (74) from the lower end of the return line, remove the retaining washer and seal and withdraw the filter.

To disassemble the back pressure valve, remove the wire retainer from lower end of valve housing. The retaining plate spring and valve can now be withdrawn. The valve housing will not need to be removed from the return line for service on the valve. Examine the valve body and valve for scoring, sticking or wear and renew as required.

HYDRAULIC PUMP

101. **PRESSURE RELIEF VALVE.** The hydraulic pump pressure relief valve is located in the pump housing. To check the pressure, remove the pressure plug in the center of the pump mounting flange immediately above the right step plate and fit at least a 3000 psi pressure gage to the opening. Due to the design of the pressure relief valve, once the relief pressure of the valve has been reached and the valve unseated, it will remain unseated at low pressure until the hydraulic control valve is returned to neutral position. For this reason, the manufacturer recommends that a pressure line including a needle - type shut off valve be installed in the jack tapping on the accessory plate, the exhaust oil being

Fig. FO567 — Suggested arrangement for pump pressure testing. Shut off valve is to assist in slowly increasing pressure to relief valve setting. Relief valve is designed so that once valve is activated, system pressure drops, making actual relief pressure difficult to read.

returned to the reservoir as shown in Fig. FO567.

NOTE: The pump relief pressure will depend upon whether the pump is fitted with the early type relief valve shown in Figs. FO568 and FO568A or the later type relief valve shown in Fig. FO568B. Although individual components are not interchangeable, the new type relief valve assembly shown in Fig. FO568B may be installed in prior production tractors to raise the pump relief pressure from 2100-2300 psi to 2450-2500 psi. Factory installation of the later type valve was effective at Fordson Dexta Tractor Serial No. 957E-49624.

To make the test, start the tractor engine and bring the speed up to 1550 rpm. Move the auxiliary control valve knob to the outer position, open the needle valve on the return hose and raise the touch control lever to top of quadrant. While observing the gage needle carefully, very slowly close the needle valve in the return hose and note the highest reading obtained before the relief valve opens. The maximum reading obtained at the time the relief valve opened should have been 2100-2300 psi with early type valve or 2450-2500 psi with late type valve. The pressure then should drop to approximately 600 psi with early type valve or 300 psi with late type valve and hold steady at that point.

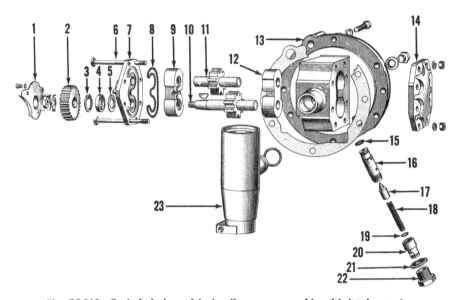

Fig. FO568—Exploded view of hydraulic pump assembly with intake strainer.

1. Shroud	5. Washer	9. Front bearing	13. Pump body	17. Relief valve	21. Sealing washer
2. Drive gear	6. Dowel bolt	10. Pump drive	14. Rear cover	18. Valve spring	22. Plug
3. Snap ring	7. Front cover	shaft & gear	15. "O" ring	19. Adjusting shim	23. Intake strainer
4. Seal	8. "O" ring	11. Driven gear	16. Relief valve body	20. Spring retainer	**housing**
		12. Rear bearing			

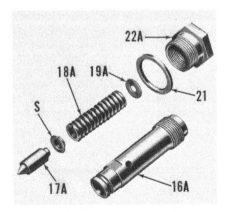

Fig. FO568A — Exploded view of early production pressure relief valve assembly. Refer to Fig. FO568 for legend.

Fig. FO568B — Exploded view of late production pressure relief valve assembly.

S. Spring seat	19A. Shims
16A. Valve body	21. Sealing washer
17A. Valve	22A. Plug
18A. Spring	

If the pressure hose and needle valve are not available, an accurate reading is very difficult to obtain. Due to the action of the unloading valve, feathering of the control valve is impossible. Once the control valve is moved to the raised position, the gage needle will flicker to the relief valve pressure and immediately drop. After observing the action several times a fairly accurate reading may be obtained.

To adjust the relief valve pressure, add or remove shims (19—Fig. FO568A or 19A—Fig. FO568B) as required. Shims are available in 0.005, 0.010, 0.015 and 0.025 thickness for early type valve and 0.010 and 0.025 for late type valve. Adding or removing 0.005 in shims will change the relief pressure approximately 50 psi. Note: Do not increase the total shim pack to more than the recommended 0.080.

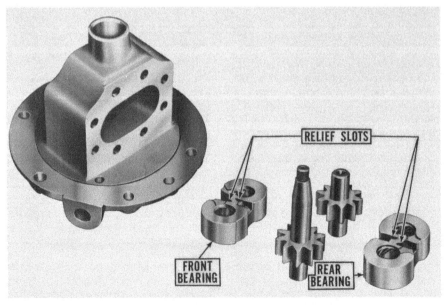

Fig. FO569 — When assembling pump, arrange parts in the order shown, so that front and rear bearings can be identified. Although bearings are very similar, position of relief slots differ in front and rear bearings.

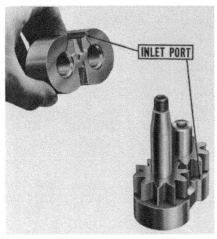

INLET PORT

Fig. FO570 — Assembly procedure of the hydraulic pump gears and bearings. Note the large cut-out area of bearings which are to be installed to the inlet port side of the pump housing.

Reinstall the pressure relief valve and recheck the pressure before releasing the tractor for service.

102. **REMOVE AND REINSTALL.** To remove the hydraulic pump for service, first drain the hydraulic system and remove the lift cover as outlined in paragraph 91.

Note: If other service work required makes it necessary to detach the transmission from the rear axle center housing, the required operations can be done from the front of the housing, without removing the lift cover.

Unbolt and remove the right hand step plate and disconnect the right brake linkage. Remove the six cap screws retaining pump mounting flange to rear axle center housing and withdraw pump assembly.

The pump inlet strainer (23—Fig. FO568) is retained to left side of rear axle center housing by a dowel screw. When reinstalling the hydraulic pump, first clean the inlet strainer and make sure the strainer

housing is properly located over the dowel screw. Renew "O" rings on inlet strainer and outlet passage of the pump mounting flange when reinstalling the pump. The design of the inlet housing makes it impossible to properly locate it in the center housing without having access through the top cover or housing front opening. Tighten the retaining cap screws to a torque of 30-35 ft.-lbs.

103. **OVERHAUL.** To disassemble the hydraulic pump, first remove drive gear shroud (1—Fig. FO568) and unbolt and remove the driving gear (2) from pump drive shaft (10) with a suitable puller. Remove the Woodruff key from the pump shaft, examine the shaft and remove any nicks or burrs which might damage the pump bearings. Remove nuts, washers and bolts retaining pump end plates (7 and 14) and remove the plates.

Note: The upper and lower bolts (6) nearest the mounting flange are special dowel bolts machined for a close fit in the plates and pump housing. These bolts are identified by a letter "D" stamped on the bolt head. During reassembly, these bolts must be reinstalled in the same two holes. If renewal is necessary, special dowel bolts must be obtained.

Remove and discard the sealing "O" rings in the two covers and renew the drive shaft seal (4) if it is worn or damaged.

Apply even pressure to the center of the rear pump bearing and slide the two bearings (9 and 12) with the pump gears from the housing as a unit. Examine the bearings for wear or scoring on their faces or shaft journals. If the bearing bores are worn out of round more than 0.001, they should be renewed. The bearings should always be renewed in pairs and must not be mixed. Although similar in appearance, bearings are not interchangeable.

Examine the pump housing (13) for wear in the gear running track. If track is worn deeper than 0.0025 on the pump inlet side, the housing, gears and bearings should be renewed.

Examine the pump gears (10 and 11) for wear or scoring on the gear faces or shafts. The gear width and diameter should be identical within 0.001 for the two gears and over the entire surface of each gear.

To correctly assemble the pump, lay the rear bearing (12) gear face up, on a clean shop towel. Position the bearing so that milled inlet (large) port is to the right and the outlet port to the left as shown in Fig. FO569. Note that the two shaft areas are relieved to the inlet side. The rear bearing can be identified by the fact that the relief slot from the gear mesh area to the inlet port will be at the bottom while the relief slot from the gear mesh area to the outlet port will be at the top of the bearing as shown. Position the driven gear, either end up in the upper bore of the rear bearing and the driving gear, long shaft up, in the lower bore, as shown in Fig. FO570. Place the front bearing over the gear shafts with inlet port to the right and insert the assembly in the pump housing so that pump drive shaft will be forward and to the bottom when pump is reinstalled on the tractor. The large inlet ports in both bearings should now align with the inlet port of the pump housing. Renew the sealing "O" rings in the cover plates and install them so that the straight side of the "O" rings is on the side nearest the pump mounting flange. Insert the two dowel retaining bolts in the upper and lower hole nearest the mounting flange and fit the remaining bolts in the other six holes and reinstall the lock washers and nuts. Tighten the nuts evenly to a torque of 40-50 ft.-lbs., reinstall the drive gear and shroud and fill the pump with a small quantity of hydraulic fluid to provide initial lubrication when starting the tractor.

NOTES

FORD

Models ■ Fordson Major Diesel (FMD) ■ Fordson Power Major (FPM)
■ Fordson Super Major (FSM)
■ New Performance Super Major (New FSM) ■ 5000 Super Major

Previously contained in I&T Shop Service Manual No. FO-21

SHOP MANUAL

FORD

MODELS
FORDSON MAJOR DIESEL (FMD)
FORDSON POWER MAJOR (FPM)
FORDSON SUPER MAJOR (FSM)
NEW PERFORMANCE FORDSON SUPER MAJOR (New FSM)
FORD 5000 SUPER MAJOR (Same As New FSM)

This manual provides service procedures and specifications on the Fordson Major series of tractors manufactured in England and imported to the United States, beginning with approximately tractor Serial No. 1260402.

On early tractors the serial (engine) number is stamped on the flange of the flywheel housing near the starter mounting.

Effective with tractor (engine) Serial No. 1380939, the serial number is located on a pad at the front right hand side of the engine cylinder block.

In November, 1961, the serial number designation was changed from a 7 digit number to a six digit number with prefix and suffix. (Note: When this new numbering system was introduced, only the six digit number was stamped on the tractor; later a "Z" was stamped in front of the prefix, but was deleted after a short time.)

As the Fordson Major Tractor Parts List indicates parts usage by month and year of tractor production instead of serial number or model range, the following list indicating month and year of production by serial numbers is presented below:

FORDSON MAJOR TRACTOR SERIES SERIAL NUMBERS

The following list of serial numbers for the Fordson Major series of tractors represents the serial number of the first tractor built on the first working day of each month from beginning of production in January, 1952, to end of production in October, 1964.

Month/Year	Starting Serial No.	Month/Year	Starting Serial No.	Month/Year	Starting Serial No.
		1/54	1276857	1/57	1412409
		2/54	1280461	2/57	1416126
		3/54	1284114	3/57	1420047
		4/54	1288616	4/57	1424724
		5/54	1292616	5/57	1429067
		6/54	1296979	6/57	1434128
		7/54	1301371	7/57	1438156
		8/54	1304721	8/57	1441861
		9/54	1308341	9/57	1444787
		10/54	1312911	10/57	1448456
		11/54	1316276	11/57	1452136
		12/54	1319466	12/57	1455496
				1/58	1458381
		1/55	1322525		
		2/55	1326304	2/58	1461911
1/52	1217101	3/55	1330197	2/58	1461811
2/52	1217854	4/55	1335206	3/58	1464968
3/52	1219501	5/55	1339093	4/58	1468222
4/52	1222168	6/55	1343610	5/58	1471551
5/52	1225184	7/55	1348338	6/58	1475102
6/52	1228560	8/55	1351565	7/58	1478284
7/52	1231013	9/55	1355435	8/58	1481013
8/52	1232538	10/55	1359668	9/58	1483139
9/52	1235064	11/55	1363538	10/58	1485781
10/52	1239010	12/55	1367817	11/58	1488927
11/52	1242232			12/58	1491814
12/52	1244823				
		1/56	1371418	1/59	1494448
1/53	1247381	2/56	1375378	2/59	1497684
2/53	1249734	3/56	1379563	3/59	1501006
3/53	1252374	4/56	1384154	4/59	1504869
4/53	1255494	5/56	1388054	5/59	1509598
5/53	1257474	6/56	1392039	6/59	1512807
6/53	1259074	7/56	1395222	7/59	1517042
7/53	1260753	8/56	1398262	8/59	1520046
8/53	1262438	9/56	1400956	9/59	1522832
9/53	1264418	10/56	1403907	10/59	1526968
10/53	1267672	11/56	1406790	11/59	1531041
11/53	1271038	12/56	1409843	12/59	1534587
12/53	1273713				

INDEX (By Starting Paragraph)

INDEX CONT.

CONDENSED SERVICE DATA

GENERAL

Engine Make	Own
No. Cylinders	4
Bore, Inches	3.937-3.938
Stroke, Inches	4.524-4.528
Displacement, Cubic Inches	220.35
Compression Ratio	16:1
Pistons Removed From:	Above
No. Main Bearings	5
Cylinder Sleeves, Type	Wet
Forward Speeds	6
Reverse Speeds	2
Starter & Generator Make	Lucas
Injection Pump & Injector Make	Simms

TUNE-UP SPECIFICATIONS

Firing Order	1, 2, 4, 3
Valve Gap Intake	0.015-H
Exhaust	0.012-H
Valve Face Angle, Degrees	29-1/2
Valve Seat Angle, Degrees	30
Injection Timing	See Index
Engine Low Idle RPM	550
Engine High Idle RPM:	
Model "New FSM"	1925
All Other Models	1900
Engine Oil Pressure, PSI	30-40

SIZES—CAPACITIES—CLEARANCES

Crankshaft Main Journal Diameter	3.0002-3.0010
Crankpin Journal Diameter	2.4997-2.5005
Camshaft Journal Diameter	2.0595-2.0600
Piston Pin Diameter:	
Prior To Serial No. 1425097	1.250
After Serial No. 1425096	1.375
Intake Valve Stem Diameter	0.373-0.374
Exhaust Valve Stem Diameter	0.3723-0.3733
Camshaft Cam Lift:	
Prior to Serial No. 1425097	0.258

Serial No. 1425097 To 1481090	0.305
Serial No. 1481091 To 1609838	0.258
After Serial No. 1609838	0.255
Piston Ring Width:	
Compression Rings	0.0933
Oil Rings	0.187
Piston Ring End Gap, All Rings	0.011-0.016
Ring Side Clearance In Groove:	
Compression Rings	0.0014-0.0034
Oil Rings	0.0015-0.0035
Crankshaft End Play	0.002-0.010
Max. Runout At Flywheel Clutch Face	0.005
Main Bearing Running Clearance	0.0025-0.004
Rod Bearing Running Clearance	0.002-0.0035
Camshaft End Play With:	
Gears Retained By 3 Cap Screws	0.003-0.008
Gears Retained By One Cap Screw	0.005-0.021
Piston Skirt Clearance	See Paragraph 77
Camshaft Bearing Running Clearance	0.002-0.0035
Cooling System Capacity	3.6 Gallons
Crankcase & Oil Filter	8 Quarts
Transmission Lubricant Capacity	5.4 Gallons
Final Drive Lubricant Capacity	10.8 Gallons
Hydraulic Fluid	Uses Final Drive Lubricant

TIGHTENING TORQUES (In Ft.Lbs.)

General Recommendations	See End of Shop Manual
Connecting Rod Nuts	55-60
Cylinder Head	85-90
Auxiliary Driveshaft Nut	60-70
Camshaft Gear Cap Screws:	
Single Cap Screw	95-100
Three Cap Screw	18-21
Main Bearing Cap Screws:	
With 1/2-in. Cap Screws	70-75
With 9/16-in. Cap Screws	95-100
With 5/8-in. Cap Screws	115-120
Flywheel Cap Screw	80-90
Front Mounting Plate	See Paragraph 54

FRONT SYSTEM, AXLE TYPE

WHEEL ASSEMBLY

1. **BEARING ADJUSTMENT.** Wheel bearings should be adjusted with clamp bolt (3—Fig. 1) removed from nut (2). Tighten nut while rotating wheel until a drag is noticeable. Then, back off until wheel turns free, insert cotter pin through nut and spindle and install and tighten the clamp bolt.

SPINDLES

2. **R&R SPINDLES.** To renew spindle, proceed as follows: Support front end of tractor and remove wheel, hub and bearing cones. Remove clamp bolt from steering arm (16 or 26—Fig. 1) and remove arm. Withdraw spindle from axle extension.

Install new thrust bearing (11) if worn or rough. Renew spindle bushings as outlined in paragraph 3 if

spindle to bushing clearance exceeds 0.013. Insert spindle in axle extension, install dust seal (15) and steering arm. While holding all end play from spindle, adjust clearance between steering arm and axle extension to 0.002-0.007 if a felt dust seal is used, or to 0.025-0.035 if late type rubber seal is installed. (Felt and rubber dust seals are interchangeable.) Note: It may be necessary to grind a wider bolt slot in steering spindle to obtain the specified clearance. Install steering arm clamp bolt and tighten securely. Reinstall wheel, hub and bearings assembly using a new felt seal (9). Adjust wheel bearings as outlined in paragraph 1.

3. **SPINDLE BUSHINGS.** Spindle bushings can be renewed after removing spindle as outlined in paragraph 2. Remove old bushings with

cape chisel and install new bushings with OTC T-810 Bushing Driver and 815 Driving Mandrel or equivalent tools. After installing bushings, ream to 1.500 with standard 1½ inch adjustable reamer. NOTE: Upper and lower bushings are not alike; install upper bushing with blind end of lubrication groove up.

Spindle diameter (new) is 1.498-1.499 providing 0.001-0.003 clearance in the 1.500-1.501 diameter bushings. Maximum allowable spindle to bushing clearance is 0.013.

AXLE EXTENSIONS

4. Axle extension (13 or 27—Fig. 1) can be renewed after removing spindle as outlined in paragraph 2. When installing extension, be sure that one retaining bolt is placed in the outer hole in center member and the other two bolts at each side of the radius rod bolt.

AXLE CENTER MEMBER, PIVOT PIN AND BUSHING

5. To renew the axle center member (beam) or pivot pin (trunnion) bushing, support tractor under front end of transmission housing and proceed as follows: Remove bolts from tie rod clamps (24—Fig. 1), unbolt axle extensions from center member and remove wheels, axle extensions, steering arms and tie rod ends as units. Remove radius rod (refer to paragraph 7), then remove pivot pin (29) and lower axle center member from front support (cross member).

On "FMD" and "FPM" models, pivot pin (29) is retained by a cotter pin at each end and thrust washers are used at front and rear of axle center member. On "FSM" models, the pivot pin is of larger diameter, no thrust washers are used and pivot pin is retained in front support by two roll pins (see Fig. 2) or by a ½-inch diameter clevis pin and cotter pin. When reinstalling pivot pin with ⅜-inch hole, be sure to install roll pins as shown in Fig. 2. Only the latest type pivot pin with ½-inch hole will be serviced; to install this pin in early "FSM," enlarge retaining pin hole in front support to ½-inch and retain with clevis pin and cotter pin.

To renew pivot pin bushing (30—Fig. 1), remove old bushing with cape chisel and drive new bushing in flush with axle center member. Bushing should not require reaming if carefully installed. Bushing is not interchangeable between "FSM" and earlier models.

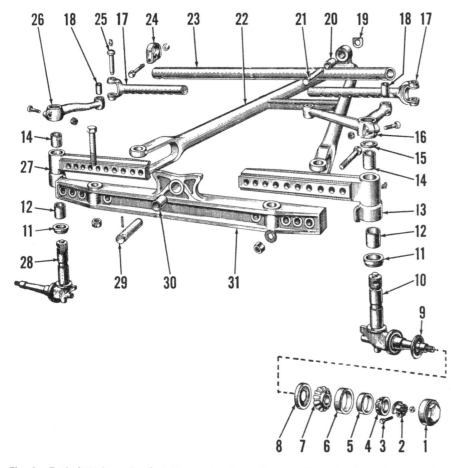

Fig. 1—Exploded view of adjustable front axle, radius rod, tie rod and steering arms for models "FSM" and "New FSM". Front axle assembly for earlier models "FMD" and "FPM" is similar except thrust washers are used on trunnion (29) at each side of axle center member (31) and thrust washer (19) is not used on radius rod pivot pin (21).

1. Hub cap	8. Seal retainer
2. Nut	9. Felt seal
3. Clamp bolt	10. Spindle, L.H.
4. Bearing cone	11. Thrust bearing
5. Bearing cup	12. Bushing, lower
6. Bearing cup	13. Axle extension, L.H.
7. Bearing cone	14. Bushing, upper

15. Dust seal	22. Radius rod
16. Steering arm, L.H.	24. End clamps
17. Tie rod ends	25. End pins
18. Bushings	26. Steering arm R.H.
19. Thrust washer	27. Axle extension, R.H.
20. Bushing	28. Spindle, R.H.
21. Pivot pin	

TIE ROD AND TOE-IN
All Models

6. The tie rod assembly, except for tread width adjustment, is non-adjustable. Toe-in should be correct at each tread width adjustment unless tie rod or steering arms are bent or excessive wear has occured. Zero toe-in (no toe-in or toe-out) is specified.

Pins (25) are a press fit in the tie rod ends (17) and bushings (18) are a press fit in the steering arms (16 and 26).

RADIUS ROD, REAR PIVOT PIN AND BUSHING

7. The radius rod (22—Fig. 1) pivots on a pin (21) retained in bosses on the bottom of engine oil pan (sump). Radius rod is fitted with a renewable, pre-sized steel bushing (20).

Due to increased diameter of pivot pin, radius rod, bushing, pivot pin and engine oil sump are not interchangeable between the "FSM" models and earlier models.

To remove radius rod, remove pivot pin and the bolts through radius rod and front axle center member. Slide radius rod to either side of tractor until clear of oil sump, then remove from tractor. Remove bushing with cape chisel and drive new bushing in until flush. Reinstall rear pivot retaining roll pins, if so equipped, as shown in Fig. 2.

FRONT SUPPORT

8. To renew the front support (cross member), first drain radiator, then proceed as follows:

Remove engine hood, disconnect headlight wiring and remove radiator shell from front support. Disconnect both radiator hoses and remove radiator from front support. Support tractor under front end of transmission and remove front axle pivot pin (trunnion). Unbolt and remove side rail from right side of tractor. Unbolt front support from left engine side rail, move support to right until clear of channel, then lift front support from tractor. Install new front support by reversing removal procedure.

Due to increased diameter of front axle pivot pin (trunnion), front support is not interchangeable between "FSM" models and earlier models.

NOTE: Due to change in radiator mounting, it may be necessary to relocate radiator mounting holes in new front support; refer to paragraph 134 and Fig. 88 in COOLING SYSTEM section.

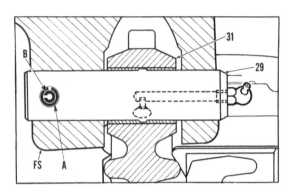

Fig 2—Cross-sectional drawing showing front pivot pin (29) installation on model "FSM" wide front axle. Note proper installation of inner (B) and outer (A) retaining roll pins. Front support is (FS), axle center member is (31).

FRONT SYSTEM, TRICYCLE

LUBRICATION

9. Oil level should be maintained at bottom of cover plate opening in front support (cross member) as shown in Fig. 3. Capacity is 4½ pints of SAE 90 gear lubricant. When refilling, remove bleed plug from upper front side of pintle housing (12) and add oil until oil flows from plug opening. Then reinstall plug and fill to proper level.

ADJUSTMENTS

10. **PINTLE BEARINGS.** Pintle shaft tapered roller bearings (11 & 14 —Fig. 3) are adjusted by tightening nuts at top end of pintle (13). Straighten tap of washer located between the two nuts and loosen top nut. Adjust lower nut to remove all end play from pintle shaft without causing binding. Then, while holding lower nut from turning, securely tighten top nut and bend tabs of washer against flats on nuts.

11. **BEVEL GEAR BACKLASH.** With pintle bearings properly adjusted as outlined in paragraph 10, vary number of shims (10—Fig. 3) located between cross shaft bearing (5) and front support (23) to remove all possible backlash between the bevel gears (3 and 4) without binding at any point in turning range of gears.

WHEEL ASSEMBLY

12. To remove wheel hub from axle, follow conventional procedure except unbolt seal retainer (20) from hub (21) before pulling hub from axle. Seal and retainer can be removed from axle after removing inner bearing cone and roller assembly. Reinstall by reversing removal procedure.

Adjust front wheel bearings on tricycle models as outlined for adjustable axle models in paragraph 1. Lubricate wheel bearings with pressure gun and multi-purpose grease; fill wheel hub with grease until grease appears from seal retainer.

AXLE SHAFT

13. To renew axle shaft (17—Fig. 3), support front end of tractor and remove front wheels, hubs, bearings and seals. Remove cotter pin and castellated nut from lower end of pintle (13) and pull axle from pintle. Renew key (18) if damaged. Install new axle and tighten retaining nut securely, then install cotter pin. Check for any end play or binding condition of pintle shaft and, if necessary, adjust pintle shaft bearings as outlined in paragraph 10.

PINTLE AND HOUSING

14. **R&R ASSEMBLY.** To remove pintle and housing with front wheels as an assembly, support front end of tractor and proceed as follows:

Remove bleed plug from upper front side of pintle housing and drain as much oil as possible. Reinstall plug, unbolt pintle housing from front support (cross member) and remove assembly from tractor.

When reinstalling, renew pintle housing to front support gasket and be sure that side of gear (3—Fig. 3) having the two punch marked teeth is to left side of tractor. Single punch marked tooth on cross shaft gear (4) must mesh between the two punch marked teeth on gear (3). Steering arm (9) should be nearly vertical when gears are properly meshed and wheels are in a straight ahead position. Refer to paragraph 9 for lubrication information.

15. **OVERHAUL PINTLE AND HOUSING UNIT.** Support front end of tractor and remove front wheels then, remove pintle, pintle housing and axle assembly as a unit following procedures outlined in paragraph 14. Turn unit upside down to drain oil from pintle housing.

Straighten tabs of washer located between the two nuts at top end of pintle and remove the nuts, washer and bevel gear (3—Fig. 3). Remove pintle housing from pintle and withdraw upper bearing cone and roller assembly from housing. Remove cotter pin and castellated nut from lower end of pintle and remove pintle from axle shaft (17). Remove key (18) from pintle, then remove pintle housing cover (16), seal (15) and lower bearing cone and roller assembly. Remove upper and lower bearing cups from housing (12) if renewal of bearings is indicated.

Renew oil seal (15) and other parts as necessary. Reassemble by reversing disassembly procedure. Align the two punch marked teeth of gear (3) with punch marked spline on pintle (13). If mark on spline is not visible, install gear with punch marked teeth 90 degrees to left of key (18). Securely tighten castellated nut and install cotter pin at bottom end of pintle, then tighten nuts at top of pintle to properly adjust pintle bearings as outlined in paragraph 10. Reinstall the assembly as outlined in paragraph 14.

CROSS SHAFT AND FRONT SUPPORT

16. RENEW CROSS SHAFT BEARING AND/OR SEAL. Remove bleed plug from upper front side of pintle shaft housing and drain oil level down below the cross shaft bearing. Remove

clamp bolt from steering arm and remove arm from cross shaft, leaving arm attached to steering drag link. Remove oil seal retainer (7—Fig. 3) and bearing (5), taking care not to lose or damage shims (10) located between bearing and front support. Remove old seal from bearing and install new seal with lip to inside. Using the removed shim pack or one of equal thickness, carefully install bearing (5) to avoid damage to seal (6). Reinstall seal retainer and steering arm. Check and readjust bevel gear backlash as outlined in paragraph 11.

17. RENEW CROSS SHAFT AND/OR FRONT SUPPORT. Drain radiator and proceed as follows: Remove pintle shaft, housing, axle and front wheels assembly as outlined in paragraph 14. Remove engine hood, radiator shell and radiator. Disconnect steering arm from cross shaft and unbolt and remove front support from engine side rails. Remove top cover plate. Remove seal retainer and "O" ring from around crank extension shaft at front of support (cross member). Remove cotter pin from crank extension shaft inside the front support and withdraw the shaft from rear of support. Unbolt and withdraw the cross shaft bearing as outlined in paragraph 16, then remove cross shaft from front support. If front support casting is to be renewed, remove the mud guard from old casting and install on new part.

Reassemble by reversing disassembly procedure. Renew cover plate gasket, crank extension shaft "O" ring and cross shaft seal. Be sure that the punch marked tooth on cross shaft gear meshes between the two punch marked teeth on pintle shaft gear.

STEERING GEAR

All Models

18. ADJUSTMENT. Steering gear adjustment is usually not required unless necessary to renew steering gear component parts. When overhauling steering gear assembly, adjust steering shaft and rocker shaft during reassembly as outlined in following paragraphs 19 and 20.

19. STEERING SHAFT. Steering gear adjustment is correct when, with steering retaining nuts tightened, steering shaft turns freely without binding and there is no noticeable end play in shaft.

To adjust steering shaft, refer to Fig. 4 and add or remove shims (20) as required to obtain proper adjustment. A gasket (19) should be placed on each side of shim stack. Be sure that holes (H) in gaskets and shims are aligned with oil holes in steering column and gear housing.

20. ROCKER SHAFT. Rocker shaft adjustment is correct when, with cover retaining bolts tightened, there is no noticeable end play in rocker shaft in mid (straight ahead) position. There will be some end play in rocker

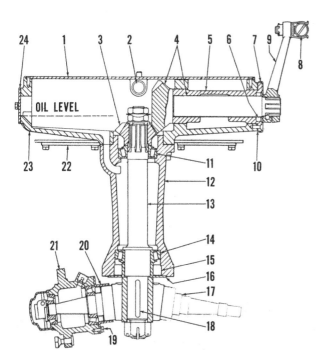

1. Top cover plate
2. Cranking extension shaft
3. Bevel gear
4. Cross-shaft
5. Cross-shaft bearing
6. Cross-shaft seal
7. Seal cover
8. Drag link
9. Steering arm
10. Shims
11. Upper pintle bearing
12. Pintle housing
13. Pintle
14. Lower pintle bearing
15. Pintle seal
16. Pintle housing cover
17. Axle
18. Key
19. Cap screws
20. Seal retainer
21. Wheel hub
22. Mud guard
23. Cross-member
24. Cover plate

Fig. 3—Drawing showing cross-sectional view of tricycle front pedestal (cross-member and pintle). Oil level is checked by removing plate (24). Wheel bearing detail is similar to that shown for adjustable front axle in Fig. 1, except that seal retainer (20) is secured to inner side of hub (21) by cap screws (19).

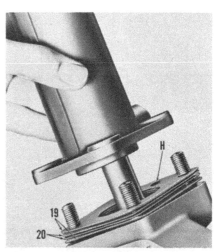

Fig. 4—Shims are used between steering column and gear housing to adjust worm-shaft bearings. A gasket (19) should be placed each side of shim stack (20). Be sure oil holes (H) in shims and gaskets are aligned with oil holes in housings.

shaft when at either side of mid position.

To adjust rocker shaft, refer to Fig. 6, be sure shaft (4) is in mid-position and that roller (6) is in place, then install cover (1) with shims as required to eliminate shaft end play. A gasket (3) should be on each side of shim stack. Note that roller (6) engages slot (S) in cover.

21. R&R STEERING GEAR UNIT. To remove the steering gear unit, first turn front wheels to mid (straight ahead) position, remove engine hood and steering wheel, then proceed as follows:

On models "FPM" and "FSM", disconnect battery ground cable and drain cooling system. Remove temperature gage bulb from engine and disconnect wiring to instrument panel at connectors on wiring harness. Remove pin or clamping screw from throttle control lever and remove the lever. Remove screws retaining instrument panel and remove the panel, disconnecting proofmeter (tachometer) cable as panel is lifted from fuel tank. On all models shut-off fuel supply valve and disconnect fuel supply and excess fuel return lines, unbolt fuel

tank from supports and lift fuel tank from supports and steering column.

Remove the nut retaining steering (drop) arm to steering gear rocker shaft and using gear pullers, remove arm from shaft. Then, unbolt and remove steering gear assembly from transmission.

When reinstalling steering gear assembly, be sure gasket surfaces are clean and apply thin coat of gasket sealer to gear housing and transmission mating surfaces. Install new gasket on top of transmission, then reinstall steering gear assembly. Turn gear unit to mid position and reinstall steering arm to rocker shaft. Complete the reassembly of tractor by reversing disassembly procedure and bleed the diesel fuel system as outlined in paragraph 107.

22. OVERHAUL STEERING GEAR UNIT. With steering gear unit removed as outlined in paragraph 21, drain lubricant from gear housing and proceed as follows:

Remove the side cover plate (1—Fig. 6) and roller (6), then withdraw rocker shaft (4) from housing. Retain shims (2) for reassembly. The rocker shaft seal can be renewed at this time. On early models, refer to Fig.

Fig. 6—Slot (S) in cover plate (1) guides roller (6) and ball nut in straight line parallel with wormshaft. Vary shims (2) as required to eliminate end play of rocker shaft (4).

7 and pry the seal (21) out of groove between bushing (13) and retainer (22). On later models, pry the lip type seal from shaft bore.

Refer to Fig. 4, and unbolt and remove steering column from gear housing and steering shaft. Lift steering shaft up far enough to remove upper bearing race (15—Fig. 5) and the ten loose steel balls (14). Remove steering shaft and ball nut assembly from side opening in housing and remove the ten loose steel balls of lower bearing from housing. If necessary to renew lower bearing race, remove plug (11) and drive race upward. Unscrew ball nut (9) fom worm on steering shaft and remove the 14 recirculating balls.

Inspect rocker shaft journals and bushings. If necessary to renew bushings on early models, refer to Fig. 7 and carefully remove the staking and seal retainer (22). On all models,

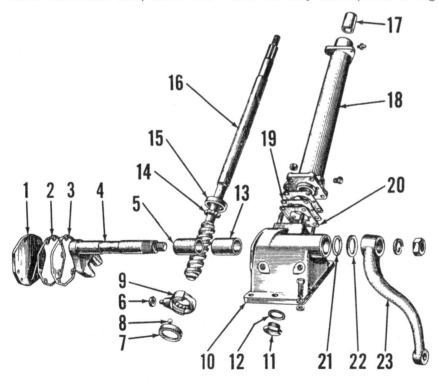

Fig. 5—Exploded view of typical steering gear assembly for adjustable front axle models. Tricycle steering gear is same as that shown except that a different drag link is used. Steering column (18) for "FMD" models does not have instrument panel support and different wormshaft (16) is required. Late production oil seal (21) does not require retaining ring (22).

1. Cover	7. Bearing race	12. Seal ring	18. Steering column
2. Shims	8. Steel balls	13. Bushing	19. Gaskets
3. Gasket	9. Main nut assy.	14. Steel balls	20. Shims
4. Rocker shaft	10. Housing	15. Bearing race	21. Oil seal
5. Bushing	11. Plug	16. Wormshaft	22. Retainer
6. Roller		17. Bushing	23. Steering arm

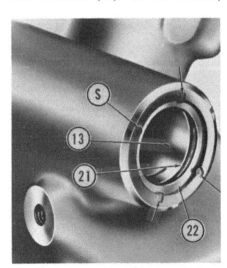

Fig. 7—Seal retainer (22) (not used on late models) is retained in gear housing by staking (S) as shown. Oil seal (21) can be renewed without removing retainer.

drive or press bushings from housing. When installing outer bushing (13) be sure open end of oil groove is to inside of housing. Bushings are pre-sized and should not require reaming if carefully installed. On early models, install oil seal retainer and carefully stake in position, then insert "O" ring in groove between bushing and retainer. On late models, drive seal into housing with lip to inside.

Renew steering shaft and ball nut assembly if shaft or nut are damaged.

To reinstall shaft and ball nut assembly, stick the 14 steel balls in nut with heavy grease and thread nut onto shaft. Install new lower race if necessary and stick ten steel balls into race. Insert shaft up through side opening and seat into lower bearing assembly. Stick ten steel balls into upper race, invert the bearing and install over steering shaft. Install new bushing (17—Fig. 5) in upper end of steering column, then install column over steering shaft with two gaskets

and proper number of shims as outlined in paragraph 19. Lubricate rocker shaft and seal, then carefully install shaft through bushings and seal. Place roller (6) on ball nut, then install side cover with two new gaskets and proper number of shims as outlined in paragraph 20. Refill steering gear housing to filler plug opening with SAE 90 gear lubricant and lubricate steering column bushing with pressure grease gun.

POWER STEERING

The power steering system consists of a belt driven hydraulic pump which furnishes pressurized oil, through flexible hoses, to a combined power cylinder and control valve assembly. The system utilizes the standard steering gear assembly which allows the operator to steer the tractor manually should the loss of power occur.

With the exception of the power cylinder, the component parts of the power steering system used on four wheel adjustable axle tractors and tricycle tractors remain the same. Since the power requirement of the tricycle models is less than that of four wheel tractors, the bore of the power cylinder for tricycle tractors is 1¾ inches compared to 2 inches for four wheel tractors.

FLUID AND BLEEDING
All Models

23. To bleed the system, turn front wheels to the right against stop. Fill reservoir to "full" mark on dip stick (see Fig. 9) with good quality SAE 10W oil. Note: If temperature is consistently below 10 degrees F. use

S.A.E. 5W oil. Start engine and run at a fast idle; then, turn front wheels full left and full right and observe return oil for bubbles and turbulence. Continue until the returning oil is free of bubbles and turbulence, then turn front wheels to the straight ahead position and refill reservoir to the "full" mark on dip stick.

SYSTEM OPERATING PRESSURE
All Models

24. A pressure test of the power steering circuit will disclose whether the pump, relief valve or some other unit in the system is malfunctioning; proceed as follows:

Connect a pressure test gage and shut-off valve in series with the pump pressure line and be sure that the pressure gage is connected in the circuit between the shut-off valve and the pump. Open the shut-off valve and run the engine at slow idle speed until the working fluid is warmed to normal operating temperature. Advance the engine speed to high idle rpm, close the shut-off valve and retain in the closed position only long

enough to observe the gage reading.

NOTE: Pump may be seriously damaged if valve is left in closed position for more than a few seconds. If the gage reading is 720-800 psi, with the shut-off valve closed, the pump and relief valve are O.K. and any trouble is located in the control valve, power cylinder and/or connections.

If the gage pressure is more than 800 psi, the relief valve is probably stuck in the closed position. If the gage pressure is less than 720 psi, renew the relief valve spring (25—Fig. 10) and recheck the pressure reading. If the gage pressure is still too low, it will be necessary to overhaul the pump as outlined in paragraph 26.

PUMP AND RESERVOIR
All Models

25. REMOVE AND REINSTALL. To remove the power steering pump proceed as follows: Remove the reservoir cover and withdraw as much oil as possible from the reservoir with a suction gun. Disconnect hoses from pump pressure and return line fittings and secure the hoses in a raised position to prevent oil drainage. Loosen

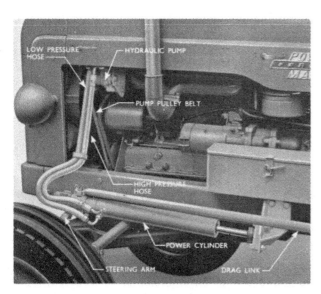

Fig. 8 — View showing power steering components on model "FPM" with wide adjustable front axle. Power steering systems for other models (including tricycle front end) are similar.

Fig. 9—Maintain fluid level to full mark on level indicator (dipstick) attached to pump reservoir filler cap.

the pump drive belt adjusting bolts and remove belt. Withdraw bolts and lift pump from tractor.

After pump is reinstalled, fill and bleed the system as in paragraph 23.

26. **OVERHAUL.** With pump removed from tractor as outlined in paragraph 25, refer to Fig. 10 and proceed as follows: Remove the reservoir spring and filter assembly, then clean the pump assembly. Remove the reservoir retaining cap screws and the reservoir stud (11), then lift reservoir and reservoir retaining plate (13) from pump body. Remove and discard the two "O" rings which are between reservoir and pump body. Remove the nut and lockwasher retaining pulley to pump shaft and using a suitable puller, remove pulley. Remove the four cap screws which retain pump cover (33) and separate pump cover and pump body (28). Remove and discard the two "O" rings from their grooves in pump body. Remove the pump rotors (23) and rotor drive key from pump shaft. Remove snap ring (16) retaining rotor shaft

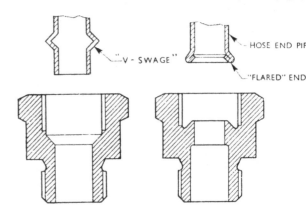

Fig. 10A—View showing early "V" swage hose end, late flared end hose fitting and related adaptors.

bearing in pump body and press bearing and shaft assembly from pump body. Bearing (18) can be removed from shaft after removing snap ring (17).

Remove the pump outlet adapter (38) from pump body. Remove flow control valve spring (37) and flow control valve (24). If necessary, use a piece of hooked wire to pull flow control valve from housing. Remove the snap ring retaining the pressure

relief valve (35) in the flow control valve and remove the pressure relief valve and spring (25).

27. Wash all parts EXCEPT the rotor shaft bearing in a suitable solvent and inspect. If the bushing and/or pump cover is worn, renew the cover and bushing assembly. If the bushing and/or the pump body is worn, renew the pump body and rotor sub-assembly. (Bushings are not available separately.) If the drive or driven rotors show evidence of wear or damage, renew the pump body and rotor assembly.

Check clearances as follows: Insert rotor shaft and bearing into housing until bearing is in position. Install key, drive rotor and driven rotor and check clearance at tooth ends as

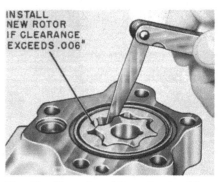

Fig. 11—Measuring drive rotor to driven rotor tooth clearance. Renew rotor set if clearance exceeds 0.006.

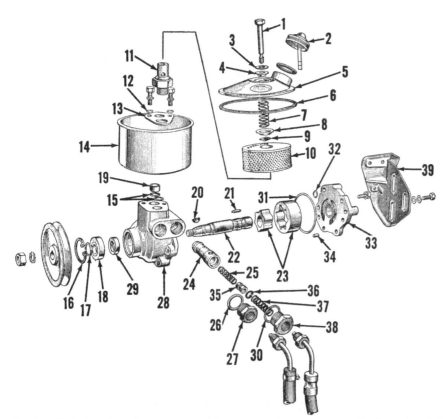

Fig. 10—Exploded view of the power steering pump and reservoir assembly. Pump bracket (39) for early models is different than shown.

1. Cover bolt	11. Stud	21. Rotor key
2. Filler cap	12. "O" rings	22. Pump shaft
3. Washer	13. Reinforcement	23. Rotor set
4. Gasket	14. Reservoir	24. Flow control
5. Cover	15. "O" rings	valve
6. Gasket	16. Snap ring	25. Relief valve
7. Spring	17. Snap ring	spring
8. Retainer	18. Bearing	26. Seal
9. Clip	19. Retainer	27. Adaptor
10. Filter	20. Woodruff key	28. Pump body

29. Oil seal	
30. Seal	
31. "O" ring	
32. "O" ring	
33. Pump cover	
34. Dowel pin	
35. Relief valve	
36. Snap ring	
37. Flow control	
valve spring	
38. Adaptor	

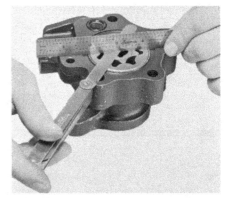

Fig. 12—Measuring rotor end clearance with straight edge and feeler gage. Maximum allowable clearance is 0.0025.

shown in Fig. 11. If clearance exceeds 0.006, renew pump body and rotor assembly. Check clearance between top of rotors and surface of pump body with a straight edge and feeler gage as shown in Fig. 12. If clearance exceeds 0.0025, renew pump body and rotor assembly. Check clearance between driven rotor and insert in pump body as shown in Fig. 13. If clearance exceeds 0.008, renew pump body and rotor assembly. The flow control valve spring should exert 16-18

Fig. 13—Measuring driven rotor to pump body clearance. If clearance exceeds 0.006, renew rotor set and/or pump body.

pounds when compressed to a height of 1.2 inches. The pressure relief valve spring should exert 30-33 pounds when compressed to a height of 1.18 inches. Renew springs if they do not meet specifications.

Thoroughly dry relief valve and bore of flow control valve; then insert relief valve and make sure it moves freely in bore of flow control valve. If necessary to remove any burrs, use crocus cloth. Check freedom of movement of flow control valve in pump cover bore in the same manner.

Before assembly, coat all parts with a light film of oil. If a new rotor shaft seal is used, coat lip of same with Lubriplate or its equivalent, and be sure it is installed with lip toward pump rotors. Always use new "O" rings. Reassemble by reversing the disassembly procedure.

CYLINDER AND CONTROL VALVE ASSEMBLY

28. **REMOVE AND REINSTALL.** Disconnect hoses from power cylinder and allow fluid to drain from

reservoir. Disconnect drag link from control valve actuating ball stud and disconnect front ball stud from steering arm on pedestal (tricycle models) or on front wheel spindle (four-wheel models). Unbolt piston rod anchor ball cap from abutment bracket taking care not to lose shims from between clamp and ball seat, then remove cylinder assembly from tractor.

When reinstalling unit, tighten ball stud nuts to a torque of 100-110 Ft.-Lbs. Reconnect piston rod anchor end with sufficient shims (61—Fig. 14 or 15) between clamp (59) and seat (62) to preven anchor ball from binding, yet be without end play. Shims are available in thicknesses of 0.002, 0.005 and 0.010. Tighten anchor ball clamp bolts to a torque of 60-65 Ft.-Lbs. Reconnect cylinder hoses as shown in Fig. 8, then refill and bleed system as in paragraph 23.

NOTE: Pump to cylinder hoses with two different type ends have been used; refer to Fig. 10A. Early "V-swage" type hose end and adapter is shown at left; later flared type hose end and adapter fitting is

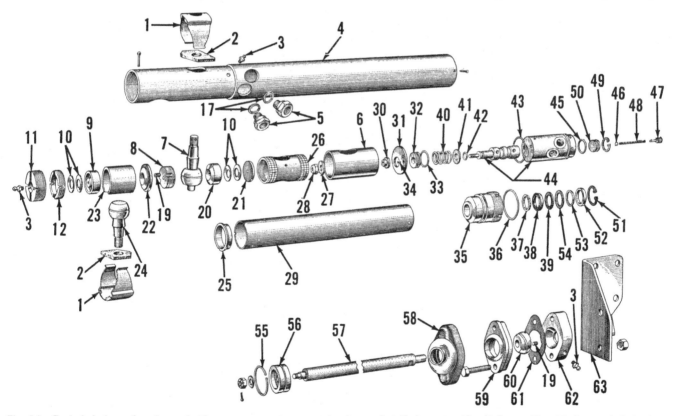

Fig. 14—Exploded view of early production power steering control valve and cylinder assembly. Refer to Fig. 15 for exploded view of later unit. Unit shown is for wide adjustable front axle models; cylinder on tricycle model is smaller in diameter.

1. Spring cover	10. Belleville washer	25. Locating collar	35. Bearing assembly	45. "O" ring	53. Scraper
2. Grease retaining pad	11. End cover	26. Operating sleeve	36. "O" ring	46. Relief valve ball	54. Washer
3. Grease fittings	12. Lock ring	27. Washer	37. Gland seal spacer	47. Relief valve plug	55. Piston ring
4. Outer tube	17. Seals	28. Self-locking nut	38. Vellumoid washer	48. Relief valve spring	56. Piston
5. Hose adapters	19. Set screw	29. Inner tube	40. Reaction spring	49. Snap ring	57. Piston rod
6. Locating sleeve	20. Manual pin ball cup	30. Collar	41. Washer	50. End plug	58. Rubber cover
7. Manual ball pin	21. Back-up washer	31. Spacer	42. "O" ring	51. Snap ring	59. Abutment cap
8. Manual ball pin cup	22. Power ball pin cup	32. Reaction ring	43. "O" ring	52. Washer	60. Anchor ball
9. Power ball pin cup	23. Spacer	33. "O" ring	44. Valve body & spool		61. Shim
	24. Power ball pin	34. Dowel pin			62. Abutment spacer
					63. Abutment bracket

shown at right. Only the late type hose with flared ends are available for service. When renewing early type hose with "V-swage" ends, it will also be necessary to install new adapter fittings in pump and cylinder.

29. OVERHAUL. The cylinder and control valve components are clamped together within the outer tube (4—Fig. 14 or Fig. 15) by the plug (11) and lock ring (12) in front end of tube and by the piston rod bearing (35—Fig. 14) (early units) or bearing retainer (35B—Fig. 15) (late production) in rear end of tube. The control valve spool is connected to inner sleeve (26—Fig. 14 or Fig. 15) and valve actuating ball stud (7) is clamped into inner sleeve by the threaded ball cap (8—Fig. 14) or plug (8A—Fig. 15).

To disassemble unit, clamp outer tube lightly in vise and remove front plug (11), spring washers (10) and ball cap (9), then remove lock ring (12), sleeve (23), ball stud (24) and ball stud seat (22). On early units, remove grub screw (19—Fig. 14) and

unscrew ball stud cap (8); on later units, remove spring clip (19A—Fig. 15), unscrew plug (8A) and remove ball stud cap (8B). Then, on all units, remove ball stud (7) and unscrew adapters (4) from unit.

On early units, refer to Fig. 14 and unscrew bearing (35) from rear end of tube, then withdraw piston rod (57) and piston assembly. With a wood dowel inserted from front end of tube against ball stud seat (20), drive the control valve unit and inner tube (29) out rear end of outer tube.

On late units, refer to Fig. 15 and unscrew bearing retainer (35B) from rear end of tube, then, insert wood dowel from front end of tube against ball stud seat (20) and drive the control valve and cylinder components from outer sleeve. Note: On late production units, an "O" ring is moulded inside outer tube; be careful not to damage this "O" ring as it is not serviced separately from outer tube.

On all units, proceed as follows: Separate inner tube from control valve body and if not already removed, remove piston rod, piston and bearing from inner tube. On four-wheel units, it is not necessary to remove collar (25) or dowel pins from rear end of control valve body. Pull inner sleeve and control valve spool from valve body and remove nut (28) to disassemble this unit. Remove snap ring (49) and push end plate (50) from rear end of valve body. Remove piston (56) from front end of piston rod then slide bearing and seal assembly from rod. Remove scraper (if fitted) and snap ring (51) from piston rod bearing, then remove seal components from bearing.

Carefully clean and inspect all parts for scoring, undue wear or other damage. Remove any burrs with crocus cloth. Take care not to remove any sharp edges from lands of control valve spool. Renew any parts which are questionable and reassemble unit as follows:

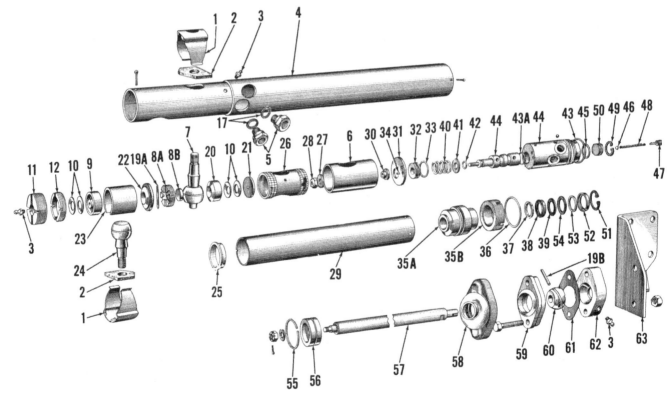

Fig. 15—Exploded view of later production power steering cylinder and control valve assembly for wide front axle models; cylinders for tricycle models are similar but are smaller in diameter. Refer to Fig. 14 for exploded view of earlier production unit.

1. Spring cover	8B. Manual ball pin cup	21. Back-up washer
2. Grease retaining pad	9. Power ball pin cup	22. Power ball pin cup
3. Grease fittings	10. Belleville washer	23. Spacer
4. Outer tube	11. End cover	24. Power ball pin
5. Hose adapters	12. Lock ring	25. Locating collar
6. Locating sleeve	17. Seals	26. Operating sleeve
7. Manual ball pin	19A. Retainer	27. Washer
8A. Cup retainer	19B. Pin	28. Self-locking nut
	20. Manual ball pin cup	29. Inner tube
		30. Collar
		31. Spacer

32. Reaction ring	43A. "O" ring	51. Snap ring
33. "O" ring	44. Valve body & spool	52. Washer
34. Dowel pin	45. "O" ring	53. Scraper
35A. Bearing assembly	46. Relief valve ball	54. Washer
35B. Bearing retainer	47. Relief valve plug	55. Piston ring
36. "O" ring	48. Relief valve spring	56. Piston
37. Gland seal spacer	49. Snap ring	57. Piston rod
38. Gland seal	50. End plug	58. Rubber cover
39. Vellumoid washer		59. Abutment cap
40. Reaction spring		60. Anchor ball
41. Washer		61. Shim
42. "O" ring		62. Abutment spacer
43. "O" ring		63. Abutment bracket

30. Install new "O" rings (33 and 42) on reaction ring (32) and control valve spool, then lubricate component parts and reassemble washer (41), centering spring (40), reaction ring, spacer (31), collar (30) and inner sleeve (26) on spool and secure with washer (27) and nut (28). Note: Nut (28) on early units is self-locking; later units have castellated nut with cotter pin and access hole in inner sleeve (26) for installing pin. Tighten the nut to a torque of 10-16 Ft.-Lbs. and install cotter pin if so fitted. Be sure that the dowel pin is properly fitted in spacer (31) and inner sleeve. Install new "O" ring (43A) inside rear end of valve body (this "O" ring is not shown in Fig. 14), lubricate valve spool and bore and insert spool through valve body. Install check valve (46), spring and screw if removed from valve body.

Renew the piston rod (57) and anchor ball (60) as a unit if either part is damaged. Slide clamp (59) and rubber boot (58) over front end of rod, then refer to Fig. 16 for correct placement of seal components in piston rod bearing and proceed as follows: Install spreader (37) in bore of bearing, then use suitable sleeve to drive seal (38) into place. Slide scraper (DS), snap ring (51), seal seat (52), square seal (53), metal washer (54) and fiber washer (39) onto piston rod in proper order. Lubricate rod and inner bore of bearing, slide bearing onto rod and install seal components into bearing. Install piston and tighten retaining nut to a torque of 35-45 Ft.-Lbs.

Reassemble components into units as they were removed and install into outer tube. Be careful not to damage "O" rings on valve body and bear-

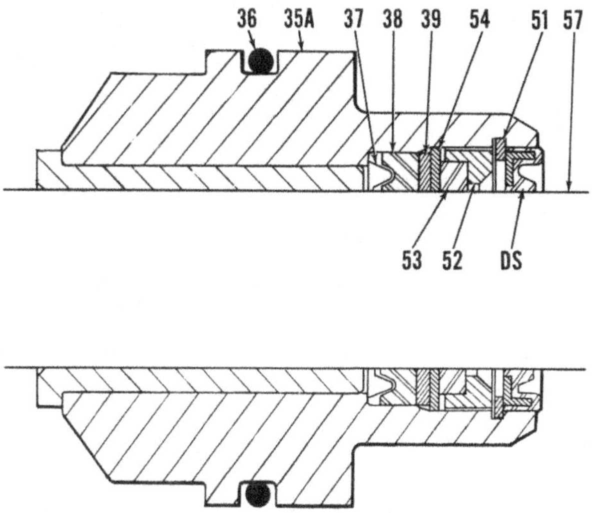

Fig. 16 — Cross-sectional view of late type bearing assembly (35A—Fig. 15) showing seal installation and also outer dust seal (DS) not shown in Fig. 15. Refer to Fig. 15 for remainder of legend.

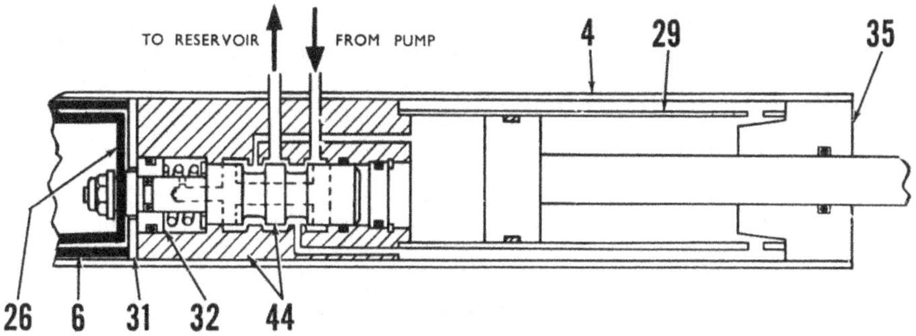

Fig. 17—Cross-sectional view of cylinder and control valve assembly. Refer to Figs. 14 and 15 for legend.

ing as they are installed past threads on inside diameter of tube. Thread piston rod bearing or bearing retainer into rear end of tube until hose ports on control valve body are aligned with holes in outer tube. Reinstall components into front end of unit by reversing removal procedure. Spring washers (10) are installed in pairs with convex (rounded out) faces to-gether. Tighten lock ring (12) to clamp all units together, then install ball stud cup, spring washers and front end plug. Secure end plug and bearing or bearing retainer with cotter pins. Note: Ball stud tension is adjusted by turning plug (11 and 8 or 8A) in tight, then backing off ¼ turn. Secure plug with cotter pin and grub screw or spring clip as fitted.

ENGINE AND COMPONENTS

R&R ENGINE AND CLUTCH ASSEMBLY
All Models

35. To remove engine, first drain cooling system and, if engine is to be disassembled, drain oil pan. Remove air pre-cleaner, vertical exhaust muffler and the rear hinge for hood, then lift hood from tractor. Disconnect radiator hose. Remove the battery or batteries. Remove tool box from left side frame and, if so equipped, remove horizontal (underneath) exhaust pipe and muffler. Disconnect air cleaner to intake manifold hose and, on later production, the air cleaner to rocker arm cover hose. Disconnect

starter motor actuating rod at starter. Disconnect wiring from starter motor, generator, oil pressure switch (late production) and headlights, remove wiring loom clips and roll wiring back out of way. Remove water temperature gage bulb from cylinder head, release tube clips and coil bulb and tube back out of way. Disconnect proofmeter drive cable, if so equipped, at auxiliary drive or from rear of fuel injection pump. Shut off fuel supply valve and disconnect line to fuel pump. Disconnect oil pressure gage tube (early production) from cylinder block, excess fuel return line from cylinder head, throttle control

rod from cross-shaft (early production) or from injection pump (late FSM) and disconnect engine stop cable from injection pump. Unbolt radiator top brace (not used on "FSM"). Remove radiator shutter operating rod on early "FMD."

If engine is being overhauled, remove engine oil filter, starter, generator, fuel filter(s), fuel lift pump, fuel injection pump, intake manifold and vacuum governor tubes and the exhaust manifold. Note: Late "FSM" has fuel injection pump with integral mechanical governor.

On manual steering models, disconnect drag link at either end. On power steering models, disconnect drag link from steering gear arm, remove steering cylinder bracket from side frame, disconnect steering cylinder from spindle arm and unbolt pump and reservoir assembly from engine. This removes the power assist steering system without disconnecting lines or draining system.

On wide front axle models, proceed as follows: Drive wedges between front axle and front support. Block up under transmission and attach hoist to engine. Remove fan blades from water pump. Unbolt engine and side frames from transmission; be sure to remove the engine to transmission bolts located under each side frame. Roll engine and front axle unit forward until clear of transmission. Support front axle and right side frame and remove left side frame from engine and front support. Remove radius rod to oil pan pivot pin. Unbolt engine front plate from right side frame, move engine to left until clear of radius rod, lift engine higher and roll the front axle, radiator and right side frame unit forward. Mount engine on work stand.

On tricycle models, proceed as follows: Block up under transmission and attach hoist to front pedestal and side frames. Unbolt side frames from

transmission and engine front plate and loosen side frame to pedestal bolts. Roll the pedestal, radiator and side frame unit forward until clear of tractor and lower rear end of side frames to floor. Attach hoist to engine, then unbolt and remove engine and clutch assembly from transmission.

CYLINDER HEAD
All Models

36. REMOVE AND REINSTALL. To remove cylinder head, first drain cooling system, remove air pre-cleaner, vertical exhaust muffler and rear hood hinge, then lift engine hood from tractor. Disconnect upper radiator hose, unbolt radiator upper brace and remove thermostat housing. Disconnect air tube from between rocker arm cover and intake manifold or inlet pipe of air cleaner. Remove decompressor lever at front end of rocker arm cover if so equipped. Note: Decompressor lever seal in rocker arm cover may be renewed at this time. Remove rocker arm cover. On any valves that are closed, push rocker arms aside and withdraw push rods. Turn engine so that remaining valves are closed and remove push rods in similar manner.

If early model "FMD" with compressor lever in rear end of cylinder head is encountered, disconnect com-

pressor lever at rear end of rocker arm shaft.

Unbolt and remove rocker arm assembly. Remove fuel filter and disconnect excess fuel return line from cylinder head and immediately cap all openings. Although not required for cylinder head removal, it is good procedure to remove fuel injector assemblies. Disconnect the excess fuel return (leak-off) line from each injector and at connection to cylinder head, then remove line. Remove injector retaining cap screws and lift fuel injectors from cylinder head with screwdriver.

Note: Be careful not to drop leak-off line or injector retaining cap screws inside engine.

On models with vacuum governor, disconnect the governor tubes from intake manifold and injection pump, then remove the tubes. Disconnect throttle control rod from throttle shaft arm on intake manifold.

On all models, unbolt and remove intake and exhaust manifolds from cylinder head, then unbolt and remove cylinder head from block. CAUTION: If fuel injectors have not been removed, be sure to lift cylinder head straight up from block and handle the removed head with care in order not to damage the injector tips that protrude through bottom face of head.

Prior to reinstalling cylinder head, check cylinder liner (sleeve) protrusion as outlined in paragraph 79 and shim sleeves to proper height if necessary. Place new rocker shaft oil passage seal ring in recess in face of cylinder block. Refer to paragraph 38 for proper cylinder head gasket to use and proceed as follows: If installing copper faced gasket, lightly coat both sides of gasket with sealer to within $\frac{3}{16}$-inch of gasket openings; do not apply gasket sealer to composition gasket. Use guide studs (S—Fig. 19) or ring dowels (D) to locate gasket, install cylinder head and tighten retaining cap screws to a torque of 85-90 Ft.-Lbs. in sequence shown in Fig. 18. Reinstall injectors, if removed, as outlined in paragraph 131. Tighten manifold bolts to torque of 20-25 Ft.-Lbs. Note: A joint sealing compound was used to seal manifolds to cylinder head during factory production of some models; however, gaskets are available for service installation of manifolds.

Complete reassembly by reversing removal procedure. Adjust valve gap cold to 0.015 on intake valves and 0.012 on exhaust valves. Bleed diesel

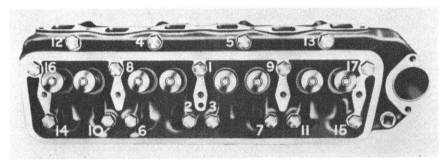

Fig. 18—View of cylinder head showing head bolt tightening sequence for all models. When removing cylinder head, bolts should be loosened in reverse of sequence shown; that is, loosen No. 17 first and No. 1 last.

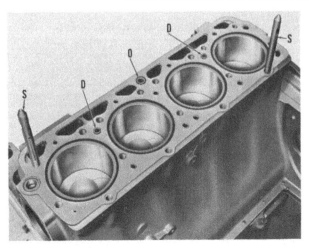

Fig. 19—Prior to fitting cylinder head gasket to cylinder block, be sure to install new sealing ring (O) in recess in top face of block at rocker arm shaft oil feed hole. Align gasket on dowels (D) and use guide studs (S) so that cylinder head will set down squarely over the dowels. Note: Dowels not used on model "FMD" and early "FPM".

fuel system as outlined in paragraph 107, start engine and run until coolant temperature is 180° F. Then, retighten cylinder head cap screws to a torque of 85-90 Ft.-Lbs. and readjust valve gap hot to 0.015 on intake valves and 0.012 on exhaust valves.

37. **RENEW CYLINDER HEAD.** Early type cylinder head with decompressor lever at rear is no longer available. To renew with later production cylinder head, a later rocker arm cover and front mounted decompressor lever must be installed. Refer to current Ford tractor parts catalog for list of necessary parts.

Only two cylinder heads will be serviced. One cylinder head, Ford part No. E1ADDN-6052-S, is used to service all early engines with sealed crankcase ventilation system and offset intake and exhaust ports in head and manifolds. Another cylinder head, Ford part No. E1ADDN-6052-J, is used to service all engines with oil bath breather on timing gear cover and suction pipe between rocker arm cover and inlet pipe of air cleaner. This later type cylinder head has intake and exhaust ports in line.

38. **CYLINDER HEAD GASKET.** Three different cylinder head gaskets are available for service. Recommended usage is as follows:

A 0.6 mm. (0.0236 in.) thick composition gasket, Ford part No. E1ADDN-6051-D, is available for use with plain top sleeves and low crown pistons (more than 0.008 below face of block).

A 1.0 mm. (0.039 in.) thick composition gasket, Ford part No. E1ADDN-6051-E, is available for use with cylinder sleeves having small diameter spigot (extension) and/or high crown pistons.

A thick copper faced gasket, Ford part No. E1ADDN-6051-C, is for installation with cylinder sleeves having large diameter spigot (extension), but can be used instead of composi-

tion gasket with any sleeve. Refer also to paragraphs 76 and 79.

ROCKER ARMS AND SHAFT
All Models

39. **R&R ROCKER ARM SHAFT ASSEMBLY.** To remove rocker arm shaft and rocker arms assembly, proceed as follows: With right side of hood lifted, disconnect vacuum tube from rocker arm cover and intake manifold on early "FMD" or from rocker arm cover and inlet pipe of air cleaner on later models. Remove decompressor lever at front end of rocker arm cover if so equipped, then remove cover from cylinder head. On all valves that are closed, push rocker arms aside and lift push rods from engine. Turn engine by hand to that remaining valves are closed and remove remaining push rods in same manner. On early production model "FMD" with decompressor handle in rear end of cylinder head, disconnect handle from lever at rear end of rocker shaft. Then, on all models, unbolt and remove the rocker arm assembly. Note: On models with decompressor lever at front of rocker assembly, be careful not to lose the detent ball and spring from between front support and the control plate while handling the assembly.

To reinstall rocker arm and shaft assembly, proceed as follows: Install the assembly with the five cap screws and lock tab washers or lock washers. Tighten the cap screws evenly and securely and if lock tab washers are used, bend tabs against cap screw head. Reconnect rear mounted decompressor lever to handle if so equipped. Push the rocker arms aside, insert push rods in engine, seat rocker arm ball socket in push rod and move rocker arm back over valve. Note: Turn engine slightly by hand, if necessary, to bring camshaft lobe to correct position to slide rocker arm over valve stem. Adjust valve gap cold to 0.015 on intake valves and 0.012 on exhaust valves. Start engine and run until coolant temperature reaches 180° F., then readjust valve gap hot to 0.015 on intake valves and 0.012 on exhaust valves. Reinstall rocker arm cover and if so equipped, front mounted decompressor lever.

40. **OVERHAUL ROCKER ARM SHAFT ASSEMBLY.** With rocker arm shaft assembly removed as outlined in paragraph 39, proceed as follows:

On models without decompressor, remove the jam nuts and set screws from intermediate rocker arm supports. On late model "New FSM",

slide the rocker arm supports, rocker arms and springs from shaft. On earlier models, first remove securing pin and plug from end of shaft.

On models with decompressor, each pair of rocker arms ride on an eccentric sleeve that is keyed to the shaft. On models with front mounted decompressor lever, set shaft assembly on rear end, push down the front rocker arm shaft support and remove the detent ball, spring and pin as shown in Fig. 21. Refer to Fig. 22 for exploded view of assembly. Early model FMD with rear mounted decompressor can be disassembled in similar manner.

Check rocker arm shaft, rocker arms, eccentrics (models with decompressor) and springs against the following values:

Models Without Decompressor
Shaft diameter 0.743-0.744
Clearance between rocker
    arm and shaft 0.001-0.003
Spring free length 1.51
Pressure @ 1.06 inches . . . . .4-5 Lbs.

Models With Decompressor
Clearance between eccentric
    and shaft 0.0005-0.0035
Clearance between rocker
    arms and eccentric . . .0.0007-0.0029
Spring free length 1.89
Pressure @ 1.06 inches . . . . . .4-5 Lbs.

Renew rocker arms, shaft and/or eccentrics if excessively worn or scored and renew the springs if rusted, distorted or if they do not meet specifications.

Rocker arms for early model "FMD" engines (prior to serial No. 1481091) are $\frac{3}{16}$-inch longer on adjusting screw end than later engines and arms are not interchangeable. Later

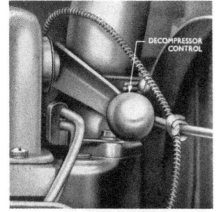

Fig. 20 — Decompressor control lever at front end of rocker arm cover must be removed before removing rocker arm cover.

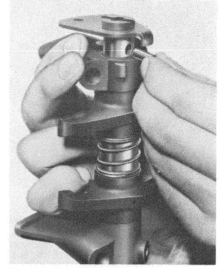

Fig. 21—To disassemble rocker arm assembly, set rocker arm on end and push front support down to allow removal of pin.

type eccentric sleeve with identification dimple on outside diameter may be used with early engines, but early type sleeves without identification dimple can be used only on engines prior to serial No. 1481091.

VALVES AND VALVE SEATS
All Models

41. Intake and exhaust valves seat on renewable seat inserts in cylinder head. Face angle for all valves on all models is 29½ degrees and seat angle is 30 degrees resulting in a ½ degree interference angle. Seat width should be 3/64 to 1/16-inch.

Standard intake valve seat outside diameter is 1.8900-1.8905 and seat thickness is 0.2175-0.2195. Intake valve seats are available in 0.010 oversize outside diameter as well as standard size.

Standard exhaust valve seat outside diameter is 1.7020-1.7025 and seat thickness is 0.2175-0.2195. Exhaust valve seats are available in 0.010 oversize outside diameter with standard seat thickness and in 0.020 oversize outside diameter with 0.010 oversize thickness as well as in standard size.

Both the intake and exhaust valve seats should have an interference fit of 0.0045-0.0060.

Valve stem diameter (new) is 0.373-0.374 for intake valves and 0.3723-0.3733 for exhaust valves. Valve stem to guide clearance should be 0.001-0.003 for intake valves and 0.0017-0.0037 for exhaust valves. Maximum alowable clearance is 0.006 for intake valves and 0.008 for exhaust valves.

When renewing exhaust valves on early models, note groove for spring retainer collets (keepers) in valve stem. If lower shoulder is tapered, new spring retainer, collets and rotor cap must be installed as only late type exhaust valve with square lower shoulder on collet groove is available for service. Refer to Figs. 23 and 24.

NOTE: At serial No. 1425097, diameter of the exhaust valve rotor cap was increased to 0.575-0.580 and exhaust valve spring retainer and collets were changed to accomodate the larger diameter rotor caps. Therefore, if necessary to renew an exhaust valve spring retainer, collets or rotor cap, the correct mating retainer, collets and/or rotor cap must be installed. Part numbers for late parts are as follows:

Rotor cap .DDN-6550-B
Retainer DDN-6514-B
Collets DDN-6518-B

After renewing valve and/or valve seats, the protrusion of the valve face above machined surface of cylinder head should be checked. If valve protrudes more than 0.034, the valve should be refaced or the valve seat reground to bring valve protrusion below 0.034.

VALVE GUIDES
All Models

42. Valve guide I.D. should be 0.375-0.376. Renew the valve guides if excessively worn. Press old guides out towards bottom of cylinder head and new guides in from top of cylinder head to dimension shown in Fig. 25 for intake and Fig. 26 for exhaust valves.

Service intake and exhaust valve guides are identical although early exhaust guides were shorter than the intake valve guides.

VALVE STEM SEALS
All Models

43. New umbrella (cup) type valve stem seals (Ford part No. EOTA-6571-C) should be installed on both the intake and exhaust valve stems before installing valve springs and retainers.

EXHAUST VALVE ROTATORS
All Models

44. Refer to Figs. 23 and 24. The exhaust valves are fitted with free spin type valve rotator caps which will not function unless there is a gap between end of valve stem and inside face of rotator cap when the open end of cap contacts the spring retainer collets (keepers) as shown. Desired gap is 0.001-0.003. The gap can be measured wtih special micrometer gages or as follows:

45. Cut from 0.010 thick shim stock, a disc that will fit into rotaor cap.

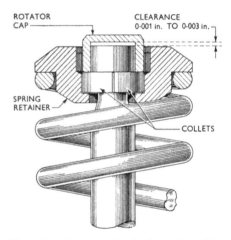

Fig. 23 — Cross-sectional view of earlier type exhaust valve spring retainer, rotator cap and collets. Clearance between end of stem and inside of rotator cap should be 0.001-0.003.

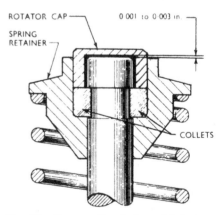

Fig. 24—Cross-sectional view of late type exhaust valve spring retainer, rotator cap and collets. Note square lower shoulder in valve stem versus tapered lower shoulder of valve stem groove in Fig. 23. Clearance between valve stem and rotator cap should be 0.001-0.003.

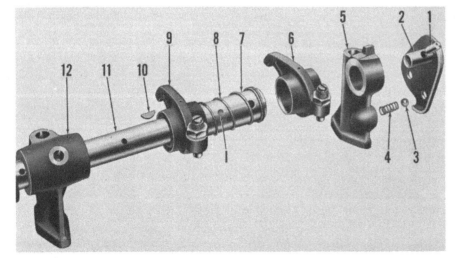

Fig. 22—Exploded view of rocker arm assembly on models with front mounted decompressor lever. Late model "New FSM" will have rocker arms with self-locking adjusting screws instead of screw and lock nut shown. Eccentrics (7) with indent (1) can be used on all models with decompressor; earlier models had higher eccentric.

1. Pin	5. Front support	9. Rocker arm, L.H.
2. Detent plate	6. Rocker arm, R.H.	10. Woodruff key
3. Detent ball	7. Eccentric	11. Rocker arm shaft
4. Detent spring	8. Spacer spring	12. Intermediate support

With disc (G) installed as shown in Fig. 24A, place the spring retainer collets (keepers) in groove of valve stem and measure the gap (x) between collets and rotor cap with feeler gage while pressing down against the collets. Subtracting the measured gap from shim thickness (0.010) will give rotor to valve stem gap. If gap is less than 0.001, grind end of valve stem squarely to obtain desired 0.001-0.003 clearance. If gap is more than 0.003, lap open end of cap to reduce the clearance.

VALVE SPRINGS

All Models

46. Intake and exhaust valve springs are interchangeable; however, two different valve springs will be encountered. Specifications are as follows:

Part No. DKN-6513-B (Early):
Spring color .... Silver gray
Number of coils 8.8
Free length, inches 2.48
Lbs. force exerted @ length
  of 1.98 inches 45-50

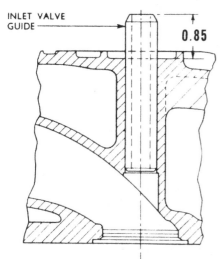

Fig. 24A—Feeler stock method of checking end clearance between rotator cap and end of valve stem. Preferred method utilizes a special micrometer.

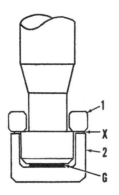

Fig. 25—Press intake (inlet) valve guide into cylinder head so protrusion above machined spring seat is 0.85 inch as shown.

Part No. DDN-6513 (Late):
Spring color .... Black
No. of coils 7.5
Free length, inches 2.31
Lbs. force exerted @ length
  of 1.98 inches 62-68

If the late spring (Part No. DDN-6513) is used for intake valves on earlier cylinder head, spring seat in head must be re-machined and a larger spring retainer installed to accommodate the larger diameter spring. Also, when used to renew earlier exhaust valve spring, larger exhaust valve spring spacer and retainer must be installed. Refer to paragraph 47.

When a camshaft with late type cam profile (refer to paragraph 56) is installed in an engine prior to serial No. 1609839, the late type valve spring and corresponding parts should also be installed.

EXHAUST VALVE SPRING SPACERS

All Models

47. Spacers are fitted under the exhaust valve springs. With smaller diameter early type spring (Ford part No. DKN-6513-B), use Ford part No. E1-CP-9. With late spring (part No. DDN-6513), a hardened spacer of larger diameter (part No. 510E-6515) is used. Spacers can be identified as shown in Fig. 27.

TIMING GEAR COVER

All Models

48. To remove timing gear cover, proceed as follows: Drain cooling system, remove vertical exhaust muffler pre-cleaner and engine hood. Disconnect radiator hoses and headlight wiring and, except on late models, disconnect radiator shell brace from engine. Disconnect radiator shutter control if so equipped. Drive wedges between front axle and front support. Unbolt radius rod from front axle, remove pivot pin from radius rod and oil pan and bump radius rod to one side until it clears pan. Disconnect steering drag link. Support front end of tractor under engine and unbolt side rails from engine front plate and at rear end, then roll the front assembly forward. Loosen the power steering drive belt on models so equipped, then remove generator and belts. Unscrew crankshaft pulley retaining cap screw (ratchet) from crankshaft, then remove crankshaft pulley. Unbolt and remove the timing gear cover.

With the timing gear cover removed, the crankshaft front oil seal can be renewed as outlined in paragraph 90.

To reinstall timing gear cover, reverse removal procedure. Tighten cover retaining cap screws to a torque of 12-15 Ft.-Lbs.

TIMING GEARS

All Models

49. The timing gear train consists of four gears; the crankshaft gear, camshaft outer gear which mates with crankshaft gear, the camshaft inner gear which drives the auxiliary driveshaft gear and the auxiliary driveshaft (fuel injection pump drive) gear. Desired backlash of any two gears is 0.003-0.004.

Due to production and service changes, it is necessary to consider several factors affecting parts procurement when renewing any timing

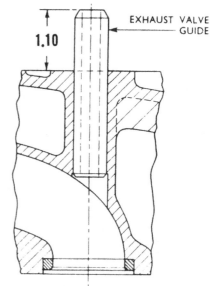

Fig. 26—Exhaust valve guide should protrude 1.10 inch above cylinder head as shown. Note: Intake and exhaust valve guides are alike but are installed to different protrusion above cylinder head.

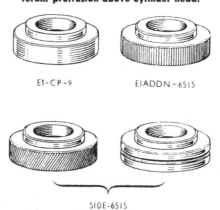

Fig. 27 — Views of the different exhaust valve spring spacers used. Latest spacer (part No. 510E-6515) is identified by diagonal knurling or by groove. Spacers without knurling or groove (part No. E1-CP-9) are used with small diameter spring. Spacers with straight knurling (DDN-6515) are no longer used.

gear. Refer to the paragraphs 50 through 53 for parts procurement information and service procedures.

CAUTION: Latest production camshaft gears are retained by a single cap screw and plate; when loosening this cap screw, torque must not be transmitted through the gear teeth. Hold the camshaft gears by applying a suitable spanner to the gear spokes while loosening or tightening the cap screw.

Also, it is very important that neither the camshaft or crankshaft are turned after removing camshaft gears on any model. The fuel pump lobe on camshaft may contact the No. 4 connecting rod causing serious damage.

To prevent damage to gears or shafts, the auxiliary driveshaft gear and the camshaft gears should be heated for installation. Gears can be heated by immersing in boiling water for approximately one minute, or by marking the gear with a 200° F. temperature indicating crayon and then heating gear by other means until crayon mark liquifies.

50. TIMING GEAR WIDTH. Two different widths of timing gears have been used. Mating gears must be of the same width; as only wider gears are available for service, it will be necessary when renewing a narrow gear to install a wider mating gear. Refer to the following for identification of timing gears by width:

TIMING GEAR WIDTH

Gear	Previous Width	Current Width
Crankshaft	0.808-0.818	0.920-0.930
Camshaft Outer	0.807-0.817	0.928-0.938
Camshaft Inner	0.810-0.815	0.935-0.940
Auxiliary Driveshaft	0.810-0.815	0.925-0.930

NOTE: Other factors besides gear width will affect parts procurement

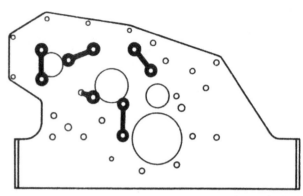

Fig. 29—When reinstalling engine front plate, place new cap screw locks in position shown. Refer to Fig. 30 for tightening torque sequence and to Fig. 31 for tightening torque values.

and service procedures, refer to the following paragraphs:

51. CRANKSHAFT GEAR. The wide crankshaft gear is available with two bore diameters: Latest crankshaft (Ford Part No. DDN-6303-F) is stepped at front end and requires large bore gear (Ford Part No. DDN-6306-B). All prior crankshafts will require smaller ID gear (Ford Part No. DDN-6306-A).

On all models, crankshaft gear is a press fit on shaft and is keyed; be sure new gear is installed with timing mark out. Refer to paragraph 50 for gear selection.

52. CAMSHAFT GEAR. Wide camshaft gears are available in two hub types. Early gears were located by a dowel pin as shown in Fig. 33, and retained to camshaft by three cap screws and a locking plate. On late models, gears are keyed to shaft and retained by a single cap screw as shown in Fig. 34. Camshaft and gear hub was changed at "FSM" serial number 1599502. Because camshaft may have been renewed in earlier models, note gear type before ordering parts.

When installing wide gear set in place of narrow outer gear, it may be

necessary on some models to install a new auxiliary driveshaft oil slinger (Ford Part No. E45-CG-9) to eliminate interference with the wide camshaft gear.

On models with three retaining cap screws, be sure hardened locking plate (Ford Part No. DDN-6258) and hardened cap screws (Part No. 118844-ESB) are used, tighten cap screws to a torque of 18-21 Ft.-Lbs. and secure with locking wire. Refer to paragraph 50 for gear selection.

On models with keyed hub, heat gears as outlined in paragraph 49 for easier installation. Hold camshaft gear from turning, using a suitable spanner, and tighten the single cap screw to a torque of 95 - 100 Ft.-Lbs. CAUTION: Do not transmit tightening torque through gear teeth.

53. AUXILIARY DRIVESHAFT GEAR. The wide auxiliary driveshaft gear is available with two hub bore diameters: The large bore gear (Ford Part No. E84-CP-9) is a service gear for use instead of the narrow production gear on models before "FSM" serial number 1599502. The small bore gear was installed in production, effective with the given serial number.

On all models, heat gear as outlined in paragraph 49 for easier installation, and tighten shaft nut to a torque of 60-70 Ft.-Lbs. Refer to paragraph 50 for gear selection.

ENGINE FRONT PLATE
All Models

Where necessary to renew the engine front plate, correct new plate must be selected by camshaft usage and serial number range.

On model "FMD" prior to serial No. 1425097, use part No. DKN-6030-C if engine is equipped with camshaft having gears retained by three cap screws and a dowel pin; if camshaft gears are retained by key and single cap screw, use front plate part No. DDN-6030-B.

On model "FPM" and model "FSM" prior to serial No. 1599502, use part No. DKN-6030-D if equipped with

Fig. 28 — View showing location of timing marks on timing gears. Early model "FMD" is shown; however, timing marks are alike for all models.

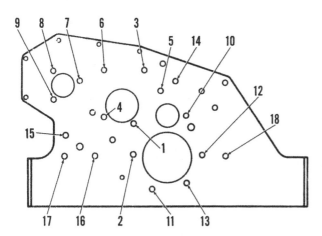

Fig. 30 — Tightening torque sequence for engine front plate retaining cap screws. Note that bolts numbered in sequence from 10 through 18 are tightened after timing gear cover is installed as they are also cover retaining cap screws. Tightening sequence of un-numbered timing gear cover retaining cap screws is unimportant after tightening 10 through 18.

camshaft having gears retained by three cap screws and a dowel pin; if camshaft gears are retained by key and single cap screw use front plate part No. DDN-6030-A.

On model "FSM" serial No. 1599502 and later and on model "New FSM," use front plate part No. DDN-6030-A.

54. R&R ENGINE FRONT PLATE. The engine front plate can be unbolted and removed after removing the timing gear cover as outlined in paragraph 48, the camshaft gears as outlined in paragraph 52, the auxiliary driveshaft gear as in paragraph 53 and the generator front bracket. It is not necessary to remove crankshaft gear.

It should be noted that there are two types of oil pan (sump) front packing strips and four types of engine front plate gaskets. A different front plate gasket is required according to type of camshaft gear retention and the latest types of camshaft gear retention and the latest gasket of each type extends below the front packing strip. Thus, packing strip used with later gaskets must be thinner as the strip contacts the rear side of the gasket instead of extending forward far enough to contact rear face of engine front plate. Only the later type thin packing strip (part No. DDN-6707) can be used with the later type gaskets (part No. DDN-6K000-A if camshaft gears are retained by three cap screws or part No. DDN-6K000-C if gears are retained by single cap screw).

When installing front plate on model "FMD" prior to serial No. 142-5097 equipped wtih camshaft having gears retained by single cap screw, note that the shouldered cap screws (133637-ES2) are used in position shown at (2—Fig. 31). On all model "FMD" tractors prior to serial No. 1425097, tighten cap screws in sequence shown in Fig. 30 and to a torque of 15-17 Ft.-Lbs.

On model "FMD" serial No. 1425097 and later, tighten cap screws at position shown at (2—Fig. 31) to a torque of 22-24 Ft.-Lbs. and all other cap screws to a torque of 15-17 Ft.-Lbs. Be sure to tighten cap screws in sequence shown in Fig. 30.

NOTE: Be sure to install new locking plates for the engine front plate retaining cap screws as shown in Fig. 29 and bend plates against cap screw heads after they are properly torqued.

CAM FOLLOWERS (VALVE TAPPETS)

All Models

55. The mushroom type tappets ride directly in unbushed bores in cylinder block and can be removed after removing engine camshaft as outlined in paragraph 59. Chilled cast tappets were used after serial No. 1358273; refer to Fig. 32 for identification of the early non-chilled tappets and the later chilled cast tappets. Tappet diameter (new) is 0.6070-0.6075; desired clearance is 0.0015-0.003 with wear limit of 0.005.

CAMSHAFT AND BEARINGS

A number of changes have been made in both production and service camshafts and related parts. As many early production engines may have been modified by

Fig. 31 — Prior to serial No. 1425097 (model "FMD"), tighten all cap screws to a torque of 15-17 Ft.-Lbs. On later models and service blocks having ⅜-inch diameter cap screws at locations marked (2), tighten the ⅜-inch cap screws only to a torque of 22-24 Ft.-Lbs. and all other (1, 3, 4 and 5) to a torque of 15-17 Ft.-Lbs. Dowel pins are located at (D); oil orifice or oil pressure relief valve is at (O). Cap screws (1) are 1⅛ inch long; (2, 3 and 4) are ¾-inch long and (5) is 1½ inches long.

installation of later parts, it is necessary to be able to identify camshaft and related parts in order to follow correct servicing procedure on a particular engine.

Model "FMD"

56. At serial No. 1358273, a camshaft with chilled lobes and chilled cast tappets (cam followers) were introduced. Refer to Fig. 32. When renewing early model "FMD" camshaft; early type tappets must be renewed using the later type. Check for interference between rounded portion of tappets and cylinder block and grind material from cylinder block if necessary. A new camshaft must not be installed with the early type tappets; only the late type chilled tappets are available for service.

At serial No. 1425097, a "high lift" camshaft was introduced.

To renew camshaft in engine prior to serial No. 1425097, use Ford part No. DDN-6251-C unless engine has been previously modified by installing camshaft with gears retained by key and single cap screw; use Ford part No. DDN-6250-D in modified engine. Refer to note following paragraph 58 regarding valve springs.

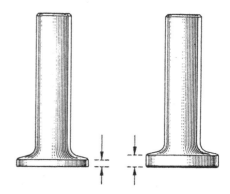

Fig. 32—If early type tappets (left) with 0.150-0.170 thick foot are encountered when renewing camshaft, the tappets must also be renewed using chilled cast tappets (right) with 0.197-0.217 thick foot.

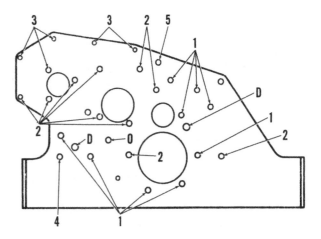

To renew camshaft in model "FMD" engine serial No. 1425097 and up, use Ford part No. 528E-6251-C unless engine has been previously modified by installing camshaft with gears retained by key and single cap screw; use Ford part No. 528E-6250-D in modified engine. Refer to note following paragraph 58 regarding valve springs.

NOTE: For a period of time, camshaft to be used with timing gears retained by three cap screws and a dowel pin (see Fig. 33) was not available for service making it necessary to modify engine by installing later camshaft (see Fig. 34) with gears retained by key and single cap screw. As a special service camshaft and improved "three bolt" camshaft gears are now available for service, it is not necessary to modify engine for later type camshaft. However, if conversion to later type camshaft and gears is desired, parts required are listed in the "Fordson Major Tractor Parts List."

Model "FPM"

57. With the introduction of the model "FPM" at serial No. 1481090, camshaft used in early model "FMD" engine was reinstated and rocker arm length was decreased approximately $\frac{3}{16}$-inch at adjusting screw end to compensate for the decreased cam lift.

To renew camshaft in model "FPM" engine, use Ford part No. DDN-6251-C unless engine has been previously modified by installation of camshaft with gears retained by key and single cap screw; use Ford part No. DDN-6250-D in modified engine. Refer to note following pararagraph 58 regarding valve springs.

Model "FSM"

58. Early model "FSM" engine was equipped with same camshaft as previous model "FPM" engine; refer to paragraph 57.

At serial No. 1599502 a new camshaft was introduced, with gears retained by key and single cap screw. At the same time, method of controlling camshaft end play was changed requiring a new engine front plate. Also, the auxiliary driveshaft, driveshaft gear, front bearing and fuel injection pump drive coupling flange were changed. The new timing gears were wider than previous gears; refer to paragraph 50.

At serial No. 1609839, the camshaft lobe profile was changed to an improved form. Camshafts installed in factory production between serial numbers 1699502 and 1609839 are either identified by the words "Old Form" stamped on rear end of shaft or by the letters "O. F." stamped on front end of shaft to denote old type cam profile. This type camshaft was never available for service installation.

To renew the camshaft in all models after serial No. 1599502, use Ford part No. DDN-6250-D. On model "FSM" prior to serial No. 1599502, use Ford part No. DDN-6251-C unless engine has been modified by installing late type camshaft with gears retained by key and single cap screw; use part No. DDN-6250-D in engines so modified. On engines prior to serial No. 1609839, refer to following note regarding valve springs.

NOTE: New valve springs (Ford part No. DDN-6513) and exhaust valve spring spacers (part No. 510E-6515) were introduced along with the new type camshaft lobe profile at serial No. 1609839. As all camshafts now available for service incorporate the late

cam profile, it is very important to check the valve springs and install the latest type, if not already installed, when renewing camshaft. On some early engines, it may be necessary to rework the cylinder head around the intake valve guides. Refer to paragraphs 46 and 47 for valve spring and exhaust valve spring spacer identification and installation.

All Models

59. **R&R CAMSHAFT.** To remove camshaft, remove the engine as outlined in paragraph 35, then, with fuel lift pump, push rods and engine front plate removed, invert engine to allow tappets (cam followers) to fall away from camshaft. Withdraw camshaft from front end of engine.

The camshaft can be removed without removing engine by the following procedure: Remove the fuel injection pump assembly as outlined in paragraph 113 or 122, then unbolt and remove push rod cover from side of crankcase. With the push rods removed as outlined in paragraph 39, lift each tappet with a magnet and retain in lifted position with a suitable clip. Remove fuel lift pump from side of crankcase and engine front plate as outlined in paragraph 54, then withdraw camshaft from front of engine.

60. With introduction of camshaft having gears retained by key and single cap screw, method of controlling camshaft end thrust was also changed. On early camshaft, end thrust was controlled by either a split type washer or a horseshoe shaped thrust plate (see Fig. 33). On new type camshaft, a thrust flange is integral with

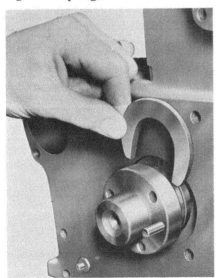

Fig. 33—Timing gears were retained to earlier type camshaft with three cap screws and dowel pin and camshaft end thrust was taken by split or horseshoe type washer fitted into groove in camshaft. Refer to Fig. 34 for later camshaft and timing gear arrangement.

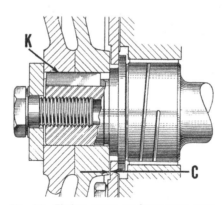

Fig. 34—Timing gears on later type camshaft are retained by cap screw, lock washer, plate and key (K). Camshaft end thrust is controlled by engine front plate which fits between rear camshaft gear and shoulder on camshaft. Clearance (C) should be 0.005-0.021.

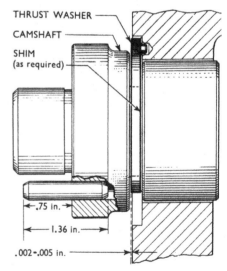

Fig. 35—Thrust washer used with early type camshaft should protrude 0.002-0.005 from front face of cylinder block; shims are available to increase protrusion. Refer also to Fig. 36.

shaft and end thrust is taken by the engine front plate which fits between hub of camshaft inner (rear) gear and thrust flange on camshaft as shown in Fig. 34.

61. On early type camshaft, the thrust plate shold protrude 0.002-0.005 from front face of crankcase as shown in Fig. 35. Shims 0.003 and 0.005 thick are available for installation behind thrust plate to obtain desired protrusion. If pre-cut shims are not available, cut shim stock to fit. Note: The latest thrust plate can be used with all early camshafts; however, if using early thrust washer or plate with currently available service camshaft, it may be necessary to remove some material from inner diameter of thrust washer or plate to fit camshaft. Desired clearance of thrust washers or plate in groove of camshaft (camshaft end play) is 0.003-0.008.

62. On late type camshaft with gears retained by key and single cap screw, camshaft end play should be 0.005-0.021. If end play is excessive,

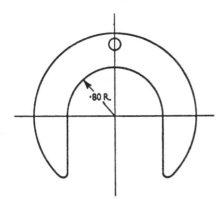

PREVIOUS TYPE

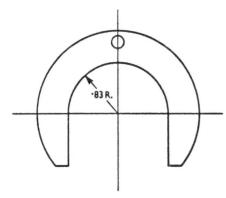

CURRENT TYPE

Fig. 36 — Currently available camshaft thrust washer (see Fig. 35) has larger inside radius. When using previous washer with currently available early type camshaft, it may be necessary to rework washer to fit shaft.

check for wear at camshaft inner (rear) gear hub, engine front plate or flange on camshaft.

63. New camshafts have a black phosphate coating to reduce wear when installed in unbushed bores in cylinder block. If camshaft is being installed in cylinder block fitted with bushings, the phosphate coating must be removed with crocus cloth or fine emery paper and the abrasive particles thoroughly cleaned from camshaft before installing the new shaft. However, **do not** remove the phosphate coating if camshaft is being installed in unbushed bores. Be sure to thoroughly lubricate the camshaft thrust surfaces before installing engine front plate as outlined in paragraph 54. Install camshaft gears as outlined in paragraph 52 and auxiliary driveshaft gear as in paragraph 53. Complete the ressembly of tractor by reversing disassembly procedure.

64. **CAMSHAFT BEARINGS.** The five journal camshaft is supported in unbushed bearing bores or renewable type bearing liners (bushings). Camshaft journal diameter is 2.0596-2.0600. Desired journal to bore bushing clearance is 0.002-0.0035 with maximum wear limit of 0.006.

If cylinder block is not fitted with bushings and camshaft bores are excessively worn, bushings can be installed after line boring all five bearing bores in crankcase to a diameter of 2.188-2.189.

Some engine originally fitted with camshaft bushings may have 0.020 oversive bores for the bushings and oversize outside diameter bushings (bearing liners) are available for this purpose.

The pre-sized bushings must be installed with a closely piloted arbor to avoid any damage to the bushings. The four rear bushings are alike; front bushing is wider and has a notch in forward end. When reinstalling front bushing, be sure the notch is aligned with oil-way in front face of bearing bore. On the four rear bushings, be sure oil holes in bushings are aligned with oil-way in front face of bearing bore. On the four rear bushings, be sure oil holes in bushings are aligned with oil passages in cylinder block. Check for proper oil hole alignment after bushings are pressed into place; alignment is satisfactory if a $\frac{3}{16}$-inch diameter steel pin can be inserted through drilling between crankshaft and camshaft bores and pass through oil hole in the bushings. Also, on number 3 and number 4 bearings, be sure that second oil

hole in bushing is aligned with rocker arm oil supply hole.

NOTE: New camshafts have a black phosphate coating; when installing camshaft in cylinder block fitted with camshaft bushings, the phosphate coating must be removed from the crankshaft journals. Do not remove the coating if installing camshaft in unbushed bearing bores.

65. **CAMSHAFT REAR BEARING PLUG.** An expansion plug is fitted in rear face of cylinder block to close the rear camshaft bearing bore. Usually, the plug does not need to be disturbed except in case of oil leakage or when renewing camshaft bushings. The expansion plug is accessible after removing flywheel. Plugs are available in three different diameters; 2 5/64 inches, 2 13/64 inches and 2 7/32 inches.

AUXILIARY DRIVESHAFT
All Models

66. The auxiliary driveshaft is mounted in two ball bearings in right front corner of engine crankcase. A gear on front of shaft is driven by the camshaft inner (rear) gear. A gear integral with shaft drives the engine oil pump and, on some models, the Proofmeter (tach-hourmeter). A coupling attached to rear end of auxiliary driveshaft drives the fuel injection pump.

67. **R&R AUXILIARY DRIVESHAFT.** To remove the auxiliary drive shaft, first remove the oil pan as outlined in paragraph 96 and the oil pump as outlined in paragraph 93, the engine front plate as outlined in paragraph 54 and fuel injection assembly as in paragraph 113 or 122.

Fig. 37—Expansion plug (P) at rear end of camshaft is accessible after removing flywheel. Plug is available in three different sizes; refer to text.

Then, with the injection pump drive coupling and key removed from rear of shaft, bump shaft forward out of engine crankcase. Drive the oil seal out to rear. Note: It is possible to renew the oil seal after removing fuel injection pump and drive coupling from rear end of shaft, then prying seal from bore in crankcase. Remove the ball bearings from each end of auxiliary driveshaft.

If necessary to renew the auxiliary driveshaft and/or bearings, refer to paragraph 68. Install the shaft and bearings as follows:

By pressing on inner bearing race only, install the small bearing on rear end of shaft and the large bearing on front (threaded) end of shaft so that bearing inner races are tight against shoulders on shaft. Insert shaft and bearing assembly in bore of crankcase from front and bump the assembly rearward until front bearing outer race is seated against shoulder in crankcase. Using a seal protector or shim stock, install seal over rear end of shaft, then remove seal protector and using suitable sleeve, drive seal into bore until seated against snap ring. Reassemble tractor by reversing disassembly procedure.

68. RENEW AUXILIARY DRIVE-SHAFT AND/OR BEARINGS. Several production changes affect the auxiliary driveshaft and related parts. As these changes may be incorporated in prior models, the following should be noted if necessary to renew the auxiliary driveshaft and/or bearings:

To renew auxiliary driveshaft on models equipped with Simms Minemec (mechanical governor) fuel injection pump, use part No. E92-CP-5.

On models with Simms fuel injection pump with vacuum governor and camshaft gears retained with key and single cap screw, use part No. E94-CP-5.

On prior models with original type engine camshaft having gears retained to shaft with three cap screws, use part No. E96-CP-9. NOTE: If drive

gear on engine oil pump does not have either part No. E1ADDN-6652 or 528E-6652 stamped on gear, it will also be necessary to renew the oil pump drive gear when installing new auxiliary drive shaft. Refer to paragraph 95.

If changing from early type camshaft having gears retained by three cap screws to late type camshaft having gears retained by key and single cap screw (refer to paragraph 52), it is also necessary to install a new fuel injection pump drive coupling flange (Ford part No. DDN-993186-B) along with the new auxiliary drive shaft (part No. E94-CP-5).

When renewing auxiliary driveshaft front bearing, use part No. E80-CP-9 if crankcase is counterbored as shown in Fig. 39, or part No. DKN-66618-A on models without counterbore.

CONNECTING RODS AND PISTONS

All Models

69. Connecting rod and piston assemblies are removed from above after removing cylinder head, oil pan, oil suction tube and inlet screen and the connecting rod caps. Note: Be sure that rod and cap have cylinder number or other marks to pair rod and cap before removing the caps.

When reassembling piston and pin to connecting rod, be sure that the valve recesses in top face of piston are to the same side of the assembly as the notches in lower end bore of rod for the bearing insert tangs. Assembly of piston pin through piston will be aided by warming the piston in boiling water.

When reinstalling connecting rod and piston assembly, be sure that word "FRONT" or arrow on top of piston is towards front of engine.

If necessary to renew connecting rod bolts and/or nuts, it should be noted that there are two different type nuts and three different type bolts that may be encountered. If the bolt hole in rod and cap has "broken through" the bearing bore when bolt

hole was drilled, a special "waisted" (necked down) bolt (Ford part. No. DKN-6215-A) must be used. At engine serial No. 1509598, length of the connecting rod bolt was increased from 2.87 inches to 2.99 inches and thickness of the nut was increased from 0.462 to 0.530. The longer bolt can be used with connecting rods originally equipped with the shorter bolts; however, the smaller nut must be used with the earlier rods as the spot face of cap will not allow the larger nut to seat. CAUTION: The shorter bolt must not be used in connecting rods originally equipped with long bolt.

Connecting rod bolt tightening torque for all models is 55-60 Ft.-Lbs. Note: Be sure that cap is installed with machined notch for bearing tang to same side of assembly as bearing tang notch in connecting rod.

PISTON PINS AND BUSHINGS

All Models

70. Piston pins are fully floating and are retained in pistons by snap rings.

Engines prior to serial No. 1425097 were equipped with 1.250 inch diameter piston pins. Later model engines are equipped with 1.375 inch diameter hollow piston pins. Refer to the following specifications for servicing piston pin and bushings:

Prior to Serial No. 1425087:
Piston pin diameter 1.2497-1.2500
Bushing inside diameter 1.2501-1.2504*
Pin to bushing clearance,
   desired 0.0001-0.0007
   Wear limit 0.0015
Pin to piston clearance,
   desired .... 0.001 interference
               to 0.0001 loose
   Wear limit 0.0010

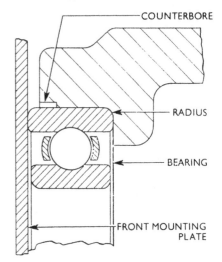

Fig. 39—Latest auxiliary driveshaft front bearing (6—Fig. 38) can be used only with cylinder block incorporating counterbore. Early bearing (part No. DKN-66618-A) can be used with all cylinder blocks.

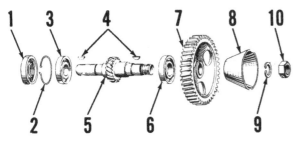

Fig. 38 — Exploded view of late production auxiliary driveshaft assembly. Outside diameter of oil slinger (8) used with wide camshaft front gear is smaller so that it will clear the gear.

1. Oil seal
2. Snap ring
3. Rear bearing
4. Woodruff keys
5. Auxiliary driveshaft
6. Front bearing
7. Driving gear
8. Oil slinger
9. Lockwasher
10. Hex nut

Serial No. 1425097 & Up:
Piston pin diameter 1.3747-1.3750
Bushing inside diameter 1.3751-1.3754*
Pin to bushing clearance,
   desired 0.0001-0.0007
   Wear limit 0.0015
Pin to piston clearance, de-
   sided....0.0001 interference to 0.001
   loose
   Wear limit 0.0010
*NOTE: Bushing must be reamed to
fit pin after installation in connecting
rod.

CONNECTING RODS AND BEARINGS
All Models

71. CONNECTING RODS. Two different connecting rods are used. Prior to serial No. 1425097, piston pin diameter is 1.375. Refer to paragraph 69 concerning connecting rod cap bolts and nuts. Connecting rod specifications are as follows:

Width at crankpin end.. 1.655-1.657
End play on crankpin,
   desired 0.003-0.009
   Wear limit 0.012

72. BEARING INSERTS. Several different factors must be considered in renewing crankpin bearing inserts; refer to the following paragraphs 73 through 75.

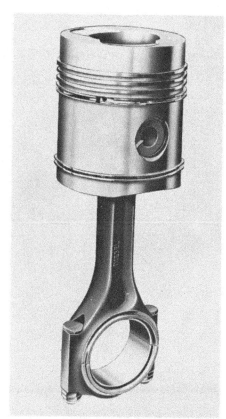

Fig. 40—When reassembling piston to the connecting rod, be sure valve recesses in piston head are to same side of assembly as crankpin bearing insert tang notches in rod.

73. BEARING TYPE. Three different types of crankpin bearing inserts have been used. Each type will be identified by stampings on the back of the insert. The letter "V.P." entwined and enclosed in a circle will indicate one type, the letter "G" enclosed in a square will indicate a seconed type and the third type will have the letter "AL" stamped in back of insert along with the letter "G" enclosed in a square. These three types of bearing inserts are interchangeable, but must be used in pairs (the top and bottom inserts of any one connecting rod must be of the same type).

74. BEARING WIDTH. Effective with serial No. 1483140, crankshafts were modified by increasing the fillet radius between the crankshaft webs and journals, and the crankpin bearing inserts were reduced in width to accommodate the increased radius. Service crankshafts and inserts also incorporate this change. Previous crankpin bearing inserts were 1.37-1.38 wide; later inserts are 1.33-1.34 wide. Narrow bearing inserts can be used with either crankshaft.

75. UNDERSIZE BEARING INSERTS. Bearing inserts are available in undersizes of 0.010, 0.020, 0.030 and 0.040 as well as in standard size. Standard crankpin bearing journal diameter is 2.4997-2.5005; desired bearing insert to journal clearance is 0.002-0.0035 with maximum allowable clearance of 0.005. Crankshaft crankpin should be reground to next standard undersize if clearance is excessive, if taper exceeds 0.001 or out-of-round exceeds 0.0015.

PISTONS AND RINGS
All Models

76. Pistons for all models are of aluminum alloy and have a combustion chamber machined in their crown. Three compression rings and one oil control ring are fitted above the piston pin and a second oil ring is located below the pin. Piston pin is fully floating and is retained by a snap ring at each end of pin bore in piston.

Engines prior to serial No. 1425097 are equipped with pistons having a 1.250 piston pin bore. Early engines in this series were equipped with pistons having combustion chambers offset from center of piston while later engines were equipped with pistons having the combustion chamber centered in top of piston. Only the pistons having centralized combustion chambers are available for service. Therefore, if an early engine with pistons

having offset combustion chambers is encountered and it is necessary to renew one or more pistons, a complete set of four later type pistons must be installed. "Spigoted" cylinder sleeves (see paragraph 79) and a copper-asbestos head gasket should be used with these pistons.

Engines after serial No. 1425097 and prior to serial No. 1518654 were equipped with pistons having a 1.375 diameter piston pin bore and a crown height (center of piston pin bore to top face of piston) of 2.783-2.785. All later engines were also equipped with pistons having a 1.375 diameter piston pin bore, but with crown height of 2.793-2.795 (0.010 higher crown than previous pistons). Both types of pistons are available for service. The earlier (low crown height) pistons should be used with non-spigoted cylinder sleeves (see paragraph 79) and a 0.6 mm thick composition head gasket. Later (high crown height) pistons should be used with spigoted cylinder sleeves and either the copper-asbetos or 1.0 mm thick composition head gasket. Low crown height pistons can be identified by part No. E1ADDN-6110E cast inside skirt (standard size piston) or by the letter "F" stamped on top of piston (0.0025 oversize piston). High crown height pistons can be identified by letter "L" stamped on top of standard piston or by letter "M" stamped on top of 0.0025 oversize piston.
NOTE: High crown height pistons should not be used in engines where more than 0.010 has been machined from top face of cylinder block. If in doubt, refer to paragraph 38 and

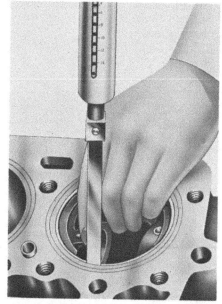

Fig. 41—Checking piston to cylinder sleeve fit with feeler gage strip and pull scale; refer to text for specifications.

measure piston height. Valve face protrusion from cylinder head should also be checked as outlined in paragraph 41.

77. FITTING PISTONS. To check piston fit, refer to Fig. 41 and proceed as follows: Position the piston in cylinder bore as shown, with top face towards crankshaft and valve recesses in top face away from camshaft side of engine, and with a ½-inch wide 0.004 thick feeler strip inserted between piston and sleeve on camshaft side of engine. Piston fit is correct when a pull of 2 to 4 pounds is required to remove the feeler strip. If pull required is less than 2 pounds with standard size piston, check fit of the 0.0025 oversize piston available for service. Cylinder sleeves should be renewed if the 0.0025 oversize piston cannot be properly fitted.

78. PISTON RINGS. Piston rings are available in standard size and 0.0025 oversize. When renewing piston rings, refer to the following specifications:

Maximum allowable cylinder
  out-of-round 0.003
Maximum allowable cylinder
  taper 0.007
Ring side clearance in groove (all rings):
  Desired 0.002-0.004
  Wear limit 0.005
Ring end gap (all rings):
  Desired 0.011-0.016
  Wear limit 0.040

Top compression ring is marked "H & T TOP" and ring must be installed with this marking up. Second and third compression rings are marked "TOP" on one face and must be installed with this mark up. The two oil control rings are interchangeable and reversible.

A "2-in-1" chrome re-ring set (Ford part No. EPN-6149-A) is also

available and can be used in some instances where cylinder sleeve wear or taper will not permit use of standard type rings. Installation instructions and specifications are packaged with each ring set.

CYLINDER SLEEVES
All Models

79. Engines for all models are equipped with wet type cylinder sleeves. On early production engines (prior to Serial No. 1591023), sleeves were sealed at bottom by an "O" ring and at the top by cylinder head gasket and also by sealer applied to under side of sleeve flange. On later engines, cylinder sleeves are sealed at top and bottom by "O" rings in grooves in cylinder block. In conjunction with change in method of sealing, cylinder sleeves are of larger outside diameter than cylinder sleeves for prior production engines.

NOTE: Some engines built during the changeover period from small to large outside diameter sleeves will require the large O.D. sleeves, but do not have an "O" ring groove at top of bore in cylinder block. As a means of identification, any engine with $\frac{9}{16}$-inch or $\frac{5}{8}$-inch diameter main bearing cap retaining bolts will require the large O.D. sleeves. Engines built prior to this change have ½-inch diameter main bearing cap bolts.

For early engines requiring sleeves with small outside diameter, two different types of cylinder sleeves (spigoted and plain) are available; refer to Fig. 42. The non-spigoted (plain top) cylinder sleeves should be used with pistons of low crown height (refer to paragraph 76) and with the 0.6 mm thick cylinder head gasket (refer to paragraph 38). The spigoted cylinder sleeves should be used with pistons of high crown height and with either the 1.0 mm thick composition type head gasket or the copper asbestos type head gasket.

For later engines requiring the large outside diameter cylinder sleeves, sleeves are available in spigoted type only and pistons of high crown height should be used. Although the 1.0 mm thick composition head gasket is recommended, the copper-asbestos gasket can be used.

NOTE: On some engines, the bores in the cylinder block for cylinder sleeves may be 0.020 oversize and require oversize outside diameter cylinder sleeves.

On early models without "O" ring at top of cylinder sleeve, the sleeve flange must protrude 0.002-0.004 above top surface of cylinder block. Fig. 43 shows one method of checking sleeve

protrusion. If sleeve does not protrude at least 0.002, install shims(s) under sleeve flange as required to bring protrusion within limits of 0.002 to 0.004. Note: Where cylinder sleeves would not otherwise be removed and are not disturbed from their installed position, a protrusion of 0.001 or more is acceptable. Shims are not available for the large outside diameter sleeves used in later model engines.

CRANKSHAFT AND MAIN BEARINGS
All Models

80. R&R CRANKSHAFT. With engine removed from tractor as outlined in paragraph 35, proceed as follows:

Remove clutch assembly, flywheel, oil pan, oil suction tube and inlet screen, timing gear cover, timing gears and the engine front support plate. Be sure the main bearing caps are identified according to location, then unbolt and remove the connecting rod caps and bearing inserts and the main bearing caps and bearing inserts. Center main bearing cap is fitted with thrust washer lower halves, do not drop or damage thrust washers as cap is being removed. Lift the crankshaft from engine. Remove the upper main bearing inserts and keep them with their respective bearing caps. Remove the two upper thrust washer halves.

To reinstall crankshaft, proceed as follows: Install new rear oil seal as outlined in paragraph 91. Using thin film of grease, stick the thrust washer upper halves to recesses at center main bearing journal making sure oil

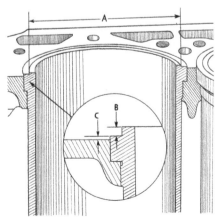

Fig. 42—Cross-sectional drawing of spigoted" cylinder sleeve installed in cylinder block. Spigot diameter is "A" and height is "B". Sleeves should protrude 0.002-0.004 (dimension "C") above cylinder block.

Fig. 43—Checking cylinder sleeve protrusion with dial indicator. A straight edge and feeler gage can also be used for this purpose. Note the four cap screws and washers used to clamp sleeves into position while checking protrusion. Refer to text.

grooves in face of thrust washers are out (away from cylinder block). Place the upper main bearing inserts in cylinder block, lubricate the inserts and carefully place crankshaft in block. Place bearing insert in center main bearing cap and using light film of grease, stick thrust washer lower halves in recesses of cap. Make sure that the oil groove faces of thrust washers are out (away from main bearing cap). Install center main bearing cap with word "REAR" embossed on cap towards rear of engine. Place bearing inserts in remaining bearing caps and install caps in their respective locations. Install front cap with flat face forward; install other caps with word "REAR" on cap towards rear of engine. Install bearing cap retaining bolts (cap screws) with new locking plates and before tightening the cap screws, bump crankshaft back and forth to be sure the center main bearing cap thrust faces are properly aligned with cylinder block.

Three different sizes of main bearing cap bolts (cap screws) have been used. Diameter of cap screws should be checked to determine proper tightening torque. Tightening torques are as follows:

MAIN BEARING TIGHTENING TORQUES

Cap Screw Size	Ft.-Lbs. Torque
½-inch	70-75
$\frac{9}{16}$-inch	95-100
⅝-inch	115-120

NOTE: On late model "New FSM" with ⅝-inch diameter main bearing cap screws, the two cap screws with 0.40 thick heads are installed in the front main bearing cap. Cap screws with 0.50 thick heads are used in the remaining bearing caps.

81. **MAIN BEARING CAPS.** Bearing caps are not interchangeable, however, service bearing caps are available and can be installed where the bearing bores are of standard size. Note: Some engines may already have oversize bearing bores, thus a service bearing cap cannot be installed and the block must be renewed. Standard bore diameter is 3.167 inches; oversize bearing bore will measure 3.182 inches.

To install a service bearing cap, install the new cap in desired position, install the remaining bearing caps in their proper locations and tighten retaining cap screws as outlined in paragraph 80; then, line bore all bearing bores to 0.0015 oversize, or to a diameter of 3.1815-3.1820. If any attempt is made to bore only one new cap, misalignment and crankshaft failure could occur.

If the center main bearing cap is being renewed, the recess in cylinder block and cap for the thrust washers must be machined to produce a continuous bearing surface (R—Fig. 45) within the limiting dimensions (W) of 1.589-1.591. Mount the center main bearing cap on block so that one of its thrust faces (F) is in line with the corresponding cylinder block thrust face (F) within 0.005. The thrust faces should then be machined square with centerline of crankshaft within 0.002 total indicator reading and the recess bore within 0.006 of being concentric with the bearing bore. An equal amount should be machined from each thrust face to give a distance between thrust faces of 1.589-1.591. File the thrust washer tab slots (S) in cap back flush with thrust surface.

82. **MAIN BEARING INSERTS.** Several different factors must be considered when renewing crankshaft main bearing inserts. Refer to the following paragraphs 83 through 87 for information concerning bearing inserts.

83. **BEARING TYPE.** Three different types of main bearing inserts have been used. Each type will be identified by stampings on the back of the insert. The letters "V.P." entwined and enclosed in a circle will indicate one type, the letter "G" enclosed in a square will indicate a second type, and the third type will have the letters "AL" stamped in the back of the insert along with the letter "G" enclosed in a square. These three types of bearing inserts must be used in pairs (the top and bottom inserts of any one bearing must be of the same type).

84. **BEARING WIDTH.** Effective with serial No. 1483140, crankshafts were modified by increasing the fillet radius between the crankshaft webs and journals, and the main bearing inserts were reduced in width to accommodate the increased radius. Service crankshafts and liners also incorporate this change. Previous front and intermediate bearing inserts were 1.12-1.13 wide; later inserts for front and intermediate bearings are 1.04-1.05 wide. Previous center and rear bearing inserts were 1.495-1.505 wide while later type inserts for center and rear bearings are 1.415-1.425 wide. The later type narrow bearing inserts can be used with all crankshafts whereas the early wide type inserts cannot be used with late crankshaft. Thus, the narrow type bearing inserts should be used for service.

85. **UNDERSIZE BEARING INSERTS.** Bearing inserts are available in undersizes of 0.010, 0.020, 0.030 and 0.040 as well as in standard size. Standard main bearing journal diameter is 3.0002-3.0010; desired bearing insert to journal clearance is 0.0025-0.004. Crankshaft main bearing journal should be reground to next undersize if clearance is exces-

Fig. 44—Crankshaft end thrust is controlled by split type thrust washers (TW) at each side of center main bearing cap. Crankshaft upper rear oil seal retainer is (SR).

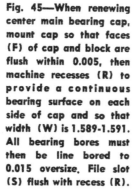

Fig. 45—When renewing center main bearing cap, mount cap so that faces (F) of cap and block are flush within 0.005, then machine recesses (R) to provide a continuous bearing surface on each side of cap and so that width (W) is 1.589-1.591. All bearing bores must then be line bored to 0.015 oversize. File slot (S) flush with recess (R).

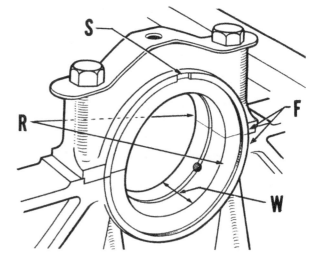

sive, if journal taper exceeds 0.001 or if out-of-round exceeds 0.0015.

86. OVERSIZE BEARING INSERTS. When a service main bearing cap has been installed and the main bearing bore diameter has been line bored to 0.015 oversize, bearing inserts of 0.015 oversize outside diameter are available for service. The oversize bearing inserts are available in inside diameter undersizes of 0.010, 0.020, 0.030 and 0.040 as well as standard size. Note: Some factory production engines will have 0.015 oversize main bearing inserts; these engines will be identified by the marking "O/S" stamped on oil pan face of cylinder block; also, later engines will have an "X" suffix to serial number (such as 08C-999999X) to indicate 0.015 oversize main bearing inserts.

As an engine may require 0.015 oversize main bearing inserts without being so marked, the main bearing bores should be measured prior to renewing the bearing inserts. Bearing bore for standard inserts will measure 3.167 inches; bore requiring oversize inserts is 3.182 inches.

87. CRANKSHAFT THRUST WASHERS. Renewable thrust washers are fitted at the center main bearing to control crankshaft end play. Desired crankshaft end play is 0.002-0.010. Thrust washers are available in kits containing two upper and two lower halves and in standard thickness as well as oversize thicknesses of 0.0025, 0.005, 0.0075, 0.010, 0.015 and 0.020. The oversize thicknesses of 0.010, 0.015 and 0.020 are for use when a new center main bearing cap has been installed and thrust faces of block and cap have been machined; refer to paragraph 81. Use of the oversize thickness thrust washers will

also allow regrinding of the center crankshaft journal thrust faces where worn or scored. Standard center journal width is 1.799-1.801.

In conjunction with the increased fillet radius of the crankshaft and decreased bearing insert width at serial No. 1483140 (refer to paragraph 84), inside diameter of the thrust washers was increased from 3.215-3.225 to 3.315-3.325. The later thrust washers with large inside diameter can be used for service in all engines.

CRANKCASE VENTILATION SYSTEM
Early Model "FMD"

88. Tractors prior to Serial No. 1425097 are equipped with a closed crankcase ventilation system. A pipe connects engine rocker arm cover to intake manifold, thus the crankcase is subjected to intake manifold vacuum. With this system, it is very important that a perfect seal be obtained at all gasket and seal points throughout the engine. Any air leakage into engine will result in entry of dirt and improper operation of the vacuum governor.

To check for proper sealing, remove oil filler cap from rocker arm cover with engine running at slow idle speed. A noticeable effect should occur on governor by change in intake manifold vacuum. With oil filler cap removed and engine running, cover filler opening with palm of hand; a definite suction should be noted. If these conditions are not encountered, air leakage into engine is occuring at one or more gasket or seal joints and must be corrected.

Late Model "FMD", Models "FPM", "FSM" and "New FSM"

89. Tractors starting with Serial No.

1425097 are equipped with an open type crankcase ventilation system. Air enters crankcase through an oil bath air cleaner located on top of timing gear cover and is withdrawn from engine through a tube connecting rocker arm cover to inlet pipe of engine air cleaner.

It is important that the breather air cleaner (refer to Fig. 46) be serviced at frequent intervals. It is recommended that the complete assembly be removed from timing gear cover, disassembled, cleaned, refilled to proper lever with clean oil and reinstalled each 50 hours of normal service. In extremely dusty conditions, service should be more frequent.

Early breather air cleaner assemblies were a push fit into tube on timing gear cover and were sealed by an "O" ring; see Fig. 47. If difficulty is encountered with unit shaking loose, a bracket should be fabricated as shown in Fig. 49 and the breather secured with a clamp (Ford part No. 204E-9628-B) as shown in Fig. 48.

Late breather assemblies are retained to tube on timing gear cover by a clamp; refer to Fig. 50. Early and late breather assemblies are not interchangeable.

CRANKSHAFT OIL SEALS
All Models

90. FRONT OIL SEAL. Crankshaft front oil seal can be renewed after removing timing gear cover as outlined in paragraph 48.

Drive old seal out towards rear (inside) of cover. If installing leather type seal, soak seal in oil for fifteen minutes prior to installing in timing

Fig. 46—The breather cap should be removed, the filter element cleaned and the cup refilled with clean oil each 50 hours of operation, or more often in extremely dusty conditions.

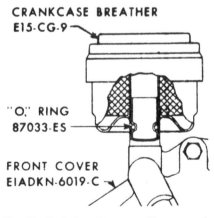

CRANKCASE BREATHER E15-CG-9

"O" RING 87033-ES

FRONT COVER EIADKN-6019-C

Fig. 47—Early breather assembly was push fit on pipe in timing gear (front) cover and is sealed with "O" ring. It is recommended that the breather be secured with clamp and bracket; refer to Figs. 48 and 49.

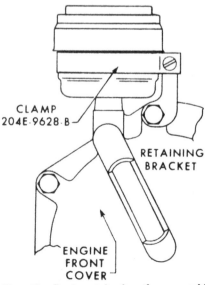

CLAMP 204E-9628-B

RETAINING BRACKET

ENGINE FRONT COVER

Fig. 48—Early style breather assembly should be secured with clamp and bracket. Refer to Fig. 49 for dimensions for fabricating bracket.

gear cover; soaking in oil is not required with rubber lip type seal. Using a suitable driver, install new seal with lip to inside of timing gear cover. NOTE: At serial No. 1308977, crankshaft pulley hub diameter was increased from 2.5 inches to 2.625 inches. Seals for the early engines with 2.5 inch diameter pulley hub will have an approximate internal sealing diammeter of 2.44 inches while seals for the later 2.625 inch diameter pulley hub will have an internal sealing diameter of approximately 2.56 inches. Be sure the correct oil seal is installed. Renew the crankshaft pulley if seal contact surface is excessively scored or rough; minor imperfections should be removed with emery cloth.

91. REAR OIL SEAL. The lower half of two piece rear oil seal is carried in a groove in the cast iron oil pan and upper half of seal is carried in a groove in retainer bolted to rear end of cylinder block. To renew the rear oil seal, first remove engine from tractor and remove the oil pan, clutch assembly and engine flywheel; then, proceed as follows:

Unbolt and remove the upper seal retainer from rear of cylinder block and remove the old seal and gasket. If stake marks are not already pres-

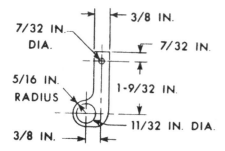

Fig. 49—Dimensions for fabricating breather bracket. Refer to Figs. 47 and 48.

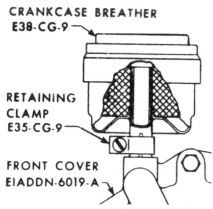

Fig. 50—Late breather element is clamped to tube in timing gear (front) cover. Early and late style breathers are not interchangeable as tubes in early and late front covers are different.

ent, use a center punch to stake bottom of seal groove in retainer in two places at either end of groove. Firmly fit a new graphite coated seal half in retainer, then trim ends of seal so that they protrude 0.015-0.025 from the retainer. Although the graphite type seal does not need to be oil soaked, the surface that contacts crankshaft should be lightly oiled. Reinstall retainer with new gasket and locking plate. Firmly fit lower half of seal in groove of oil pan, trim the seal ends so they protrude 0.015-0.025 from oil pan, then reinstall oil pan as outlined in paragraph 96.

ENGINE OIL FILTER

All Models

92. A number of different makes and types of oil filter assemblies have been used. To service the different filters, two special oil filter kits are available through Ford Tractors parts departments. For all model "FMD" tractors, use filter element kit Ford part No. EPN-6731-A. For all model "FPM", "FSM" and "New FSM" tractors, use element kit Ford part No. EPN-6731-B. Note: These part numbers are not listed in the Fordson Major Tractor Parts List as they are produced and available only in the U.S.

CAUTION: **Be sure** to closely follow instructions packaged with the filter element kit and install filter according to filter make; trade name (make) of filter will be found on filter head (base).

OIL PUMP AND RELIEF VALVE

All Models

93. R&R OIL PUMP. To remove oil pump, first remove oil pan (sump) as outlined in paragraph 96, remove the oil suction tube and filter screen, then unbolt and remove oil pump from lower face of cylinder block.

To reinstall, reverse the removal procedure and reinstall oil pan as outlined in paragraph 96.

94. RELIEF VALVE. On engines prior to serial No. 1425097, the oil pressure relief valve assembly is threaded into engine front plate and can be removed after removing timing gear cover as outlined in paragraph 48. The relief valve is serviced as a complete assembly only. When installing valve assembly, tighten to a torque of 20-25 Ft.-Lbs.; valve can be distorted causing malfunction if overtightened. On all later models, the oil pressure relief valve is located in the oil pump cover; refer to paragraph 95. Oil pressure on

all models should be 40-50 psi with engine running at 1500 RPM.

95. OVERHAUL PUMP. On later type pumps with relief valve in cover, remove the cotter pin, spring retainer, spring and relief valve plunger. Unbolt and remove cover on all pumps and withdraw idler gear from shaft and pump body. Press helical drive gear from upper end of driveshaft, then withdraw shaft and driven gear from pump body. If necessary to renew driven gear and shaft is serviceable, press gear from shaft. Note: If necessary to renew shaft, a new driven gear should be assembled to shaft as the sintered metal gear tends to lose its interference fit once it is removed from shaft. If idler shaft is worn and pump body is serviceable, press shaft from body. Carefully inspect all parts and compare with the following specifications and information:

Driveshaft to pump body bore clearance and idler gear to idler shaft clearance should be 0.001-0.0035. Gear end clearance in pump body should be 0.001-0.0045.

At serial No. 1595085 changes were made in the material used in the oil pump helical drive gear and the gear on auxiliary driveshaft to increase life of the gears. Later changes were made wherein the width of the oil pump drive gear and auxiliary driveshaft gear were increased from ½ to $\frac{19}{32}$-inch, the upper end of oil pump driveshaft was scrolled to permit oil passage to the helical drive gear and

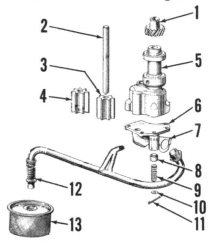

Fig. 51—Exploded view of typical oil pump assembly. Early model "FMD" pump did not have oil pressure relief valve (8) as valve was fitted in engine front plate; refer to Fig. 53. Refer to Fig. 52 for cross sectional view of latest pump.

1. Drive gear	8. Relief valve
2. Driveshaft	9. Valve spring
3. Driven gear	10. Retainer
4. Idler gear	11. Cotter pin
5. Pump body	12. Suction tube
6. Cover	13. Suction screen
7. Lock plate	

the gear was drilled to allow oil to lubricate the gear teeth. Refer to cross-sectional view of latest type oil pump body, shaft and gear assembly in Fig. 52.

Ther latest type wide, drilled gear (Ford part No. 528E-6652) and scrolled shaft (part No. 528E-6609) may be installed in previous pump assemblies to gain benefit of the oil feed to the gears. However, in installing the gear in engines prior to serial No. 1595085, a new auxiliary driveshaft of the improved material must also be installed; refer to paragraph 68. Note: Oil pump helical drive gears of the improved material can be identified by a spot of white paint or the part No. E1ADDN-6652 or part No. 528E-6652 stamped on the gear; part No. E1ADDN-6652 does not incorporate the increased width and oil drilling, however.

In late production oil pumps, the relief valve plunger is longer than in earlier pumps, a cylinder type spring retainer is used instead of a flat disc and the relief valve spring is of an improved material. All three parts are individually interchangeable, but should be installed in prior pumps as a set to improve relief valve performance. The new spring (Ford part No. 528E-6654) can be identified by a yelow paint band on the spring coils. Part number of longer valve plunger is DDN-6663-B; cylinder type retainer part number is DDN-6653-B.

Refer to paragraph 94 for relief valve on model "FMD" prior to serial No. 1425097.

To improve oil pump efficiency, late pump gears have chamfer at one end of teeth only instead of at both ends as on previous gears. If installing gear with chamfer at one end only, be sure the gear is installed with chamfer up and with square end of teeth towards pump cover. The early and late gears are completely interchangeable.

To reassemble pump, proceed as follows: Press idler shaft into pump body so that lower end of the shaft is 0.005 below flush with lower face of pump body. Press driveshaft into chamfered end of gear (early type gears have chamfer at each end) so that lower end of shaft is 0.080-0.090 below flush with lower face of gear if scrolled shaft is being installed, or 0.200 below flush with lower face of gear if early type plain shaft is being installed. Insert driveshaft and driven gear assembly into pump body and install helical drive gear as follows:

On early "FMD" with plain pump cover (relief valve in front support plate), support lower end of driveshaft and press helical gear onto upper end of shaft so that distance from machined lower face of pump body to lower edge of helical gear teeth is 5.200-5.202 inches; minimum clearance between helical gear hub and pump body (shaft end play) is 0.010.

On later type pumps with relief valve in pump cover, press helical drive gear onto drive shaft so that clearance between gear hub and pump body (shaft end play) is 0.010-0.036 if plain shaft and gear are being installed, or 0.007-0.012 if scrolled shaft and drilled gear are being installed.

Insert idler gear, chamfered end first, into pump body and install pump cover. Install relief valve plunger,

spring, spring retainer and cotter pin in cover of pumps so equipped. Be sure that pump turns freely and is well lubricated before reinstalling in engine.

OIL PAN (SUMP)
All Models

96. On tricycle front wheel models, the engine oil pan can be unbolted and removed without removing any other components.

To remove oil pan on wide front axle models, unbolt radius rod from front axle center member and remove the pivot pin from radius rod and oil pan. Bump the radius rod to one side until clear of pan, then remove radius rod from tractor. Remove the connecting rod from front wheel spindle arms. The oil pan can then be unbolted and removed from tractor.

On model "FMD" prior to serial No. 1425097, oil pan was fitted with removable plate on bottom of pan so that oil pump suction screen could be cleaned without removing oil pan. On all models after "FMD" serial No. 1425096, a one-piece oil pan is used. On model "FSM" and "New FSM", bore for radius rod pivot pin is larger than for prior models and on late production "New FSM", oil pan is retained to crankcase by 20 cap screws of ⅜-inch diameter and two cap screws of $\frac{5}{16}$-inch diameter instead of the 22 cap screws of $\frac{5}{16}$-inch diameter used on all prior production engines.

When renewing oil pan on model "FMD" prior to serial No. 1425097, it will be necessary to modify the oil pump suction screen cover as shown in Fig. 55 as only the later type one-piece oil pan is available for service.

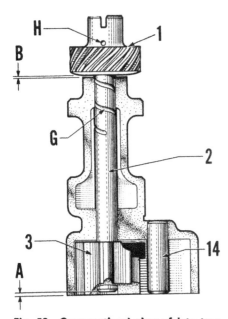

Fig. 52—Cross-sectional view of late type pump showing driveshaft with spiral groove (G) at upper end and drive gear (1) with oil hole (H). Refer to text for installation of idler shaft (14) and for drive shaft to driven gear assembly dimension (A) and drive gear to body clearance (B).

Fig. 53—Camshaft front (outer) gear removed to show oil pressure relief valve (RV) location on model "FMD" prior to serial No. 1425097.

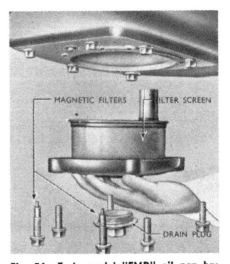

MAGNETIC FILTERS FILTER SCREEN

DRAIN PLUG

Fig. 54—Early model "FMD" oil pan has removeable plate, thus filter screen can be removed and installed without removing oil pan. As this type pan is no longer available, renewing pan requires modification of screen cover; refer to Figs. 55 and 56.

To install oil pan, proceed as follows: If installing later type oil pan on early "FMD, retain suction screen cover on oil pump tube with a spring clip (Ford part No. E120-Z-9), then install screen in cover and secure with spring as shown in Fig. 56. Be sure that retaining spring is parallel with crankshaft to avoid interference between oil pan and cover. On all later models, install oil screen assembly on oil pump tube and turn screen ¼-turn to retain it on tube. Note: On early "FMD" with original type oil pan, the oil pump screen can be installed after pan is reinstalled on engine.

Using gasket sealer, fit the two side gaskets and the sump front packing strip to engine. Note: Type of sump front packing will depend upon type of front support plate to engine gasket used; refer to paragraph 54. Firmly fit new graphite coated rear oil seal lower half in groove at rear of oil pan and trim ends of seal so that they protrude 0.015 to 0.025 from pan. Lubricate the rear oil seal, then lift pan into place with floor jack. Install the retaining cap screws, then tighten cap screws alternately and evenly. Reinstall connecting rod and radius rod on wide front axle models.

FLYWHEEL

All Models

97. **R&R FLYWHEEL.** Flywheel can be unbolted and removed after splitting tractor between engine and flywheel housing as outlined in paragraph 149 and removing the single clutch assembly as outlined in paragraph 154 or 155, or the dual ("Live PTO") clutch assembly as outlined in paragraph 157. Refer to paragraph 98 for starter ring gear and to paragraph 99 for clutch (transmission input) shaft pilot bearing.

On model "FMD" prior to serial No. 1308977, flywheel is positioned by two equally sized dowel pins, thus the flywheel can be installed in either of two positions. On later models the

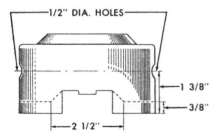

Fig. 55—When renewing early type oil pan (see Fig. 54), oil filter screen must be modified by cutting off material as indicated by dotted line and cutting two holes as shown above.

two dowel pins are of different size and the flywheel can be installed in one position only. Three different flywheels are currently available for servicing models with 12 inch single plate clutch (flywheel part No. DDN-6375-D), 13 inch single plate clutch (flywheel part No. DDN-6375-F) or with dual ("Live PTO") clutch (flywheel part No. DDN-6375-E). To install the later type flywheel (part No. DDN-6375-D) on early tractor, a new dowel pin (Ford part No. DKN-6387-B) must be obtained and installed as follows:

Position the crankshaft so that No. 1 throw (crankpin) is up and position flywheel to crankshaft so that timing marks on outside diameter of flywheel are at the "4 O'clock" position as viewed from rear; then install dowel pins in crankshaft to correspond with proper size dowel holes in flywheel.

When installing flywheel on all models, use new locking plates for cap screws, tighten the cap screws to a torque of 80-90 Ft.-Lbs. and bend locking plates up against flats on cap screw heads.

Flywheel runout should not exceed 0.005 at either the clutch friction surface on single disc clutch models or at outside diameter (periphery) of flywheel.

98. **STARTER RING GEAR.** Starter ring gear is attached to front face of flywheel with six countersunk head cap screws, thus it is necessary to remove flywheel to renew starter ring gear.

On late models, starter ring gear cap screws are secured with "Loctite" instead of with lock washers. "Loctite" can be used on prior models if desired. To install starter ring gear, be sure that chamfered edge of gear teeth is towards front side of flywheel. If using "Loctite", be sure threads in flywheel and on cap screws are clean, apply a drop or two of "Loctite" on threads of each cap screw and tighten the cap screws alternately and evenly. Proper tightening torque is 12-15 Ft.-Lbs.

99. **PILOT BEARING.** Two different types of clutch (transmission input) shaft pilot bearings have been used and are interchangeable.

Early type ball bearing (Ford part No. DKN-7600) should be packed with a high melting point grease prior to assembly and installed with shielded side to rear (out).

When installing the later type sintered bronze pilot bearing (Ford part No. DDN-7600), be careful not to

damage it in any way. Note: If necessary to remove a sintered bronze bearing from flywheel, it should be discarded and a new bearing installed on reassembly. Lightly coat surface of the bronze bearing prior to reassembling tractor.

DIESEL FUEL SYSTEM

The diesel fuel system consists of three basic components; the fuel filter(s), fuel injection pump and fuel injection nozzles. When servicing any unit associated with the fuel system, the maintenance of absolute cleanliness is of utmost importance. Of equal importance is the avoidance of nicks or burrs on any of the working parts.

Probably the most important precaution that service personel can impart to owners of diesel powered tractors is to urge them to use an approved fuel that is absolutely clean and free from foreign material. Extra precaution should be taken to make certain that no water enters the fuel storage tanks.

TROUBLE SHOOTING

All Models

100. If the engine will not start, is hard to start or does not run properly, refer to the following appropriate paragraph:

101. **ENGINE WILL NOT START.** If the engine will not start or is difficult to start, check the following:
1. Fuel tank empty.
2. Fuel supply valve closed.
3. Filters clogged or air in filters.
4. Fuel lift pump faulty.
5. Air in fuel injection pump or fuel injector lines.
6. Air cleaner dirty.
7. Low engine compression.
8. Pump installed out of time.

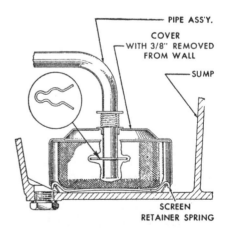

Fig. 56—When installing late type oil pan on early "FMD" engine, modify screen cover as shown in Fig. 55 and position cover and screen with retainer spring lengthwise with pan (parallel with crankshaft) as shown above.

9. Pump control rod stuck in closed position.

10. Throttle plate in intake manifold not opening (vacuum governed models only).

11. Starter not turning engine fast enough.

12. Faulty injectors.

102. ENGINE STARTS, THEN STOPS. If the engine will start, but will not continue running, check the following.

1. Air in filters or in fuel injection pump.

2. Filter clogged

3. Lift pump faulty.

4. Fuel injection pump control rod sticking.

5. Governor not properly adjusted (vacuum governed models only).

103. ENGINE LACKS POWER. If the engine will not develop sufficient (rated) power, check the following:

1. Air in fuel injection system.

2. Fuel filters clogged.

3. Air cleaner dirty or restricted hose connections.

4. Low engine compression.

5. Pump out of time.

6. Throttle plate in intake manifold not fully opening (vacuum governed models only).

7. Faulty governor.

8. Engine governed speed too low.

9. Engine overheating.

10. Engine not running hot enough.

11. Faulty injectors.

104. ENGINE EMITS EXCESSIVE SMOKE. If the engine emits excessive smoke, check the following:

1. Excess fuel delivery (starting) device stuck in excess fuel position.

2. Pump out of time.

3. Engine compression low.

4. Air cleaner dirty.

5. Faulty injectors(s).

6. Engine not running hot enough.

7. Injection pump not properly calibrated for correct fuel delivery under load.

8. Faulty vacuum governor.

105. ENGINE KNOCKS. In addition to normally encountered knocking due to engine wear, check the following:

1. Faulty injectors(s).

2. Pump out of time.

3. Injection pump not properly calibrated for correct fuel delivery under load.

FILTERS AND BLEEDING
All Models

106. FILTER MAINTENANCE. A single stage fuel filter was used on early models. Late models are fitted with a water separator filter in addi-

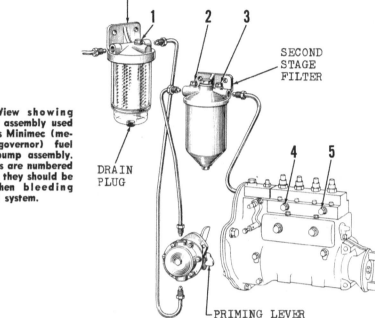

Fig. 57—View showing dual filter assembly used with Simms Minimec (mechanical governor) fuel injection pump assembly. Bleed plugs are numbered in rotation they should be opened when bleeding fuel system.

tion to the fuel oil filter previously used. Prior model "FSM" and "New FSM" may be converted to the dual filter system if desired.

On models with single stage filter, the filter element should be renewed after each 600 hours of operation.

On models with dual filter unit, the water separator element should be renewed after each 600 hours of use and the second stage element should be renewed after each 1000 hours of use. In addition, any water noted in the clear shell of the primary (water separator) filter should be drained. To drain the water, shut off fuel supply valve, then open filter element drain plug until filter is drained. Close the drain plug, open bleed screw on top of filter head, then open fuel supply valve. When fuel runs free of bubbles from the bleed hole, close the bleed screw and then bleed remainder of system in normal manner.

A special filter element kit (Ford part No. EPN-99162-A) is available for renewing element in single stage filter or for renewing element in sec-

Fig. 58—Bleeding fuel system with vacuum governor fuel injection pump and single stage filter. In some instances, it may be necessary to crank engine over to different position so that fuel lift pump can be operated with primer lever.

ond stage filter on models with dual filter unit. As several different makes of filter units have been used, instructions packaged with filter kit should be closely followed to install the filter element according to trade name of filter which will be found on filter head (base).

Another element kit (Ford part No. DDN-99162-C) is available for renewing the element in the first stage (clear shell) filter unit on models with dual filter.

107. BLEEDING. On models with dual filter unit, refer to Fig. 57, open the fuel supply valve at fuel tank, then open the bleed screw (1) on first stage filter. When fuel flows freely without air bubbles, close the bleed screw. Procedure for further bleeding of models with dual filter unit and complete bleeding procedure

for models with single filter is as follows:

With fuel supply valve open, open the filter inlet bleed screw (Fig. 58 or 2—Fig. 57) and actuate the primer lever on fuel lift pump until fuel flows from bleed screw opening free of any air bubbles. Note: If primer lever will not actuate the fuel lift pump, turn engine so that fuel pump cam on camshaft is away from fuel pump lever. When fuel flows freely without air bubbles, close the filter inlet bleed screw and open the filter outlet bleed screw (Fig. 58 or 3—Fig. 57). Again actuate the primer lever on fuel lift pump until fuel flows from bleed screw free of any air bubbles and close the screw. Open the fuel pump bleed screw (Fig. 58 or 4 and 5—Fig. 57) and continue to actuate the fuel lift pump primer lever until fuel flows freely without air bubbles from the open bleed screw(s). Close the bleed screws, then loosen the injector pressure line connecting nuts at the injectors, be sure fuel shut-off button is pushed in and crank engine until fuel flows free of air from the loosened connections. Tighten the fuel injector line connections and crank engine.

FUEL LIFT PUMP

All Models

108. Refer to Fig. 59 for exploded view of fuel lift pump used on models "FMD", "FPM" and "FSM" with fuel injection pump having vacuum governor and to Fig. 60 for exploded view of fuel lift pump used on late model "FSM" and "New FSM" having Minemec (mechanical governor) fuel injection pump.

Overhaul of the fuel lift pump is conventional. Fuel delivery pressure should be from 1½ to 3½ psi. Bleed the fuel system as outlined in paragraph 107 after reinstalling pump assembly to engine.

FUEL INJECTION PUMP

Models With Vacuum Governor

109. Models "FMD" and "FPM" and model "FSM" prior to serial No. 08B-756398 were equipped with Simms model SPE-4A fuel injection pumps having a vacuum governor. The different pumps used can be identified by a number on the injection pump cambox cover. Refer to the following cross reference chart for the different pumps used:

Ford Part Number	Simms Pump Number
DDN-993100-A	SPE 4A 70S 296
DDN-993100-B	SPE 4A 70S 380
DDN-993100-C	SPE 4A 75S 498
DDN-993100-D	SPE 4A 75S 527
DDN-993100-F	SPE 4A 75S 647

On model "FMD" tractors prior to serial No. 1425097, the fuel injection pump (Part No. DDN-993100-A or B) had 7.0 mm diameter plungers; all later pumps have 7.5 mm diameter plungers. In an emergency, pumps with 7.0 mm. plungers may be used in later model tractors providing the pump to engine timing is properly adjusted; refer to paragraph 110. However, pumps with 7.5 mm. diameter plungers must not be installed on model "FMD" engines prior to serial No. 1425097.

In some instances when renewing a fuel injection pump, it will be necessary to install a different pump than original equipment. When this oc-

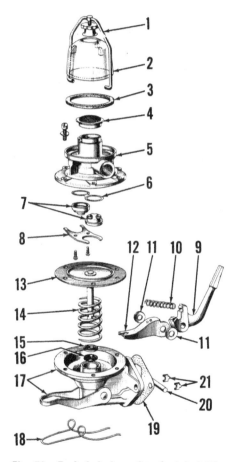

Fig. 59—Exploded view of typical fuel lift pump used with vacuum governor type fuel injection pump. On earlier lift pump, pivot pin (20) extended through pump body (17) and was staked in place.

1. Clamp	13. Diaphragm
2. Bowl	14. Spring
3. Gasket	15. Retainer
4. Screen	16. Seal
5. Valve body	17. Body & priming
6. Gasket	lever assy.
7. Valves	18. Priming lever
8. Clamp	spring
9. Cam lever	19. Gasket
10. Spring	20. Pivot pin
11. Washers	21. Retainers
12. Diaphragm lever	

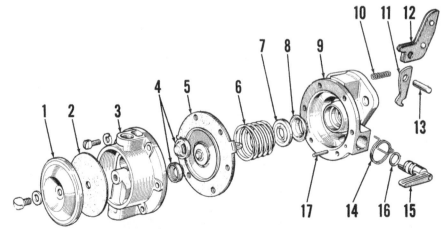

Fig. 60—Exploded view of fuel lift pump assembly used on models "FSM" and "New FSM" with Simms Minimec fuel injection pump.

1. Cover	7. Retainer
2. Pulsator	8. Seal
diaphragm	9. Pump body
3. Valve body	10. Spring
4. Valves	11. Diaphragm lever
5. Diaphragm	12. Cam lever
6. Spring	
	13. Pivot pin
	14. Primer lever
	spring
	15. Primer lever
	16. "O" ring
	17. Lever retaining pin

curs, it may be necessary to change other related parts which will be listed in the Fordson Major Tractor Parts List.

110. **PUMP TIMING.** Pump to engine timing will depend upon tractor model, serial number range and, in some instances, engine modifications. Refer to the following paragraphs:

MODEL "FMD" PRIOR TO SERIAL NO. 1308977. To check pump timing, turn engine until No. 1 piston is coming up on compression stroke (this can be checked by removing oil filler cap and observing rocker arm action), then continue turning engine slowly until the notch in rear flange of crankshaft pulley is aligned with pointer attached to engine front plate. Refer to Fig. 61. The chisel mark on pump drive coupling should then be aligned with mark on timing plate attached to fuel injection pump as shown in Fig. 62. If mark on coupling is not aligned with mark on plate, loosen the two cap screws (7), turn pump camshaft until marks are aligned, then retighten the cap screws. Note: If engine has been modified by installation of later type oil pan with timing hole and flywheel with timing marks, refer to following paragraph and timing chart.

ALL MODELS EXCEPT "FMD" PRIOR TO SERIAL NO. 1308977. Remove the timing plate from lower right rear corner of oil pan (see Fig. 63). Remove the oil filler cap so that rocker arm action can be noted and actuate the compression release if engine is so equipped. Turn engine until No. 1 piston is coming up on compression stroke, then continue to turn engine slowly until the proper degree timing mark (see TIMING CHART) is aligned with pointer. The chisel mark on fuel injection pump

drive coupling flange should then be exactly aligned with mark on timing plate as shown in Fig. 62. If not, loosen the two cap screws (7) and turn pump camshaft to align the marks and retighten the cap screws.

TIMING CHART (VACUUM GOVERNOR)

Tractor Model	Serial No. Range	Degrees BTDC On Flywheel
"FMD"	Prior to 1308977	29 (1)
"FMD"	1308977-1425096	26
"FMD"	1425097-1481090	19 (2)
"FPM"	1481091-1578885	23 (3)
"FSM"	1578886-08B756397	23 (4)

Fig. 62—View showing timing mark on fuel injection pump drive coupling flange (2) aligned with mark on timing plate (both marks are circled). If necessary to adjust timing, loosen the two cap screws (7) and rotate pump camshaft.

2. Drive coupling flange
3. Fiber drive block
4. Driving flange
5. Clamp bolt
7. Cap screws (claw bolts)
23. Dampening valve
24. Lock nut

Note: Refer to paragraph 119 for models with mechanical governor.

(1) Use timing pointer on engine front plate and notch in crankshaft pulley unless engine has been modified by installation of later type oil pan and flywheel.

(2) When fitted with original fuel injection pump (Part No. DDN-993100-C) and thin steel cylinder head gasket (no longer available). If earlier type fuel injection pump (Part No. DDN-993100-A or -B) is installed with thin steel cylinder head gasket, set timing at 22 degrees BTDC. If thick cylinder head gasket is installed with original type pump, adjust timing to 23 degrees BTDC. If both thick head gasket and earlier type pump are installed, set timing to 26 degrees BTDC.

(3) If earlier type fuel injection pump (Part No. DDN-993100-A or -B) has been installed, set timing at 26 degrees BTDC.

(4) Equipped with vacuum governor type fuel injection pump (Simms SPE 4A).

111. **EXCESS FUEL (STARTING) CONTROL.** The fuel injection pump is equipped wtih an excess fuel delivery control (see Fig. 64) which will permit excess fuel to be injected for easier engine starting. Pushing the control button in (or pulling fuel shut-off lever out on early model "FMD") allows governor spring to push fuel metering rack in pump past the maximum fuel delivery stop. As soon as engine starts, the excess fuel control button (or fuel stop lever on early model "FMD") should return to normal position. If control remains in excess fuel delivery position after engine starts, the pump should be removed and sent to authorized service shop for repairs. CAUTION: Do not fasten the control button or fuel stop lever in excess fuel delivery position.

112. **ENGINE SPEED ADJUSTMENTS.** Engine slow idle speed should be 540-560 RPM and high idle no-load speed should be 1900 RPM.

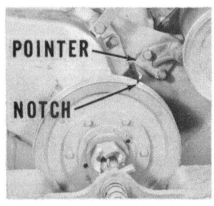

Fig. 61—To time early model "FMD" (prior to serial No. 1308977), turn engine so that No. 1 cylinder is on compression stroke and pointer can be pushed down into notch in crankshaft pulley. Pump timing marks should then be aligned as shown in Fig. 62.

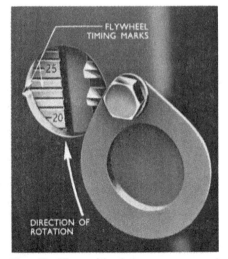

Fig. 63—View with timing cover moved aside showing timing marks on engine flywheel and notch in engine pan. Refer to text for timing specifications and to Fig. 62 for fuel injection pump timing marks.

Fig. 64—View of fuel injection pump showing the excess fuel control button. On early model "FMD", the fuel stop lever is pulled out to engage the excess fuel device for starting.

To check and/or adjust engine speed, first start engine and bring to normal operating temperature, then proceed as follows:

NOTE: Anything that interferes with normal intake manifold vacuum will cause governor malfunction. Do not attempt to adjust engine governed speed with the pre-cleaner removed from top of air cleaner or with the air cleaner disconnected. Renew air pre-cleaner if the spiral fins have been removed or damaged. Be sure that the air cleaner (including pre-cleaner and hose connections) is properly serviced and is in good condition. On early model tractors with closed crankcase ventilation system, a leaking engine gasket will cause governor malfunction; refer to paragraph 88.

Move throttle lever to slow idle speed position and check engine speed. If not within the range of 540-560 RPM, refer to Fig. 65 and adjust idling screw as required. If engine surges at idle speed, refer to Fig. 66 and adjust governor dampening valve to eliminate or reduce surge as much as possible. Readjust engine slow idle speed if necessary. Note: If engine surges excessively, refer to paragraph 116.

Move throttle lever to wide open throttle position and check engine high idle, no load speed. If high idle speed is not approximately 1900 RPM, refer to Fig. 65 and adjust the maximum speed screw as required.

NOTE: If tractor is not equipped with Proofmeter (tachometer), or to check accuracy of Proofmeter, engine speed can be checked at belt pulley or PTO output shaft if tractor is so equipped. At 1900 engine RPM, belt pulley speed with transmission in high range will also be 1900 RPM; PTO output shaft speed will be 855 RPM when equipped

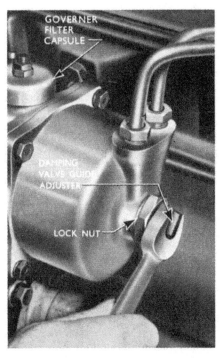

Fig. 66—If engine surges at idle speed, or at a slow engine speed at which engine must be run for certain operations, loosen lock nut and slowly turn damping valve guide adjuster back and forth to find position that will most effectively reduce surge.

with standard PTO or 641 RPM if equipped with raised PTO.

113. R&R FUEL INJECTION PUMP. To remove the fuel injection pump, first thoroughly clean the pump and surrounding area, then shut off the fuel supply valve and proceed as follows:

Remove the two vacuum pipes from between pump governor housing and engine intake manifold. Remove the fuel injector pipes and the fuel supply pipe from filter to pump and immediately cap all openings. Disconnect the stop control wire and then unbolt and remove pump assembly from mounting bracket.

To reinstall pump, set timing position as outlined in paragraph 110, then reinstall pump with timing mark on pump drive coupling aligned with mark on timing plate as shown in Fig. 62. Reconnect stop wire so that

stop button is out about ⅛-inch when stop control lever on fuel injection pump is fully forward. Reinstall the governor pipes, fuel supply pipe and fuel injector pipes, leaving the injector pipes loose at upper (injector) ends. Using a pump type oil can, fill pump cam box to level of leak-off pipe opening with clean, new engine oil, then install leak-off pipe. Bleed the diesel fuel system as outlined in paragraph 107. CAUTION: Do not attempt to start engine without the governor pipes installed.

114. PUMP OVERHAUL. Other than servicing the governor as outlined in paragraph 116 or renewing drive coupling parts as outlined in paragraph 115, overhaul and/or adjustment of the fuel injection pump should not be attempted unless in a fully equipped injection service shop.

115. PUMP DRIVE COUPLING. Refer Fig. 67 for exploded view of the fuel injection pump drive coupling. Do not attempt to renew the timing plate (1) or drive coupling flange (2) on fuel injection pump as timing marks must be placed on the flange in a diesel pump service shop.

Renewal of the fiber drive coupling block (3), driving flange (4) or adapter (6) can be accomplished after removing fuel injection pump as outlined in paragraph 113. Remove the two cap screws (7) to unbolt flange from adapter. Remove the clamping bolt (5), then pull adapter from accessory shaft.

At serial No. 1599502, width of keyway in auxiliary driveshaft and pump drive adapter was increased from $\frac{5}{32}$-inch to $\frac{3}{16}$-inch. As some earlier models may have the later shaft and adapter, it will be necessary to measure key width before obtaining a new adapter and key.

When reinstalling adapter, leave clamping bolt loose until after fuel injection pump has been reinstalled. With the adapter to driving flange cap screws snug, but not tight, move adapter on accessory driveshaft to obtain a 0.010 end float of the fiber

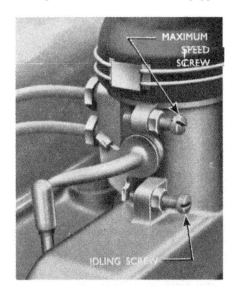

Fig. 65—Engine governed speed is adjusted by the maximum speed screw (top) and idling screw (bottom).

Fig. 67—Exploded view of pump drive coupling. The timing plate (1) and drive coupling flange (2) should not be removed from the pump as timing marks must be affixed to new parts at pump repair station.

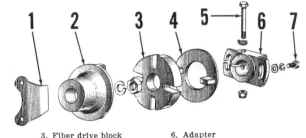

1. Timing plate	3. Fiber drive block
2. Drive coupling flange	4. Driving flange
	5. Clamp bolt

6. Adapter	
7. Cap screws (claw bolts)	

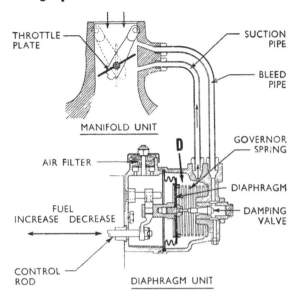

THROTTLE PLATE

SUCTION PIPE

BLEED PIPE

MANIFOLD UNIT

AIR FILTER

GOVERNOR SPRING

DIAPHRAGM

D

DAMPING VALVE

FUEL INCREASE DECREASE

CONTROL ROD

DIAPHRAGM UNIT

Fig. 68—Schematic view of the vacuum governor unit incorporated in the Simms fuel injection pump used prior to model "FSM" serial No. 08B-756397. Suction pipe is connected to intake manifold at point below throttle plate and bleed pipe is connected to intake manifold above the throttle plate. Vacuum in diaphragm chamber (D) acts against governor spring to move the pump control rod and increase or decrease fuel injected in engine according to relative vacuum above and below throttle plate in intake manifold.

drive block, then tighten adapter clamping bolt. Set pump to engine timing accurately as outlined in paragraph 110, then tighten the adapter to driving flange cap screws.

116. VACUUM GOVERNOR. Refer to Fig. 68 for schematic view showing vacuum governor operation and to Fig. 69 for exploded view of governor unit. Service of the governor unit can be performed without use of special equipment. Refer to the following procedures.

The governor filter element (see Fig. 70) should be removed and cleaned after each 200 hours of operation. The filter element should be lightly oiled with clean engine oil before being reinstalled on top of fuel injection pump housing.

To check governor diaphragm for air leaks, proceed as follows: Remove the governor housing to intake manifold air pipes, rotate stop lever at rear end of pump as far to rear as possible and tightly cover the two openings in governor housing. If the governor spring returns the pump

rack to normal position, the diaphragm is leaking.

To renew the governor diaphragm and/or the dampening valve, proceed as follows: Refer to Fig. 70 and remove the governor assembly from fuel injection pump, then refer to Fig. 69 and disassemble governor unit as follows: Hold the governor spring compressed by pushing diaphragm plate inward and remove the wire ring (10) and retaining plate (11); then, remove the diaphragm and dampening valve assembly and the governor spring. Unscrew the diaphragm guide (12) and lock nut (13), then separate the diaphragm plate (15), diaphragm (14), spring retainer plate (16) and dampening valve (17).

Inspect dampening valve for wear at ball joint or at front end of valve where it rides in the bushing (22) in dampening valve adjusting screw (23). Renew the valve and bushing if wear is noted. Note: Bushing is available separately from adjusting screw, but new adjusting screw includes new bushing. Refer to Fig. 71

for identification of two of the dampening valves and bushings that have been used. A later valve has a 7 degree taper and is used with the same "flush bushing" as the 20 degree taper valve; later 7 degree taper valve can be identified by annular groove cut on valve.

If difficulty with engine surging at about 1000 RPM is encountered on model "FPM" and surging cannot be satisfactorily corrected by turning the dampening valve adjusting screw, use of later 7 degree taper dampening valve (Ford part No. DDN-993113-C) along with smaller ($\frac{3}{16}$-inch O.D.) governor to intake manifold suction pipe (part No. DDN-993139-D) is recommended. Note: Fuel injection pumps with Simms identification "SPE 4A-75S-647" on cambox cover will have the later 7 degree taper valve.

Later type governor spring has free length of 6 inches. Free length of early

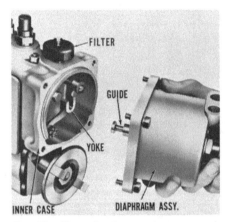

FILTER

GUIDE

YOKE

INNER CASE

DIAPHRAGM ASSY.

Fig. 70—View showing vacuum governor (diaphragm) assembly being removed from fuel injection pump. Note governor breather filter which should be serviced regularly. Lubricate guide and yoke with multi-purpose grease prior to reassembly.

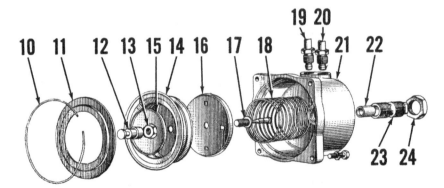

10 11   12 13 15 14 16   17 18   19 20   21 22

23 24

Fig. 69—Exploded view of the fuel injection pump vacuum governor assembly. Gap in retaining ring (10) should be installed adjacent to ejector hole in housing (21).

10. Retaining ring	14. Diaphragm	17. Dampening valve	21. Governor housing
11. Retainer plate	15. Diaphragm plate	18. Governor spring	22. Bushing
12. Diaphragm guide	16. Spring locating	19. Suction pipe	23. Adjusting screw
13. Lock nut	plate	20. Bleed pipe	24. Lock nut

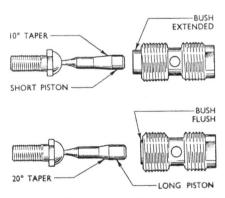

10° TAPER

BUSH EXTENDED

SHORT PISTON

BUSH FLUSH

20° TAPER

LONG PISTON

Fig. 71—View showing 10 degree taper and 20 degree taper dampening valves and their mating bushings. Latest 7½ degree taper valve is not shown, but can be identified by an annular groove on valve. Latest valve is used with same bushing as 20 degree taper valve.

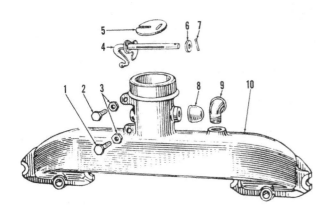

Fig. 72—View showing throttle plate and shaft removed from intake manifold. Wear on throttle shaft and in shaft bore of intake manifold is possible cause of dust entry and excessive engine wear.

1. Slow idle speed screw
2. Maximum speed screw
3. Lock nuts
4. Throttle shaft
5. Throttle plate
6. Washer
7. Cotter pin
8. Cover
9. Ventilation connector (early "FMD" only)
10. Intake manifold

spring was 7.16 inches. Early (long) spring should exert a force of 3 lbs., 8 ounces to 3 lbs., 13½ ounces when compressed to a length of 1.97 inches; later spring should exert force of 3 pounds to 3 pounds, 5 ounces, when compressed to same length. In conjunction with change in governor spring, the external diameter of diaphragm plate and internal diameter of spring retainer plate were reduced from 2.48 inches to 2.09 inches. The springs and plates are interchangeable; however, both the latest diaphragm plate (15—Fig. 69) and spring retainer plate (16) must be installed as a pair.

When reassembling governor, lightly oil the ball joint of dampening valve. Using Fig. 69 as a guide, reassemble diaphragm and dampening valve assembly. Set spring in housing being sure outer end is properly located in housing, then set the valve and diaphragm assembly on inner end of spring so that the inner end of spring is seated in the spring retainer plate. Push unit down into housing making sure dampening valve enters bushing. Fit outer edge of diaphragm in housing and secure with retaining plate and wire ring. Gap in ring should be adjacent to ejector hole in housing. Lightly grease the groove in end of dampening valve guide, then reinstall the governor unit as in Fig.

70. Be sure governor air pipes are correctly installed before attempting to start engine.

117. INTAKE MANIFOLD THROTTLE. Refer to exploded view of intake manifold in Fig. 72. Throttle shaft (4) can be removed after removing cover (8), cotter pin (7) and flat washer (6). Check throttle shaft and shaft bore in intake manifold for wear. Renew shaft and/or intake manifold if wear is sufficient to allow dust entry into air intake system.

Crankcase ventilation connector (9) is not used except on early "FMD" prior to serial No. 1425097 with closed crankcase ventilation system. Refer to paragraph 88.

Models With Mechanical Governor Type (Minimec) Fuel Injection Pump

118. At serial No. 08B-756398 of model "FSM" production, a fuel injection pump with integral mechanical governor (Simms "Minimec") was introduced. Other than adjusting engine governed speed as outlined in paragraph 121 or renewing the fiber drive coupling block, no service to the pump assembly should be attempted outside an authorized Simms or Ford diesel pump repair station.

119. **PUMP TIMING.** To check fuel injection pump to engine timing, first

Fig. 73—View of Simms Minimec (mechanical governor) fuel injection pump on Fordson Super Major. Refer to Fig. 74 for view showing oil filler, level and drain plugs and to Fig. 75 for speed adjustment screws. Timing mark on drive coupling flange (F) must be aligned with mark on timing indicator (I) when No. 1 piston is coming up on compression stroke and 21 degree BTDC mark on flywheel (see Fig. 63) is aligned with notch in timing hole.

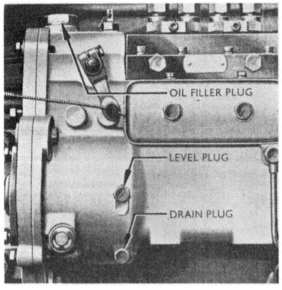

Fig. 74—Every 200 hours of operation, fuel injection cambox should be drained and refilled to level plug with clean engine oil. Oil drain plug, level plug and filler plug locations are shown.

actuate the compression release, remove timing cover from right rear side of oil pan and remove oil filler cap from rocker arm cover so that action of rocker arms can be observed. Turn engine until No. 1 piston is coming up on compression stroke, then continue turning until the 21 degree BTDC timing mark on flywheel is aligned with notch in timing hole of oil pan. The timing mark on fuel injection pump drive coupling flange should then be exactly aligned with mark on pointer attached to pump. If not, loosen the two cap screws clamping driving coupling front flange to adapter, turn pump camshaft until marks are aligned, then tighten the two cap screws. Turn engine through two revolutions and recheck timing.

120. LUBRICATION. Refer to Fig. 74 for location of fuel injection pump oil filler, oil level and oil drain plugs.

When installing a new or rebuilt fuel injection pump, be sure to fill pump to level plug with new, clean engine oil. After each 200 hours of operation, the pump should be drained and refilled wih new, clean oil. Use same weight oil as used in engine crankcase. As there will be some fuel leakage past the pump plungers, the oil level will rise and some of the fuel-oil mixture will be lost out the overflow pipe (O—Fig. 73). This is only normal and the greatest overflow will be noted after the engine has been stopped.

121. ENGINE SPEED ADJUSTMENT. Engine governed speed is controlled by the movement of governor control lever; refer to Fig. 75. The maximum speed screw is usually adjusted on a fuel injection pump calibration stand; however, it should be readjusted if engine high idle (no-load) speed is not within the range of 1800-1825 RPM on model "FSM" or 1925 RPM on model "New FSM".

Note: The maximum speed screw is wired and sealed; breaking of the seal by other than Ford authorized service personnel may void the warranty on new tractors.

Engine slow idle speed should be approximately 550 RPM on all models equipped with Simms Minimec fuel injection pump.

122. R&R FUEL INJECTION PUMP. To remove the fuel injection pump, first shut off the fuel supply valve and thoroughly clean pump and surrounding area, then, proceed as follows:

Disconnect the proofmeter (tachometer) drive cable and the stop control cable from injection pump. Remove the fuel filter to pump supply pipe and the four fuel injector pipes and immediately cap all openings, then, unbolt and remove the pump assembly from engine.

When reinstalling pump, remove oil filler cap from rocker arm cover so that rocker arm action can be observed, then turn engine until No. 1 piston is coming up on compression stroke. Refer to Fig. 63, remove the timing cover and slowly turn engine in normal direction of rotation until the 21 degree BTDC mark on flywheel is aligned with notch in timing hole in oil pan. Turn pump camshaft so that timing mark on coupling flange (F—Fig. 73) is aligned with timing indicator (I), then, attach pump to engine in reverse of removal procedure.

123. PUMP OVERHAUL. Other than renewing the drive coupling fiber block, driving flange and/or adapter as outlined in paragraph 115, overhaul of the fuel injection pump should not be attempted. Return the pump to authorized Simms or Ford fuel injection pump repair station for adjustment, rebuilding or exchange.

INJECTION NOZZLES
All Models

All models are equipped with Simms fuel injection nozzles. The nozzles have four orifices each with a spray angle of 150 degrees. Engines between serial No. 1425-097 and 1481090 were equipped with nozzles having 0.011 diameter orifices. All other models were equipped with nozzles having 0.010 diameter orifices. The two different nozzles can be identified after disassembly of the fuel injector. The nozzles having 0.010 diameter orifices will have "NL 123" etched on upper part of nozzle body and nozzles having 0.011 diameter orifices will have "NL 141" etched on upper part of body. The different size orifice nozzles should not be individually interchanged, but can be interchanged as set

Fig. 75—View of fuel injection pump removed from engine to show maximum speed screw and idling stop screw. Adjustment of the maximum fuel stop screw should not be attempted unless pump is mounted on fuel injection pump calibrating machine.

of four without seriously affecting engine operation.

CAUTION: Fuel leaves the injector nozzle with sufficient force to pentrate the skin. When testing nozzles, keep your person clear of the spray.

124. TESTING AND LOCATING A FAULTY NOZZLE. If engine does not run properly and a faulty injection nozzle is indicated, such a faulty nozzle can be located as follows: With engine running, loosen the high pressure line fitting on each nozzle holder in turn, allowing fuel to escape at the union rather than enter the injector. As in checking for faulty spark plugs in a spark ignition engine, the faulty unit is the one which, when its line is loosened, least affects the running of the engine.

125. NOZZLE TESTER. A complete job of testing and adjusting the fuel injection nozzle requires use of a special tester such as shown in Fig. 76. The nozzle should be tested for opening pressure, spray pattern, seat leakage and leak back.

Operate the tester until oil flows and then connect injection nozzle to tester. Close the valve to tester gage and operate tester lever to be sure nozzle is in operating condition and

Fig. 76—Checking nozzle seat leakage on injector nozzle test stand. Refer to text for procedure.

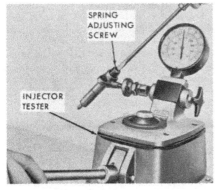

Fig. 77—Adjusting injector opening pressure.

not plugged. If oil does not spray from all four spray holes in nozzle, if tester lever is hard to operate or other obvious defects are noted, remove nozzle from tester and service as outlined in paragraph 132. If nozzle operates without undue pressure on tester lever and fuel is sprayed from all four spray holes, proceed with following tests:

126. OPENING PRESSURE. While slowly operating tester lever with valve to tester gage open, note gage pressure at which nozzle spray occurs. Pressure should be 2720-2794 psi. If gage pressure is not within these limits, remove cap nut and turn adjusting screw (See Fig. 77) as required to bring opening pressure within specified limits. If opening pressure is erratic or cannot be properly adjusted, overhaul nozzle as outlined in paragraph 132. If opening pressure is within limits, check spray pattern as outlined in following paragraph.

127. SPRAY PATTERN. Operate the tester lever slowly and observe nozzle spray pattern. All four (4) sprays must be similar and spaced at approximately 90° to each other in a nearly horizontal plane. Each spray must be well atomized and should spread to a 3 inch diameter cone at approximately 8 inches from nozzle tip. If spray pattern does not meet these conditions, overhaul nozzle as outlined in paragraph 132. If nozzle spray is satisfactory, proceed with seat leakage test as outlined in following paragraph:

128. SEAT LEAKAGE. Close valve to tester gage and operate tester lever quickly for several strokes. Then, wipe nozzle tip dry with clean blotting paper, open valve to tester gage, push tester lever down slowly to bring gage pressure to 200 psi. below nozzle opening pressure and hold this pressure for one minute. Apply a piece of clean blotting paper (see Fig. 76) to tip of nozzle; the resulting oil blot should not be greater than one-half inch in diameter. If nozzle tip drips or blot is excessively large, overhaul nozzle as outlined in paragraph 132.

If nozzle seat leakage is not excessive, proceed with nozzle leak back test as outlined in following paragraph.

130. NOZZLE LEAK BACK. Operate tester lever to bring gage pressure to approximately 2300 psi., release lever and note time required for gage pressure to drop from 2200 psi to 1500 psi. Time required should be from ten to forty seconds. If time required is less than ten seconds, nozzle is worn or there are dirt particles between mating surfaces of nozzle and holder. If time required is greater than forty seconds, needle may be too tight a fit in nozzle bore. Refer to paragraph 132 for disassembly, cleaning and overhaul information.

NOTE: A leaking tester connection, check valve or pressure gage will show up in this test as excessively fast leak back. If, in testing a number of injectors, all show excessively fast leak back, the tester should be suspected as faulty rather than the injectors.

131. REMOVE AND REINSTALL INJECTORS. To remove the injectors, first remove the rocker arm cover, then proceed as follows: Disconnect the fuel leak off pipe from the top of each injector and at rear end of cylinder head and remove the pipe. Remove the fuel injector pressure pipes, being sure to hold delivery valve from turning while removing nut at injection pump end of pipe. Remove the injector after unscrewing the two retaining cap screws. CAUtion: Take care not to drop any of the return line or injector cap screws down into the engine. If injector is tight in its bore, remove with pry bar or injector lifting bar. Immediately cap all openings.

Prior to reinstalling injectors, check the injector seats in cylinder head to be sure they are clean and free of any carbon deposit. Old copper sealing washers, if stuck in bore, can be removed by tapping a screwdriver blade down through washer, then twisting washer loose. Install a new copper washer in each seat, then insert injector assemblies into cylinder head. Tighten the retaining cap screws of each injector alternately and evenly

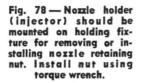

Fig. 78 — Nozzle holder (injector) should be mounted on holding fixture for removing or installing nozzle retaining nut. Install nut using torque wrench.

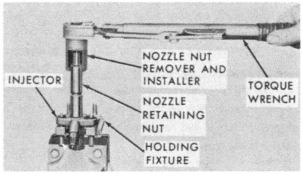

to a torque of 10-15 Ft.-Lbs. Reinstall leak off pipe to tops of injectors and reconnect pipe at rear end of cylinder head. Check fuel injector pipe connections to be sure they are clean and reinstall pipes, tightening at injection pump end only. Crank engine until a steady stream of fuel is pumped out each pipe, then tighten the connections. Start and run engine to be sure there are no leaks, then reinstall rocker arm cover.

132. OVERHAUL INJECTORS. Unless complete and proper equipment is available, do not attempt to overhaul diesel nozzles. Equipment recommended by Ford is Kent-Moore J-8666 Injector Nozzle Tester and J-8537 Injector Nozzle Service Tool Set. This equipment is available from the Kent-Moore Organization, Inc., 28635 Mound Road, Warren, Michigan.

Refer to Fig. 78 and proceed as follows: Secure injector holding fixture (J8537-11) in a vise and mount injector assembly in fixture. **Never** clamp the injector body in vise. Remove the cap nut and back-off adjusting screw, then lift off the upper spring disc, injector spring and spindle. Remove the nozzle retaining nut using nozzle nut socket (J8537-14), or equivalent, and remove the nozzle and valve. Nozzles and valves are a lapped fit and must never be interchanged.

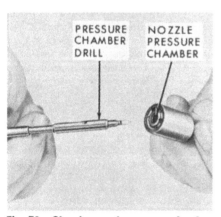

Fig. 79—Cleaning nozzle pressure chamber.

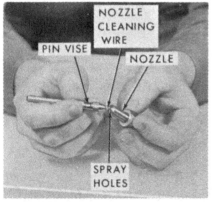

Fig. 80—Cleaning nozzle spray hole.

Place all parts in clean fuel oil or calibrating fluid as they are disassembled. Clean injector assembly exterior as follows: Soften hard carbon deposits formed in the spray holes and on needle tip by soaking in a suitable carbon solvent, then use a soft wire (brass) brush to remove carbon from needle and nozzle exterior. Rinse the nozzle and needle immediately after cleaning to prevent carbon solvent from corroding the highly finished surfaces. Clean the pressure chamber of nozzle with a 0.043 reamer (J8537-4) as shown in Fig. 79. Clean spray holes with the proper size wire probe held in a pin vise (J4298-1) as shown in Fig. 80. To prevent breakage of wire probe, the wire should pro-

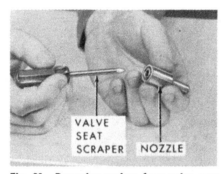

Fig. 81—Removing carbon from valve seat in nozzle.

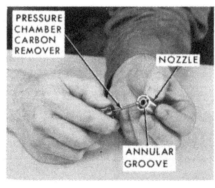

Fig. 82—Clean annular groove in top of nozzle using pressure chamber tool as shown.

Fig. 83 — View showing proper positioning of nozzle in reverse flush adapter for back-flushing nozzle on injector test stand.

trude from pin vise only far enough to pass through the pin holes. Rotate pin vise without applying undue pressure. Use a 0.009 wire probe (Kent-Moore part No. J8537-2) with nozzle having "NL 123" etched on upper part of nozzle body or a 0.010 diameter wire probe with nozzle having "NL 141" etched on upper part of body.

The valve seats in nozzle are cleaned by inserting the valve seat scraper (J-8537-18) into nozzle and rotating scraper. Refer to Fig. 81. The annular groove in top of nozzle and the pressure chamber are cleaned by using the pressure chamber carbon remover tool (J-8537-15) as shown in Fig. 82.

When above cleaning is accomplished, back flush nozzle and needle by installing reverse flushing adapter (J-8537-6) on nozzle tester and inserting nozzle and needle assembly tip end first into the adapter. Secure with knurled nut. Rotate the needle in nozzle while operating tester lever. After nozzle is back flushed, the seat can be polished by using a small amount of tallow (J-8537-28) on end of a polishing stick (J-8537-21), rotating stick in nozzle as shown in Fig. 84.

If the leak back test time was greater than 40 seconds (refer to paragraph 130), or if needle is sticking in bore of nozzle, correction can be made by lapping the needle and nozzle assembly. This is accomplished by using a polishing compound (Bacharach No. 66-0655 is suggested) as follows: Place small diameter of nozzle in a chuck of a drill having a maximum speed of less than 450 RPM. Apply a small amount of polishing compound on barrel of needle taking care not to allow any compound on tip or beveled seat portion, and insert needle in rotating nozzle body. Refer to Fig. 85. Note: It is usually necessary to hold upper pin end of needle with vise-grip pliers to keep the needle from turning with the nozzle. Work the needle in and out a few

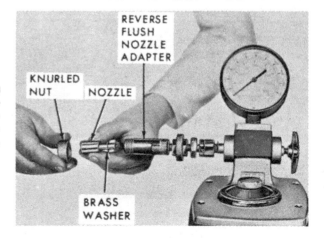

times taking care not to put any pressure against seat, then withdraw the needle, remove nozzle from chuck and thoroughly clean the nozzle and needle assembly using back flush adapter and tester pump.

Prior to reassembly, rinse all parts in clean fuel oil or calibrating fluid and assemble while still wet. The injector inlet adapter (Simms only) normally does not need to be removed. However, if the adapter is removed, use a new copper sealing washer when reinstalling. Position the nozzle and needle valve on injector body and be sure dowel pins in body are correctly located in nozzle as shown in Fig.

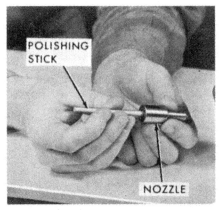

Fig. 84—Nozzle seat can be polished by using a small amount of tallow on a polishing stick and rotating stick in nozzle as stand.

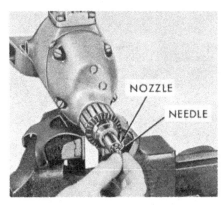

Fig. 85—Nozzle and needle can be lapped by using a slow speed electric drill. Refer to text.

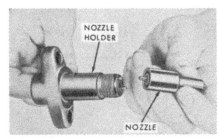

Fig. 86 — When positioning nozzle and needle on injector body, be sure that dowel pins are correctly aligned in nozzle.

86. Install the ⅜-inch shim washer (See Fig. 87) and nozzle retaining nut and tighten nut to a torque of 50 Ft.-Lbs. Note: Place injector in holding fixture (J8537-11) and tighten nut with socket (J8537-14). Install the spindle, spring, upper spring disc and spring adjusting screw. Connect the injector to tester and adjust opening pressure as in paragraph 126. Use a new copper washer and install cap nut. Recheck nozzle opening pressure to be sure that installing nut did not change adjustment. Retest injector as outlined in paragraphs 127 through 130; renew nozzle and needle if still faulty. Note: If the injectors are to be stored after overhaul, it is recommended that they be thoroughly flushed with calibrating fluid prior to storage.

COOLING SYSTEM

RADIATOR

All Models

133. Three different types of radiators may be encountered in service. Early model "FMD" were equipped with radiators having 10 fins per inch. All later radiators have five fins per inch but can be used for service installation to renew the earlier type radiator. Late production model "FSM" (Serial No. 08B767403 and later) and model "New FSM" have radiator mounting bolt holes 10¼

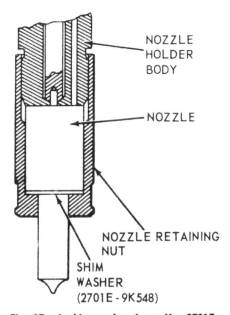

Fig. 87—A shim washer (part No. 2701E-9K548) installed between nozzle and retaining nut will help in preventing dowels from being sheared as nozzle retaining nut is tightened.

inches apart instead of 18⅛ inches apart as on earlier production. To install late production radiator on earlier tractor, it will be necessary to drill new mounting holes in tractor front support; refer to Fig. 88. Also, if installing later type front support on earlier tractor, it will be necessary to drill radiator mounting holes in the front support as shown in Fig. 88 to adapt the earlier radiator. A kit (Ford part No. DDN-8K004) is available for mounting the late type radiator (with mounting holes 10¼ inches apart) in earlier type front support where new mounting holes have been drilled.

134. **R&R RADIATOR.** To remove radiator, proceed as follows: Drain cooling system, remove air pre-cleaner and vertical exhaust muffler, then remove engine hood. Disconnect radiator hoses and, when so equipped, the flexible filler hose. Remove the two half-sections of radiator grille and disconnect headlight wiring. On "FSM" and "New FSM", unbolt and

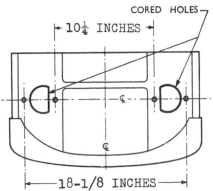

Fig. 88 — When installing early radiator with late front support or late radiator with early production front support, new mounting holes must be drilled according to above dimensions.

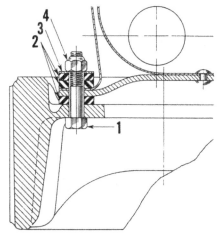

Fig. 89—Cross-sectional view showing early type radiator mounting.

1. Bolt   3. Washer
2. Insulators   4. Self-locking nut

remove the headlight assemblies. Unbolt the radiator support brace (to cylinder head or radiator shell). Unbolt and remove radiator shell, then unbolt and remove radiator assembly from front support.

When reinstalling radiator, refer to Fig. 89 or to Fig. 90 for placement of rubber insulating washers on radiator mounting bolts. Note: If renewing radiator or front support, it may be necessary to modify the front support; refer to Fig. 88.

THERMOSTAT
All Models

135. Two different types of thermostats have been used; refer to Fig. 91. The "non-shrouded" type was used in early model "FMD" (prior to Serial No. 1425097). Cylinder heads for later model "FMD" (Serial No. 1425-097 and up) and all model "FPM", "FSM" and "New FSM" require a "shrouded" type thermostat. Although the later "shrouded" type thermostat may be used in earlier engines, it is very important that only the "shrouded" type (Ford part No. 204-E-8575) be used in the later engines. The design of the shrouded thermostat is such that as the engine reaches operating temperature, the thermostat opens and gradually closes off the by-pass opening; refer to Fig. 92. A

restricted flow of hot coolant is then circulated through the radiator until, when the thermostat is fully open, an unrestricted flow is obtained.

Either type of thermostat should start to open at 170-179° F., and be fully open at 199° F. When installing thermostat, it is important that the thermostat flange protrude 0.005-0.028 above cylinder head as shown in Fig. 92. If protrusion is less than 0.005, install a 0.018-0.021 shim (Ford part No. E9-CJ-9) between thermostat flange and cylinder head.

WATER PUMP
All Models

136. **R&R PUMP ASSEMBLY.** Drain cooling system and disconnect radiator hose from pump. If equipped with power steering, loosen power steering pump mounting bolts and slide pump downward to loosen drive belt. Loosen generator mounting bolts and push generator inward to loosen fan belt. Unbolt and remove the fan blades and belts, then unbolt and remove water pump assembly.

NOTE: At serial No. 1425097, water pump passages were changed in pump body and cylinder block. If installing new pump assembly, be sure correct pump and mounting gasket are obtained.

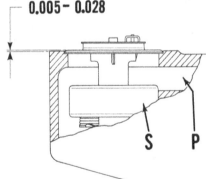

Fig. 92—On all models, thermostat must protrude 0.005-0.028 from cylinder head as shown. Except on early "FMD," thermostat shroud (S) controls flow of water through by-pass passage (P) in cylinder head.

To reinstall pump assembly, reverse removal procedures.

137. **OVERHAUL PUMP.** Using suitable pullers, remove pulley hub and, on power steering models, fan pulley from front end of water pump shaft. Pry the bearing retaining clip from slot in pump housing. Adequately support the water pump housing and press the shaft and bearing assembly with impeller out towards rear of housing. Using suitable pullers, remove impeller from shaft.

Carefully inspect all parts and install new seal or repair kit as follows: Press slinger sleeve, flange end first, onto rear (long) end of shaft so that flange is same distance from bearing as slinger flange was on shaft and bearing assembly removed from pump. Press shaft and bearing into pump body from front until retainer groove in bearing race is aligned with slot in pump housing, then install the retaining clip. Place new seal assembly over shaft with carbon ring out (towards impeller) and fit the seal into proper position. Press impeller onto rear end of shaft so that there is 0.030 clearance between impeller vanes and rear face of pump body. Supporting rear end of shaft, press the pulley hub onto front end of shaft (pulley and hub assembly on models with power steering). Recess in front of hub should be flush with front end of shaft.

NOTE: Due to changes in the by-pass passages of both the water pump and cylinder block, early "FMD" pump assemblies (prior to serial No. 142097) require a different pump body than later models. However, all other parts and service procedures are the same for all models.

FAN BLADES
All Models

138. Various sizes of fan blades have been used and are available for service. The original production installed fan blade for tractors shipped to U.S. are for temperatures below 90° F.

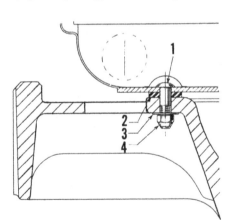

Fig. 90—Cross-sectional view showing late type radiator mounting.

1. Bolt
2. Insulator
3. Washer
4. Self-locking nut

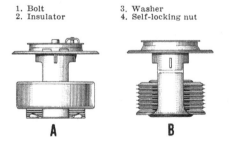

Fig. 91—A shrouded type thermostat (A) must be used in all models except early production "FMD" which can be equipped with non-shrouded thermostat (B).

Fig. 93—Exploded view of typical water pump assembly.

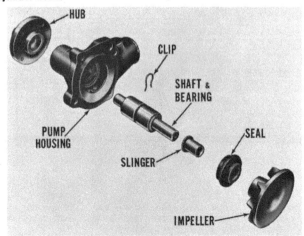

If overheating is experienced in hot weather, the original type fan blade should be removed and the "tropical" fan installed. Conversely, if operating in cold weather and trouble is experienced in not being able to keep the engine up to normal operating temperature, a smaller fan should be installed.

Model "FMD" engines were originally equipped with a 15 inch diameter two blade fan (Ford part No. DDN-8607-A). Alternate "hot weather" fan for the model "FMD" engine is a two blade 18 inch diameter fan (Ford part No. 508E-8607).

Models "FPM", "FSM" and "New FSM" engines were originally equipped with a 17 inch diameter two blade fan (Ford part No. DDN-8607-B). Alternate "hot weather" fan for these engines is a four blade 17 inch diameter fan made by adding a second two blade fan (Ford part No. DDN-8606) to the original fan installation.

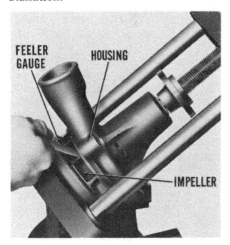

Fig. 94—Using feeler gage to check clearance between pump housing and impeller; refer to text.

ELECTRICAL SYSTEM

GENERATOR

All Models

139. Two different Lucas generators (Ford parts Nos. E27N-10001-D and E1ADN-10001) have been used and only the latest type is available for service. Note: The later generator (E1ADN-10001) is equipped with blade type connectors whereas the early unit (E27N-10001-D) had screw type post connectors. When installing new generator, it will be necessary to obtain two blade type connectors (part Nos. 114777-ESA and 114775-ESA) and two insulators (part Nos. 105E-14454-B and -C) to connect wiring to generator. The latest type

generator will develop maximum current at lower engine speed than previous generator. Refer to following generator test procedure:

140. **GENERATOR TEST PROCEDURE.** Disconnect leads marked "D" and "F" from voltage regulator terminals and connect these leads together. Connect negative lead of voltmeter, which is calibrated to at least 30 volts, to the connected leads and connect positive lead of voltmeter to a good ground connection. Start engine and gradually increase engine speed to approximately 1000 RPM;

voltmeter reading should rapidly rise to above 24 volts and remain steady. If reading is low, or if no reading can be obtained, check to see that generator leads are in good condition and that generator is polarized. Note: To polarize (magnetize) generator, disconnect wire from field ("F") terminal of regulator and momentarily touch this wire to the battery ("A") terminal of regulator. If generator will not develop the required voltage, follow normal overhaul procedures to repair or renew the generator. If generator checks OK, check voltage

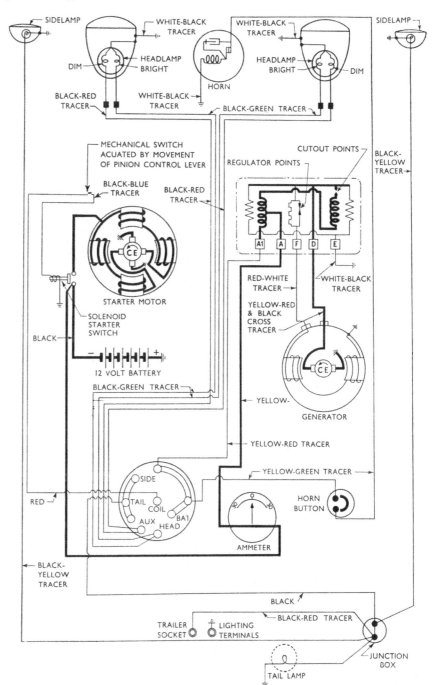

Fig. 95—Wiring diagram for model "FMD"; for other models, refer to Fig. 96. Note: Late production model "FMD" will have two 6-volt batteries connected in series instead of single 12-volt battery shown.

regulator as outlined in paragraph 142.

VOLTAGE REGULATOR
Model "FMD" and Early Model "FPM"

141. All model "FMD" and model "FPM" prior to serial No. 1542263 were equipped with Lucas adjustable voltage regulators with removable removable plastic covers. A Lucas model RB 106/1 (Ford part No. E27N-10505-E) regulator was used prior to serial No. 1426221 of model "FMD" production. Later model "FMD" and early model "FPM" prior to serial No. 1542263 were equipped with a Lucas model RB 106/2 regulator (Ford part No. E27N-10505-F). These regu-

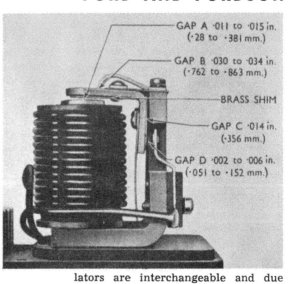

Fig. 97 — Adjustment points for Lucas model RB 106/1 regulator cut-out relay; refer to text.

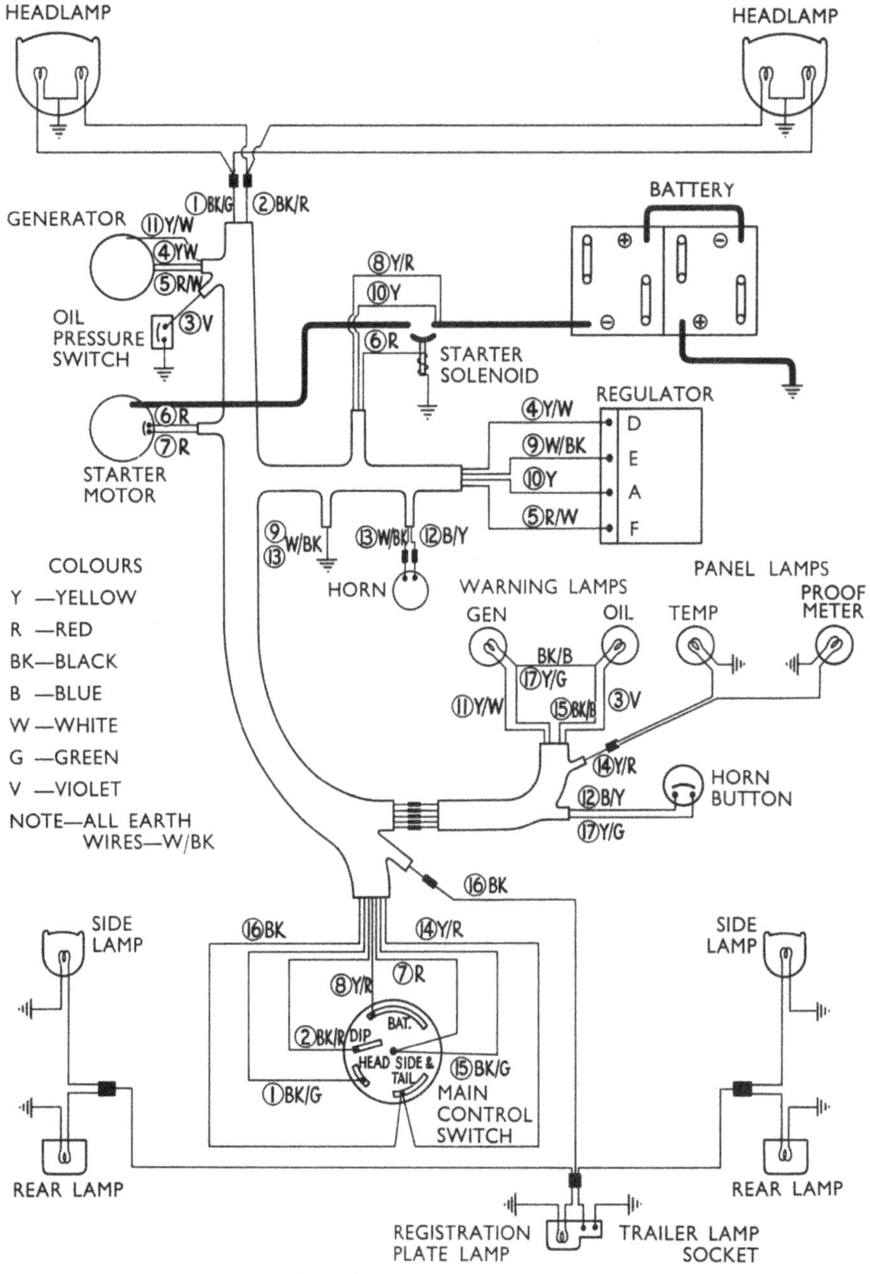

GAP A ·011 to ·015 in.
(·28 to ·381 mm.)

GAP B ·030 to ·034 in.
(·762 to ·863 mm.)

BRASS SHIM

GAP C ·014 in.
(·356 mm.)

GAP D ·002 to ·006 in.
(·051 to ·152 mm.)

lators are interchangeable and due to service installations, may not be installed according to tractor serial number. However, they can be identified by Lucas identification (model) numbers stamped on bottom of regulator. Refer to following test and adjustment procedures:

142. REGULATOR TEST PROCEDURE. To check the voltage regulator, disconnect the leads from regulator terminals marked "A" and "A1" and connect them together. Connect the negative lead of a test voltmeter to "D" terminal of voltage regulator and connect positive lead of test voltmeter to a good ground connection. Start engine and gradually increase speed until voltmeter needle "flicks", then steadies. The voltmeter reading should then be within the limits given in the VOLTAGE REGULATOR SETTING CHART according to approximate temperature of the regulator unit (air temperature if engine has been started cold).

VOLTAGE REGULATOR SETTING CHART

Lucas Model RB 106/1

TEMPERATURE, DEGREES F.	VOLTAGE READING
50°	15.9-16.5
68°	15.6-16.2
86°	15.3-15.9
104°	15.0-15.6

Lucas Model RB 106/2

TEMPERATURE, DEGREES F.	VOLTAGE READING
50°	15.7-16.1
68°	15.6-16.0
86°	15.5-15.9
104°	15.0-15.6

If the reading is not between the limits given in the VOLTAGE REGULATOR SETTING CHART, the regulator is in need of adjustment. Increase engine speed to maximum speed; the voltmeter reading should not rise more than one-half volt above the tabulated readings at 1000 RPM.

HEADLAMP

HEADLAMP

GENERATOR

①BK/G ②BK/R
⑪Y/W
④Y/W
⑤R/W
③V

OIL PRESSURE SWITCH

⑥R
⑦R

STARTER MOTOR

⑧Y/R
⑩Y
⑥R

STARTER SOLENOID

BATTERY

REGULATOR

④Y/W
⑨W/BK
⑩Y
⑤R/W

D
E
A
F

COLOURS

Y —YELLOW
R —RED
BK—BLACK
B —BLUE
W —WHITE
G —GREEN
V —VIOLET

NOTE—ALL EARTH WIRES—W/BK

⑨W/BK
⑬

⑬W/BK ⑫B/Y

HORN

WARNING LAMPS
GEN   OIL

BK/B
⑰Y/G
⑪Y/W   ⑮BK/B ③V

PANEL LAMPS
TEMP   PROOF METER

⑭Y/R

HORN BUTTON

⑫B/Y
⑰Y/G

⑯BK

SIDE LAMP

⑯BK   ⑭Y/R

SIDE LAMP

⑧Y/R   ⑦R
BAT.
②BK/R/DIP
HEAD SIDE & TAIL   ⑮BK/G
①BK/G

MAIN CONTROL SWITCH

REAR LAMP

REAR LAMP

REGISTRATION PLATE LAMP   TRAILER LAMP SOCKET

Fig. 96—Wiring diagram for all models except "FMD". Refer to Fig. 95 for "FMD" wiring diagram.

If voltmeter reading continues to rise as engine speed is increased, renew the regulator.

To adjust voltage regulator, shut off engine and remove regulator cover. Loosen lock nut on regulator adjusting screw (see Fig. 97 or 99) and turn screw clockwise to raise setting or counter-clockwise to lower setting. Turn the screw only a fraction of a turn at a time, then start engine and test as before. Note: Adjustment of regulator open-circuit voltage should be completed in 30 seconds; otherwise, heating of windings will cause false settings to be made. A generator run at high speed on open circuit will build up a high voltage, thus when adjusting regulator, increase speed slowly until regulator operates. After adjusting regulator, reconnect wires to regulator "A" and "A1" terminals. Leave voltmeter connected between "D" terminal and ground and check cut-out relay as follows: Gradually increase engine speed and note voltage reading immediately before the cut-out relay points close. This voltage should be between 12.7 and 13.3. If not, loosen locknut on cut-out adjusting screw and turn screw clockwise to increase cut-out closing voltage or counter-clockwise to decrease closing voltage.

143. CLEAN AND ADJUST REGULATOR CONTACT POINTS. The voltage regulator can be disassembled and the contact points cleaned if not burned. Use carborundum paper or crocus cloth in a circular movement and wash away all dirt or abrasive with alcohol. Reassemble, but prior to tightening screws, loosen the locknuts on the voltage regulator and cut-out adjusting screws and back the screws out. Then, adjust as follows:

144. ADJUST LUCAS MODEL RB 106/1. Refer to Fig. 97 and clamp a 0.011-0.015 thick gage between armature and core (GAP A) and a 0.014 thick gage between armature and frame (GAP C) of cut-out relay and tighten armature spring retaining screws. With the 0.011-0.015 gage still clamped between armature and core (GAP A), bend armature stop so that it clears armature (GAP B) from 0.030 to 0.034 and adjust point gap (GAP D) to 0.002-0.006. Remove the gage from cut-out relay and adjust regulator points as follows:

Refer to Fig. 98 and insert a 0.018 thick gage between armature and frame (GAP A) and clamp a 0.012-0.020 thick gage between armature and coil core (GAP B), then tighten armature spring retaining screws. Remove

Fig. 98 — Adjustment points for voltage regulator armature on Lucas model RB 106/1 regulator; refer to text.

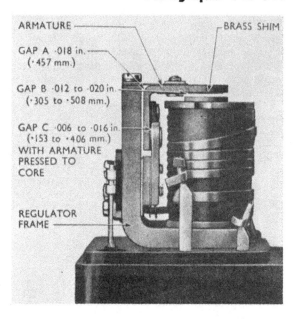

Fig. 99—Adjustment on Lucas model RB 106/2 cut-out relay; refer also to Fig. 100 and text.

Fig. 100—After making adjustments shown in Fig. 99, refer to text and continue with cut-out relay adjustment shown.

the feeler gages and push armature down against coil core; regulator point gap should then be 0.006-0.016; if not, add or remove shims between contact point base and regulator frame as required. Take care that the shims do

not short out the contact base to frame.

With armature and point gap adjusted for both the regulator and cut-out relay, adjust the voltage regulator setting and cut-out closing voltage as

outlined in paragraph 142.

145. ADJUST LUCAS MODEL RB 106/2. Refer to Fig. 99 and push the cut-out armature down against coil core, then tighten armature retaining screws. While still holding armature down against coil core, bend armature stop so that it clears armature by 0.025-0.040. Insert a 0.010-0.020 thick gage between armature and coil core (see Fig. 100), clamp down on gage and set the fixed contact by bending contact arm so that points are just touching.

Refer to Fig. 101 and insert a 0.015 thick feeler gage between armature and core face shim and press armature down against gage. Tighten the two armature retaining screws while holding down on armature and with gage still in position, turn the point (fixed contact) screw down until points just touch and tighten locknut.

With voltage regulator and cut-out points adjusted, refer to paragraph 142 and adjust the voltage regulator setting and cut-out closing voltage.

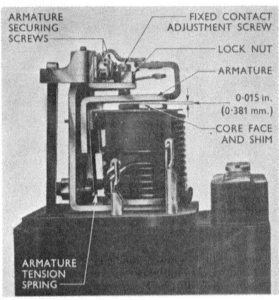

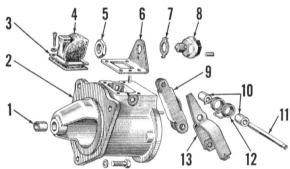

Fig. 101—View of voltage regulator of model RB 106/2 Lucas regulator assembly; refer to text for adjustment procedure.

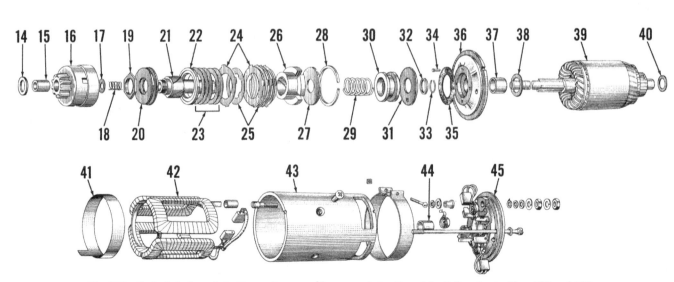

Fig. 102—Exploded view of starting motor assembly as used on all models. Refer also to Figs. 103 and 104.

1. Bushing
2. Housing
3. Retainer
4. Cover
5. Nut
6. Bracket
7. Lockwasher
8. Relay switch
9. Actuating lever
10. Spacers
11. Pin
12. Spring
13. Lever & shoe assy.
14. Thrust washer
15. Spring guide
16. Pinion
17. Washer
18. Spring
19. Nut
20. Plate
21. Sleeve
22. Ring
23. Shims
24. Driven plates
25. Driving plates
26. Sleeve
27. Washer
28. Snap ring
29. Spring
30. Collar
31. Brake plate
32. Retaining clip
33. Lock ring
34. Rivets
35. Brake lining
36. Bearing plate
37. Bushing
38. Washer
39. Armature
40. Washer
41. Insulator
42. Field coils
43. Frame
44. Bushing
45. End plate assy.

Late Model "FPM" and Models "FSM" and "New FSM"

146. At serial No. 1542263 of model "FPM" production, and on all later models, the voltage regulator is a sealed unit. If the regulator does not meet the test specifications in paragraph 142 for voltage regulator setting (as given for Lucas Model RB 106/2) or cut-out closing voltage, the unit should be renewed.

STARTING MOTOR

All Models

147. Refer to Fig. 102 for exploded view of the Lucas starter motor used on all models. The starter motor incorporates a manually shifted drive pinion assembly which includes both an over-running and a torque limiter clutch. When the drive pinion is in engaged position, the actuating lever contacts the relay switch which engages the solenoid switch. Releasing starter lever engages the brake plate (31) with friction disc (35) attached to center bearing support plate (36) to stop motor.

Current draw at normal cranking speed of 200 engine RPM is 450 amperes. Lock torque is 28 Ft.-Lbs.

Use Fig. 102 as disassembly and reassembly guide. The torque limiting clutch should be reassembled with sufficient thickness of shims (25) so that clutch will slip when a torque of 65-80 Ft.-Lbs. is applied; refer to Fig. 103 for suggested method for checking clutch slipping torque. Shims are available in thicknesses of 0.004,

0.005, 0.006, 0.009 and 0.020. Adding shim thickness will increase slip torque.

When starter is reassembled, refer to Fig. 104 and proceed as follows to check starter relay switch adjustment: Connect a battery and test light in series to the relay switch terminals, then slowly move starter actuating lever forward until test light just goes on (switch contacts close). At this time, distance between mounting face of starter housing and rear face of pinion should be exactly $1\frac{7}{16}$ inches as shown in Fig. 104. If not, loosen switch bracket retaining screws and shift the bracket on housing so that relay switch makes contact when starter drive pinion is in position shown, then tighten the bracket retaining screws.

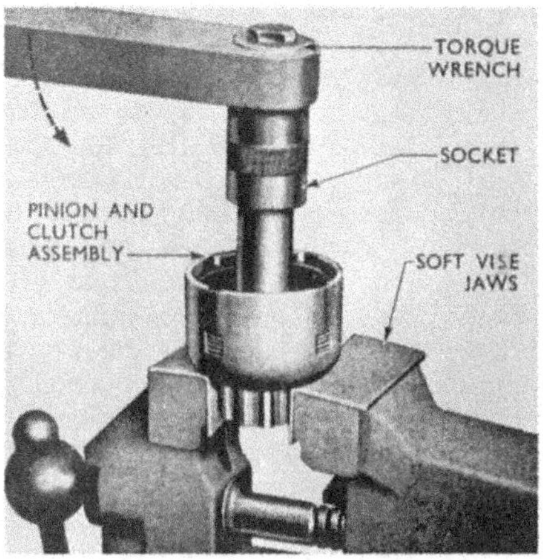

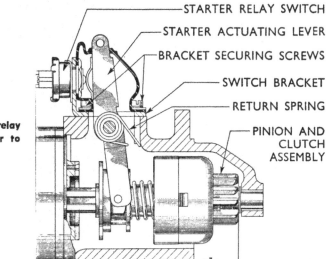

Fig. 103—Checking clutch assembly for proper slip torque; refer to text for procedure and specifications.

Fig. 104—Adjusting relay switch position; refer to text.

— STARTER RELAY SWITCH
— STARTER ACTUATING LEVER
— BRACKET SECURING SCREWS
— SWITCH BRACKET
— RETURN SPRING
— PINION AND CLUTCH ASSEMBLY

←$1\frac{7}{16}$ in.→

CLUTCH

148. Tractor may be equipped with either a 11 or 13 inch single clutch or with a dual disc clutch having 12 inch friction discs for both the transmission and PTO input shafts. To service the clutch, first split tractor between flywheel and transmission housings as follows:

TRACTOR SPLIT

All Models

149. To separate tractor for service of clutch assembly, proceed as follows: Drain cooling system, remove air precleaner and vertical exhaust muffler, then remove engine hood. Disconnect proofmeter drive cable at engine end. Disconnect stop control cable at fuel injection pump. Disconnect throttle

control rod from cross-shaft through the engine (early "FMD") or at cross-shaft at rear of engine (later models). Turn off the fuel supply valve and disconnect fuel line to fuel lift pump. Disconnect excess fuel return line at rear end of cylinder head. Disconnect hose from air cleaner to intake manifold and, except on early "FMD", disconnect air cleaner to rocker arm cover hose. Disconnect battery ground cable and wiring from starter relay switch, generator, oil pressure sending switch on models with electric pressure gage, and wiring from headlights. Disconnect starter operating rod at starter and, on early "FMD", disconnect radiator shutter operating rod. Remove temperature gage bulb from

cylinder head and coil tube back out of way. Remove tool box and disconnect drag link from steering gear arm. Remove horizontal exhaust if so equipped. Drive wood wedges between front axle and stops on front axle support. Place a jack or stand under front end of transmission and support rear end of engine with moveable hoist or rolling floor jack. Unbolt side rails and flywheel housing from transmission housing. Note: There are flywheel housing to transmission housing bolts located behind the side rails. Roll engine forward away from transmission.

To reconnect tractor at flywheel housing to transmission housing, reverse procedure used to separate tractor. After reconnecting fuel lines, bleed the diesel fuel system as outlined in paragraph 107.

LINKAGE ADJUSTMENT
Single Clutch, All Models

150. Clutch pedal free travel should be 1½ to 2 inches and adjustment is made by varying the length of clutch operating rod; refer to Fig. 105.

Dual Clutch, Models "FMD" and "FPM"

151. Free pedal adjustment is made by turning the adjusting screw in clutch pedal arm (see Fig. 106) in or out to obtain a free pedal of ½-inch. Note: Do not adjust length of clutch operating rod to obtain free pedal; if adjustment cannot be made by turning adjusting screw, proceed as follows:

Loosen locknut on pedal adjusting screw and turn screw into pedal (shorten screw) as much as possible. Check length of clutch operating rod; if not 15 inches from center-to-center of hole at each end of rod, remove rod from balance arm, loosen locknut and turn clevis end as required to obtain this adjustment. Reconnect operating rod, but do not tighten locknut or secure clevis pin at this time. With pedal stop pin in lower hole in bracket, engage PTO, start engine and slowly depress clutch pedal. Note: If equipped with raised PTO, be sure both PTO shift levers are engaged. The PTO shaft should stop turning just before the clutch pedal contacts stop pin. If this condition is not obtained, shorten the clutch operating rod as required. Note: Take care not to shorten rod to extent that release mechanism in clutch assembly bottoms before pedal contacts stop pin. Move stop pin to top hole in stop bracket, start engine and check to see that transmission can

be shifted satisfactorily when clutch pedal is depressed against stop pin. If transmission cannot be shifted without "clashing" gears, continue to shorten the clutch operating rod until transmission can be shifted satisfactorily. Move stop pin back to the bottom hole of stop bracket and check to be sure that clutch assembly release mechanism does not bottom before clutch pedal contacts stop pin in the lower position. Note: If release mechanism bottoms at this point, overhaul clutch assembly. When the adjustment of operating rod is completed, tighten operating rod locknut and secure clevis pin with cotter pin. Turn adjusting screw to obtain ½-inch free pedal and tighten adjusting screw locknut. If operation of both the PTO and transmission clutch is

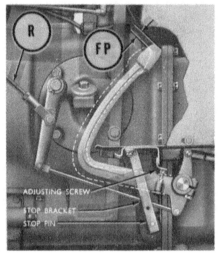

Fig. 106—Adjustment points for models "FMD" and "FPM" with dual clutch. Refer to text for adjustment procedure and specifications.

desired, install stop pin in lower hole and secure with cotter pin. Place stop pin in upper hole in stop bracket when operation of PTO clutch is not desired.

Dual Clutch, Model "FSM"

152. Clutch pedal free travel should be 1½ to 2 inches and adjustment is made by varying the length of clutch operating rod; refer to Fig. 105. After adjusting operating rod, make the following checks:

153. Remove stop pin from stop bracket, engage PTO, start engine and depress clutch pedal to bottom of stop bracket. If PTO shaft continues to turn with pedal fully depressed, shorten operating rod as required so that depressing clutch against bottom of stop bracket will stop shaft. Install the stop pin in lower hole of stop bracket, depress the clutch pedal against stop pin and check to see that transmission can be shifted. If transmission cannot be satisfactorily shifted with pedal against pin, continue to shorten operating rod until this condition is met. Check to see that sufficient free pedal travel remains and that with stop pin removed, clutch mechanism does not bottom before clutch pedal contacts bottom of stop bracket; overhaul clutch if either of these conditions are not found. Store the stop pin in top hole of stop bracket if operation of both the PTO and transmission clutches is desired. If operation of PTO clutch is not desired, install stop pin in lower hole of stop bracket.

Fig. 105—Clutch adjustment points for all models with single clutch and models "FSM" and "New FSM" with dual clutch. Refer to Fig. 106 for models "FMD" and "FPM" with dual clutch. Refer to text for adjustment procedure and specifications.

C. Clevis end
R. Clutch release rod
N. Nut

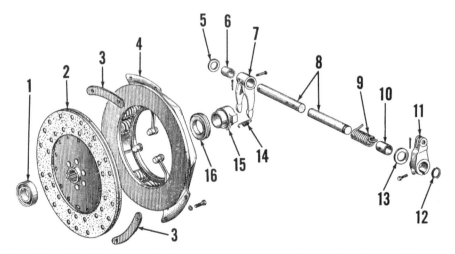

Fig. 107—Exploded view of 11-inch diameter single clutch assembly. Spacers (3) are used between cover (4) and flywheel when heavy duty friction disc assembly is installed.

1. Pilot bearing	10. Bushing
2. Friction disc	11. Cross-shaft arm
3. Spacers	12. Snap-ring
4. Cover & pressure plate assy.	13. Washer
5. Washer	14. Return spring
6. Bushing	15. Release bearing hub
7. Release fork	16. Release bearing assy.
8. Cross-shaft	
9. Return spring	

R&R AND OVERHAUL
Models With 11 Inch Single Clutch

154. With tractor split between flywheel and transmission housings as outlined in paragraph 149, alternately and evenly loosen the clutch cover to flywheel cap screws to remove the cover and pressure plate assembly. Note: Most models will have three spacers (3—Fig. 107) located between cover and flywheel; be careful not to lose spacers where used.

The clutch cover and pressure plate assembly is serviced as a complete unit only (Ford part No. 8MTH-7563-A) which is interchangeable with Ford truck assembly of same part number. Original production assembly and/or service parts are not available in the U.S. Renew the cover and pressure plate assembly if inspection reveals any part unfit for further service.

The clutch friction disc (2) is serviced as either a standard duty (Ford part No. DKN-7550-A) or a heavy duty (Ford part No. DKN-7550-B) assembly and either unit may be used. However, if renewal of friction disc assembly is indicated, use of the heavy duty disc assembly is recommended.

To install clutch assembly, proceed as follows: Using suitable pilot, position friction disc assembly on flywheel with long hub away from flywheel. If installing standard friction disc, attach cover and pressure plate assembly to flywheel with 25/32-inch long shouldered cap screws (Ford part No. 350433-S) and lockwashers. If installing heavy duty friction disc, in-

sert a spacer (Ford part No. DKN-77596) at each of the three bolting points between cover and flywheel and attach cover and pressure plate assembly to flywheel using 0.85 long shouldered cap screws (Ford part No. 355599-ES) and lockwashers. Tighten cap screws to a torque of 12-15 Ft.-Lbs.

NOTE: Model FPM tractors after Serial No. 1418861 originally equipped with 11 inch clutch can be converted to 13 inch clutch by installing new flywheel, friction disc, cover assembly, release bearing and related parts. Tractors before this Serial number cannot be converted to 13 inch clutch due to location of release (cross) shaft in transmission housing. Consult Ford tractor parts department for all necessary conversion parts.

Models With 13 Inch Single Clutch

155. **REMOVE AND REINSTALL.** With tractor split between flywheel and transmission housings as outlined in paragraph 149, alternately and evenly loosen cover to flywheel cap screws to remove the clutch assembly. Note: Some models may have spacers located between clutch cover and flywheel; be careful not to loose these spacers, if so equipped, when removing clutch assembly.

Service parts for the 13-inch clutch cover and pressure plate assembly are available for service. Refer to paragraph 156 for overhaul procedure.

Two different 13 inch diameter friction disc assemblies or a 12 inch friction disc assembly may be used. A 0.33 thick 12 inch diameter cushioned disc (Ford part No. DDN-7550-C) or

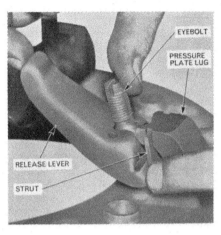

Fig. 109—Assembling release lever, strut, lever pin and eyebolt to clutch pressure plate.

a 0.33 thick 13 inch diameter solid disc (Ford part No. DDN-7550-B) can be used with early type 13 inch cover and pressure plate assembly (Ford part No. DDN-7563-B) which can be identified by drive straps (S—Fig. 108) of two laminations (layers). A 0.52 thick disc (Ford part No. DDN-7550-D) is used with later type cover and pressure plate assembly (Ford part No. DDN-7563-C) which can be identified by drive straps (S) made of three laminations.

To install clutch and friction disc assembly, proceed as follows: Using a suitable pilot, position the friction disc on flywheel with long hub of disc away from flywheel. If disc is 0.33 thick, align dowel holes in cover with proper size dowels in flywheel and attach cover and pressure plate assembly with 1 inch long shouldered cap screws (Ford part No. 20388-S2) and lockwashers. If installing a 0.52 thick friction disc, align dowel holes in cover with proper size dowels in flywheel, insert a spacer (Ford part No. E117-GC9) at each bolting point between cover and flywheel and attach cover and pressure plate assembly with 1⅛ inch long cap screws (Ford part No. 20408-S7) and lockwashers. Tighten cap screws to a torque of 12-15 Ft.-Lbs. Note: If installing a new clutch cover and pressure plate assembly, red painted spacers inserted between release levers and cover for shipping purposes will fall out of the assembly as the retaining cap screws are tightened. Be sure all of the spacers are removed from clutch and discarded.

156. **OVERHAUL.** To disassemble cover and pressure plate assembly, proceed as follows: Mark the plate, cover, release levers and release lever plate so that they can be reinstalled in the same relative positions. Remove

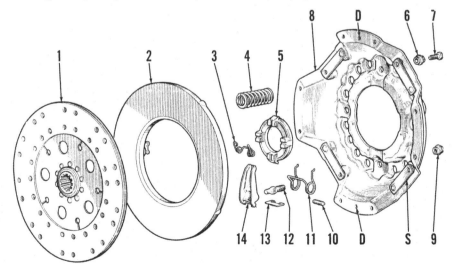

Fig. 108—Exploded view of 13-inch single clutch assembly. Refer also to Figs. 109 and 110.

D. Dowel holes	5. Lever plate	10. Lever pins
S. Straps	6. Ferrules	11. Anti-rattle springs
1. Friction disc	7. Drive strap	12. Eyebolts
2. Pressure plate	screws	13. Struts
3. Retaining springs	8. Cover	14. Release levers
4. Pressure springs	9. Adjusting nuts	

flywheel from engine and lay flywheel on bench with clutch friction surface up. Position three 0.329-0.331 thick by 3½ inch long flat spacers evenly around friction surface. Place cover and pressure plate assembly over the spacers with dowel holes in cover aligned with proper size dowel pins in flywheel and attach to flywheel with four of the cover retaining cap screws at alternate bolting positions. Note: Do not install clutch cover to flywheel spacers; it may be necessary to add flat washers under cap screw heads on models having spacers and 1⅛ inch cover retaining cap screws. In remaining four bolting points, install ⅜-inch—16 x 1½ inch long cap screws and turn these cap screws in until they bottom in flywheel.

Remove the release lever plate (5—Fig. 108) and the retaining springs (3). Using a punch, drive edge of ferrules (6) back away from drive strap cap screws (7), then remove the cap screws. Unscrew and remove adjusting nuts (9) from eyebolts (12); these nuts are staked and may be hard to turn. Evenly loosen the four short cover retaining cap screws until cover rises against heads of the long cap screws, then remove the four short cap screws and evenly loosen the long cap screws until all spring pressure is released. Remove cover from pressure plate. Remove the sixteen pressure springs. Holding release lever inner ends upward against eyebolts, move the struts until eyebolts and release levers can be removed from pressure plate. Withdraw the eyebolts from release levers, then remove the pivot pins.

Examine all parts thoroughly and renew any that are cracked, scored, excessively worn or that show signs of overheating. Note: Two different clutch cover and pressure plate assemblies have been used and it is important that the assembly being serviced is correctly identified. Inspect the four laminated drive straps (S) riveted to inner side of clutch cover (8). Drive straps of early assembly (Ford part No. DDN-7563-B) are made of two laminations (layers) whereas late cover and pressure plate assembly (Ford part No. DDN-7563-C) have drive straps made of three laminations.

The same pressure springs are used in both the early and late clutch assemblies. Check the sixteen springs against the following specifications:

Spring color .... Yellow/Green

Free length ...... 2.665-2.711 inches
Pounds pressure @
   1.69 inches 135-145

To reassemble, proceed as follows: Place three 0.329-0.331 thick spacers approximately 3½ inches long evenly around friction surface of engine flywheel. Place pressure plate on top of the spacers. Install pivot pins (10) through eyebolts (12) and insert threaded ends of eyebolts through release levers (14). Refer to Fig. 109, place struts under lugs on pressure plate and while holding inner ends of release levers against eyebolts, insert plain ends of eyebolts in bores of pressure plate and outer ends of release levers under the struts. Place the pressure springs on bosses of pressure plate. Install anti-rattle springs (11—Fig. 108) in clutch cover, then place cover down over the springs, release levers and pressure plate so that anti-rattle springs are over the release levers. Rotate the assembly so that dowel holes in clutch cover are properly aligned with correct size dowel pins in flywhel and move the spacers so that they are under each release arm. Also adequately support the pressure plate. Install the four 1½ inch long cap screws at alternate bolt holes and tighten them evenly until they bottom in flywheel. Install four of the cover retaining cap screws in the remaining bolt holes and tighten them evenly until cover is pulled down against flywheel. Use washers if necessary. Be sure that dowel pins in flywheel and eyebolts in levers enter proper holes in cover as cover is being pulled down. Remove the four long cap screws, install the remaining cover retaining cap screws and tighten all eight cap screws to a torque of 12-15 Ft.-Lbs.

Install **new** adjusting nuts on the release lever eyebolts, but do not tighten the nuts at this time. Insert **new** ferrules through the drive straps into counterbores in pressure plate and securely install the drive strap cap screws. Using a punch, drive the outer ends of drive strap ferrules against cap screw heads. Install release lever plate with the four retaining springs.

If possible, operate release levers several times to be sure all parts are fully seated. This can be accomplished by placing the flywheel and clutch unit in a press. CAUTION: Be sure not to bend release levers. Using a depth gage, measure distance (D—Fig. 110) between face of release lever plate and spacers. Turn the adjusting nuts down so that the measured dis-

tance is equal within 0.015 at all measuring points and is within the following specifications:

Ford Part No.	Release Plate Height
DDN-7563-B	2.449-2.521
DDN-7563-C	2.150-2.185

If possible, operate levers and recheck adjustment. With release levers properly adjusted, firmly stake adjusting nuts to eyebolts. Remove clutch assembly from flywheel, reinstall flywheel as outlined in paragraph 97 and the clutch assembly as outlined in paragraph 155.

Models With Dual ("Live PTO") Clutch

157. **REMOVE AND REINSTALL.** With tractor split between flywheel and transmission housings as outlined in paragraph 149, the dual clutch assembly can be unbolted and removed from engine flywheel. Support clutch assembly as it is being unbolted. Note: The clutch cover may move away from the center drive plate as the retaining cap screws are removed.

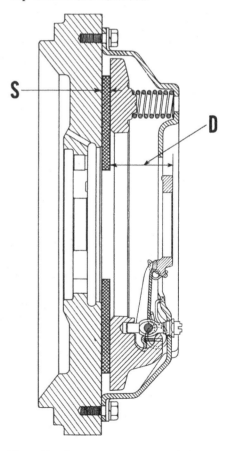

Fig. 110—Cross-sectional view of 13-inch clutch cover and pressure plate assembly mounted on flywheel with three spacers (S) between plate and flywheel for adjustment. Refer to text for thickness of spacer (S) and lever height dimension (D).

When reinstalling clutch assembly, tighten retaining cap screws evenly until clutch cover is pulled down against center drive plate, then tighten all cap screws securely.

158. OVERHAUL. To service the dual clutch assembly, a special adjusting fixture (Nuday tool No. NE-7502) is required for clutch disc alignment and release lever adjustment after clutch unit is reassembled. Note: Late production clutches will also require a new transmission clutch disc pilot (Nuday tool No. NE-7502-Detail 5B). If the special adjusting fixture is available, proceed as follows:

With clutch assembly on bench with cover up, place correlation marks on the cover (11—Fig. 111), PTO clutch pressure plate (7), center drive plate (5) and the main (transmission) pressure plate (2) so that if not renewed, the parts may be reinstalled in same relative position. Remove the pins that retain actuating struts (13) to main pressure plate (2). Remove snap rings from pins that retain PTO clutch release (short) levers (18) in pivots (20) on clutch cover, then remove the pins and release lever springs (19). Remove the adjusting screws from PTO release levers, move levers to vertical position and lift off clutch cover with transmission release levers and struts attached; refer to Fig. 112. Compress the pressure springs with an overhead valve spring compressor and remove spring retainers (11—Fig. 111) as shown in Fig. 113. Then lift off the spring seats, springs and insulating washers. CAUTION: Do not attempt to release springs by unscrewing the self-locking nuts from

spring retainer pins. The pressure plates, center drive plate and the two clutch friction discs can now be separated.

Check all parts for signs of scoring, distortion or cracking due to overheating and for excessive wear. Clutch disc linings should be tight on disc and free from oil. If necessary to renew one disc, the other disc even though serviceable should also be renewed. Installing one new and one worn disc may cause difficulty in adjusting relative position of PTO and transmission clutch release levers.

If any of the spring retainer pins in transmission pressure plate are bent or broken, they must be renewed. Pressure plate (Ford part No. DKN-7566-B) in early production clutch assembly is equipped with pins having $\frac{5}{16}$-inch diameter threaded end. Improved retainer pins (Ford part No. E54-GC-9) are available for servicing

this early pressure plate. To install the improved pins, a 0.040 deep 45 degree chamfer must be cut in friction side of pressure plate; refer to Fig. 114. Transmission pressure plate (Ford part No. DKN-7566-C) for late

Fig. 113—Compress pressure springs with valve spring compresser to remove retainers.

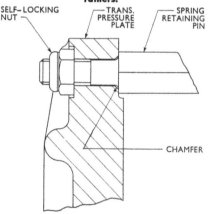

Fig. 114—Note chamfer required in transmission clutch pressure plate to allow installation of late type spring retaining pins.

Fig. 112—Removing cover from dual clutch assembly.

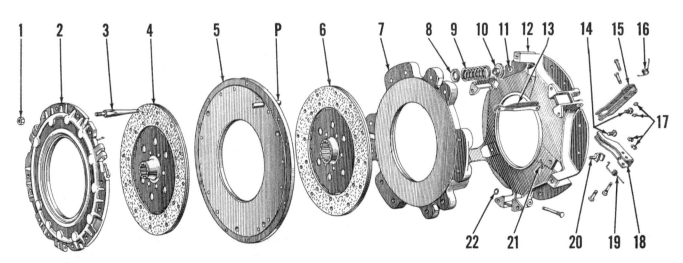

Fig. 111—Exploded view of typical dual ("live PTO") clutch assembly. When servicing unit, be sure that pins (P) are securely fastened in center plate (5).

P. Pins	4. Transmission disc	8. Insulating washer	12. Clutch cover
1. Nuts	5. Center drive plate	9. Pressure springs	13. Struts
2. Pressure plate	6. PTO disc	10. Spring retainers	14. Adjusting screws
3. Pins	7. Pressure plate	11. Retainer locks	15. Transmission release

levers	19. Anti-rattle springs
16. Anti-rattle springs	20. Yokes
17. Snap rings	21. Pins
18. PTO release levers	22. Snap rings

production clutch assembly is fitted with spring retainer pins having ⅜-inch diameter threaded ends (Ford part No. E97-GC-9). When installing pins in pressure plate, **always** install **new** self-locking nuts. Tighten the $\frac{5}{16}$-inch nuts (Ford part No. 34443-ES2C) to a torque of 25-28 Ft.-Lbs. or the ⅜-inch nuts (Ford part No. 34420-ES2C) to a torque of 30-35 Ft.-Lbs.

Renew the center drive plate (5—Fig. 111) if warped, scored or over-heated or if any of the three driving pins (P) are loose in plate.

Two different clutch pressure springs have been used. The early spring (Ford part No. DKN-7552-A) is no longer available for service. If springs are required for early clutch, the complete set of twelve springs should be renewed using the latest type spring (Ford part No. DDN-7552). Test specifications and spring identifications are as follows:

Ford part No. DKN-7552-A:

Free length ....3.23 inches
Lbs. pressure @
  2 inches 92.5

Ford part No. DDN-7552:
  Color ....Red
  Free length ....3.23 inches
  Lbs. pressure @
    2 inches 109.5

The adjusting screws in clutch release levers should require a minimum torque of 5 Ft.-Lbs. to rotate screws in levers. Renew adjusting screws and/or levers as required to obtain the minimum turning torque.

To reassemble clutch, reverse disassembly procedure noting that long hub of transmission clutch disc (4) is towards pressure plate (2) and that long hub of PTO clutch disc (6) is towards PTO clutch pressure plate (7). Note: PTO clutch disc has larger splined hole in hub than does the transmission clutch disc. When clutch is assembled, bolt the cover to center drive disc using three bolts at evenly spaced bolting points and proceed as follows to adjust the assembly.

Engine flywheel has two $\frac{5}{16}$-inch tapped holes 180° apart which are used to attach clutch fixture base plate and spindle as shown in Fig. 116. Remove the pilot bearing, one pair of flywheel retaining cap screws and the locking plate covering one of the tapped holes, then install spindle as shown. Install the six risers (R) spaced as shown in Fig. 116. Mount assembled clutch unit on the six risers as shown in Fig. 117 and secure clutch to risers with six of the clutch to flywheel retaining cap screws. Refer to Fig. 118 and slide the transmission clutch disc pilot (T), PTO clutch disc pilot (P) and the gage block (B), with flat face of gage block towards release levers, onto the adjusting fixture spindle. Then install washer (W) and nut (N), tightening nut until the clutch discs are released. Refer to Fig. 119 and center the clutch discs with one screwdriver while pushing pilots

into clutch disc hubs with second screwdriver inserted through opening in gage block. Install the three plate spacers (see Fig. 120) at equally spaced points around the clutch assembly and be sure that the spacers contact edges of both clutch discs. It may be necessary to tighten the spindle nut slightly in order to insert the spacers between the clutch pressure plates and center drive disc. Loosen and remove the nut and washer from adjusting fixture spindle, making sure that spacers stay in place as nut is loosened. Remove gage block from spindle and reinstall it in reversed position with stepped gage surfaces towards release levers. Using a 0.005 thick feeler gage, adjust the transmission clutch (long) release lever screws to the low step on gage block and the PTO clutch (short) release lever screws to the high step on gage block. CAUTION: In order to obtain proper adjustment, the transmission

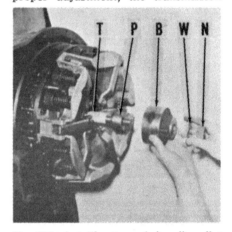

Fig. 118—Installing transmission disc pilot (T), PTO disc pilot (P), gage block (B), washer (W) and nut (N) to compress release levers.

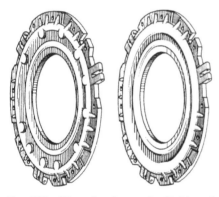

Fig. 115—View showing early (left) and late (right) production transmission clutch pressure plate (2—Fig. 111).

Fig. 116—Attaching spindle and risers (R) of clutch assembly and adjustment fixture to engine flywheel.

Fig. 117—Mounting dual clutch assembly on spindle and risers; refer to Fig. 116. Cover is bolted (arrow) to center driving plate at three equally spaced points.

Fig. 119—With release levers compressed, align clutch friction discs with one screwdriver while pushing pilots (T and P—Fig. 118) into place with second screwdriver.

Fig. 120—With release levers compressed (Fig. 119), install plate spacers at equally spaced points making sure that blocks contact the friction discs before releasing levers.

clutch disc pilot must contact shoulder of spindle, the PTO disc pilot must contact transmission clutch disc pilot and the gage block must be held firmly against the PTO disc pilot.

With the release finger screws adjusted, remove clutch and fixture by reversing installation procedure. Reinstall pilot bearing and the two flywheel retaining cap screws and lock plate, tighten the cap screws to a torque of 80-90 Ft.-Lbs. and bend locking plate tabs against cap screw heads. NOTE: If late type sintered bronze pilot has been removed from flywheel, the pilot must be renewed. A sintered pilot bearing cannot be reused. Reinstall the serviced clutch assembly as outlined in paragraph 157.

AUTOMATIC CLUTCH
RELEASE (ACR)

All Models So Equipped

159. **ACR OPERATING PRINCIPLES.** The automatic clutch release is a hydraulic cylinder with adjustable relief valve that is an integral part of the three-point hitch top link and connected through linkage to the clutch pedal release rod. When compression on the top link establishes a pressure within the cylinder that exceeds the relief valve setting, the top link and cylinder collapses and, through the connecting linkage, disengages the engine clutch. The purpose of the ACR is to stop the tractor when an implement mounted on the three point hitch strikes an obstruction such as a stump or rock.

To reset the collapsed cylinder (top link), proceed as follows: Shift transmission to neutral, depress the clutch pedal, actuate the release hand lever,

re-engage the clutch and place the hydraulic lift in raised position. Tension on the top link will then extend and reset the hydraulic cylinder.

Refer to Fig. 121 for early release linkage, to Fig. 122 for later type release linkage and to Fig. 123 for exploded view of typical top link and hydraulic release cylinder assembly.

160. **ADJUST ACR RELIEF PRESSURE.** To adjust the automatic clutch release, refer to exploded view of the assembly in Fig. 123 and proceed as follows:

Disconnect the front end of the top link from tractor, then turn the swivel ball in front end (1) so that a screwdriver can be inserted through the end and swivel ball unit. Turn the adjusting screw (not shown) in or clockwise to increase pressure on relief valve spring (5), or out to decrease pressure.

161. **BLEED TOP LINK CYLINDER.** Air in the pressure side of the top link cylinder will result in spongy action of the unit and partial release of the engine clutch. Free movement of the cylinder should not exceed $\frac{1}{8}$-

inch after use or $\frac{1}{16}$-inch after being bled. To bleed the cylinder, proceed as follows:

Remove the filler plug (17) and fill reservoir with SAE 10W oil, then loosely reinstall plug. Tilt rear of top link up about 30° and loosen the bleed plug (25). Push in or tap on rear end (30) of cylinder to expel air, then tighten the bleed plug. Pull on rear end of link to reset the cylinder, then repeat bleeding operation as required to expel all air from cylinder. After bleeding is completed, hold the top link level and remove the oil level plug (31) and allow excess oil to drain from reservoir, then, tighten the filler and bleed plugs.

162. **OVERHAUL ACR CYLINDER ASSEMBLY.** Refer to Fig. 123 and proceed as follows:

Thoroughly clean outside of cylinder, remove filler plug and drain excess oil from unit. Remove top cover (19), front end and piston guide (22) and rear end (26). Remove set screw (11) and unscrew piston (12) from front end (1), then withdraw front end from guide (22). Further dis-

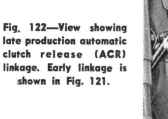

Fig. 121—View showing early production automatic clutch release (ACR) linkage. Refer to Fig. 122 for late production linkage.

Fig. 122—View showing late production automatic clutch release (ACR) linkage. Early linkage is shown in Fig. 121.

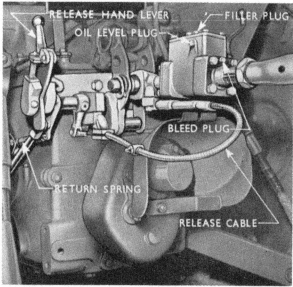

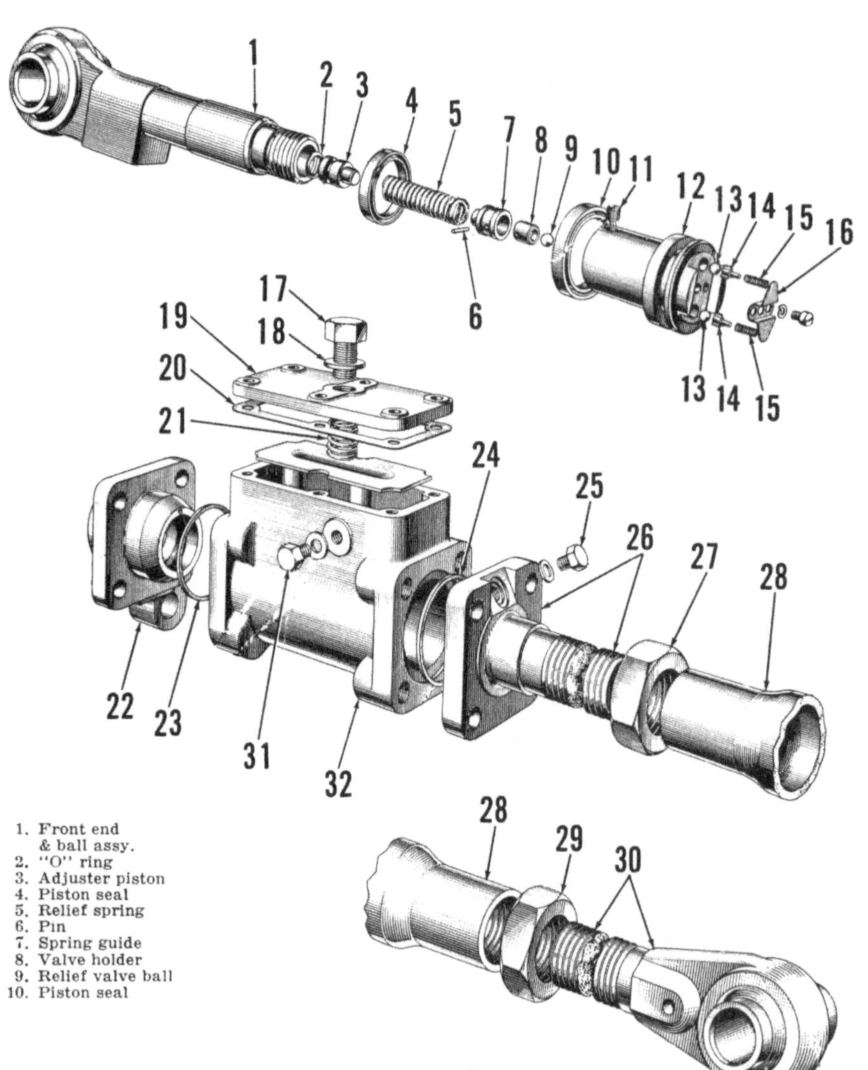

Fig. 123—Exploded view of typical automatic clutch release cylinder and three-point hitch top link assembly. Early units used leaf springs instead of coil springs (15) shown.

1. Front end & ball assy.
2. "O" ring
3. Adjuster piston
4. Piston seal
5. Relief spring
6. Pin
7. Spring guide
8. Valve holder
9. Relief valve ball
10. Piston seal

11. Set screw	23. "O" ring
12. Piston	24. "O" ring
13. Check valves	25. Bleed plug
14. Valve guides	26. End plate &
15. Springs	screw
16. Retainer	27. Lock nut
17. Filler plug	28. Turnbuckle
18. Gasket	29. Locknut
19. Cover	30. End & ball assembly
20. Gasket	31. Oil level plug
21. Spring	32. Cylinder
22. End plate	

LUBRICATION

All Models

163. Transmission lubricant capacity is approximately 20½ quarts for models without PTO or 21½ quarts for models with PTO. It is recommended that the transmission be drained, flushed and refilled with new lubricant after each six months of service. Recommended refill lubricant is SAE 80 Mild E.P. lubricant (Ford specification M-4864-A) for temperatures below 10° F., or SAE 90 Mild E.P. lubricant (Ford specification M-4864-B) for temperatures above 10° F.

TRANSMISSION BREATHER

All Models

164. A special long-headed bolt securing the front left hand fuel tank bracket to transmission housing functions as the transmission breather. If leakage of oil around transmission gaskets is experienced, this bolt should be checked to be sure the breather hole is not plugged.

SHIFTER MECHANISM

All Models

165. **PRIMARY (HIGH-LOW) SHIFT MECHANISM.** Refer to exploded view of primary gearbox shifter mechanism for single clutch models in Fig. 124. Mechanism for dual clutch models is similar except for shift rail (24—Fig. 134) and fork (23). Primary gearbox is in high range when shift lever (97—Fig. 124) is in "up" position on single clutch models or in "down" position on dual clutch models.

The shift arm (99) and rail (98) are mounted in the shifter housing (100) which also functions as the fuel tank rear support. To remove the shifter housing assembly, first remove the steering gear housing as outlined

assembly is obvious from inspection of unit and reference to Fig. 123. Note: Early units may have leaf spring for the reset valves (13) instead of coil springs (15) shown; however, the coil spring can be used in all units.

Inspect cylinder bore and valve seats in piston and renew any broken or extensively worn parts. Reassemble by reversing disassembly procedure, refill with SAE 10W oil and bleed the unit as outlined in paragraph 161. Set relief pressure as outlined in paragraph 160 so that the top link will not collapse from normal operating compression on top link.

TRANSMISSION

The transmission for all models consists of a two-speed primary gearbox installed in front of the main three-speed transmission providing six forward and two reverse speeds. On models with standard power take-off, the PTO drive is taken from the primary gearbox lower gear. On models with "live PTO", the transmission and power-take-off input shafts are separate and PTO drive train gears are separate from transmission gears. A number of production changes have been introduced with greatest change at introduction of the New Performance Super Major ("New FSM"). Where service procedures or specifications vary because of type of transmission (standard or "live" PTO) or because of production changes, it will be noted in the text or will be covered by separate paragraphs.

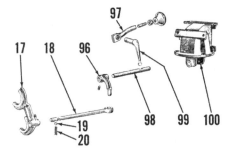

Fig. 124—Exploded view of primary gearbox shift mechanism for single clutch models. Dual clutch shift mechanism is similar except for shift rail and fork; refer to Fig. 134 (items 23 and 24).

17. Shift fork
18. Shift rail
19. Detent ball
20. Spring
96. Connector
97. Lever
98. Upper rail
99. Arm
100. Lever housing

Fig. 125—View of rear end of primary gearbox assembly. When reinstalling main shift cover, be sure lower fork (70—Fig. 127) engages both gear flanges (F). Connector (96—Fig. 124) engages hole in shift rail (18).

in paragraph 21. The shift housing can then be unbolted and removed. When reinstalling the unit, be sure that connector arm (96) enters hole in rear end of gearbox shift rail (18); this can be observed through opening in transmission where steering gear unit was mounted.

To remove the primary gearbox shift rail, fork and detent, refer to paragraph 177 for single clutch models or to paragraph 182 for dual clutch models.

166. MAIN TRANSMISSION SHIFT MECHANISM. Refer to exploded view of the main transmission shifter mechanism in Fig. 126.

The shift lever housing (79) and lever assembly can be unbolted and removed from top of rear axle center housing without removing other components. To remove lever from housing, remove snap ring (82) and withdraw lever from lower side of housing. When reinstalling assembly, be sure lower end of shift lever enters socket in connector (77).

To remove shift cover (64) assembly from left side of transmission, first drain lubricant to below level of cover and disconnect clutch linkage from balance lever. Then, unbolt and remove cover assembly. Refer to Fig. 127 for view showing inner side of cover. To disassemble unit, remove the rivets or clevis pins retaining shift forks to rails. Push top shift rail out of cover and remove the shift fork, interlock pin (69), detent plungers and spring (see Fig. 128); then, remove lower shift rail and fork. To reassemble unit, reverse disassembly procedure. Note: Top and bottom shift rails are interchangeable; end of rails with single notch are installed to rear (interlock plunger) end of cover. The

later clevis pins and cotter pins can be used to secure forks to rails on all models. Also, the later type detent plungers and spring can be used as a set to renew early type plungers and spring. When reinstalling assembly, be sure the flanges of the sliding gears are together and aligned, then reinstall assembly so that connector arm (72—Fig. 129) engages slots in forks and forks engage the flanges (F—Figs. 125 and 129) of both gears on main transmission upper and lower shifts. Note that the PTO shift lever pivots on the lower right cover retaining bolt.

To remove the connector (77—Fig. 126), shift rail (74) or bearing (75), the tractor must first be split between transmission housing and rear axle

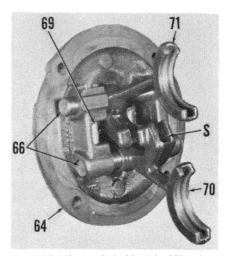

Fig. 127—View of inside of shift plate (64—Fig. 126). When installing assembly, be sure slot (S) engages selector arm (72—Fig. 126) and that shift forks (70 and 71) engage flanges (F—Figs. 125 and 129). Refer to Fig. 126 for legend.

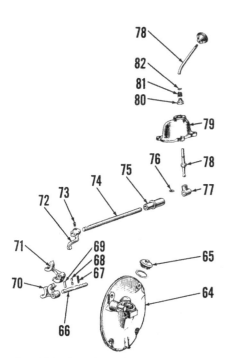

Fig. 126—Exploded view of main transmission shift mechanism. Refer to Fig. 127 for view showing inner side of plate (64).

64. Shift plate	73. Set screw
65. Oil filler & level plug	74. Rail
66. Shift rail	75. Bearing & guide assy.
67. Detent spring	76. Set screw
68. Detents	77. Socket
69. Interlock pin	78. Shift lever
70. Lower shift fork	79. Shift housing
71. Upper shift fork	80. Retainer
72. Selector arm	81. Spring
	82. Snap ring

Fig. 128—View showing latest production main shift rail detents (left) and early production detents (right). Refer also to Figs. 126 and 127.

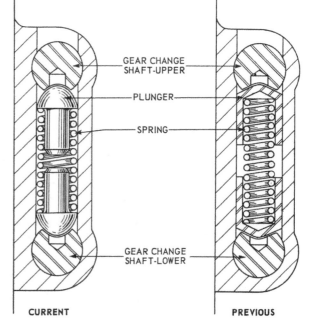

GEAR CHANGE SHAFT-UPPER

PLUNGER

SPRING

GEAR CHANGE SHAFT-LOWER

CURRENT　　　　　　PREVIOUS

center housing as outlined in paragraph 197 and the shift cover assembly removed as outlined in preceding paragraph. Remove the set screw (73) and front connector arm (72), then slide rail out of rear end of transmission. The bearing (75) can then be removed and a new bearing installed if necessary. Early model "FMD" was fitted with an oil seal located in rear end of bearing; however, use of the seal has been discontinued for all models.

TRANSMISSION INPUT (CLUTCH) SHAFT

Models With Single Clutch

167. Transmission input (clutch) shaft and housing can be removed after splitting the tractor between engine and transmission housing as outlined in paragraph 149. To remove shaft and housing, proceed as follows: Disconnect clutch release rod from cross-shaft arm and remove clutch release bearing and hub assembly. Note: On models with 13 inch clutch, disconnect lubrication hose from release bearing hub or from housing. Remove the pins retaining clutch release fork to cross-shaft, withdraw shaft and remove fork and return spring. Cut cap screw locking wire, then unbolt and remove the transmission input (clutch) shaft housing (1—Fig. 131) and seal (3) assembly. Withdraw input shaft (7) and bearing (5) assembly from gearbox. Remove seal from housing. Remove snap ring (4), then remove bearing from input shaft. Remove outer race for roller bearing (8) from rear bore of input shaft if necessary to renew bearing.

Install new bearing outer race in rear bore of input shaft, then rein-

Fig. 129—View of transmission with planetary gearbox assembly (Fig. 125) and main shift plate assembly (Fig. 127) removed. Upper shift fork (71—Fig. 127) must engage flange (F) and slot (S—Fig. 127) must fit connector (72) when reinstalling shift plate assembly.

stall shaft so that it engages primary upper shaft dog coupling. Install new oil seal in shaft housing, then install the housing with new gasket and with clutch spring lug to bottom. Install the drilled head cap screws, tighten them securely, then install locking wire through the drilled heads.

Models With Dual ("Live PTO") Clutch

168. After splitting tractor between engine and transmission housing as outlined in paragraph 149, disconnect clutch release rod from cross-shaft arm and remove clutch release bearing and hub assembly. Remove the pins retaining clutch release fork to cross-shaft, withdraw the shaft and remove fork and return spring. Cut locking wire and remove cap screws retaining input shaft housing (1—Fig. 134) to primary gear box, then remove the housing and seal (5) assembly. Withdraw transmission input shaft (10) and PTO input shaft (9) as a unit.

Early production units were fitted with a composition sealing ring and metal retainer instead of the oil seal (8); however, the oil seal (Ford part No. DDN-7052-A) can be used in all models. A new seal should be installed whenever the transmission input shaft is removed from PTO input shaft. Be sure that the bushing in front end of PTO input shaft is driven back flush with step in shaft bore when renewing early composition seal and metal sealing ring. Drive seal into shaft with spring loaded lip to rear and lightly bottom seal against front end of bushing. Liberally grease the seal and transmission input shaft and carefully insert transmission input shaft through the PTO input shaft and oil seal assembly.

R&R COMPLETE TRANSMISSION ASSEMBLY

Model "FMD"

169. With tractor split between engine and transmission as outlined in paragraph 149, proceed as follows: Remove steering wheel and the instrument control box. Unbolt fuel tank from supports and lift tank off over steering gear column. Remove the steering gear assembly from top of transmission. Drain lubricant from rear axle center housing and remove the PTO output shaft as outlined in paragraph 224 or raised PTO unit and driveshaft as outlined in paragraph 225. Disconnect wiring to rear lights. Remove right foot rest (step plate), disconnect brake cables and remove brake locking lever guide and the

right foot rest bracket. Remove the left foot rest, clutch pedal stop and foot rest bracket. Disconnect clutch release rod at balance lever. Remove brake outer pedal, loosen clamp bolt on inner pedal and slide inner pedal out on shaft far enough to clear transmission rear flange. Remove main shift lever housing from top of rear axle center housing. Attach hoist to transmission, support rear unit under front end of rear axle center housing, then unbolt and remove the transmission assembly.

To reinstall transmission, reverse removal procedure.

Models "FPM", "FSM" and "New FSM"

170. With tractor split between engine and transmission as outlined in paragraph 149, proceed as follows:

Remove steering wheel, throttle control lever and the grease fitting from upper end of the steering column. Remove instrument panel retaining screws and disconnect wiring loom from horn (if so equipped), generator and oil pressure warning lights, starter and front lights. Disconnect proofmeter (tach-hourmeter) drive cable at upper end, free the water temperature gage tube and bulb, then remove the instrument panel assembly. Disconnect stop control cable housing from clamp on battery tray and the wiring loom from main switch, unscrew primary gearbox shift lever knob, then unbolt and remove shroud (sheet metal cover) from rear of fuel tank. Remove the fuel tank and the starter control lever. Unbolt and remove steering gear unit from top of transmission.

Drain lubricant from rear axle center housing and remove PTO output shaft as outlined in paragraph 224 or raised PTO unit and driveshaft as

Fig. 130—Removal and installation of primary gearbox (housing) assembly is made easier by use of locating studs.

outlined in paragraph 225. Disconnect wiring to rear lights. Remove right foot rest (step plate), disconnect brake cables (on "FPM"), then remove brake locking lever guide ("FPM" only) and the right foot rest bracket. Remove the left foot rest, clutch pedal stop and foot rest bracket. Remove outer brake pedal and loosen clamp bolt on inner pedal, then slide inner pedal out far enough on shaft to clear transmission rear flange. Unbolt and remove the main shift lever assembly from top of rear axle center housing. Attach hoist to transmission, support rear unit under front end of rear axle center housing, then unbolt and remove the transmission assembly.

Reinstall transmission by reversing removal procedures.

R&R PRIMARY GEARBOX

All Models

171. First, split tractor between flywheel housing and transmission housing as outlined in paragraph 149, then proceed as follows:

Drain oil from transmission and if equipped with PTO, also drain rear axle center housing. Disconnect clutch release rod from cross-shaft arm and remove clutch release bearing and hub assembly. Remove clutch release fork to cross-shaft retaining pins and withdraw cross-shaft from transmission housing, then remove fork and return spring. Remove raised PTO unit if so equipped and partially withdraw extension shaft from rear of rear axle center housing. On models with standard PTO, unbolt PTO output shaft bearing retainer and partiall withdraw output shaft. Disconnect PTO engagement lever and remove PTO gearbox from bottom of transmission housing. Remove belt pulley unit if so equipped.

Disconnect clutch pedal to balance lever rod at balance lever, then unbolt and remove main transmission shifter plate assembly (see Fig. 126) complete with balance lever and release rod. Place primary gearbox lever in lower position on model with single

clutch or in upper position on model with "live PTO" (dual clutch). Working through main transmission shifter plate opening, remove lockwire and square headed set screw retaining selector lever to primary gearbox upper shift rail. Move primary gearbox shift lever to opposite position and with locating studs threaded into right and left center gearbox retaining cap screw holes (see Fig. 130), slide primary gearbox forward about 2 inches. Slide connector lever from upper shift rail, remove lever from primary gearbox shift rail, then remove primary gearbox from transmission housing.

172. To reinstall primary gearbox, place new gasket on locating studs. Note: Gasket sealer should be applied to both sides of gasket between two lower bolt holes. Guide primary gearbox onto locating studs, fit connector lever to primary gearbox shift rail and slide over end of upper shift rail, then slide gearbox into position and secure with retaining cap screws. Wire the cap screw heads to prevent loosening. Complete reassembly of tractor by reversing disassembly procedure.

OVERHAUL PRIMARY GEARBOX

Models With Single Clutch

173. With primary gearbox removed as outlined in paragraph 171 and transmission input shaft and housing removed from gearbox as in paragraph 167, proceed as follows:

174. **REVERSE IDLER.** Early model "FMD" was fitted with transmission brake unit as shown in Fig. 133 although tractors shipped to U.S. were not equipped with the hand lever and camshaft (95) required to operate the brake. On later model "FMD" and all subsequent models, a revised idler gear (27—Fig. 131), shaft (26), bolt (24) and retainers (25 and 28) are used, eliminating the brake components.

If early "FMD" with transmission brake unit is encountered, remove cotter pin retaining the nut (93—Fig. 133), then remove nut, spring (92), retainer plate (88), spring (86), the five rotating discs (91), the seven stationary discs (87) and the idler gear (90). Note: Seventh stationary disc is located between idler gear and primary gearbox housing. The bolt (84) can now be removed from inside of housing.

On models without transmission brake, unscrew nut (29—Fig. 131) and remove retainer (28) and idler gear (27), then remove bolt (24) and sec-

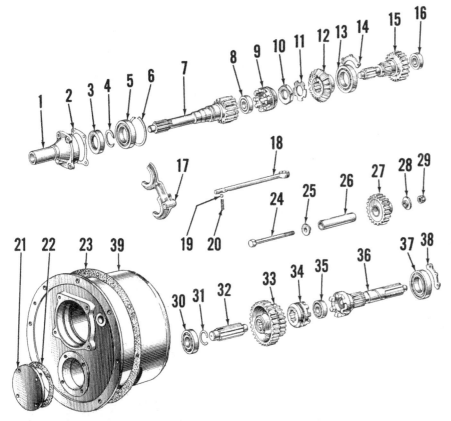

Fig. 131—Exploded view of primary gearbox assembly for single clutch models. Refer to Fig. 134 for gearbox used with dual clutch (Live PTO) models. Fig. 132 shows assembled view of gearbox.

1. Input shaft housing	10. Nut	19. Detent ball	29. Nut
2. Gasket	11. Lockwasher	20. Spring	30. Bearing
3. Oil seal	12. Bevel gear	21. Plate	31. Snap ring
4. Snap ring	13. Bearing	22. Gasket	32. Lower shaft
5. Bearing	14. Locking plate	23. Gasket	33. Cluster gear
6. Snap ring	15. Shaft	24. Bolt	34. Coupling
7. Input shaft	16. Bearing	25. Retainer	35. Bearing
8. Bearing	17. Shift fork	26. Idler shaft	36. Lower main shaft
9. Coupling	18. Shift rail	27. Reverse idler	37. Bearing
		28. Retainer	38. Locking plate

ond retainer (25) from inside of housing.

Idler shaft outside diameter (new) is 1.122-1.123 and idler gear inside diameter (new) is 1.124-1.125 resulting in a shaft to bushing clearance of 0.001-0.003. Renew shaft (26—Fig. 131 or 85—Fig. 133) if worn to a diameter of 1.120 or smaller and gear if inside diameter is worn to 1.127 or larger. Note: On early models with transmission brake unit, the stationary disc retaining pins (89—Fig. 133) can be removed, the brake components discarded and the later type reverse idler and components (items 24 through 29—Fig. 131) installed if so desired.

If necessary to renew idler gear shaft, press old shaft from housing and press new shaft into place. If early transmission brake components are being reinstalled, the shaft (85—Fig. 133) should protrude 1.20-1.23 inches from rear machined (brake) surface of housing. On models without brake, the shaft (26—Fig. 131) should protrude 1.30-1.31 inches from machined face of housing.

On early models with brake, reassemble as follows: Insert special bolt (84—Fig. 133) from inside of housing. Place a stationary disc on idler shaft so that it engages the retaining pins (89), then install idler gear with recess towards housing. Alternately place the stationary and rotating discs on idler shaft, install spring (86) on bolt, install retainer plate (88) and spring (92) and secure with nut (93). Adjust nut so that distance from mounting face of primary gearbox housing to rear face (end) of nut is 13.870-13.895 inches, then retain nut with cotter pin.

On models without transmission brake unit, insert bolt (24—Fig. 131) with retainer (25) through idler gear shaft from inside of housing, place idler gear on shaft with hub away from housing, then secure with second retainer (28) and self locking nut (29). Note: Be sure that small diameter of retainers are towards each other and enter bore of shaft. Tighten the nut securely and measure end play of gear on shaft. If end play is not within 0.010-0.025, press idler shaft into or out of housing as required to bring idler gear end play within limits.

175. LOWER SHAFT AND GEARS. The main transmission lower pinion gear (46—Fig. 140), reverse pinion (45) and lower shaft gear (43) are carried on the primary gear box lower rear shaft (36—Fig. 131). To remove

the shaft and gears as an assembly, remove locking plate (38), then withdraw the unit from rear of primary gear box. Refer to paragraph 185 for service of the removed shaft and gears.

With shaft (36) removed, use suitable pullers to remove inner race and roller assembly of bearing (35) and remove front plate (21) from front face of primary gearbox housing. Select a pipe that will fit over rear end of shaft (32) and into recess of coupling gear (34). With unit supported on this pipe, use suitable drift punch to drive the shaft (32) out to rear of the assembly and into the pipe. With shaft out, remove the cluster gear (33) and thrust washer (31). Drive the front bearing (30) from bore in housing. The coupling gear cannot be removed from shifter fork (17); remove fork and gear as a unit as outlined in paragraph 177, then remove gear from fork.

To reassemble, be sure that shift fork (17) and gear (34) are in place with clutch teeth of gear to rear of housing. Then, reinstall shaft, thrust washer, cluster gear and bearing by reversing removal procedure. Reinstall front plate with new gasket and secure the cap screws with locking wire. NOTE: With introduction of model "New FSM" at serial No. 08C-960337, diameter of bearing journal on front end of primary gearbox lower shaft (32—Fig. 131) and inside diameter of bearing (30) were decreased to provide a larger capacity bearing. The new shaft (Ford part No. DKN-77112-B) and bearing (part No. DDN-7A452) can be used as service replacements for the prior shaft (part No. DKN-77112-A) and bearing (part No. BB-7065) in all prior models. When installing the late shaft and bearing in all models, a thrust washer (part No. 113435-ES) is placed between bearing (30) and gear (33).

Also, with introduction of model "New FSM", number of teeth of primary gearbox upper shaft gear (15—Fig. 131) was changed

from 23 to 24 and number of teeth on lower mating gear (43—Fig. 140) was changed from 24 to 23. The new upper shaft (Ford part No. DDN-7024) and lower gear (part No. DKN-77103-B) can be installed as a pair in all prior models as a service replacement for prior upper shaft (part No. DKN-7024-A) and lower gear (part No. DKN-7103-A).

176. UPPER SHAFT AND GEARS. With reverse idler and lower shaft removed as outlined in paragraphs 174 and 175, proceed as follows:

Using suitable pullers, remove bearing (8—Fig. 131) inner race and roller assembly from front end of upper shaft (15). Straighten tab on locking washer (11), then hold shaft from turning while unscrewing nut (10), or hold nut and turn shaft as practical with available tools. Note: A suggested method of holding or turning shaft is to place clutch disc on front end of input shaft (7), engage rear end of input shaft with coupling (9), then hold or turn with the clutch disc.

With nut (10) off, remove locking plate (14) from rear face of gearbox and withdraw shaft (15) and bearing (13) from rear. Then, remove gear (12), tab washer, nut and coupling from the gearbox.

If necessary to remove outer race of bearing (16) from rear end of upper shaft, first remove the ball bearing (13), then drive outer race from rear end of shaft with punch inserted through holes in gear part of shaft. If necessary to renew upper shaft, refer to note following paragraph 175.

Reinstall by reversing removal procedure. Be sure to bend tab of washer against nut (10) and secure cap screws for locking plate (14) with wire.

177. SHIFT RAIL AND FORK. With the upper and lower primary shafts and gears removed as outlined in paragraphs 175 and 176, proceed as follows:

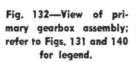

Fig. 132—View of primary gearbox assembly; refer to Figs. 131 and 140 for legend.

Cut the locking wire and remove set screw from fork (17—Fig. 131) and rail (18). Withdraw the shift rail, catching the detent ball and spring as rail is removed. The shift fork along wtih the lower shaft coupling gear (34) can be removed at this time. Remove plug from shift rail bore at front side of primary gearbox housing by driving plug forward out of bore.

To reinstall, insert coupling gear (34) in lower end of fork with coupling teeth to rear, then place fork and coupling gear in housing. Insert shift rail through rear bore and into shift fork. Insert detent spring and ball and hold the detent ball in depressed position as the shift rail is moved forward.

Coat outside of new shift rail bore plug (Ford part No. 115422-ES) with sealer, then drive the plug into bore with open end out. Note: Previous installation of plug was with open end towards shift rail. After plug is installed, stake around the plug bore in three positions to keep plug from moving forward out of bore.

Models With Dual Clutch

178. With primary gearbox removed as outlined in paragraph 171 and the transmission and PTO input shafts removed as in paragraph 168, proceed as follows:

179. **REVERSE IDLER.** Unscrew the nut (32—Fig. 134) and from bolt (27) and remove the retainer (31) and reverse idler gear (30) from shaft (29). Bolt and inner retainer (28) can be removed from inside of primary gearbox housing.

Outside diameter of idler shaft (29) is 1.122-1.123 (new) and idler gear inside diameter (new) is 1.124-1.125 resulting in a gear to shaft clearance of 0.001-0.003. Renew shaft if worn to a diameter of 1.120 or smaller and gear if inside diameter is worn to 1.127 or larger. Gear end play on shaft of 0.010-0.025 can be adjusted by pressing idler shaft into or out of primary gearbox housing as required.

When installing new idler shaft, press shaft into housing until it protrudes 1.30-1.31 inches from machined rear face of housing. Insert bolt with retainer (28) (small diameter away from bolt head) into shaft from inside housing. Place idler gear on shaft with hub away from housing, then install retainer (31) and self-locking nut (32). Note: Be sure small diameter of both retainers enter bore of shaft. Tighten the nut securely and check idler gear end play; press shaft into or out of housing as required to bring end play within limits of 0.010-0.025.

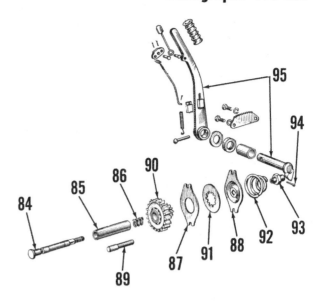

Fig. 133—Early model "FMD" primary gearbox is fitted with transmission brake assembly (items 84 through 94) although lever and actuating shaft (95) were not fitted on tractors shipped to United States.

84. Bolt
85. Idler shaft
86. Spring
87. Stationary discs
88. Retainer plate
89. Retaining pins
90. Reverse idler
91. Rotating discs
92. Spring
93. Nut
94. Cotter pin
95. Lever & shaft (not used)

180. **R&R MAIN TRANSMISSION LOWER SHAFT AND GEARS.** The main transmission lower pinion gear (46—Fig. 140), reverse pinion (46) and lower shaft gear (43) are carried on the primary gear box lower rear shaft (main transmission lower shaft) (40—Fig. 134). To remove the shaft and gears as an assembly, remove locking plate (42), then withdraw the unit from rear of primary gearbox. Refer to paragraph 184 for service information on the removed shaft and gears.

Reinstall shaft and gears as a unit as shown and secure the locking plate cap screws with wire.

181. **PRIMARY GEARBOX UPPER SHAFT AND GEARS.** With reverse idler removed as outlined in paragraph 179 and the transmission lower

shaft and gears removed as in paragraph 180, proceed as follows:

Straighten the upper shaft locknut retainer (20—Fig. 134), then unscrew the nut (21). Note: A tool for removing and tightening the nut can be made by welding four pieces of square key stock to a suitably sized socket wrench.

Place shift rail (24) in outermost position and remove the locking plate (18). Withdraw the gear (19) and bearing (17) assembly from rear of gearbox. Place gearbox on its side and push shift rail to inner (forward) detent position. Separate the shaft (14) and sliding gear (12) from bevel gear (16), remove shaft and sliding gear through opening in side of gearbox, then remove the bevel gear.

The bearing (17) can be removed

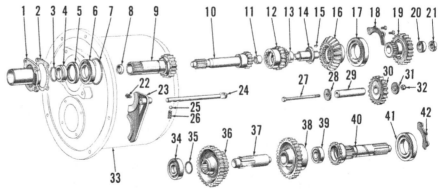

Fig. 134—Exploded view of primary gearbox assembly used on models with dual clutch (Live PTO). Refer to Fig. 131 for gearbox used with single clutch models.

1. Input shaft housing
2. Gasket
3. Snap ring
4. Sleeve
5. Oil-seal
6. Bearing
7. Snap ring
8. Oil seal
9. PTO input shaft
10. Transmission input shaft
11. Bearing
12. Sliding gear
13. Bushing
14. Upper shaft
15. Woodruff key
16. Bevel gear
17. Bearing
18. Locking plate
19. Upper gear
20. Retainer
21. Nut
22. Set screw
23. Shift fork
24. Shift rail
25. Detent ball
26. Detent spring
27. Bolt
28. Retainer
29. Reverse idler shaft
30. Reverse idler gear
31. Retainer
32. Nut
33. Housing
34. Bearing
35. Snap-ring
36. Cluster gear
37. Lower shaft
38. Sliding gear
39. Bearing
40. Main shaft
41. Bearing
42. Locking plate

from gear (18) with suitable pullers. Remove bearing cup from rear face of gear with a punch inserted through holes in gear. The sliding gear and bushing (13) are not serviced separately.

NOTE: At model "FMD" serial No. 1435545, the dog teeth on rear end of sliding gear (12) were increased in length and the distance between detent notches on shift rail (24) was decreased to compensate for longer gear teeth. Only the later type gear with longer teeth and the shift rail with shorter distance between detent notches are available for service. Thus, when servicing model "FMD" prior to serial No. 1435545, gear and shaft must be renewed as a matched pair. Late type gear can be identified by letter "B" stamped on external dog teeth (rear) end of gear and late type rail can be identified by letter "S" stamped on circular end of rail.

With introduction of model "New FSM" at serial No. 08C-960337, number of teeth on primary gearbox upper gear (19) was changed from 23 to 24 and number of teeth on main transmission lower gear (43—Fig. 140) was changed from 24 to 23. The new upper gear (part No. DDN-7102-B) and new lower gear (part No. DKN-77103-B) can be installed as a pair to renew the previous upper gear (part No. E84-GA-9) and lower gear (part No. DKN-77103-A) in prior models.

To reinstall upper shaft and gears, proceed as follows: Be sure that Woodruff key (15—Fig. 134) is placed in slot of shaft (14) and position sliding gear (12) on shaft with external dog teeth towards threaded end of shaft. Place bevel gear (16) in gearbox and push shift rail to inner position (forward). Place sliding gear and shaft in gearbox so that shift fork engages slot in gear and enter threaded end of shaft through the bevel gear. With bearing (17) installed on hub of gear (19), insert splined end of gear into splines of

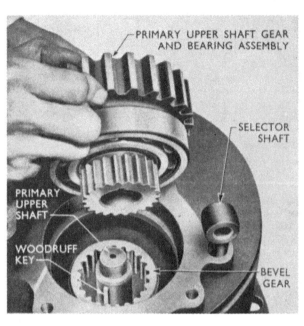

Fig. 136—Removing primary upper shaft gear (19—Fig. 134) and bearing (17) assembly from primary gearbox.

bevel gear so that slot in gear hub is correctly aligned with key (15) in shaft. Take care that the key is not pushed out of place as gear is pushed onto shaft. Install nut (21) with new retainer (20) and tighten nut to a torque of 100 Ft.-Lbs. Bend edges of retainer into all four slots in nut.

182. SHIFT RAIL AND FORK. With the primary gearbox upper shaft and gears removed as outlined in paragraph 181, proceed as follows:

Cut the locking wire and remove set screw (22—Fig. 134) from shift fork (23). Rotate the rail (24) away from detent ball (25) and withdraw rail from gearbox, catching the detent ball (25) and spring (26) as rail is removed. Rotate the shift fork (23) up out of slot in sliding gear (38), then remove fork from opening in side of gearbox. Drive the plug from

shift rail bore at front side of gearbox.

To reinstall shift rail and fork, position the fork in slot of lower shaft sliding gear and insert rail from rear through gearbox and fork. Insert the detent spring and ball in their bore and hold detent ball depressed while pushing shift rail into position. Install set screw and secure with wire. Coat new shift rail bore plug with sealer and install plug with hollow side out. Stake the plug at three points to retain it in bore.

183. LOWER SHAFT AND GEARS. With upper shaft and gears removed as outlined in paragraph 181 and the shift rail and fork removed as in paragraph 182, proceed as follows:

Using suitable pullers, remove bearing (39—Fig. 134), inner race and rollers from rear end of lower shaft (37). Remove plate (see 21—Fig. 131) from front face of gearbox. Support unit on a piece of pipe that will fit over rear end of lower shaft and into the recess in sliding gear (38—Fig. 134). Then, using a suitable drift punch, drive shaft out of bearing (34)

Fig. 135—Removing upper shaft locking plate (18—Fig. 134) from primary gearbox housing.

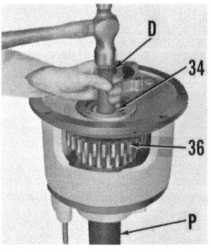

Fig. 137—Using suitable drift (D), drive lower shaft from bearing (34) and cluster gear (36). Use hollow pipe (P) to support gear.

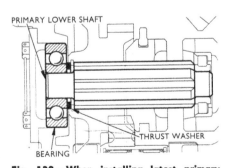

Fig. 138—When installing latest primary gearbox lower shaft and front bearing, a thrust washer is placed between bearing inner race and gear locating snap ring.

and gears (36 and 38) and into the pipe. Remove the gears out through hole in side of gearbox and remove bearing (34) from bore.

NOTE: At model "New FSM" serial No. 08C-960337, diameter of bearing journal on front end of lower shaft (37) and inside diameter of bearing (34) were decreased to provide a larger capacity bearing. The new shaft (Ford part No. DKN-77112-C) and bearing (part No. DDN-7A452), along with new thrust washer (part No. 113435-ES) can be used as service replacements for the prior shaft (E81-GA-9) and bearing (BB-7065) in all prior models.

To reinstall lower shaft and gears, proceed as follows: Install bearing (34) and retaining plate in front bore of gearbox, then turn gearbox front face down. Be sure that snap ring (35) is properly located in groove inside front hub of gear (36) and if late type lower shaft and bearing are being installed, stick thrust washer to front side of snap ring with heavy grease. Position sliding gear (38) inside gearbox, then place the gear (36) in gearbox with front face (or thrust washer) against bearing inner race and lower the sliding gear onto the cluster gear. Insert the shaft (37) through the gears and enter front journal of shaft into bearing. Tap rear end of shaft until shoulder of shaft contacts snap ring. Remove the retaining plate from in front of bearing (34) and support front end of shaft, then drive bearing (39) inner race and roller assembly onto rear end of shaft. Reinstall front plate with new gasket and secure retaining cap screws with wire.

MAIN TRANSMISSION

All Models

184. The main transmission lower shaft and gears can be serviced after being removed from the primary gearbox assembly as outlined in paragraph 175 for models with single clutch or in paragraph 180 for models with dual clutch (live PTO). For service of other main transmission components, the complete transmission assembly must be removed as outlined in paragraph 169.

185. **LOWER SHAFT AND GEARS.** The main transmission lower pinion gear (46—Fig. 140), reverse pinion (45) and lower shaft gear (43) are carried on the primary gearbox lower rear shaft (main transmission lower shaft (36 or 40). Remove the shaft and gears as in paragraph 175 for models with single clutch or as in paragraph 180 for models with dual clutch, then disassemble as follows:

Fig. 139—Cut-away view of typical single clutch transmission. Refer to Fig. 131 for exploded view of single clutch primary gearbox; to Fig. 134 for exploded view of dual clutch primary gearbox; and to Fig. 140 for exploded view of main transmission gears.

Using suitable pullers, remove the inner race and roller assembly of bearing (47—Fig. 140) from rear end of shaft, then slide the lower pinion (46) and reverse pinion (45) from shaft. Remove the snap ring (44) from rear side of gear (43), then press the gear and bearing (37 or 41) assembly from the shaft. Using suitable pullers, remove bearing from gear and outer race of bearing (35—Fig. 131 or 39—Fig. 134) from the bore in front end of shaft.

NOTE: With introduction of model "New FSM" at serial No. 08C-960337, number of teeth on lower shaft gear (43—Fig. 140) was changed from 24 to 23 and number of teeth on the mating primary gearbox upper shaft gear was changed from 23 to 24. The latest gears can be installed as a pair in earlier model tractors; refer to note in paragraph 175 or 181.

To reassemble lower shaft and gears, proceed as follows: Press outer race of bearing (35—Fig. 131 or 39—Fig. 134) into front bore of shaft. Place

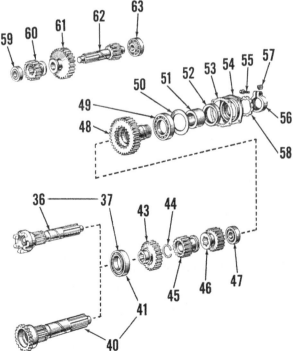

Fig. 140—Exploded view of main transmission unit. Refer to Fig. 139 for cross-sectional view.

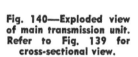

36. Main-shaft (single clutch)
37. Bearing
40. Main shaft (dual clutch)
41. Bearing
43. Lower shaft gear
44. Snap ring
45. Reverse pinion
46. Lower pinion
47. Bearing
48. Output gear
49. Bearing
50. Oil baffle
51. Sleeve
52. Oil seal
53. Gasket
54. Retainer
55. Clamping bolt
56. Nut
57. Locking tab
58. Locking washer
59. Bearing
60. Upper pinion
61. Low gear
62. Upper shaft
63. Bearing

roller bearing assembly (37 or 41—Fig. 140) on shaft, then start gear (43) front hub onto large diameter splines on shaft. Place the unit in a press and push shaft into gear making sure that bearing starts onto hub of gear. Continue pressing shaft into gear until shoulder of shaft forces bearing inner race against shoulder on gear hub, then install snap ring (44). Install the reverse pinion (45) and lower pinion (46) on shaft, then press inner race and roller assembly of bearing (47) onto rear end of shaft. Reinstall the shaft and gear assembly as outlined in paragraph 175 or 180.

186. UPPER SHAFT AND GEARS AND OUTPUT GEAR. After removing primary gearbox from transmission as outlined in paragraph 171, proceed as follows to remove the upper shaft and gears and the transmission output (large) gear:

Remove the clamp bolt (55—Fig. 140 or 141) and the locking tab (57) from nut (56), then unscrew and remove nut and locking washer (58). Unbolt and remove the oil seal retainer (54), seal and sleeve (51). Push the output (transmission large) gear forward. As the gear is pushed forward, the upper shaft (62) and gear assembly must move forward until rear bearing (63) is clear of bore in transmission housing; otherwise, the output gear (48) will be forced against upper shaft gear (61). Remove the upper shaft and gears as soon as rear bearing is free of housing, then continue pushing output gear forward until free of bearing (49) in transmission. Remove bearing from rear of transmission.

NOTE: On model "FMD" and model "FPM" prior to serial No. 1531530, upper shaft rear bearing (63—Fig. 140) was a ball bearing; all later models are equipped with a roller bearing at this location which is the only bearing now used for service of all models.

Fig. 141—View of rear end of main transmission showing plug (P) located behind upper shaft bearing (63—Fig. 140). Refer also to Fig. 143. Refer to Fig. 140 for legend.

Also, with introduction of model "New FSM" at serial No. 08C-960337, a new upper main shaft (62) with 11 teeth was introduced; previous shaft had 13 gear teeth. Along with the change in gear teeth, a new output (transmission large) gear with same number of teeth but having a different tooth form was introduced. The new gear can be identified by overall diameter of 7.368 inches compared with diameter of 7.135 inches for old gear. The internal dog teeth on rear of upper shaft low gear (61) was also changed from 13 to 11 to mesh with the new upper shaft gear. The new upper shaft (62), upper shaft low gear (61) and new output (transmission large) gear (48) can be installed in earlier models, but the gears are not individually interchangeable.

Remove the expansion plug (P—Fig. 141) and bearing cup for upper shaft rear (roller type) bearing (63—Fig. 140) if necessary. Note: Three different plugs have been used to seal bore at rear of upper shaft rear bearing (63). Early model "FMD" was fitted with a $3\frac{7}{32}$ inch diameter expansion plug (Ford part No. 119781-ES) which is installed from front after removing bearing. Later model "FMD", model "FPM" and early "FSM" are fitted with a cup type plug (part No. 119782-ES) installed with open side of cup towards bearing (front) as shown in Fig. 143. Later model "FSM" and model "New FSM" are also equipped with a cup type plug, but the bore in housing is 0.045 smaller than prior models and a different plug (part No. 115422-ES) is required. If in doubt as to which plug diameter is required, measure bore diameter in housing; if bore measures 2.880-2.882 inches, install large plug (part No. 119782-ES). Install small diameter plug (part No. 115422-ES) if bore measures 2.835-2.837.

Inspect the outer race of bearing (47—Fig. 140) in bore of output (large) gear (48); remove the bearing cup and expansion plug from bore of gear if necessary. Install new expansion plug, then install new bearing cup.

Using suitable pullers, remove upper shaft front bearing (59) inner race and rollers. Slide the pinion (60) and upper shaft low gear (61) from shaft. On latest shaft (62), shoulder is too large to permit removal of bearing (63) inner race and roller assembly with pullers. If necessary to renew bearing (63) and shaft is otherwise serviceable, remove old bearing as follows: Break the roller retainer and grind a flat on bearing inner race (track), then break inner race with chisel as shown in Fig. 142. The inner race can then be easily removed from shaft rear journal. Press new bearing

inner race and roller assembly onto shaft.

To reassemble, proceed as follows: Install low gear (61) and pinion gear (60) on upper shaft with shifter flanges together and install front bearing (59) inner race and rollers. Install the new oil seal (52) in retainer (54) with lip to front (machined) side of retainer. Position the bearing (49) and oil baffle (50) in bore of transmission housing and secure with retainer (54) and new gasket (53). Start rear hub of output (large) gear (48) into bearing and bump gear rearward until about halfway through bearing. Position the upper shaft, gears and bearing assembly into rear bearing race, then continue driving output gear rearward until seated against bearing inner race. Insert sleeve (51) through oil seal, then install locking washer (58) and nut (56). Tighten nut securely so that slot in nut is aligned with notch in washer, then install locking tab (57) and clamping bolt (55).

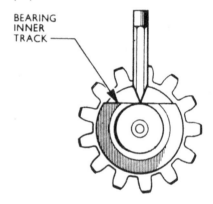

Fig. 142—Removing rear bearing from late type shaft (62—Fig. 140); refer to text.

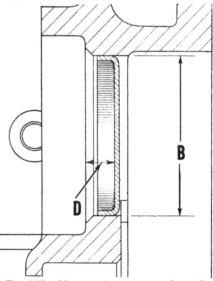

Fig. 143—Measure bore (B) to determine proper size of plug; refer to text for sizes. Plug depth (D) measured from shoulder of bearing bore to front face of plug should be 0.53 inch.

DIFFERENTIAL

CARRIER BEARING ADJUSTMENT
All Models

187. The differential carrier bearings will be properly adjusted when the total thickness of shims located between the rear axle center housing and both differential carrier bearing supports (bull pinion housings) is 0.016. Shims from one side of center housing may be shifted to opposite side to adjust bevel pinion backlash as outlined in paragraph 195; however, total installed thickness of 0.016 must be maintained.

R&R DIFFERENTIAL

Models "FMD" and "FPM"

188. To remove the differential and bevel ring gear assembly, first remove the left rear axle shaft as in paragraph 206, the left bull gear as in paragraph 205 and the left pull pinion and brake assembly as outlined in paragraph 202. Unbolt and remove ring gear thrust pad support (Fig. 152) and the oil deflector assembly (Fig. 144). Thread jack screws into the two tapped holes in differential carrier bearing support and remove the support taking care not to lose or damage shims located between support and rear axle center housing. The differential and bevel ring gear assembly can now be lifted from rear axle center housing.

When reinstalling differential assembly, be sure to reinstall the same shim thickness as was removed from between carrier bearing support and the rear axle center housing. Then, reinstall bull pinion and brake assembly

and check bevel gear backlash. If backlash is not within limits of 0.004-0.018, remove both bull pinion and brake assemblies (refer to paragraph 202) and transfer shims from under right bearing carrier to under left bearing carrier to increase backlash, or from under left bearing carrier to under right bearing carrier to decrease backlash. Total thickness of shims under both bearing carriers must be 0.016.

When bevel gear backlash is properly adjusted, reinstall ring gear thrust pad and support assembly with proper number of gaskets so that clearance between thrust pad and back face of ring gear is 0.004-0.014. Reinstall oil deflector assembly and, if necessary, adjust clearance between scraper and ring gear to 0.004.

Reinstall bull gear, rear axle and hydraulic pump and refill rear axle center housing with lubricant. Reinstall hydraulic lift cover assembly.

Models "FSM" and "New FSM"

189. To remove the differential and bevel ring gear assembly, proceed as follows: Remove the left rear axle shaft as outlined in paragraph 206 and the left bull gear as in paragraph 205. Remove both brake assemblies and brake housings as in paragraph 216, then remove the left bull pinion and the right bull pinion and differential lock assembly. Unbolt and remove the ring gear thrust pad and support assembly (Fig. 152) and the oil deflector (Fig. 145). Thread jack screws into the left differential carrier bearing support (bull gear housing) and remove the support taking care

that the shims located between bearing support and rear axle center housing are not lost or damaged. The differential and bevel ring gear assembly can now be lifted from rear axle center housing.

To reinstall differential and bevel ring gear assembly, proceed as follows: Position differential assembly in rear axle center housing and reinstall left carrier bearing support with same thickness of shims as was removed and reinstall both brake housings without the bull pinions. Check bevel ring gear to bevel pinion backlash and if not within the limits of 0.004-0.018, remove the brake housings and both carrier bearing supports and transfer shims from under right support to under left support to increase backlash, or from under left support to under right support to decrease backlash.

When bevel gear backlash is correct, remove the brake housings taking care not to disturb the differential carrier bearing supports (bull pinion housings), then install the bull pinions, brake housings and brake assemblies. Install the ring gear thrust pad and support assembly with proper number of gaskets to provide 0.004-0.014 clearance between thrust pad and back face of bevel ring gear. Reinstall the oil deflector and, if necessary, adjust deflector scraper to clear ring gear a distance of 0.004.

Reinstall bull gear, rear axle and hydraulic pump and refill rear axle center housing with lubricant. Reinstall hydraulic lift cover assembly.

OVERHAUL DIFFERENTIAL
Models "FMD" and "FPM"

190. Place correlation marks on differential case halves if not already present, then cut locking wire and

Fig. 145 — On models "FSM" and "New FSM", oil deflector (D) mounts on left differential carrier bearing support and delivers oil to bevel pinion bearings. Adjust deflector scraper (S) to ring gear clearance to 0.004. Hydraulic pump pressure tube is (HP); return tube is (HR).

Fig. 144—On "FMD" and "FPM" models, oil deflector (D) attaches to both differential carrier bearing supports. Adjust scraper (S) to clear ring gear by 0.004. Oil tube (T) delivers oil from scraper to differential.

Fig. 146—Be sure mating marks are placed on differential case halves before disassembling unit.

the rivets out of gear and case.

When installing ring gear to differential case, be sure the mating surfaces are clean and free of any burrs, then install ring gear with special bolts (Ford part No. 115605-ES), spacers (part No. E21-EB-9) and self-locking nuts (part No. 34447-ES). Tighten the nuts to a torque of 40-45 Ft.-Lbs.

When reassembling differential, be sure the correlation marks on case halves are properly aligned, install the eight cap screws and tighten them to a torque of 65-75 Ft.-Lbs. Wire the drilled heads of the cap screws securely.

Models "FSM" and "New FSM"

191. With the differential and bevel ring gear assembly removed as outliner in paragraph 189, proceed as follows:

Place correlation marks on the the differential case halves if not already present, then cut locking wire and remove cap screws retaining right case half to left half. Separate the case halves as shown in Fig. 146 and remove the spider, gears and thrust washers.

remove cap screws retaining right case half to left half. Separate the case halves as shown in Fig. 146 and remove the spider, gears and thrust washers.

NOTE: On early "FMD", bores for the differential side gear hubs were unbushed. Diameter of the side gear hubs was later reduced and bushings installed in the differential case bores. At serial No. 1481091, number of teeth on side gears was changed from 18 to 20 and number of teeth on differential pinions was changed from 9 to 11. Only the latest type side gears and pinions are available for service. Thus, both new side gears and four new pinions must be installed if differential assembly is fitted with the early type gears. Also, if the differential case is unbushed, bushings must be installed or if the bores are worn excessively, a new differential assembly must be installed.

Inspect the differential spider, gears, thrust washers and bushings for excessive wear or breakage. If neces-

sary to install new bushings in differential case, a closely fitting bushing driver should be used and the inside ends of bushing should be flush with bottom of chamfer on inside of housing. The bushings are pre-sized. The differential pinions and their bushings are serviced as assemblies only.

The bevel ring gear is riveted to the left differential case half during factory production; however, special bolts, spacers and nuts are available for service installation of gear to case. If necessary to remove a riveted ring gear, center punch the upset ends of rivets on ring gear side of the assembly, then using a ⅛-inch diameter drill, drill pilot holes in each of the rivet heads. Using a ½-inch diameter drill bit, drill into the pilot holes until the hardened face of the ring gear is reached; then, using a ⅜-inch diameter drift punch, drive

The bushings in the differential case halves for the differential side gear hubs are renewable. New bushings should be driven into bores from inside of differential case so that inner ends of bushings are flush with the chamfer in case. The bushings are pre-sized and reaming should not be necessary if bushings are carefully installed.

Holes are provided in the differential case halves so that a punch may be inserted through the holes to drive the carrier bearing cone and roller assemblies from the case. When renewing bearing cones, the cups in

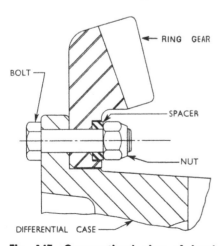

Fig. 147—Cross-sectional view of bevel ring gear showing proper installation of retaining bolts, spacers and nuts.

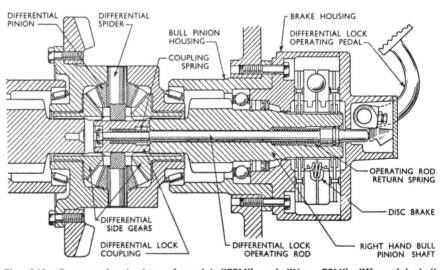

Fig. 148—Cross-sectional view of model "FSM" and "New FSM" differential, bull pinion and right brake unit showing operation of differential lock. Depressing operating pedal locks both differential side gears together.

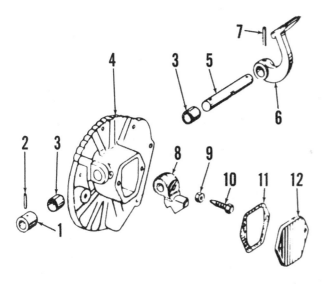

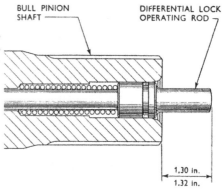

Fig. 149—Exploded view of differential lock operating pedal and disc brake cover assembly.

1. Collar
2. Pin
3. Bushings
4. Brake cover
5. Shaft
6. Operating pedal
7. Pin
8. Operating lever
9. Lock-nut
10. Set screw
11. Gasket
12. Cover plate

Fig. 151—Adjust nut (2—Fig. 150) so that differential lock operating rod protrudes 1.30-1.32 from bull pinion shaft.

carrier bearing supports (bull pinion housings) should also be renewed.

If necessary to renew the bevel ring gear or differential case separately, unbolt and remove the ring gear from left case half. When installing ring gear on differential case, be sure the mating surfaces are clean and free of any burrs, then install the ring gear with special bolts (Ford part No. 43068-ES), spacers (part No. E21-EB-9) and self-locking nuts (part No. 34447-ES). Note: Clean the bolts and nuts thoroughly in solvent, allow to dry, then apply a thin line of "Loctite" sealer along the threads of each bolt before installing nuts. Tighten the nuts to a torque of 50-55 Ft.-Lbs. and allow two hours for "Loctite" to set after tractor is assembled before driving the tractor.

Assemble differential case halves with correlation marks properly aligned, tighten cap screws to a torque of 65-75 Ft.-Lbs. and secure with locking wire through the drilled cap screw heads.

DIFFERENTIAL LOCK

Models "FSM" and "New FSM"

192. Refer to cross-sectional view in Fig. 148 for operation of the differential lock. Repair procedures are as follows:

193. **OPERATING PEDAL.** To disassemble the operating pedal and brake cover assembly, refer to exploded view of unit in Fig. 149 and proceed as follows: Remove cover (12), loosen locknut (9) and remove the set screw (10). Drive pin (2) from collar (1) and shaft, then withdraw the pedal and shaft from brake cover and remove operating lever (8). Drive pin (7) from pedal and remove pedal from cross-shaft. Remove cross-shaft bushings from brake cover.

To reassemble operating pedal and brake cover assembly, reverse disassembly procedure. Be sure that operating pedal is installed on correct end of shaft; this can be checked by noting position of dimple in shaft for the set screw (10). New pins should be used to secure pedal and collar to shaft.

194. **DIFFERENTIAL LOCK AND BULL PINION ASSEMBLY.** Refer to paragraph 204 to remove and reinstall bull pinion assembly. The differential lock and bull pinion assembly can be disassembled by removing the cotter pin and castellated nut from inner end of differential lock operating rod. Refer to exploded view in Fig. 150.

Install a new "O" ring on operating rod, lubricate "O" ring and reassemble the unit. Turn nut onto inner end of rod so that outer end protrudes 1.30-1.31 inches from outer face of bull

pinion shaft as shown in Fig. 151 then secure nut in this position with new cotter pin.

MAIN DRIVE BEVEL GEARS

BEVEL GEAR ADJUSTMENT
All Models

195. Bevel gear mesh position is non-adjustable and is maintained by installation of matched bevel ring gear and bevel pinion whenever renewal of either part is required.

Bevel gear backlash should be 0.004-0.018 and is adjustable by transfering shims located between rear axle center housing and differential carrier bearing supports (bull pinion housings). For proper adjustment of differential carrier bearings, the total shim thickness of both locations must be 0.016. Shims are available in thicknesses of 0.003 and 0.005; thus the total shim thickness must include two 0.003 thick and two 0.005 thick shims. Bevel gear backlash should be checked whenever installing differential assembly is outlined in paragraph 189.

RING GEAR THRUST PAD
All Models

196. A thrust pad (see Fig. 152) is provided to prevent bevel ring gear and pinion separation under heavy thrust loads. Clearance of 0.004-0.014 between thrust pad and back face of bevel ring gear should be maintained by varying number of the gaskets used between thrust pad support and rear axle center housing.

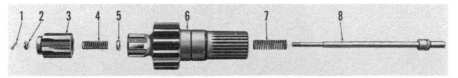

Fig. 150—Exploded view of right bull pinion and differential lock assembly. Protrusion of rod (8) is adjusted by nut (2); refer to Fig. 151.

1. Cotter pin
2. Nut
3. Coupling
4. Spring
5. Spring seat
6. Bull pinion
7. Spring
8. Operating rod

MAIN DRIVE BEVEL PINION

All Models

197. REMOVE AND REINSTALL. The bevel pinion assembly can be unbolted and removed from front of rear axle center housing after splitting the tractor between transmission and rear axle center housings as follows:

Drain the rear axle center housing lubricant and remove PTO output shaft as outlined in paragraph 224 or the raised PTO unit and driveshaft as outlined in paragraph 225. Disconnect wiring to rear lights. Remove the foot rests (step plates) from each side of tractor. On models "FMD" and "FPM", disconnect brake cables and remove the brake locking lever guide. Remove the outer brake pedal, loosen clamp bolt on inner pedal and slide inner brake pedal out far enough to clear the transmission flange. Disconnect clutch release rod at balance lever. Remove the main shift lever assembly from top of rear axle center housing.

If differential has been removed, attach suitable hoist to rear axle center housing and place a support under the transmission housing. On four wheel models, drive wood wedges between front axle and front support; adequately brace tricycle front end to keep from tipping.

Unbolt the rear axle center housing from transmission and separate the

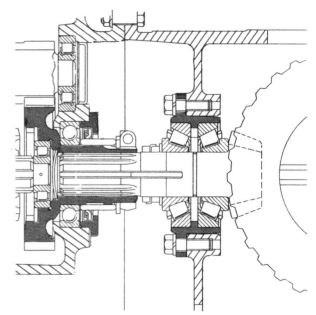

Fig. 153—Cross-sectional view showing bevel pinion mounting. Refer to Fig. 141 for view showing rear face of transmission housing and to Fig. 152 for view showing front face of rear axle center housing.

units. Remove retaining cap screws and withdraw bevel pinion assembly from front of rear axle center housing. If necessary to drive assembly from rear axle center housing, use proper size drift and drive on rear end of bearing carrier only; driving against rear end of pinion could damage pinion bearings.

If necessary to renew bevel pinion, refer to paragraph 190 or 191. Pinion and ring gear must be installed as a matched pair to maintain bevel pinion mesh position.

With differential and bevel ring gear assembly removed, reinstall bevel pinion assembly first and then adjust pinion bearing preload as outlined in paragraph 199. If differential and bevel ring gear assembly is not re-

moved, clamp flange of bearing carrier in vise and adjust pinion bearing preload prior to reinstalling the assembly.

Using guide studs, install bevel pinion assembly in front face of rear axle center housing, then remove guide studs and securely install carrier bearing retainer cap screws. Rejoin the tractor by reversing split procedure.

198. OVERHAUL BEVEL PINION ASSEMBLY. With bevel pinion assembly removed as outlined in paragraph 197, proceed as follows: Lift up the bent sides of the locking washer, then loosen and remove the locking nut, locking washer, adjusting nut and thrust washer from front end of pinion. Withdraw the pinion and rear

Fig. 152—View showing bevel ring gear thrust pad and support assembly.

1. Expansion plug
2. Support
3. Gaskets
4. Thrust pad
5. Rivet

Fig. 154—Cut-away view of bull pinion shaft and housing; refer to Fig. 155 for exploded view.

bearing cone and roller assembly from rear of carrier and remove front bearing cone and roller assembly from front of carrier. NOTE: The front and rear pinion bearing assemblies are alike; however, if bearings are to be reused, the bearing cone and roller assemblies must be kept with their mating cups due to established wear patterns.

Remove rear bearing cone and roller assembly from bevel pinion and remove the bearing cups from the carrier.

Drive rear bearing cone and roller assembly firmly against shoulder on

bevel pinion and be sure that bearing cups are firmly seated in carrier. Install pinion in carrier, install front bearing cone and roller assembly and the thrust washer and adjusting nut. Adjust bearing preload as follows:

199. **ADJUST PINION BEARING PRELOAD.** Gradually tighten the adjusting (rear) nut until torque required to turn pinion in bearings is 12-16 inch-pounds. Turning torque may be measured with a torque wrench or with cord wrapped around the pinion shaft, then using pull scale and cord to steadily rotate shaft. Pull scale reading should be 12 to 16

pounds when bearings are properly adjusted. When correct torque or pull scale reading is obtained, place new locking washer on shaft and hold the adjusting nut while installing and tightening locknut. Recheck adjustment after locknut is tight, then bend locking washer down against flats on adjusting nut and locknut.

200. **RENEW BEVEL RING GEAR.** The bevel ring gear is either riveted or bolted to the differential housing. Refer to differential overhaul in paragraph 190 for models "FMD" and "FPM", or in paragraph 191 for models "FSM" and "New FSM".

FINAL DRIVE AND REAR AXLES

BULL PINIONS

Models "FMD" and "FPM"

201. Refer to cut-away view of the bull pinion and related parts in Fig. 154 and to the exploded view of bull pinion, pinion housing and differential carrier bearing support in Fig. 155. Bull pinion inner bearing (7) seats in outer end of differential carrier bearing support (8) and outer bearing (3) is located in outer end of brake support (4).

202. **R&R BULL PINION AND BRAKE ASSEMBLY.** To remove the bull pinion for access to differential or to adjust axle bearing preload, the bull pinion and brake assembly should be removed as a unit as follows:

If rear wheel is set close to fender, support rear end of tractor and remove wheel. Remove the fender and footrest as an assembly. Disconnect the brake cable from operating lever. Cut the locking fire and remove cap screws retaining brake support to rear axle center housing. Taking care not to

Fig. 156—Cut-away view of model "FSM" and "New FSM" disc brake, bull pinion and differential lock. Exploded view of the right bull pinion and differential lock assembly is shown in Fig. 150.

dislodge the differential carrier bearing support (8—Fig. 155) and shims (9), remove the brake, brake support and bull pinion as an assembly.

When reinstalling the brake support and bull pinion assembly, tighten retaining cap screws securely, then install locking wire through holes in cap screw heads. Note: The two lower cap screws also retain the brake cable conduit support.

203. **RENEW BULL PINION.** To remove the bull pinion, remove the brake support and pinion assembly as outlined in paragraph 202, then proceed as follows: Remove brake drum cover and brake drum from outer end of bull pinion shaft. Cut the locking wire, then unbolt and remove brake backing plate assembly from outer end of brake support housing.

Press the bull pinion shaft out towards inner end of brake support housing. Remove snap ring (6—Fig. 155) from bull pinion, then press pinion from bearing (7). Pry oil seal (5) from inner end and oil seal (1) from outer end of housing (4). Remove snap ring (2), then remove bearing (3) from housing.

Reassemble by reversing disassembly procedure. Install oil seals (5 and 1) with lips toward inner end of brake support housing.

Models "FSM" and "New FSM"

204. **R&R BULL PINION.** To remove either bull pinion, first remove the disc brake housing as outlined in paragraph 216. Remove the oil baffle at outer side of bull pinion bearing.

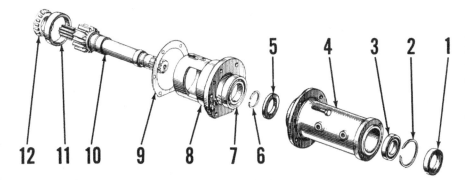

5 4 3 2 1

12 11 10 9 8 7 6

Fig. 155—Exploded view of model "FMD" and "FPM" bull pinion shaft and housings. Cut-away view is shown in Fig. 154.

1. Oil seal	6. Snap ring
2. Snap ring	7. Bearing
3. Bearing	8. Bull pinion
4. Brake support	housing
5. Oil seal	9. Shims
	10. Bull pinion

11. Differential carrier bearing race
12. Bearing cone & rollers

The bull pinion and ball bearing assembly can then be withdrawn from bull pinion housing (differential carrier bearing support). To disassemble left bull pinion, remove the snap ring, then press bull pinion from ball bearing. On right bull pinion, the bearing is removed in same manner; however, refer to paragraph 192 for information concerning the differential lock assembly incorporated in the right bull pinion.

To reinstall bull pinion, reverse removal procedure.

BULL GEARS
All Models

205. To renew a bull gear or rear axle inner bearing, first remove axle shaft as outlined in paragraph 206. Remove hydraulic pump assembly from pedestal in bottom of rear axle center housing. The bull gear (or gears) can then be lifted from housing. Remove rear axle bearing cone and roller from outer hub of bull gear as shown in Fig. 158 and the bearing

Fig. 157—Removing nut from inner end of rear axle. Note that both rear axles and bull gear retaining nuts have left hand threads. Axle shaft bearing preload is adjusted by nut; refer to paragraph 207.

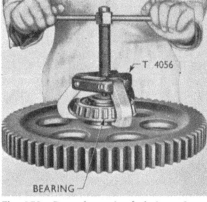

Fig. 158—Removing axle shaft inner bearing from hub of bull gear.

cup from inner end of axle housing; refer to Fig. 160.

NOTE: Bull gears used in late models have a different. tooth form. and are smaller in diameter than bull gears used in early models and the gears are not interchangeable. If in doubt, the gears can be identified by measuring their outside diameter. Bull gear (part No. E27N-4258-C) for models "FMD" and "FPM" measures 17.65-17.66 inches; whereas bull gear (part No. DDN-4258) for models "FSM" and "New FSM" measures 17.52-17.53 in diameter.

Fig. 159—Rear axle oil seal retainer (13) is staked into groove in rear axle housing (7). Axle outer bearing is lubricated from grease fitting (F). Refer to Fig. 160 for exploded view of rear axle (15), housing and related parts.

Drive bearing cone firmly against shoulder of bull gear and install cup in axle housing making sure cup is fully seated. Position bull gear in housing, then reinstall hydraulic pump and rear axle.

REAR AXLES AND HOUSINGS
All Models

206. **AXLE SHAFT, OUTER BEARING AND OIL SEAL.** To renew the axle shaft, outer bearing and/or axle shaft seal, proceed as follows:

Remove the hydraulic lift cover as outlined in paragraph 233 or 261. Drain rear axle center housing lubricant. Support rear of tractor and remove wheel. Refer to exploded view of the assembly in Fig. 160 and pry edge of seal retainer (13) out of groove (G) in outer end of rear axle housing. Refer also to Fig. 159. Remove the cotter pin (1—Fig. 160) and unscrew the axle shaft nut (2) at inner side of bull gear. Note: The axle shaft nut has left-hand threads. Withdraw axle, bearing and oil seal assembly from axle housing and bull gear.

Bend or cut oil seal retainer so that a bearing puller attachment can be fitted at outer side of bearing cone (9) and pull bearing from shaft. Remove bearing cup (8) from outer end of housing and oil seal retainer assembly from axle shaft.

Soak new felt (12) and oil seal (11) in oil, then install them in a new retainer (13) and place the retainer and seal assembly on axle shaft. Pack

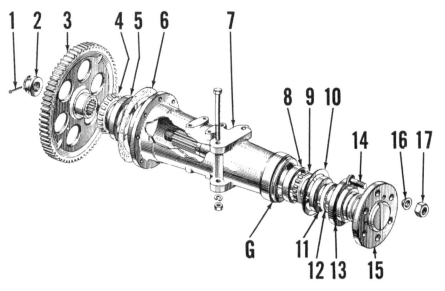

Fig. 160—Exploded view of rear axle, housing and related parts. On late production, rear wheel nuts (17) and tapered retainers (16) are made in one piece. Oil seal retainer fits over outer end of axle housing; refer to Fig. 159.

1. Cotter pin	4. Axle inner bearing	8. Bearing cup	12. Felt
2. Nut (L.H. threads)	5. Bearing cup	9. Axle outer bearing	13. Seal retainer
3. Bull gear	6. Gasket	10. Gasket	14. Wheel bolts
	7. Axle housing	11. Oil seal	15. Axle shaft
			16. Tapered retainers
			17. Nuts

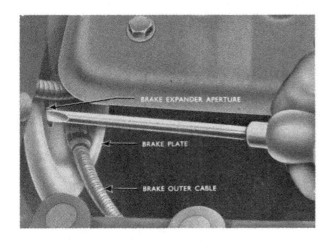

Fig. 161—Minor adjustment of model "FMD" and "FPM" brake shoes. Refer to text.

BRAKES

Models "FMD" and "FPM"

210. Models "FMD" and "FPM" are equipped with expanding shoe type brakes with brake drums mounted on outer ends of the bull pinion shafts.

211. **MINOR ADJUSTMENT.** Remove the adjusting hole cover from backing plate (see Fig. 161) and rotate notched adjuster wheel towards front of tractor until shoes drag on drum, then back adjuster wheel off until shoes are free. If brakes do not hold satisfactorily after completing minor adjustment, proceed with major adjustment as outlined in following paragraph.

212. **MAJOR ADJUSTMENT.** To adjust the brake shoe anchor pin, proceed as follows: Remove brake drum cover. Loosen the anchor pin nut and turn adjuster wheel (see paragraph 211) forward until shoes are tightly expanded against drum. Tap against anchor pin nut to be sure shoes are centralized, then, if possible, tighten adjuster wheel further. Securely tighten anchor pin nut, then back off adjuster wheel until there is 0.008 clearance between drum and shoes around complete circumference of shoes. Clearance can be measured with feeler gage inserted through slots in brake drum.

213. **STEADY POST ADJUSTMENT.** The brake shoe hold down springs hold the shoes against adjustable steady posts; refer to Fig. 163. The steady posts should not require adjustment unless renewing or relining the brake shoes; refer to paragraph 214 for procedure.

bearing cone and roller assembly with grease, then using a length of pipe, drive cone firmly against shoulder on shaft. Install bearing cup in outer end of axle housing.

207. To properly adjust axle shaft bearing preload, the bull pinion should first be removed as outlined in paragraph 202 or 204, then, install axle shaft and adjust bearing preload as follows:

Place new gasket (10—Fig. 160) in oil seal retainer and install axle shaft through housing and into bull gear. Drive the oil seal retainer over outer end of axle housing as shaft is moved into place, then stake edge of retainer down into groove in housing at diametrically opposite points. Install axle shaft nut and tighten nut so that a torque of 40-45 inch-pounds is required to turn axle shaft in bearings and oil seal. Secure the nut with new cotter pin when bearing preload is correct. Reinstall bull pinion as outlined in paragraph 202 or 204 and refill rear axle center housing with lubricant. Reinstall hydraulic lift cover as outlined in paragraph 233 or 261 and reinstall rear wheel.

208. **AXLE INNER BEARING.** The axle shaft inner bearing cone and roller assembly is pressed onto outside hub of the final drive bull gear and bearing cup is in inner end of axle housing. Refer to paragraph 205 for bull gear and/or bearing renewal.

209. **AXLE HOUSING.** With axle shaft removed as outlined in paragraph 206, disconnect wiring to fender mounted light, remove fender from axle and disconnect hydraulic lift lower link check chain. The axle housing can then be unbolted and removed from center housing.

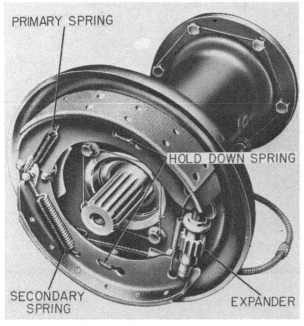

Fig. 162—View of model "FMD" and "FPM" brake assembly with brake drum cover and drum removed. Refer also to Fig. 163.

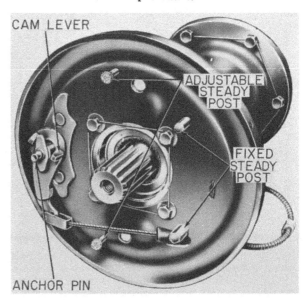

Fig. 163—View of brake backing plate assembly with brake shoes removed. Refer also to Fig. 162.

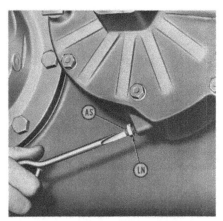

Fig. 164 — Adjusting model "FSM" and "New FSM" brakes. Refer to text and also to Fig. 165.

214. OVERHAUL. Remove tractor rear wheels, brake drum covers and brake drums. Unhook the shoe hold down springs from shoes and the two return springs from anchor post. Then, lift off the brake shoes, adjuster and retractor spring.

Brake linings and rivets are available for relining early type shoes with riveted linings or bonded shoe and lining assemblies are available for servicing all models with shoe type brakes. Upper and lower brake shoes are interchangeable.

Inspect the return, retracting and hold down springs and renew any that are broken, cracked, rusted or distorted. Inspect brake cables, cam levers, steady posts and anchor pins and renew as required. Note: Hole in backing plate for anchor pin is elongated to permit centralizing the brake shoes.

To install new or relined shoes, proceed as follows: Fit square ends of each shoe in adjuster slots, connect the retracting (orange) spring at adjuster ends of shoes, then place the assembly on backing plate as shown in Fig. 162. Connect the shoe hold down springs and anchor cups with long ends of springs in brake shoes. Connect the red (secondary) return spring with long end to anchor pin and short end in second hole in lower shoe. Connect the black (primary) return spring with long end over anchor pin and short end in first hole of upper shoe. Loosen the anchor pin nut, back off the adjusting screw (retract shoes) and install the brake drum. Turn adjuster forward to expand shoes tightly against the drum. Tap on anchor pin nut to center the shoes, then tighten adjuster further if possible. Tighten anchor pin nut, then back off adjuster until shoes just clear the drum. Loosen nuts on adjustable steady posts and slowly back the posts out until brake drum just drags on the brake shoes. Counting the turns, turn the

Fig. 165—Adjusting disc brake free pedal. Refer also to Fig. 164 and to text.

steady posts in until brake drum just drags on shoes, then back steady posts out one-half number of turns counted and tighten the lock nuts.

Reinstall brake drum covers and rear wheels when adjustment is completed.

Models "FSM" and "New FSM"

215. ADJUSTMENT. To adjust the disc type brakes, first refer to Fig. 164 and loosen the locknut (LN) on adjusting screw (AS) protruding from lower side of brake housing. Turn the adjusting screw in until brake is fully locked, then back screw out 1½ turns

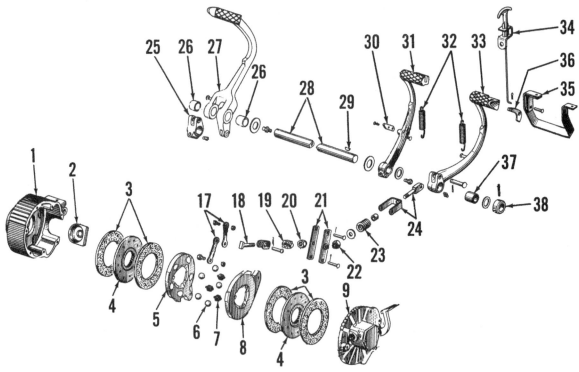

Fig. 166—Exploded view of model "FSM" and "New FSM" right brake assembly and actuating pedals and linkage. Left brake assembly is similar except that cover (9) does not contain differential lock operating linkage. Refer also to Figs. 163, 164 and 165.

1. Brake housing	6. Steel balls	19. Pivot block
2. Grommet	7. Return springs	20. Adjusting nut
3. Friction linings	8. Actuating disc	21. Operating levers
4. Brake disc	9. Brake cover	23. Spring
5. Actuating disc	17. Links	24. Rod and yoke
	18. Pull rod	

25. Cross-shaft lever	30. Latch	33. Right brake pedal
26. Bushings	31. Left brake pedal	34. Parking lock
27. Clutch pedal	32. Pedal return springs	35. Pedal stop
28. Cross-shaft		36. Latch
29. Woodruff key		37. Bushing
		38. Collar

and while holding screw in this position, tighten locknut. Then, remove the brake return spring (32—Fig. 166). Refer to Fig. 165; tighten the pull rod nut (20) until brake pedal contacts upper stop, then back nut off 1½ turns. This should give a brake free pedal of 1½ inches measured at pedal pad. Reconnect brake pedal return spring and adjust opposite brake in similar manner.

216. **OVERHAUL DISC BRAKES.** To gain access to brakes, unbolt and remove the platform (foot rest) from brake housing and support bracket. On the left hand platform, it will also be necessary to disconnect clutch pedal return spring.

Unbolt and remove the brake housing cover; right hand brake housing cover (9—Fig. 166) and differential lock pedal, shaft and cam are removed as an assembly. Withdraw the outer friction disc assembly (3 & 4), disconnect and remove the actuating plate assembly complete with links, operating rod and dust cover. Then, withdraw the inner friction disc assembly. If necessary to remove brake housing, be careful not to disturb the bull pinion housing and shims.

To disassemble the actuating plate assembly, remove rubber dust cover (2), remove the bolts securing the links (17) to plates (5 & 8), then remove the three springs (7), separate the plates and remove the five steel balls (6). Thoroughly clean and inspect all parts and renew any parts which are excessively worn or damaged. To reassemble, lay one actuating plate with inner face up on bench, place the five steel balls in pockets on plate, then place the other plate face down on top of the steel balls. NOTE: **Do not** lubricate either the steel balls or their pockets in the actuating plates. Install the three plate return springs making sure that they

are correctly located on lugs of plates. Otherwise the bull pinion shaft may rub against them when the assembly is reinstalled. Reconnect the links and pull rod (18). Stake the nuts to bolts that secure the links to actuating plates.

Brake disc linings (3) are available separately from discs (4) and the disc and linings are also available as an assembly. Renew the linings or disc assembly if excessively worn or oil soaked.

Check the oil seal in brake housing and if in doubt, renew the seal. Renew brake housing if friction surface is excessively worn or scored. Using a petroleum solvent, flush out the bolt holes in rear axle center housing and wash the brake housing retaining cap screws, then air dry the bolt holes and cap screws. Be sure that holes in shims behind the bull pinion housing flange are aligned with holes in hous-

ings, then reinstall brake housing. Apply a thin line of "Loctite" to each of the cap screw threads, then install and tighten the cap screws to a torque of 65-70 Ft.-Lbs.

Install inner friction disc, actuating plate assembly, outer friction disc and brake cover or cover and differential lock operating pedal assembly. Adjust brakes as outlined in paragraph 215.

If necessary to remove or overhaul the brake pedals, cross shaft and linkage, refer to exploded view in Fig. 166 as disassembly and reassembly guide. Note: In order to remove cross shaft, it is first necessary to drain the rear axle center housing lubricant to below shaft level and remove right rear wheel (unless wheel is mounted in extended position). It is possible to renew the cross shaft oil seals in rear axle center housing without removing the cross shaft.

BELT PULLEY

All Models So Equipped

217. **REMOVE AND REINSTALL.** First, either drain transmission to below level of belt pulley opening or tilt tractor so that level is below opening. Unbolt and remove belt pulley assembly from transmission housing.

Before reinstalling pulley, engage shift collar and check backlash in drive coupling by holding the bevel gear and turning pulley back and forth. Note backlash present at pulley rim, then reinstall the belt pulley assembly with sufficient number of gaskets between pulley housing and transmission to obtain an additional ⅛-inch backlash at pulley rim. Usually, one to three gaskets are required. The belt pulley shift collar should be kept in neutral position except when required to transmit power.

218. **OVERHAUL.** To disassemble belt pulley unit, refer to exploded view in Fig. 168 and proceed as follows: Remove pulley guard (6) and the shift lever (10). Remove nut from outer end of shaft (17), then remove pulley and hub assembly. Bump outer end of shaft to remove shaft, gear (20) and bearing (19) assembly from inner end of housing. Shift collar (18) can then be removed from housing. Remove pin (14) from lever (12), then remove lever, spring (15) and block (13) from inside of housing. Remove snap ring (22) and washer (21), then withdraw shaft (17) from gear (20). Bearing (19) can be removed by inserting punch through holes in inner end of gear and driving bearing from gear hub. Remove seal (7) and bearing (8) from outer end of housing. Remove hub (3) from pulley.

Fig. 167—Installing brake housing.

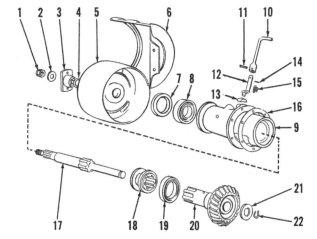

Fig. 168—Exploded view of belt pulley assembly.

1. Nut
2. Washer
3. Adapter
4. "O" ring
5. Pulley
6. Guard
7. Oil seal
8. Bearing
9. Housing
10. Shift lever
11. Pin
12. Shift arm
13. Thrust block
14. Pin
15. Spring
16. Gasket
17. Pulley shaft
18. Shift collar
19. Bearing
20. Bevel gear
21. Washer
22. Retaining clip

Inspect all parts and renew any that are excessively worn or damaged. Gear (20) is fitted with two renewable bushings. Shaft (17) should turn freely in bushings and there should be some end play of gear on shaft with thrust washer and retaining snap ring installed.

Reassemble unit as follows: Install outer bearing (8) in housing, then install new oil seal with lip of seal towards bearing. Install inner bearing on bevel gear, then insert shaft through gear and secure with thrust washer and snap ring. Install shift lever (12) and spring (15) in housing and secure with pin (14). Place block (13) on inner end of shift lever, then locate shift collar in housing with block engaged in groove of collar. Install shaft through collar and bump inner end of gear until bearing (19) is seated against shoulder in housing. Lubricate inner hub of belt pulley and install pulley on shaft. Install hub (3) with new gasket (4) and securely tighten nut (1) and hub retaining cap screws. Check to see that with shift collar disengaged, shaft will turn freely in bevel gear.

POWER TAKE-OFF

PTO DRIVE UNIT (GEARBOX)

All Models

219. REMOVE AND REINSTALL. Drain transmission and rear axle center housing and remove PTO output shaft as outlined in paragraph 224, or if so equipped, remove the raised PTO unit and driveshaft as outlined in paragraph 225. Disconnect shift lever link from shift rail (23—Fig. 170). Support the drive unit while unbolting it from bottom of transmission housing, slide the unit forward until "O" ring (13) boss is clear of bore in transmission housing, then lower drive unit from tractor.

When reinstalling drive unit, fit a new "O" ring (13) on rear end of housing and a new gasket on top face of housing. Lubricate the "O" ring, raise unit to bottom face of transmission housing, slide unit rearward into bore of housing and secure with the two dowel type cap screws in left and right center holes. Install plain cap screws in the remaining holes, then tighten all cap screws evenly and securely.

220. OVERHAUL. With unit removed as outlined in paragraph 219, proceed as follows:

221. IDLER GEAR. Remove cotter pin retaining idler gear shaft (12—Fig. 170) in housing (11), push shaft from housing and remove the idler gear and, on model "New FSM", the two "D" shaped thrust washers.

Idler gear for model "FMD" was fitted with two bushings. Renew bushings and/or shaft if clearance is excessive. Bushings are pre-sized and should not require reaming if carefully installed.

Model "FPM" and early model "FSM" idler gears were fitted with two caged needle roller bearing assemblies (8—Fig. 171). Late model "FSM" idler gear was fitted with retaining snap rings (7) to keep bearings from moving out of the gear. Model "New FSM" idler gear is fitted with thrust spacer (10) between the bearing cages. When renewing the caged needle bearings, drive or press on lettered end of bearing cage only; if lettering is not visible after installing bearing, renew the bearing. On idler gear without retaining snap rings, lettered end of bearing cage should be 13/64-inch below flush with end of gear. For gears with retaining rings, press bearing cages in just far enough to install the snap ring in groove of gear.

Latest (model "New FSM") gear and bearing assembly may be used to renew all model "FPM" and "FSM" gears, but the "D" shaped thrust washers cannot be used in the earlier housings. To install needle roller bearing idler gear and shaft in model "FMD" in place of the bushing type idler requires installing new housing.

To reinstall idler gear, place gear and bearing (or bushing) assembly in housing with long hub to rear. On model "New FSM", insert thrust washer at each side of idler gear. Insert shaft through the housing and gear and secure shaft with cotter pin. Note: If installing new idler gear, check for proper clearance between front hub of idler gear and cluster gear on primary gear box lower shaft after PTO drive housing assembly is reinstalled. Remove material from hub as necessary to obtain clearance at (C —Fig. 171).

222. SELECTOR COVER, SHAFT AND FORK. Unbolt and remove cover (22—Fig. 170) from housing. Cut locking wire and remove set screw (20), then remove shaft (23) from cover and fork. Take care not to lose detent ball (25) or spring (26) as shaft is being removed. Remove oil seal (24) from cover.

Shifter fork (19) is not interchangeable between model "New FSM" and earlier models. Inspect all parts and

Fig. 169—Cross-sectional view showing PTO drive gear train. Single clutch model is shown; on dual clutch model, PTO input shaft is separate from transmission input (main drive) shaft.

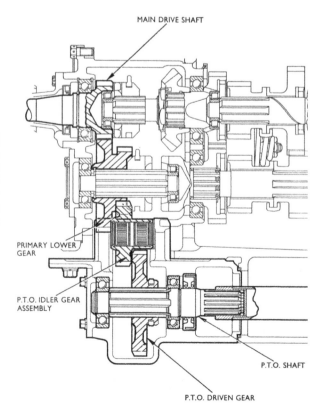

MAIN DRIVE SHAFT

PRIMARY LOWER GEAR

P.T.O. IDLER GEAR ASSEMBLY

P.T.O. SHAFT

P.T.O. DRIVEN GEAR

renew any that are excessively worn or damaged. Reassemble using new oil seal as follows:

Drive new oil seal into bore of cover with metal face out. Position shifter fork with boss to front of cover (away from seal). Insert shaft through cover and fork, insert detent spring and ball in bore of shaft, then push shaft into place. Rotate shaft so that detent ball drops into place, then install set screw in fork and shaft and secure screw with lock wire. Reinstall cover, shaft and fork assembly with new gasket.

223. DRIVEN GEAR, SHAFT AND BEARINGS. With idler gear removed as outlined in paragraph 221 and selector cover assembly removed as in paragraph 222, proceed as follows:

Remove front bearing plate (1—Fig. 170) and bump shaft (17) forward until front bearing (4) is exposed. Remove snap ring (3) from front end of shaft and remove front bearing using suitable bearing puller. Bump the shaft, rear bearing (15) and oil seal (18) out towards rear of housing and remove driven gear (6). Remove oil seal from shaft. If necessary to renew shaft or bearing, remove snap ring (14) at front side of bearing and press bearing from shaft. Note: Model "New FSM" has retaining snap ring (16) at rear of bearing; earlier models retain

bearing between snap ring (14) and shoulder on shaft.

With early production shaft (one snap ring), either a double lip seal (18) or two single lip seals may be used. When installing two single lip seals, lip of each seal must face away from other seal. The late type shaft (with two snap rings) may be used to renew early type shaft if corresponding oil seal is used. Model "New FSM" driven gear has 32 teeth; earlier models have gear with 26 teeth.

To reassemble, proceed as follows: On late type shaft, install rear snap ring (16). Then, on either shaft, press rear bearing (15) on from front end of shaft until it is tight against shoulder or rear snap ring. Install snap ring (14) at front side of bearing. On all models except "New FSM", install front bearing snap ring (5) in groove of bearing bore in housing. Press front bearing (4) into place against snap ring or, on model "New FSM", against shoulder in housing. Place driven gear (6) in housing with shifter groove rearward, then insert shaft through housing and gear. While supporting front bearing, press shaft and rear bearing into place and install snap ring (3) and cover (1) with new gasket (2). Using a seal protector or shim stock and suitable seal driver, install shaft seal or, on early models, the alternate two single lip seals.

STANDARD PTO OUTPUT SHAFT
All Models

224. REMOVE AND REINSTALL. First, drain rear axle center housing lubricant, then remove cap screws retaining output shaft bearing support (40—Fig. 172) in rear axle center housing and withdraw the output shaft assembly.

Output shaft (34) for model "New FSM" has 2 ⅛ inch O.D. for increased torque capacity and can be used as service replacement for earlier shaft having 2 inch O.D.

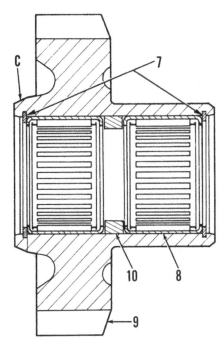

Fig. 171 — Cross-sectional view of latest type PTO idler gear and bearing assembly. Refer to Fig. 170 for exploded view. When installing gear in earlier models, it may be necessary to remove some material from gear hub at (C) for clearance. Refer to Fig. 170 for legend.

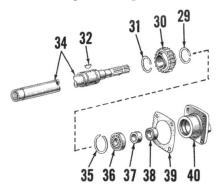

Fig. 172—Exploded view of standard PTO output shaft assembly.

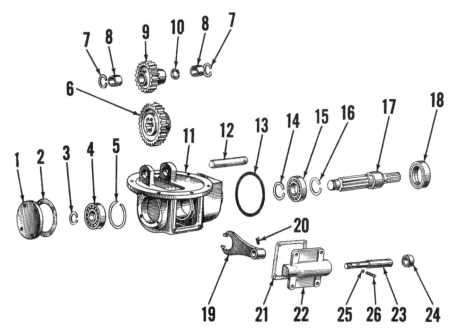

Fig. 170—Exploded view of PTO gearbox. Refer to Fig. 171 for cross-sectional view of latest type idler gear and bearing assembly and to Fig. 169 for cross-sectional view of complete drive gear train.

1. Cover plate	8. Bearing	14. Snap ring	20. Set screw
2. Gasket	9. Idler gear	15. Bearing	21. Gasket
3. Snap ring	10. Spacer	16. Snap-ring	22. Shift cover
4. Bearing	11. Housing	17. PTO drive shaft	23. Shift rail
5. Snap ring	12. Idler shaft	18. Oil seal	24. Seal
6. Driven gear	13. "O" ring	19. Shift fork	25. Detent ball
7. Snap ring			26. Detent spring

29. Snap ring	35. Snap ring
30. Hydraulic pump drive gear	36. Bearing
	37. Sleeve
31. Snap ring	38. Oil seal
32. Woodruff key	39. Gasket
34. Output shaft	40. Bearing support

Hydraulic pump drive gear (30) for model "New FSM" has 34 teeth whereas gear for earlier models has 33 teeth. To remove gear, remove front snap ring (31) and bump gear off front end of shaft.

To renew oil seal (38), remove snap ring (35) from front end of the bearing retainer (40) and bump retainer off rear end of shaft. Install new oil seal in retainer, then reinstall retainer and secure with snap ring.

If necessary to renew bearing (36), remove bearing retainer and oil seal. Grind the collar (37) thin enough that it can be cracked with chisel, then pull bearing and collar from rear end of shaft. Install new bearing, then heat collar to dull red (approximately 800° F.) and install on shaft next to bearing. After allowing collar to cool slowly, install bearing retainer.

When reinstalling the assembled output shaft, use new gasket (39). With PTO shift lever in engaged position, rotate shaft until splines are aligned and hydraulic pump gears are in mesh. then push shaft into place. Install and securely tighten the four retaining cap screws.

RAISED PTO OUTPUT UNIT

Purpose of the raised PTO unit is to raise the position of the PTO output shaft to meet ASAE specifications for distance of PTO output shaft above tractor drawbar. The unit also incorporates an additional engagement lever which allows the hydraulic pump to be driven by the PTO extension shaft from PTO gearbox without also turning the PTO output shaft. On model "New FSM", raised PTO output shaft turns at same speed as the PTO extension shaft (or standard PTO output shaft); on all earlier models, raised PTO output shaft speed to extension shaft speed ratio is 1:1.33.

All Models So Equipped

225. REMOVE AND REINSTALL. First, drain rear axle center housing lubricant and remove PTO output shaft guard. Remove the two lower retaining cap screws and lock washers, then unscrew the two upper cap screws and remove unit from tractor. NOTE: Do not remove the upper right cap screw (19—Fig. 174) as it also functions as the rail for shifter fork (17). If unit is not to be disassembled, retain cap screw with nut until unit is to be reinstalled.

With the raised PTO unit removed, the extension shaft (33), bearing and drive gear assembly (see Fig. 175) can be withdrawn from rear axle center housing.

To reinstall, proceed as follows: Insert extension shaft, bearing and drive gear assembly into rear axle center housing and rotate shaft to align splines and hydraulic pump drive gears, then push the assembly into place. Install the raised PTO unit using a new gasket (1—Fig. 174) and securely tighten the retaining cap screws. NOTE: Special rubber bonded washers must be used on the two upper cap screws.

226. OVERHAUL. With the raised PTO unit and extension shaft assembly removed as outlined in paragraph 225, proceed as follows:

To disassemble raised PTO unit, push down on shift lever and remove lever pivot bolt (7—Fig. 174), then withdraw the lever. Remove upper right cap screw (shift rail) (19) taking care not to lose detent ball (15) or spring (16) and remove shift fork (17). Drive idler gear shaft (21) out rear of housing and remove idler gear (20). Unbolt and remove output shaft bearing retainer (13) with gear (9) and bearings (8 and 10). Bump retainer from rear bearing and remove output shaft seal from retainer. Remove bearings from output shaft using suitable bearing puller.

On model "New FSM", extension shaft differs from that shown in Fig. 175 in that a bearing and retaining snap ring are fitted on rear end of shaft instead of the flat washer, nut and cotter pin. To disassemble extension shaft, remove the snap ring or cotter pin, nut and washer and pull drive gear from rear end of shaft. Remove the retaining snap ring (28—Fig.

Fig. 175—View of PTO drive shaft and drive gear with raised PTO assembly removed. Latest shaft has gear retaining snap ring instead of nut.

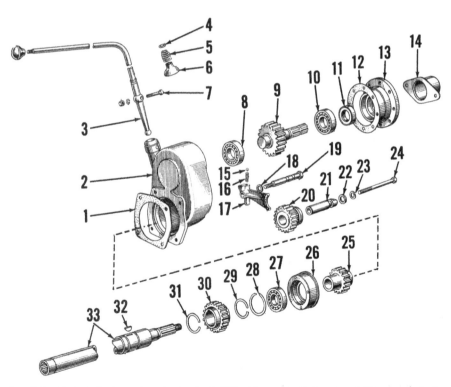

Fig. 174—Exploded view of typical raised PTO drive unit. Unit raises PTO output shaft to within ASAE specificatons as measured from drawbar hitch point.

1. Gasket	9. Output shaft	16. Detent spring	25. Drive gear
2. Housing	& gear	17. Shift fork	26. Bearing support
3. Shift lever	10. Bearing	18. Seal	27. Bearing
4. Snap ring	11. Seal	19. Shift rail	28. Snap ring
5. Spring	12. Gasket	20. Idler gear	29. Snap ring
6. Cover	13. Bearing support	21. Idler shaft	30. Hydraulic pump
7. Pivot bolt	14. Output shaft	22. Seal	drive gear
8. Bearing	cover	23. Lock washer	32. Woodruff key
	15. Detent ball	24. Cap screw	33. PTO drive shaft

174) and remove retainer (26), then remove snap ring and pull bearing (27) from rear end of shaft. The hydraulic pump drive gear (30) can be removed by first removing snap ring (31) and driving gear off front end of shaft without disturbing PTO drive gear or bearings if so desired.

NOTE: Late production service installed raised PTO units for earlier models (prior to "New FSM") have features of the later ("New FSM") unit except the 1:1.33 gear ratio is retained.

To reassemble, reverse disassembly procedure.

HYDRAULIC SYSTEM
(Models "FMD" & "FPM")

Hydraulic system for model "FMD" and "FPM" incorporates only lift, lower and hold control of the three-point hitch. An attachment is available to provide "position control" where any leak-down of the lift system will move the control lever to "raise" position until the lift arms are returned to set position. A remote cylinder port ("jack tapping") is provided in the control valve housing for remote control of single acting cylinder. To operate a cylinder from this port, the hydraulic system control lever is used and the three-point hitch lift arms are tied in raised position.

Differential and final drive lubricant is utilized for hydraulic fluid. Gear type hydraulic pump is mounted in rear axle center housing and is driven by a gear on the PTO output shaft. Production changes have been made from time to time and where changes affect parts procurement or service procedure, it will be noted in the text.

FLUID AND FILTERS
Models "FMD" and "FPM"

227. Rear axle and final drive lubricant is used for hydraulic fluid; capacity is 10.8 gallons. SAE 80 Mild E. P. (Ford specification M-4864-A) or SAE 90 Mild E. P. (Ford specification M-4864-B) gear lubricant is recommended. After each 12 months of operation, the rear axle center housing should be drained and flushed, the pump suction screen removed and cleaned and the system refilled with proper new lubricant. A dipstick located at left rear corner of the transmission housing is for checking final drive lubricant level.

TROUBLE-SHOOTING
Models "FMD" and "FPM"

228. **WILL NOT LIFT.** First, check oil level in final drive housing and be sure that the PTO output shaft is engaged and turning. Be sure that the lift arms will move up and down. Check the hydraulic system relief pressure as outlined in paragraph 231; if pressure is OK, check for broken ram cylinder piston rod or broken ram lift arm on lift shaft. If system relief pressure is low, or there is no pressure, check for worn or broken hydraulic pump, clogged pump suction filter, broken pressure pipes, broken lift cylinder, broken check (non-return) valve spring, broken pressure relief (unloading) valve spring or sticking valve.

229. **WILL NOT HOLD.** If the lift arms settle with control lever in neutral, or if implement gains depth when working in field, a leak is occuring in the lift cylinder or control valve unit. If the cylinder leaks down rapidly check for ruptured piston seal or for check (non-return) valve and/or pressure relief (unloading) valve not seating. If the cylinder leaks down slowly, piston seal may be nicked or worn, check (non-return) valve or seat may be damaged, pressure relief (unloading) valve or seat may be damaged or the control valve spool may be worn or fit too loosely.

230. **SLOW IN LIFTING.** If the lift arms raise slowly, but will remain in raised position with control valve in neutral, a worn or damaged hydraulic pump is indicated. Note: Leakage that will allow lift arms to fall rapidly will also cause slow rate of lift; refer to paragraph 228.

ADJUSTMENTS
Models "FMD" and "FPM"

231. **SYSTEM RELIEF PRESSURE.** Operate the tractor until final drive lubricant is at normal operating temperature. With the lift arms in lowered position, remove the remote cylinder (jack tapping) plug from left side of control valve body and connect a "Tee" fitting, 0-3000 psi pressure gage, shut-off valve and hose to the open port. Insert other end of hose in rear axle center housing filler cap at rear of hydraulic lift cover. With shut-off valve open and engine running at fast idle speed, hold the hydraulic control lever in raising position and slowly close the shut-off valve. The lift arms will raise to their uppermost position and pressure gage reading will raise. Continue to slowly close the shut-off valve until the relief (un-loading) valve unseats and the pressure gage reading drops. The highest gage reading obtained is the system relief pressure.

On model "FMD" (except very late production), the check valve will seat as the pressure relief valve opens and the lift arms will be held in raised position. However, flow of oil from the pump will hold the pressure relief (unloading) valve open as long as the control valve is in raising position and the pump pressure will drop to approximately 40 psi. In order to "reset" the relief valve, it will be necessary to return the control valve to neutral position.

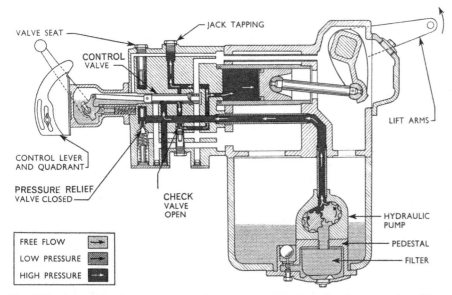

VALVE SEAT

JACK TAPPING

CONTROL VALVE

LIFT ARMS

CONTROL LEVER AND QUADRANT

PRESSURE RELIEF VALVE CLOSED

CHECK VALVE OPEN

HYDRAULIC PUMP

PEDESTAL

FILTER

FREE FLOW	→
LOW PRESSURE	→
HIGH PRESSURE	→

Fig. 176—Schematic diagram of model "FMD" and "FPM" hydraulic system in raising position.

On model "FPM" and some late production model "FMD" tractors, the design of the pressure relief valve and valve housing were changed and the pressure drop will be very slight when the relief valve is in open position. Therefore, pressure will be maintained in the lift cylinder circuit and the check (non-return) valve will not close until the control valve is returned to neutral position. At that time, the pressure gage reading will drop to approximately zero.

Recommended pressure relief setting for both models "FMD" and "FPM" is 2000-2200 psi. However, where operating conditions are abnormal or heavier than average implements are used, it is permissible to increase the pressure relief setting to 2500-2700 psi. This will be especially advantageous with model "FMD" tractors where shock loading will cause the relief (unloading) valve to "pop-off" before the implement is fully lifted. Note: On model "FMD" tractors, it will also help to reduce engine speed before attempting to lift heavy implements on the three-point hitch.

To adjust pressure relief setting, remove the left (opposite from control lever) plate from top side of control valve housing and remove the valve, spring, shim and adjuster assembly; refer to Figs. 178 and 180 for the three types of assemblies that may be encountered. Add or remove shims at location shown as necessary. Shims are available in thicknesses of 0.016 and 0.028. When reinstalling late type valve with ball at tip, stick ball to valve with heavy grease. Install new "O" ring on adjuster ("FMD") or on relief valve stop ("FPM"). Recheck relief pressure and readjust if necessary. If adding shims does not increase pressure reading, service the hydraulic pump. Refer to paragraph 242 for overhaul data.

232. RATE OF LOWERING (DROP) ADJUSTMENT. The control valve spool incorporates a series of five cylinder exhaust ports (early "FMD") or an annular recess on inner end of spool and a series of three cylinder exhaust ports (late "FMD" and all "FPM") as shown in Fig. 183. The farther the control valve lever is pushed down, the greater the exhaust port area and thus a faster implement drop is obtained by pushing the lever down farther past neutral position.

To adjust rate of lowering for a particular implement, start with the implement in raised position and the control lever in neutral position. Then,

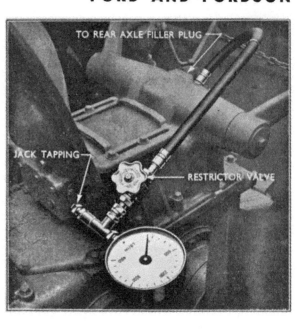

Fig. 177—Pressure gage, shut-off (restrictor) valve and return hose installed to check hydraulic relief pressure on model "FMD" or "FPM".

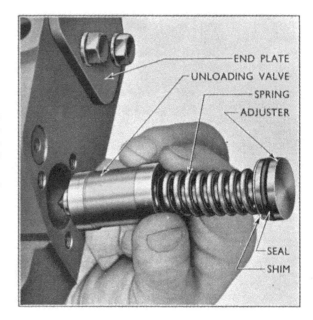

Fig. 178 — Early type pressure relief (unloading) valve assembly. Refer to Fig. 179 for relief valve seat and to Fig. 180 for later type valve assemblies.

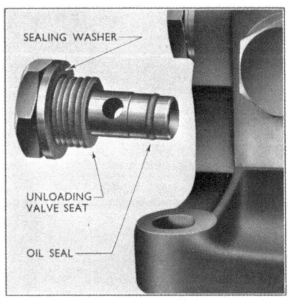

Fig. 179—View showing pressure relief (unloading) valve seat removed from bottom of valve housing.

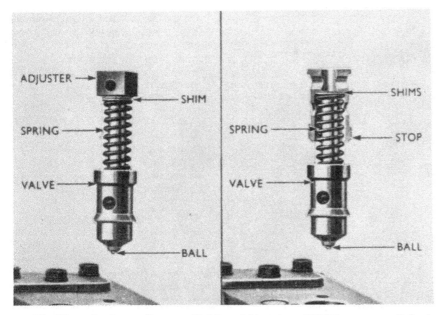

Fig. 180—Views showing earlier type (Left) and latest type (Right) pressure relief valve assemblies.

slowly push lever downward until desired rate of lowering is obtained and adjust the stop (43—Fig. 181) on control lever quadrant to this position.

OVERHAUL

Models "FMD" and "FPM"

233. **R&R LIFT COVER & CYLINDER ASSEMBLY.** Remove operator's seat and disconnect lift arms from lift links. Unbolt cover from rear axle center housing and remove with suitable hoist.

When reinstalling cover, renew the "O" rings on pump pressure pipe, then reinstall pipe in pump. Carefully

lower cover into position so that pipe enters bore in cover.

Note: Most overhaul work can be accomplished without removing lift cover from tractor. Refer to following paragraphs:

234. **R&R CONTROL VALVE UNIT.** Lower the lift arms, then unbolt and remove control valve unit from front of hydraulic lift cover. Lift cylinder and piston may be removed with valve or may remain in lift cover.

To reinstall, first remove horizontal pressure pipe and fit a new "O" ring at each end of pipe, lubricate the "O" rings and reinstall pipe in lift cover.

Reinstall lift cylinder and piston if removed with valve unit, fit a new "O" ring in inside diameter of cylinder and lubricate the "O" ring. Reinstall valve unit with new gasket, carefully placing unit casting over pressure pipe and into lift cylinder. Tighten retaining cap screws securely.

235. **CONTROL VALVE SPOOL AND LINKAGE.** To remove control valve, unbolt control lever support from valve housing and withdraw link and spool with support. The valve spool can then be removed from the link and the lever return spring and cam roller can be withdrawn from the lever support; refer to Fig. 181.

To disassemble lever, quadrant and support assembly, refer to exploded view in Fig. 184. Unbolt quadrant (33) and cover (42) from support (40) and remove set screw from valve lever (39). Then remove hand lever (31) and valve lever from support. When reassembling, renew the "O" rings (32) on hand lever shaft and install cover with new gasket (41).

The control valve spool (piston valve) is a select fit to the valve bore in housing. Valves are available in five different size ranges which are color coded as follows:

Color Code	Valve Diameter
Red	0.8685-0.8687
Yellow	0.8687-0.8689
Blue	0.8689-0.8691
Green	0.8691-0.8693
White	0.8693-0.8695

The largest diameter spool that will fit in the valve bore without binding should be selected. Note: When checking valve fit, the front cover of the valve housing should be installed and retaining cap screws tightened to a torque of 25-39 Ft.-Lbs. When installing a new valve housing, the valve bore in housing will be color coded to indicate correct size range of valve spool to be installed.

With proper size valve selected, attach the valve to control lever link, insert cam roller in bore of lever support and insert return spring in

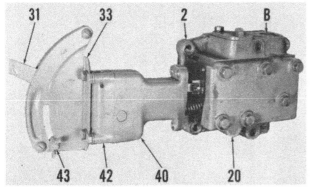

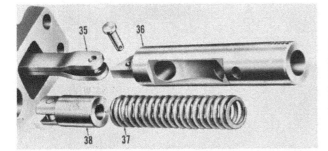

Fig. 181—View showing control valve lever support split from valve housing.

B. Relief valve bore
2. Valve housing
20. Cover plate
31. Control lever
33. Quadrant
40. Support
42. Cover
43. Rate of lower stop

Fig. 182—Control valve spool (36) can be removed from link (35) after removing lever support as shown in Fig. 181. Lever return spring (37) and follower (38) fit in bore in support.

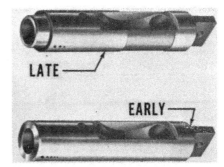

Fig. 183—View showing late and early type control valve spools.

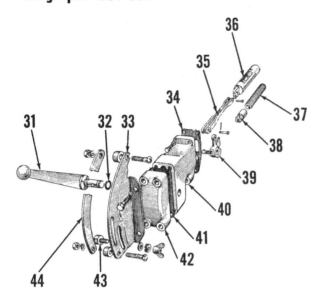

Fig. 184—Exploded view of control lever, support and control valve unit; refer to Fig. 181 for view showing unit split from valve housing.

31. Control lever
32. "O" ring
33. Quadrant
34. Gasket
35. Control link
36. Control valve spool
37. Lever return spring
38. Cam follower
39. Valve lever
40. Lever support
41. Gasket
42. Cover
43. Rate of lower stop
44. Lever guide

On late type relief valve, stick steel ball in tip of valve with heavy grease, then insert valve in bore. Install spring, adjusting shims and adjuster or stop. Install retaining plate with new gasket. Note: Refer to paragraph 231 for shim selection.

237. CHECK (NON-RETURN) VALVE. To remove the check (non-return) valve, first remove valve housing as outlined in paragraph 234, then remove the check valve assembly from bottom of housing. Refer to Fig. 187.

With check valve removed, carefully inspect valve seat. Seat should be renewed if excessively worn or damaged. Renew the check valve if grooved or damaged in any way. Renew spring if cracked, rusted or distorted, or if free height is not approximately equal to that of new spring.

Reinstall check valve assembly using a new sealing washer and securely tighten guide in valve housing.

238. LIFT CYLINDER AND PISTON. To remove lift cylinder and piston, first remove the valve housing as outlined in paragraph 234. If not removed with valve housing, withdraw cylinder and piston unit from

bore of valve housing. Then, reinstall the lever, support and valve assembly with new gasket between lever support and valve housing.

236. RELIEF (UNLOADING) VALVE AND SEAT. Three different types of pressure relief (unloading) valves may be encountered. Except for some late production units, model "FMD" systems were fitted with valve having tapered tip and adjuster with "O" ring seal as shown in Fig. 178. Early production model "FPM" and some late model "FMD" systems were fitted with a valve having a loose steel ball at the tip and a square adjuster as shown in left view in Fig. 180. To prevent the valve from raising high enough to allow the ball to become dislodged from the tip of valve, the adjuster was discontinued and a hollow valve stop was introduced in later production model "FPM" systems. If the square type adjuster is encountered, it should be renewed using the later type stop and small diameter adjusting shims. To renew the early "FMD" valve with tapered tip using the new steel ball tip valve, a new type valve housing is also required. Note: Only the late type valve housing will be available for service of both "FMD" and "FPM" systems.

For adjusting relief pressure, valve can be removed without removing valve housing from tractor. Remove the plate (7—Fig. 185) from left side of top face of valve housing, then remove the adjuster or stop, adjusting shims, spring and valve. Note: On models with steel ball at tip of valve, be careful not to lose the ball as it may stick to valve. If not removed with valve, extract the steel ball from valve bore.

With valve removed, carefully inspect valve bore for any deep score marks that would cause valve to stick and inspect valve seat for excessive wear or damage. To remove seat, first remove valve unit as outlined in paragraph 234, then unscrew seat from housing. To install seat, place new sealing ring on seat and install new "O" ring in groove on seat, then securely install seat in valve housing.

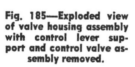
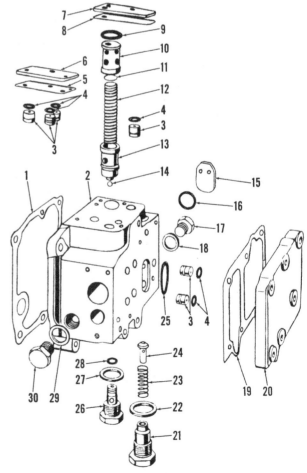

Fig. 185—Exploded view of valve housing assembly with control lever support and control valve assembly removed.

1. Gasket
2. Housing
3. Sealing plugs
4. "O" rings
5. Gasket
6. Retainer plate
7. Retainer plate
8. Gasket
9. "O" ring
10. Spring guide
11. Shims
12. Relief valve spring
13. Valve retainer
14. Relief valve ball
15. Retainer plate
16. "O" ring
17. Jack tapping (remote cylinder port)
18. Seal
19. Gasket
20. Cover plate
21. Check valve guide
22. Seal
23. Check valve spring
24. Check valve
25. "O" ring
26. Relief valve seat
27. Seal
28. "O" ring
29. Seal
30. Plug

lift cover. If difficulty is encountered, the rear access cover can be removed from lift cover and a drift can be used to bump cylinder forward out of bore in lift cover.

With cylinder and piston unit removed, push piston from rear end of cylinder. Renew cylinder if cracked, deeply scored or worn. Renew piston if deeply scored. To renew piston seal,

straighten lock tab washer on front end of piston and remove cap screw, washer and seal retainer. Fit new seal over front end of piston and check to see that seal retainer fits evenly; install new retainer if bent or otherwise damaged. Using new lock tab washer, secure retainer to front end of piston with cap screw and bend tab of washer against screw head. Install sleeve in bore of lift cover, then lubricate piston and with "O" ring removed from bore of cylinder, install piston as shown in Fig. 188. Install new "O" ring in cylinder bore and reinstall valve housing assembly; refer to paragraph 234.

239. **LIFT SHAFT AND ARMS.** To remove the lift shaft with lift cover installed, remove rear fender from one side of tractor. Also, depending on tire

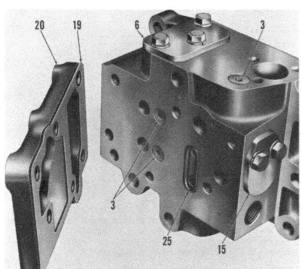

Fig. 186—View showing cover (20) removed from valve housing. Refer to Fig. 185 for legend.

Fig. 187—Removing check valve and guide assembly from bottom of valve housing.

Fig. 188—View showing lift cylinder piston partially installed in cylinder. The "V" shaped seal is clamped to front end of piston by the retainer, bolt and lock washer.

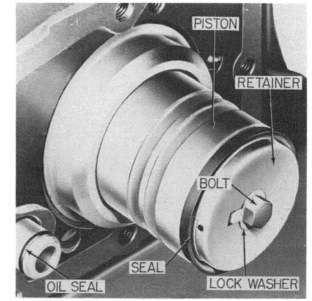

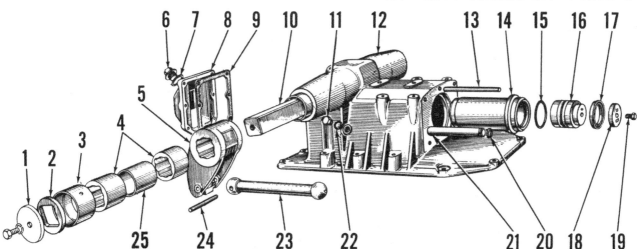

Fig. 189—Exploded view of model "FMD" and "FPM" hydraulic lift cover, cylinder, lift shaft and related parts. Refer to Fig. 184 and Fig. 185 for exploded views of valve housing and related parts that bolt to front face of lift cover (12).

1. Flat washer	5. Lift cylinder arm	8. Cover	13. Return oil pipe
2. Seal	6. Filler plug	9. Gasket	14. Cylinder
3. Bushings	7. Gasket	10. Lift shaft	15. "O" ring
4. Bearings		11. Oil passage plug	16. Piston
		12. Lift cover	17. Seal

18. Retainer	22. Gasket
19. Cap screw	23. Piston rod
20. "O" ring	24. Pin
21. Pressure pipe	25. Spacer

size and/or wheel width setting, it may be necessary to remove rear wheel. Disconnect lift links from lift arms and remove lift arm from side of tractor opposite removed fender. Remove rear access cover from lift cover and remove the lift shaft and remaining lift arm from cover. Remove ram cylinder arm and connecting rod from rear opening and extract the two lift (cross) shaft bearings and spacers from each side of cover.

Inspect the four lift shaft bearing liners in lift cover and renew liners if excessively worn. Install new bearing liners using suitable bushing driver, making sure holes in liners are aligned with grease holes in lift cover. Bearing liners should not require reaming if carefully installed.

Place new shaft seal, outer bearing, spacer and inner bearing on lift shaft next to arm not removed. Place ram cylinder arm and piston rod assembly in lift cover through rear opening, then insert lift shaft through cover and ram cylinder arm. Insert inner bearing, spacer, outer bearing and lift shaft seal into lift cover bore over end of lift shaft, then reinstall lift arm and reconnect lift links. Reinstall rear cover on lift cover.

POSITION CONTROL ATTACHMENT
Models "FMD" and "FPM"
So Equipped

240. A "position control" attachment is available for installation on models "FMD" and "FPM." Refer to exploded view of attachment in Fig. 190. The unit is mounted to right end of lift shaft instead of lift arm retaining washer and cap screw. Flange on plate (1) is placed below right hand lift arm and front end of spring loaded arm (8) is inserted through slot in quadrant below the control lever. Note: Early production units do not have slot in quadrant for lever; either install new quadrant or cut hole in quadrant according to directions packaged with attachment.

In operation, the lever (2) is clamped to plate (1) in position that spring loaded arm (8) will return control lever to neutral position when desired depth of implement is reached. Any required repairs are evident from inspection of unit and reference to Fig. 190.

HYDRAULIC PUMP
Models "FMD" and "FPM"

241. REMOVE AND REINSTALL PUMP. The hydraulic pump can be unbolted and remove from the pedestal after removing lift cover as outlined in paragraph 233, or can be removed as follows:

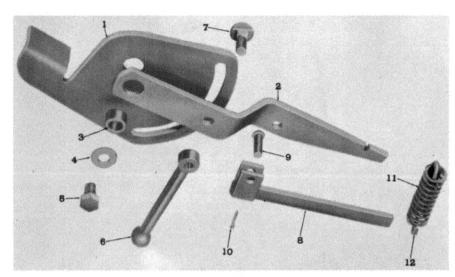

Fig. 190—Exploded view of position control attachment. Plate (1) is mounted on right end of lift shaft (10—Fig. 189) with projection on plate placed under lift arm. Downward movement of lift arm then moves spring loaded arm (8) upward returning control lever from lowering position to neutral position at desired lift arm height. Adjustment is made by clamping arm (2) to desired positon on plate (1).

1. Plate	7. Carriage bolt
2. Arm	8. Spring loaded
3. Bushing	arm
4. Flat washer	9. Pin
5. Cap screw	10. Cotter pin
6. Handle nut	11. Spring
	12. Retainer

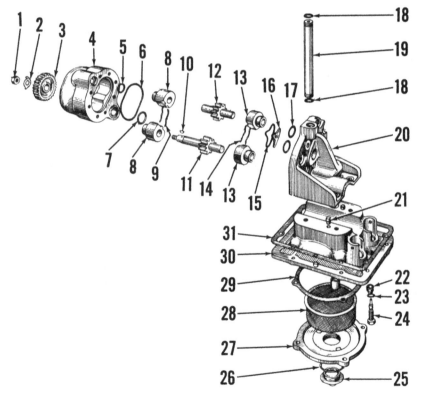

Fig. 191—Exploded view of model "FMD" hydraulic pump assembly. Pump is no longer available as a complete assembly; thus, some "FMD" or "FPM" tractors may be fitted with later type pump as shown in Fig. 216.

1. Nut	9. Locking springs	17. "O" rings
2. Locking washer	10. Woodruff key	18. "O" rings
3. Gear	11. Drive gear	19. Pressure pipe
4. Housing	12. Driven gear	20. Mounting bracket
5. "O" ring	13. Front bearings	21. Dowel pins
6. "O" ring	14. Locking springs	22. Check ball
7. "O" ring	15. Seal ring	(usually omitted)
8. Rear bearings	16. Relief plate	23. "O" ring

24. Cap screw	
(magnet tip)	
25. Drain plug	
26. Gasket	
27. Screen cover	
28. Suction screen	
29. Gasket	
30. Pump base	
31. Gasket	

Drain rear axle center housing and remove PTO output shaft assembly as outlined in paragraph 224 or raised PTO unit and drive shaft as outlined in paragraph 225. Then, unbolt pump pedestal from bottom of rear axle center housing and lower the pedestal and pump assembly from tractor. Remove vertical pressure pipe from pump or lift cover. Unbolt and remove pump from pedestal.

To reinstall pump, proceed as follows: Install new "O" rings on the vertical pressure pipe, then insert pipe into pump front cover (bracket). Reinstall pump by reversing removal procedure and refill the rear axle center housing with proper lubricant as outlined in paragraph 227.

242. OVERHAUL PUMP. Note: A later model "FSM" pump may be installed; if pump has removable cover at drive gear end, refer to paragraph 275. If pump body is as shown in Fig. 191, proceed as follows:

Straighten lock tab washer on pump drive shaft, remove nut and washer, pull gear from shaft and remove the Woodruff key. Unbolt and remove bracket (front cover) from pump, then remove the "O" ring seals and relief plate. Bump pump drive shaft against wood block to remove front bearings. Note: Bearings are "locked" by two wire springs and tension is removed from springs when one bearing is moved about ¼-inch forward of other bearing. Remove the drive gears, then using suitable tool, push rear drive gear bearing forward to loosen spring tension and remove rear bearings.

Carefully inspect pump body for cracks, deep score marks or excessive wear. Maximum wear allowance for gear track at intake side of pump body is 0.0025. If body is not suitable for further service, the complete pump must be renewed using the later type pump used for model "FSM" tractors. To adapt the later type pump, a conversion kit is also required as the pressure outlet is located in different positions on the two pumps.

If pump body is suitable for further service, carefully inspect removed parts and renew as required. Lubricate all parts and using all new sealing rings, reassemble as follows:

The rear pump bearings can be identified by their wider flange. With the two rear bearings connected by their locking springs, insert them into pump body with one bearing slightly ahead of the other and push them into pump body. Turn the bearings clock-

wise, then push trailing bearing down into place to "lock" the bearings. Insert the drive and driven gears. With the two front bearings connected by their locking springs, insert them into pump body with one bearing slightly ahead of the other. When leading bearing is against gear, turn the bearings clockwise, then push trailing bearing down into place to "lock" the bearings.

If, when installed, the flanges of the front bearings are below flush with machined recess in pump body, measure distance between bearing flange faces and machined face of pump body with depth gage at several points, then install relief plate of thickness 0.003-0.0055 less than minimum measured distance. If bearing flange faces are above flush with machined recess, se-

lect a relief plate of thickness that will provide a minimum clearance of 0.003-0.0055 between plate and machined surface of pump body when measured with straight edge and feeler gage as shown in Fig. 192. Relief plates are available in four thicknesses: 0.1125-0.1130, 0.1150-0.1155, 0.1175-0.1180 and 0.120-0.1205.

Insert new body and oil passage sealing rings and place new sealing ring around relief plate. Taking care not to dislodge "O" rings, install pump cover (bracket) to bearings and pump body. Evenly and securely tighten cover retaining cap screws. Install Woodruff key, drive gear, new lock tab washer and nut, tighten nut and secure by bending tab of washer against nut. Fill pump with oil and turn it by hand to be sure it is free.

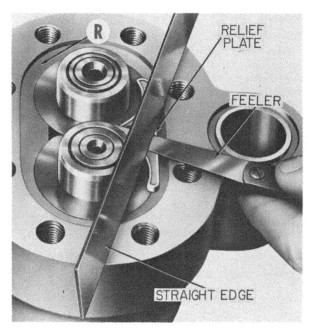

Fig. 192—Checking with feeler gage and straight edge for proper relief plate thickness. Refer to text.

HYDRAULIC SYSTEM
(Models "FSM" & "New FSM")

The Model "FSM" hydraulic lift system incorporates automatic draft control with compression on the top link of the three-point hitch; the Model "New FSM" provides automatic draft control with either tension or compression on the top link. Both models incorporate automatic implement position control. Fluid for either system is common with the rear axle final drive, but is separated from the transmission by oil seals. Hydraulic power is supplied by a gear type hydraulic pump mounted in the rear axle center housing and driven by a gear on the PTO shaft. Production changes have been made from time to time as well as a change in the control linkage between the models "FSM" and "New FSM". Where changes affect parts procurement and service procedure, it will be noted in the text.

FLUID AND FILTERS

Models "FSM" and "New FSM"

243. The rear axle differential and final drive lubricant is utilized for the hydraulic system fluid. Capacity is 10.8 gallons. Lubricant from factory is SAE 30 H.D. oil; however, SAE 80 Mild E.P. (Ford specification M-4864-A) or SAE 90 Mild E.P. (Ford specifictaion M-4864-B) gear lubricant is recommended for refill when the factory installed lubricant is drained. CAUTION. Do not mix SAE 80 or SAE 90 gear lubricant with the SAE 30 H.D. oil as the different type lubricants are not compatible.

A wire mesh screen suction filter is located in the pump pedestal and is accessible after draining rear axle center housing and removing cover from bottom of pedestal; refer to Fig. 191.

The rear axle center housing should be drained, the suction filter screen removed and cleaned, the housing flushed and the system refilled with new oil of the correct type after each 12 months or 2000 hours of service.

A replaceable element type filter (120—Fig. 210) is located on the hydraulic system sump return line. It is recommended that this filter be renewed whenever performing major overhaul of the differential, final drive or hydraulic system.

TROUBLE-SHOOTING

Models "FSM" and "New FSM"

244. Trouble in the hydraulic lift system will usually show up as (a), failure to lift; (b), inability to hold implement in raised position without excessive "corrections" (up and down bobbing motion); (c), over-correction (inability to maintain desired depth) in draft control; or (d), erratic action of the system. The possible causes of trouble and methods of checking to locate source are outlined in the following paragraphs.

245. **WILL NOT LIFT.** First, check to be sure the PTO output shaft is engaged and that the shaft turns when clutch is engaged. Check to see that system (rear axle center housing) contains proper amount of oil; a dipstick is located at left rear corner of transmission. Move the control lever to top of quadrant and check with the selector lever (see Fig. 194) in both the up (draft control) and forward (position control) positions. If the lift still fails to operate, move the auxiliary service control (selector valve) knob to out position, remove plug from front side of valve housing and turn engine with starter. If no oil flows from plug opening, remove lift cover and cylinder assembly, then again turn engine with starter. If no oil flows from pump pressure tube, remove and overhaul hydraulic pump. If oil flows from open pressure tube and pressure can be obtained by blocking tube opening, overhaul lift cover and cylinder assembly.

Possible causes of failure to lift within the pump assembly are shearing of drive key or shaft, broken or extremely worn pump body or gears, plugged intake screen, or ruptured seals in pump.

Possible causes of failure to lift within the cover assembly are improper adjustment of linkage, bent or broken linkage, sticking or binding of valves or broken ram cylinder, piston or connecting rod.

246. **OVER-CORRECTION IN DRAFT CONTROL.** Under some conditions, uneven depth control and over-correction may result from excessive oil flow from the hydraulic pump. Models "FSM" and "New "FSM" are equipped with an adjustable flow control valve which is used to regulate hydraulic pump output to meet varying field conditions and implement requirements.

If adjusting the flow control valve to minimum flow position does not change rate of hydraulic lift, check the flow control valve plunger (74—Fig. 210) to be sure that it is not sticking or that the plunger spring (75) is not damaged or broken.

247. **EXCESSIVE CORRECTIONS ("BOBBING" OR "HICCUPS" IN RAISED POSITION).** Leakage of oil in the hydraulic lift circuit will allow lift arms to fall, then automatically correct to raise back to set position. This is referred to as corrections in raised position. With hydraulic oil at normal operating temperature, engine running at 1600 RPM and approximately 1500 pounds on the lift arms, three or less corrections in two minutes is considered normal. The system will operate satisfactorily with up to

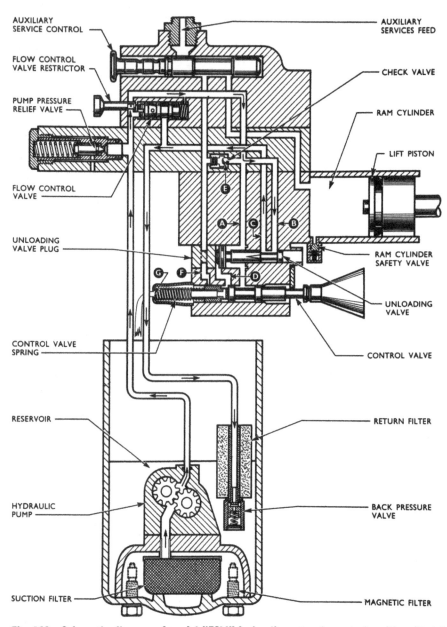

Fig. 193—Schematic diagram of model "FSM" hydraulic system in neutral position. Model "New FSM" system is similar except that a restrictor valve is incorporated in the return circuit (see Fig. 202) to control rate of lowering of the three-point hitch.

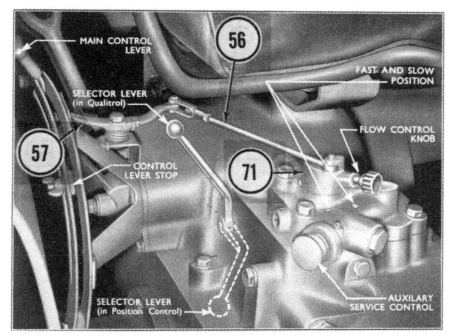

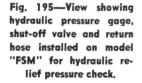

Fig. 194—View of controls on model "FSM" hydraulic system. Model "New FSM" controls are similar except for drop control knob shown in Fig. 201. Refer to Fig. 210 for legend.

30 corrections in a two minute period, however, if there is any sudden increase in number of corrections, service of the hydraulic lift is indicated.

To determine cause of leakage, mount a heavy implement on the three-point hitch and raise the implement, then, shut off engine. If implement falls steadily all the way to ground, a leaking piston seal, check valve or safety valve is indicated. Renew check valve as outlined in paragraph 263; if this does not correct trouble, remove lift cover and renew piston seal and, if necessary, the lift cylinder safety valve.

If implement falls only part way, then stops or rate of fall decreases noticeably, a leaking control valve is indicated and the valve and bushing should be renewed. Refer to paragraph 267.

Additional points to check are the "O" rings located between lift cylinder and cover and the fit of unloading valve bore plug in lift cylinder. While lift cover is off, make sure that control linkage operates without binding.

248. ERRATIC ACTION. Usually caused by binding of the control valve, back pressure valve, unloading valve, flow control valve or linkage. Before removing top cover, check to be sure that lift arms can be moved up and down.

SYSTEM RELIEF PRESSURE

Models "FSM" and "New FSM"

249. To check system relief pressure, first operate tractor until hydraulic

fluid (differential and final drive lubricant) is at operating temperature, then proceed as follows:

Stop engine and remove filler cap from rear of lift cover and pressure port plug from right side of lift cover. Connect a 0-3000 psi pressure gage, shut-off valve and return tube to pressure port with return tube inserted in filler opening as shown in Fig. 195. With test gage shut-off valve open, start engine, lower the 3-point hitch lift arms, pull selector (auxiliary control) valve out and move lift control lever to top of quadrant. With engine running at high idle speed, gradually close test gage shut-off valve while observing pressure gage.

With early type pressure relief valve assembly (see Fig. 196), gage reading should gradually increase to system relief pressure as shut-off valve is

Fig. 195—View showing hydraulic pressure gage, shut-off valve and return hose installed on model "FSM" for hydraulic relief pressure check.

closed, then drop to approximately 300 psi when relief valve opens.

On systems fitted with late type pressure relief valve (see Fig. 196), gage reading should gradually increase to slightly above system relief pressure as shut-off valve is closed, then drop to relief pressure when valve opens.

System relief pressure should be 2450-2500 psi with either pressure relief valve assembly. The only difference is that late valve will maintain system pressure while early valve will drop system pressure to approximately 300 psi when valve opens and control valve must be returned to neutral to reset the valve. Late type relief valve assembly may be installed in early production lift cover if desired.

To adjust system relief pressure, remove the relief valve assembly, refer to Fig. 196, and disassemble the unit.

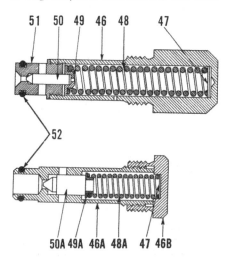

Fig. 196—Cross-sectional views of early type (bottom) and late type (top) pressure relief valve assemblies. Refer to text.

46. Valve housing
47. Shims
48. Relief-valve spring
49. Spring seat
50. Relief valve
51. Valve seat
52. "O" ring

Inspect valve (50 or 50A) and seat (51 or 46A) for wear or damage and spring (48 or 48A) for cracks or distortion. If valve appears serviceable, add shims (47) to increase relief pressure or remove shims to decrease pressure. Shims are available in thicknesses of 0.010 and 0.025. Adding or removing shim thickness of 0.010 should change relief pressure approximately 100 psi. CAUTION: Do not install a total shim thickness of more than 0.080. If adding shims does not increase system relief pressure, overhaul pump assembly as outlined in paragraph 275.

SYSTEM ADJUSTMENTS
Model "FSM"

250. **ADJUST MAIN CONTROL SPRING.** To adjust main draft control spring, disconnect top link rocker from yoke, then turn yoke in until shoulder on yoke is flush with lift cover as shown by arrows in Fig. 197. As a functional check of main control spring adjustment, proceed as follows: Place selector lever in draft control and attach an implement to the lower links of the three-point hitch. With engine running at approximately 1600 RPM, move control lever slowly upward until the lift arms raise, then move lever back down the quadrant a distance of 1 inch. The lift arms should lower, then raise again when a force of 110 pounds is applied directly to the main control spring yoke. If required pressure exceeds 110 pounds, disconnect top link rocker from yoke and turn yoke out ½-turn at a time until adjustment is correct.

If required pressure is considerably less than 110 pounds, turn yoke in.

251. **ADJUST DRAFT CONTROL LINKAGE.** Adjustment of the hydraulic lift linkage should be made using special gages as follows:

With lift cover assembly removed as outlined in paragraph 261 and main draft control spring adjusted as in paragraph 250, attach locating arm (Nuday tool No. N-503-C) to lift cover and position lift arms with pin as shown in Fig. 198. Move selector lever to draft control position and control lever to ½-inch from top stop on quadrant. Using adjusting gage (Nuday tool No. N-502-B) (same gage as used for model "NAA" Ford tractor), measure gap between shoulder on control valve spool and machined face of valve housing with small (draft control) end of gage. Gage should just enter gap without any binding or side clearance. If adjustment is not correct, loosen jam nut on turnbuckle (102) and lengthen or shorten linkage as necessary. Recheck adjustment after tightening jam nut.

If special gages are not available, proceed as follows: With selector lever in draft control position and control lever at bottom of quadrant, move lift arms in lowering direction until piston contacts front end of lift cylinder; then, back upward ½-inch measured at pin hole in end of lift arms. Lock lift shaft in this position by tightening cap screws in ends of lift shaft. Move control lever to ½-inch from top stop on quadrant and measure gap between shoulder on control valve spool and machined face of valve housing. If distance is not 0.342, lengthen or shorten linkage as required to provide this measurement.

Remeasure gap after tightening jam nut.

With draft control linkage adjusted, refer to paragraph 252 and adjust position control linkage.

252. **ADJUST POSITION CONTROL LINKAGE.** Before adjusting position control linkage, adjust draft control main spring and linkage as outlined in paragraphs 250 and 251, then proceed as follows:

With lift arms positioned as in paragraph 251, move control lever to bottom of quadrant and selector lever to position control; refer to Fig. 199. Using large (position control) end of adjustment gage (Nuday tool No. N-502-B), measure gap between shoulder of control valve spool and machined face of valve housing. If gage binds or fits loosely, refer to Fig. 210 and loosen jam nut (37) on position control adjustment screw (32), then turn screw in or out until gage fits without binding or side clearance. Recheck adjustment after tightening jam nut.

If special adjustment gages are not available, proceed as follows: With lift shaft locked in position described in paragraph 251, move control lever to bottom of quadrant and move selector lever to position control. Gap between shoulder on control valve spool and machined face of valve housing should then measure 0.432. If not, loosen jam nut (37—Fig. 210) and turn adjustment screw (32) as necessary to obtain this measurement. Remeasure gap after tightening jam nut. With position control linkage properly adjusted, loosen cap screws in ends of lift shaft until lift arms will fall of their own weight, then bend locking tabs against cap screw heads.

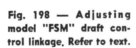

Fig. 198 — Adjusting model "FSM" draft control linkage. Refer to text.

Fig. 197—Model "FSM" main spring adjustment requires that shoulder on yoke (89) be flush with rear face of lift cover as indicated by arrows. Filler plug is (87).

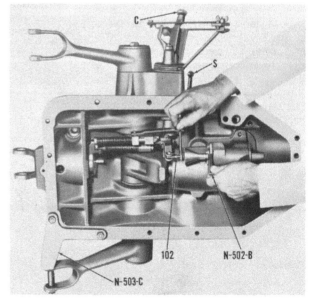

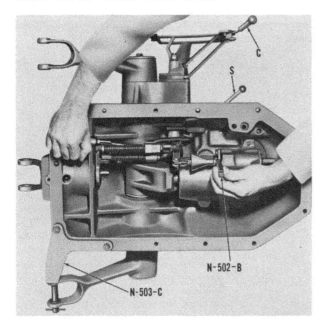

Fig. 199 — Adjusting model "FSM" position control linkage. Refer to text.

Model "New FSM"

253. ADJUST MAIN CONTROL SPRING. To adjust main draft control spring, disconnect top link rocker from spring yoke, then turn yoke in until snug against spring, back yoke out until pin hole in yoke is horizontal and reconnect top link rocker. NOTE: If any end play of yoke and spring can be noted when yoke is turned in against spring, the spring retainer nut should be re-shimmed as outlined in paragraph 254.

As a functional check of main control spring adjustment, proceed as follows: With implement or weight attached to the two lower links and with engine running at approximately 1600 RPM, move control lever upward until lift arms raise, then move lever back down quadrant a distance of 1 inch. The lift arms should lower, then raise again when a force of 250 pounds

is applied directly to main control spring yoke.

254. SHIM MAIN CONTROL SPRING NUT. If the main control spring retainer nut and spring have been removed from lift cover, or if end play of spring can be noted when yoke is turned in tight against spring, re-shim retainer nut as follows:

With the spring inner seat (IS— Fig. 200), spring (88A) and outer seat (OS) installed in lift cover, install retainer nut without any shims between nut and lift cover and turn nut in until all end play is removed from spring and seats. Then, using a

feeler gage, measure gap between flange on nut and face of lift cover. Remove the nut, install shims of thickness equal to measured gap and reinstall nut. Tighten nut to a torque of 80-85 Ft.-Lbs. Note: Each shim has a thickness of 0.008-0.012 and in selecting number of shims to be used, total thickness must be as near as possible to measured gap without exceeding this measurement. Adjust main control spring yoke as outlined in preceding paragraph 253.

255. ADJUST DROP CONTROL RESTRICTOR. The model "New FSM" is fitted with a restrictor valve in the exhaust passage from the hydraulic lift cylinder and this restrictor valve should be adjusted according to the weight of the implement mounted on the three-point hitch.

To adjust the restrictor valve, refer to Fig. 201 and turn drop control knob fully in. Raise implement to fully raised position, then move control lever to lowering position. If the implement lowers too slowly, back drop control knob out until desired rate of lowering is obtained.

256. ADJUST DRAFT CONTROL LINKAGE. To adjust draft control linkage, remove lift cover assembly as outlined in paragraph 261, refer to Fig. 203 and proceed as follows:

Insert a 0.010 feeler gage between draft control plunger (110) and face of housing, then tighten main control spring yoke (89A) so that gage is held tightly. Move lift arms to raised position so that thrust pad on lift cylinder arm (97A) contacts draft control plunger. Move the control lever (20) beyond upper stop to extreme top end of quadrant. Move selector lever (29) up to draft control position. Loosen jam nut (37) on position control rod (32) and screw rod back out of way of control valve lever (99). Insert the thick (position control) end of adjustment gage (Nuday tool No.

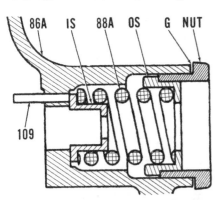

Fig. 200 — Cross-sectional view showing model "New FSM" main control spring. Refer to text for procedure to adjust spring.

G. Gap
IS. Inner spring
   spring
OS. Outer spring
   seat
86A. Lift cover
88A. Main control
   spring
109. Guide pin

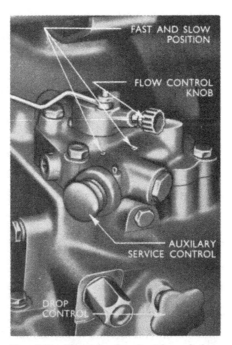

Fig. 201—View showing location of drop control valve knob on model "New FSM". Refer also to Fig. 202.

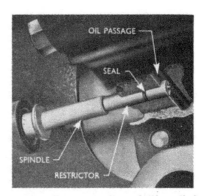

Fig. 202—Cut-away view of lift cylinder showing drop control restrictor valve. Refer also to Fig. 201.

N-502-B) or a 0.432 thick spacer between shoulder on control valve spool and face of valve housing. Loosen jam nut on turnbuckle (102) and lengthen or shorten turnbuckle as required to obtain a 0.200 gap between snap ring on front end of draft control plunger (107) and fork (105). Recheck gap after tightening jam nut on turnbuckle. With draft control linkage properly adjusted, refer to following paragraph 257 and adjust "knock-off" pin.

257. ADJUST "KNOCK-OFF" PIN. After the draft control linkage is properly adjusted as outlined in paragraph 256, it is important that the eccentric "knock-off" pin (see Fig. 204) in control valve actuating lever is properly adjusted.

With the control lever, selector lever and lift arms positioned as for draft control adjustment, with feeler gage inserted between draft control plunger and housing and with the 0.432 thick gage (Nuday tool No. N-502-B) inserted between shoulder of control valve spool and valve housing as outlined for draft control adjustment in paragraph 256, proceed as follows: Loosen lock nut on the "knock-off" pin and turn the pin so that eccentric is as far away from rear end of lift cylinder as possible. Remove the ram cylinder safety valve from front end of cylinder and using a suitable curved rod, push ram cylinder piston rear-

ward as far as possible. Be sure that piston rod is properly seated in center of piston. Turn the eccentric "knock-off" pin so that it contacts skirt of piston and while holding pin in this position, tighten lock nut.

Remove the feeler gage from between draft control plunger and housing and proceed with position control linkage adjustment as outlined in following paragraph 258.

258. ADJUST POSITION CONTROL LINKAGE. First, adjust draft control linkage as outlined in preceding paragraph 256 and "knock-off" pin as in paragraph 257 then proceed as follows:

Refer to Fig. 199 and position lift arms with locating arm (Nuday tool No. N-503-B) or, if tool is not available, move lift arms until piston contacts front end of cylinder, move lift arms back up ½-inch from this position and tighten cap screws in each end of lift arm shaft to hold arms in this position. Move control lever to bottom of quadrant and selector lever to position control. Turn position control rod screw (32—Fig. 203) as necessary to obtain a gap of 0.432 between shoulder on control valve spool (17) and face of valve housing. Tighten jam nut (37) and recheck adjustment. If special adjustment gage (Nuday tool No. N-502-B) is available, use thick (position control) end to check gap between valve spool shoulder and housing.

Models "FSM" and "New FSM"

259. ADJUST FLOW CONTROL VALVE LINKAGE. With hydraulic lift cover and component parts fully assembled, proceed as follows: Disconnect adjustable rod (56—Fig. 205) from lever (57). Move stop on main control lever to front side of lever and move the control lever against top stop on the quadrant. Turn flow control fully out, hold lever (71) against slow ("S") stop and slowly turn knob in until lever starts to move away from "S" stop. Note: It is important that the knob not be turned past this position. Hold lever (57) against rear side of control lever (not against stop on lever) and lengthen or shorten adjustable rod (56) so that rod can be reconnected to lever without moving either lever (57 or 71). Move stop on control lever to rear side of lever and check operation of linkage.

FLOW CONTROL AND SELECTOR VALVE ASSEMBLY

Models "FSM" and "New FSM"

260. Models "FSM" and "New FSM" are equipped with a flow control and selector valve assembly. Refer to exploded view in Fig. 206.

A single acting remote cylinder can be connected to the hydraulic system by removing plug (81) from the valve housing (85) and connecting hose to that port. When the selector valve spool (65) is pushed in, hydraulic oil under pressure is directed to the three-point hitch lift cylinder; pulling selector valve spool out will direct oil pressure to the port in valve housing and operate the single acting remote cylinder with the hydraulic system control valve.

To regulate hydraulic pump output, turn knob (70) in or out to move

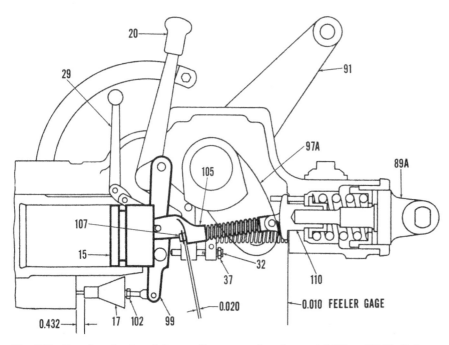

Fig. 203—Drawing showing linkage adjustment points for model "New FSM". Refer to text for procedure.

15. Lift piston	32. Adjusting screw	91. Lift arms	105. Yoke
17. Control valve spool	37. Lock nut	97A. Lift cylinder arm	107. Draft control rod
20. Control lever	89A. Main spring yoke	99. Control valve lever	110. Main spring plunger
29. Selector lever		102. Turnbuckle	

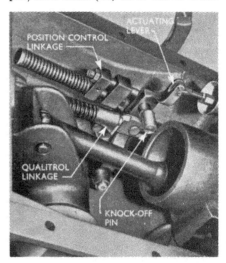

Fig. 204—View of model "New FSM" control linkage showing knock-off pin.

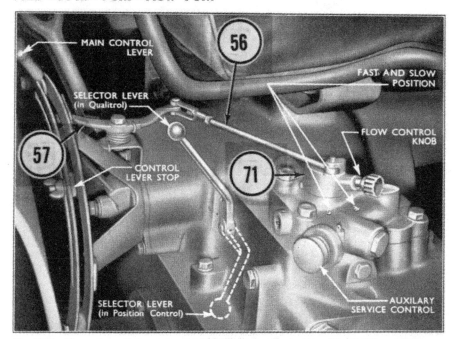

Fig. 205—View of controls on model "FSM" hydraulic system. Model "New FSM" controls are similar except for drop control knob; refer to Fig. 201. Refer to text for adjustment of linkage (56).

out binding when lubricated with hydraulic fluid. Usually, the required size will be of the same color code as the valve spool removed. Valve sizes are indicated by color code as follows:

Flow Control Valve Spool

Color Code	Valve Diameter
Red	0.6670-0.6672
Yellow	0.6672-0.6674
Blue	0.6674-0.6676
Green	0.6676-0.6678
White	0.6678-0.6680

Selector Valve Spool

Color Code	Valve Diameter
Green	0.7482-0.7485
White	0.7485-0.7488
Blue	0.7488-0.7491
Yellow	0.7491-0.7494
Orange	0.7494-0.7497

Lubricate all valve parts prior to reassembly and reassemble using all new "O" rings. Reinstall valve assembly using new gasket and "O" rings and tighten retaining bolts to a torque of 40-45 Ft.-Lbs.

LIFT COVER AND CYLINDER
Models "FSM" and "New FSM"

The lift cover assembly includes the lift (rock) shaft, control quadrant, lift cylinder, main control valve, unloading valve, system relief valve, check valve and safety valve. The system back pressure valve is located at the lower end of system fluid return line.

261. R&R LIFT COVER AND CYLINDER ASSEMBLY. To remove lift cover, first exhaust all oil from lift cylinder by placing control valve lever at bottom of quadrant, selector lever in draft control position and forcing oil from cylinder by pushing lift arms to their lowest position. Then, proceed as follows:

Disconnect lift links from lift shaft arms and top link rocker from draft control yoke. Remove operator's seat

restrictor valve (82—Fig. 210) to minimum or maximum flow positions. Movement of the flow control valve (74) is related to the position of the restrictor valve; the flow control valve bypasses a certain amount of oil back to the hydraulic sump and thereby regulates amount of oil flowing to the tractor lift cylinder or remote cylinder. The main control lever (20) is equipped with a moveable spacer which, when moved to rear side of lever, contacts arm (57) when control valve is in raised position causing the restrictor valve to return to maximum flow position regardless of setting of knob (70).

To remove the flow control and selector valve assembly, first push selector valve (65) in, move hydraulic control lever to bottom of quadrant, selector lever (29) to draft control position and force lift arms downward to exhaust all oil from lift cylinder. Unbolt and remove valve assembly from top of hydraulic lift cover.

Disassembly procedure is evident from inspection of unit and reference to exploded view in Fig. 206. The flow control valve (74) and the selector valve spool (65) are a selective fit in bores of valve housing. If necessary to renew either valve spool, select the largest size that will fit in bore with-

Fig. 206—Exploded view of flow contol and selector valve assembly; refer also to Fig. 210.

58. Knob
59. Cover
60. Washer
61. Detent ball
62. Detent spring
63. Cap
64. "O" ring
65. Selector valve
70. Adjuster knob
74. Flow control valve
75. Flow control spring
76. "O" ring
77. Plug
78. Retaining pin
81. Remote cylinder port plug
85. Valve body

Fig. 207 — Removing lift cylinder from model "FSM" lift cover assembly.

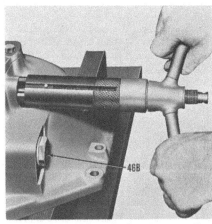

Fig. 208—Removing check valve seat; English tool is shown. Refer to text for procedure. Early type relief valve is (46B).

and the cap screws and nuts retaining lift cover to rear axle center housing. Attach suitable hoist to lift cover and remove the assembly from tractor.

When reinstalling lift cover, use new gasket and "O" rings and tighten the retaining cap screws and nuts to a torque of 40-45 Ft.-Lbs.

262. R&R LIFT CYLINDER ASSEMBLY. First, remove lift cover and cylinder assembly as outlined in paragraph 261, then proceed as follows: Unbolt and remove the flow control and selector valve assembly from top of lift cover, then remove the bolts retaining lift cylinder to cover and remove cylinder from bottom side of

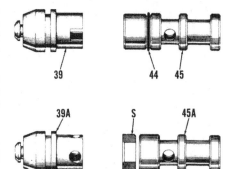

Fig. 209—View showing early type (top) and late type (bottom) check valve seat and pilot. Refer to text for interchangeability.

S. "Delrin" seat
39. Pilot
44. "O" ring
45. Check valve seat

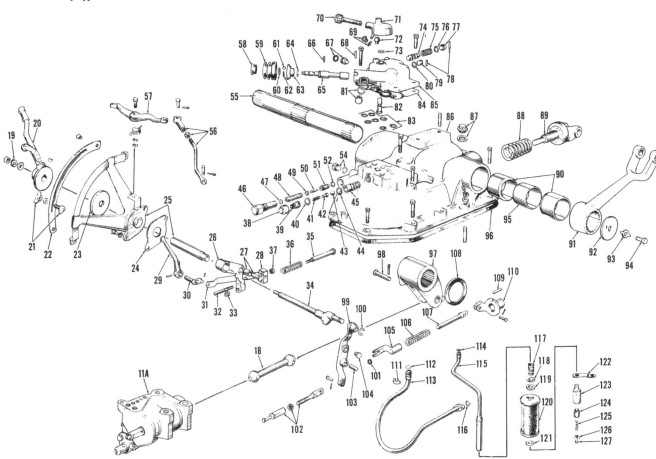

Fig. 210—Exploded view of model "FSM" lift cover assembly. Refer to Fig. 211 for exploded view of lift cylinder (11A). Refer to Fig. 203 for model "New FSM" linkage. Figs. 196 and 209 show production changes in pressure relief valve and check valve seats applicable to both models "FSM" and "New FSM."

11A. Cylinder assembly	37. Lock nut	57. Actuating arm	74. Flow control valve	91. Lift arm	110. Draft control plunger
18. Piston rod	38. Retainer plug	58. Knob	75. Spring	92. Retaining washer	111. Retainer
19. Tension spring	39. Pilot	59. Cover	76. "O" ring	93. Locking washer	112. "O" ring
20. Control lever	40. "O" ring	60. Washer	77. Plug	94. Cap screw	113. Pump pressure pipe
21. Stop bolt	41. Spring	61. Detent ball	78. Pins	95. Spacer	114. "O" ring
23. Friction disc	42. Spring guide	62. Detent spring	79. Plug	96. Gasket	115. Return pipe
24. Gasket	43. Check valve ball	63. Cap	80. "O" ring	97. Lift-cylinder arm	116. "O" ring
25. Quadrant	44. "O" ring	64. "O" ring	81. Plug & seal	98. Pin	117. Spring
26. Position control arm	45. Check valve seat	65. Selector valve	82. Restrictor valve	99. Draft control arm	118. Washer
27. Pin and roller	46. Relief valve body	66. Pin	83. "O" rings	100. Snap ring	119. Seal
28. Block	47. Shims	67. Plug & seal	84. Gasket	101. Snap ring	120. Filter element
29. Selector lever	48. Spring	68. Pin	85. Flow control valve housing	102. Turnbuckle	121. Seal
30. Selector arm	49. Spring seat	67. Plug & Seal	86. Lift cover	103. Pin	122. Bracket
31. Link	50. Relief valve	68. Pin	87. Filler plug & gasket	104. Pin.	123. Valve housing
32. Adjusting screw	51. Valve seat	69. Plug & seal	88. Main control spring	105. Yoke	124. Back pressure valve
33. Locknut	52. "O" ring	70. Flow control knob	89. Yoke	106. Spring	125. Spring
34. Control lever shaft	54. Plug and seal	71. Flow control lever	90. Bushings	107. Draft control rod	126. Spring retainer
35. Spring guide	55. Lift shaft	72. Stop pin		108. Spacer	127. Snap ring
36. Spring	56. Flow control rod	73. "O" ring		109. Guide pin	

cover. Discard all "O" rings and gasket from between cylinder, flow control and selector valve assembly and cover.

Install cylinder to bottom side of lift cover using all new "O" rings and tighten retaining bolts to a torque of 50-55 Ft.-Lbs. Install flow control and selector valve assembly with new gasket and "O" rings and tighten retaining bolts to a torque of 40-45 Ft.-Lbs.

263. CHECK VALVE AND SEAT. The check valve and seat can be removed without removing lift cover and cylinder assembly. Unscrew the plug (38—Fig. 210) from front of lift cover. Using needle nose pliers, pull the pilot (39) from lift cover and extract spring (41), spring seat (42) and check valve ball (43).

Two different types of check valve seats have been used; refer to Fig. 209. Early production model "FSM" systems were equipped with a one-piece hardened steel seat (44). Late production models have a "Delrin" material seat (S) with an inner bushing (45A). A different pilot (39A) is required with the "Delrin" seat. The late pilot, seat and inner bushing are interchangeable as a group with the earlier pilot (39), "O" ring (44) and seat although the parts are not interchangeable individually.

To remove either the late or early seat, proceed as follows: Using special puller, Nuday tool No. NCA-997-A or

equivalent, thread puller rod into check valve seat or inner bushing with late type seat and pull the bushing and/or seat from lift cover as shown in Fig. 208.

To install new early type seat, install new "O" ring (44—Fig. 209) on seat, lubricate bore, seat and "O" ring and install seat by threading puller into seat and driving the seat into housing until it contacts shoulder in bore.

To install new late type seat, thread inner bushing on puller rod, lubricate bushing and bore and drive bushing into bore until seated against shoulder. Tap the "Delrin" seat into place with chamfered seat out (away from inner bushing).

Insert check valve ball, spring seat and spring in bore. Install pilot with new "O" ring, then install retaining plug. With early type hardened steel check valve seat, tighten plug to a torque of 45-55 Ft.-Lbs. On late type "Delrin" seat, tighten plug to a torque of 25-30 Ft.-Lbs. CAUTION: Do not exceed recommended torque value when tightening plug with new type "Delrin" seat.

264. SYSTEM RELIEF VALVE. The system relief valve can be removed from front end of lift cover without removing cover from tractor; refer to (46B—Fig. 208). The relief valve is removed as an assembly.

Two different type relief valve assemblies have been used; refer to Fig.

196. Both type valves are adjusted as outlined in paragraph 249.

265. CYLINDER SAFETY VALVE. To remove the cylinder safety valve (9—Fig. 211), first remove the lift cover as outlined in paragraph 261. The valve can then be unscrewed from front end of cylinder.

The cylinder safety valve is adjusted at the factory to open at a pressure of 2750-2850 psi. If service is indicated, renew the complete valve assembly. Renew sealing ring (10) when installing valve in cylinder.

266. DROP CONTROL VALVE (Model "New FSM" Only). After the lift cylinder has been removed from lift cover as outlined in paragraph 262, the drop control restrictor valve (see Fig. 202) can be withdrawn from bore in front of lift cylinder casting. To remove spindle from lift cover, drive pin from knob and spindle, remove knob and unscrew spindle from inside of lift cover.

Install new "O" ring on restrictor valve, lubricate "O" ring and insert restrictor in bore of cylinder casting. Install new spindle seal in lift cover with lip outward, screw the spindle into cover from inside and install knob and retaining pin.

267. CONTROL VALVE AND BUSHING. With the lift cylinder removed as outlined in paragraph 262, proceed as follows: Remove baffle plate (1—Fig. 211) and control valve spring (7) from front end of cylinder casting. Unbolt plate (16) from rear (open) end of cylinder and remove the control valve spool (17) and plate. Using a suitable sleeve, press bushing (8) out towards front (closed) end of cylinder.

Inspect lands on control valve and renew valve spool **and** bushing if erosion or scoring is visible. Note: Late production control valve spools have an annular recess (see Fig. 212) in front land on spool. Early valve spools had a solid land and the change was made to correct valve sticking in raising position under heavy hydraulic loading. Early and late type spools are interchangeable, and the late type spool should be installed with new bushing if sticking of con-

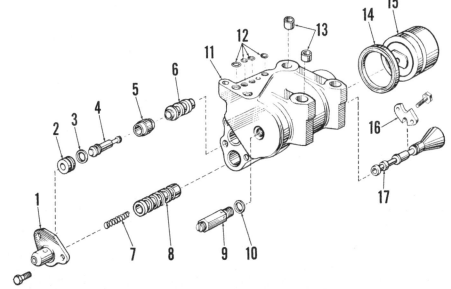

Fig. 211—Exploded view of lift cylinder assembly for model "FSM". Model "New FSM" is similar except for restrictor valve shown in Fig. 202.

1. Baffle	6. Rear unloading valve bushing
2. Plug	7. Valve return spring
3. Valve "O" ring	8. Control valve bushing
4. Unloading valve	
5. Front unloading valve bushing	
9. Safety valve	14. Piston seal
10. Seal	15. Piston
11. Lift cylinder	16. Retaining plate
12. "O" rings	17. Control valve spool
13. Hollow dowels	

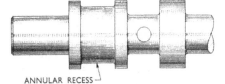

ANNULAR RECESS

Fig. 212—Latest type control valve spool has annular recess in front land; refer to text.

trol valve with solid land is encountered.

The control valve bushing is available in eight different outside diameter size ranges which are color coded to indicate size. Always renew the bushing with new bushing having same color code. Color codes for each size range are as follows:

Control Valve Bushings

Color Code	Outside Diameter
Blue/White	1.0000-1.0002
White	1.0002-1.0004
Blue	1.0004-1.0006
Yellow	1.0006-1.0008
Green	1.0008-1.0010
Orange	1.0010-1.0012
Green/White	1.0012-1.0014
Red/White	1.0014-1.0016

To install new control valve bushing, insert end opposite that having widest annular recess into front (closed) end of cylinder, then press bushing in until front end of bushing is flush with front machined face of lift cylinder casting.

After bushing is installed, select fit a new control valve spool to bushing. Note: Valve cannot be fitted prior to installing bushing in cylinder casting. Lubricate valve spool and insert spool in normal position. A light drag should be felt as spool is moved within its normal range of travel. Valve spools are color coded for diameter size range as follows:

Control Valve Spools

Color Code	Diameter
White	0.5917-0.5919
Blue	0.5919-0.5921
Yellow	0.5921-0.5923
Green	0.5925-0.5926
Orange	0.5927-0.5928

It should be noted as that the color code indicates a size range only, a valve of one color code (size range) may fit correctly while another valve spool of the same color code may fit too tight or too loose. After installing valve, retaining plate, spring and baffle plate, check to see that spring will return valve quickly when valve is depressed and released; if not, valve is either too tight or foreign material is causing valve to bind.

268. UNLOADING VALVE AND BUSHING. After control valve spool has been removed as outlined in paragraph 267, thread an impact puller (slide hammer) adapter into bore plug (2—Fig. 211) and pull plug from front end of cylinder. Insert small rod or screwdriver into rear bushing (6) and push unloading valve out towards front. Remove and discard "O" ring (3) from valve (4). Using suitable sleeve or driver, press both bushings (5 and 6) out towards front end of cylinder.

The unloading valve bushings and bore plug are available in eight different size ranges and each size range is color coded. Always install the same size bushings as were removed from a lift cylinder casting. Usually, the bore plug will not need to be renewed but, in some instances, it may be necessary to fit a larger size bushing if one of the same color code does not fit tightly. Color codes and size ranges for the unloading valve bushings and plug are the same as listed for control valve bushing in paragraph 267.

When installing unloading valve bushings, note that one end of the front bushing (5) has a single notch while opposite end has two notches. End of bushing with single notch must be forward placing end with two notches against rear bushing (6). To install bushings, start front bushing into bore from rear (open) end of lift cylinder with single notched end forward, then press both bushings into place by pressing against rear bushing only. The rear face of rear land on bushing (6) must be flush with machined face of lift cylinder.

Renew the unloading valve if scored, excessively worn or otherwise damaged. Valve is available in one size only. Lubricate valve and insert in bushings without the "O" ring (3). Valve should be a free sliding fit in the bushings. If not, check to be sure that no foreign material is present, that front bushing is concentric with rear bushing and that the valve or bushings are not damaged. With a free sliding fit obtained, install new "O" ring on bushing, lubricate both parts and reinstall valve. The "O" ring should impart a slight drag when moving valve back and forth but should not cause valve to stick or bind. Install another "O" ring if valve moves freely as without "O" ring, or if binding occurs. CAUTION: Never install an "O" ring of unknown quality at this location as any shrinking or swelling of "O" ring will cause malfunction of hydraulic system.

Reinstall unloading valve bore plug with threaded hole in plug out. If plug fits loosely, it should be renewed.

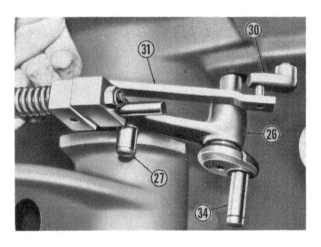

Fig. 213—Removing position control linkage and control lever shaft. Refer to Fig. 210 for legend.

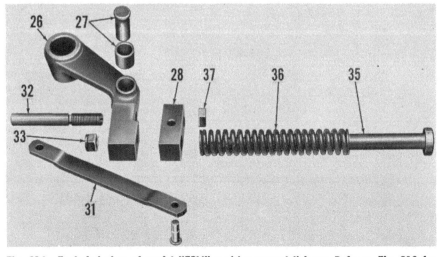

Fig. 214—Exploded view of model "FSM" position control linkage. Refer to Fig. 210 for legend.

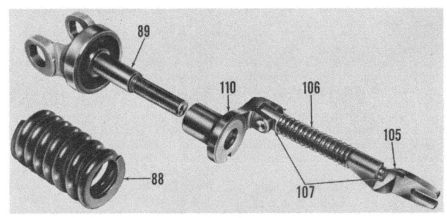

Fig. 215—Exploded view of model "FSM" main control spring and linkage. Refer to Fig. 200 for cross-sectional view of model "New FSM" main control spring assembly. Refer to Fig. 210 for legend.

cover has step cut at each side at lift shaft bushing bore, install new "O" rings around shaft and into stepped bore. Install the lift arms, spring washers, lock tabs and cap screws. On models with "O" ring seal, tighten one cap screw, back it out slightly and secure with lock tab. Tighten opposite cap screw to remove all end play from lift shaft, but so that lift arms and shaft can be rotated, and secure cap screw with lock tab. On models without "O" rings, tighten one cap screw fully, back this cap screw out one turn and secure with lock tab. Tighten opposite cap screw securely, back cap screw out, then retighten so that lift arms will just fall of their own weight and secure with lock tab.

In some instances, it may be necessary to select a plug one or two size ranges larger than original plug to obtain tight fit.

269. **LIFT CYLINDER PISTON.** After removing lift cylinder as outlined in paragraph 261, remove ram cylinder safety valve and insert rod through opening to push piston from cylinder. Inspect piston and cylinder for scoring, cracks or excessive wear and renew if necessary. Remove sealing ring from piston and install new ring with lip forward (toward closed end of piston). Lubricate piston, ring and cylinder bore and carefully install piston to avoid damaging lip of sealing ring. Reinstall ram cylinder safety valve with new sealing washer.

270. **LIFT ARMS, SHAFT AND BUSHINGS.** The lift arms, shaft and bushing can be removed from lift cover after removing cover and cyl-

inder assembly as outlined in paragraph 261. Refer to Fig. 210 and proceed as follows: Straighten the locking tabs (93) and remove cap screws (94), tabs, spring washers (92) and lift arms (91) from each end of shaft. Then, withdraw shaft with the two bushings (90) and bushing spacer (95) from right (control lever) side of lift cover. Remove the two bushings and bushing spacer from opposite side of lift cover. To remove the ram cylinder arm, first remove control linkage as outlined in paragraph 271 or 272, then lift ram arm (97) and spacer (108) from cover.

To install shaft, be sure ram cylinder arm is properly placed, insert spacer washer (108) at left side of arm and insert lift shaft through cover and arm. Insert bushing, bushing spacer and second bushing into lift cover over shaft at each side. If lift

HYDRAULIC CONTROL LINKAGE
Model "FSM"

271. With the lift cylinder assembly removed as outlined in paragraph 262, proceed as follows: Remove quadrant (22—Fig. 210), then remove self-locking nut, spring (19) and flat washer from control lever shaft (34). Pull control lever (20) from shaft, remove the Woodruff key and friction disc (23), then unbolt and remove quadrant support and tube (25) from lift cover. Remove lift arms and cross shaft as outlined in paragraph 270. Unscrew main draft control spring yoke (89) and remove the spring (88). Slide fork (105) from pin on control valve actuating lever (99) and remove the fork, rod, spring and plunger as a unit (Fig. 215). Remove snap ring (100—Fig. 210) retaining actuating lever to control lever shaft (34) and remove actuating lever and turnbuckle (102). Remove pin from selector lever shaft (30) and remove the position control linkage and control lever shaft as a unit; refer to Fig. 213.

If necessary to perform further disassembly of linkage, refer to Fig. 210 for guide to disassembly and reassembly. After control linkage, lift shaft, lift arms and lift cylinder assembly are reinstalled on lift cover, adjust linkage as outlined in paragraphs 250, 251 and 252.

Model "New FSM"

272. Control linkage for Model "New FSM" is similar to that used on Model "FSM". Procedure as outlined in paragraph 271 for Model "FSM" can be followed for disassembly of linkage; however, refer to Figs. 200 and 203 as guide to reassembly.

After reinstalling linkage, refer to paragraphs 253, 256, 257 and 258 for adjustment procedure.

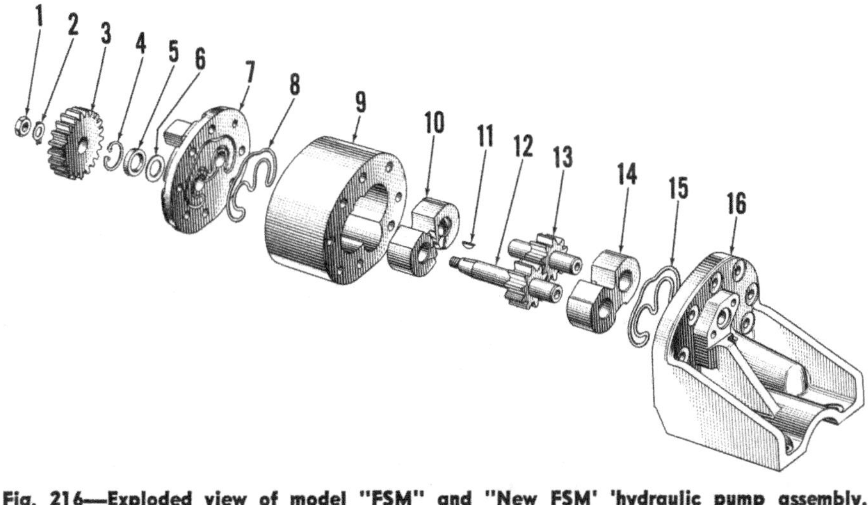

Fig. 216—Exploded view of model "FSM" and "New FSM' hydraulic pump assembly. An adapter kit is required to install this pump in model "FMD" or "FPM" due to different location of pressure outlet.

1. Nut	7. Rear cover	12. Drive-shaft & gear
2. Locking washer	8. Sealing ring	13. Driven gear
3. Drive gear	9. Pump body	14. Front bearing
4. Snap ring	10. Rear bearing	15. Sealing ring
5. Seal	11. Woodruff key	16. Mounting bracket
6. Washer		

RETURN LINE, FILTER AND BACK PRESSURE VALVE

Models "FSM" and "New FSM"

273. The return line (115—Fig. 210), complete with filter (120) and back pressure valve (123 through 127) can be removed after removing lift cover and cylinder assembly as outlined in paragraph 261. Unbolt return line bracket, then push return line down far enough to clear lift cover flange of rear axle center housing, then remove unit from tractor.

The return filter will not need to be renewed except at times of major overhaul unless the oil has become contaminated. To renew the filter, unscrew the back pressure valve from lower end of return line, remove the retaining washer, filter and seals and install new filter element and seals.

To disassemble back pressure valve, remove snap ring (127), spring retainer (126), spring (125) and valve (124). The valve body (123) can be unscrewed from lower end of return line. Inspect valve body and valve for scoring, sticking or excessive wear and renew if necessary. Spring should exert a pressure of 2.64 to 2.92 pounds when compressed to a length of 0.74 inch.

HYDRAULIC PUMP

Models "FSM" and "New FSM"

274. **REMOVE AND REINSTALL.** To remove the hydraulic pump, first drain the hydraulic sump (rear axle center housing) and remove the hydraulic lift cover as outlined in paragraph 261. Unbolt the pressure line from pump front cover (mounting bracket), then unbolt and remove the pump from pump pedestal. Note: Pump front cover is secured to mounting bracket with three cap screws and two dowel pins.

When the pump is removed, pedestal cover should be removed from lower side of tractor and the pump suction screen cleaned.

To reinstall hydraulic pump, position the pump on the two dowel pins and securely install the three retaining cap screws. Use a new "O" ring and reconnect pressure tube to pump front cover. Reinstall lift cover as outlined in paragraph 261 and refill hydraulic system as outlined in paragraph 243.

275. **OVERHAUL PUMP.** With the hydraulic pump removed as outlined in paragraph 274, proceed as follows: Straighten the lock tab washer (2—Fig. 216) and remove nut (1), washer and pump gear (3) and extract Woodruff key (11) from shaft. Note: Use of a special gear remover (Nuday tool No. N-6306-B) is recommended to prevent damage to pump cover or other parts when removing gear. Remove the nuts, washers and bolts retaining the pump end covers. Note that the two bolts in line with the bolt holes for pressure line are machined dowel bolts and are identified by a "D" on the bolt head. Remove the covers and discard the sealing "O" type rings (8 and 15). Remove snap ring (4), seal (5) and washer (6) from shaft bore in rear cover (7). Remove the pump gears and bearing blocks as an assembly as shown in Fig. 217, then separate the gears and blocks.

Carefully inspect pump gears, bearing blocks and pump body (9—Fig. 216) for signs of seizure or scoring; light score marking can be removed by careful lapping with "O" grade emery paper and kerosene. Examine body for wear in gear running track; if track is worn deeper than 0.0025 on inlet side of body, renew the body. Run-out across gear face to tooth edge should not exceed 0.001. Face widths

of each pair of gears must be within 0.001.

Early production bearing blocks had "run-out slots" (R—Fig. 219) and front bearing block (14—Fig. 216) is different from rear bearing block (10). When assembling early type pump, refer to Fig. 219 for view of rear bearing block; "run-out slots" in front block will be opposite to that shown. On later production pumps, the "run-out slots" have been eliminated and front and rear bearings are alike. Later type bearings can be used as a pair to renew early type front and rear bearings.

The groove for the "O" type sealing rings (8 and 15—Fig. 216) in pump covers (7 and 16) have been changed from that shown, requiring a differently shaped and heavier "O" ring. The early and late type covers are interchangeable providing the correct "O" type sealing rings are used.

To reassemble the pump, refer to Fig. 216 and lay out the component parts as shown. Note: Late type bearing blocks do not have the "run-out slots". Refer also to Fig. 218. Set the pump gears in the rear bearing, then invert front bearing over gear shafts. Be sure the intake side of both bearing blocks (sides with oil slot into bearing bore) are to the same side of the assembly. Insert the assembled bearing blocks and gears into pump body as shown in Fig. 217; refer to Fig. 216 for view of rear face of pump body.

Place washer (6) in drive shaft bore of rear cover (7), then using suitable driver, install new seal (5) with lip forward (to inside of pump). Lubricate the seal and install snap ring (4). Be careful when installing rear cover over drive gear shaft so as not to damage the seal. Install the two dowel bolts (identified with "D" on bolt heads) in holes that are in

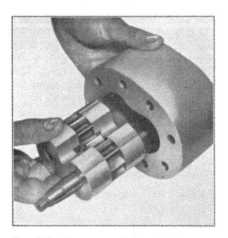

Fig. 217—Remove and install the pump gears and bearings as an assembly as shown.

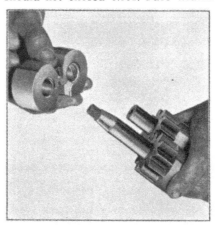

Fig. 218—Assembling rear bearing to pump gear shafts. On later pumps, both bearings are alike.

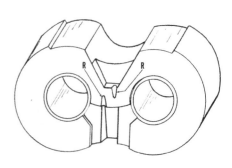

Fig. 219—On early bearings, front and rear bearings differ due to position or run-out slots (R). Rear bearing is shown. Latest bearings do not have the run-out slots and both bearings are alike.

line with pressure line adapter bolt holes. Install remaining cap screws, then evenly tighten all bolts until a final torque of 40-45 Ft.-Lbs. is reached. It is important that this torque specification be strictly adhered to.

Insert Woodruff key in pump drive shaft and install pump drive gear. Note: Drive gear for model "New FSM" has 18 teeth while gear for earlier models has 21 teeth. Tighten the gear retaining nut and secure with lock tab washer.

Pour oil into pump suction opening and turn gear until pump is thoroughly lubricated. Reinstall pump as outlined in paragraph 274.

NOTES

NOTES

FORD

Models ■ 6000 ■ Commander 6000

Previously contained in I&T Shop Service Manual No. FO-22

SHOP MANUAL

FORD

SERIES 6000

INDEX (By Starting Paragraph)

CONDENSED SERVICE DATA

GENERAL

	Series 6000 Commander 6000 Non-Diesel	Series 6000 Commander 6000 Diesel
Torque Recommendations	———————See End of Shop Manual———————	
Engine Make	Own	Own
Cylinders, Number of........................	6	6
Bore—Inches...............................	3.62	3.62
Stroke—Inches.............................	3.60	3.90
Displacement—Cubic Inches..................	223	242
Compression Ratio..........................	8.40	16.5
Pistons Removed From	Above	Above
Main & Rod Bearings Adjustable?	No	No
Generator & Regulator Make..................	Own	Own
Starter Make...............................	Own	Delco-Remy
Carburetor Make	Zenith	None
Distributor Make	Own	None

TUNE-UP

Firing Order	1-5-3-6-2-4	1-5-3-6-2-4
Distributor Poin Gap	0.024-0.026	......
Valve Tappet Gap—Intake....................	0.015 Hot	0.015 Hot
Valve Tappet Gap—Exhaust	0.015 Hot	0.015 Hot
Valve Face Angle—Degrees	———————See. Par. 28———————	
Vavle Seat Angle—Degrees..................	———————See. Par. 28———————	
Ignition Timing	See Par. 82	......
Injection timing	......	See Par. 63
Spark Plug Make	Auto-Lite	......
Spark Plug Size	18MM	......
Spark Plug Electrode Gap (Gasoline)	0.028-0.032	......
(LPG)...................................	0.018-0.022	......
Engine Low Idle—RPM	650-700	800-850
Engine High Idle—RPM (Series 6000)	2575-2625	2475-2525
(Commander 6000)	2675-2725	2645-2695
Engine Rated Speed—RPM (Series 6000)..........	2300	2230
(Commander 6000)	2430	2430
Battery Terminal Grounded	Negative	Negative

SIZES—CAPACITIES—CLEARANCES

Crankshaft Journal Diameter	———————See. Par. 43———————	
Crankpin Diameter..........................	2.2980-2.2988	2.2980-2.2988
Camshaft Journals Diameter	1.9255-1.9265	1.9255-1.9265
Piston Pin Diameter	0.9120-0.9123	1.1240-1.1243
Valve Stem Diameter, Intake	———————See. Par. 28———————	
Valve Stem Diameter, Exhaust.................	———————See. Par. 28———————	
Main Bearings Running		
Clearance	See Par. 43	0.0019-0.0040
Rod Bearings Running		
Clearance	0.0006-0.0025	0.0006-0.0025
Piston Skirt Clearance	———————See. Par. 40———————	
Crankshaft End Play.........................	0.004-0.008	0.004-0.008
Camshaft Bearing Running		
Clearance	0.001-0.003	0.001-0.003
Cooling System—Quarts (Series 6000)	21	21
(Commander 6000)	15-1/2	15-1/2
Crankcase—Quarts (With Filter)	5	5
Transmission—Quarts	12	12
Rear Axle, No PTO—Quarts	14	14
Rear Axle, With PTO—Quarts..................	12	12
Hydraulic System Reservoir—		
Quarts.................................	14	14
Steering Gear Housing—Lbs.	1/2	1/2

FRONT SYSTEM

The series 6000 and Commander 6000 tractors are available with any of four front end assemblies; a single wheel tricycle, a dual wheel tricycle, a wide adjustable front axle and an all purpose front axle.

The most obvious difference between the two adjustable front axles is that the all purpose axle (Fig. F012A) is mounted to the tractor with the stay rod portion of the axle toward the front of the tractor rather than rearward as with the wide adjustable axle. Mounting the axle this way shortens the tractor wheelbase to 85 inches as compared to the 104.24 inch wheelbase when tractor is equipped with the wide adjustable axle. Service procedures will not differ greatly and will be obvious upon examination.

Power steering is standard equipment for all models.

TRICYCLE TYPE

Single Wheel Models

1. The single wheel tricycle wheel spindle is bolted to the power steering motor shaft and is installed so that the arm is on left side as shown in Fig. FO10. When installing wheel spindle, tighten the four retaining cap screws to 150-180 ft.-lbs. torque.

To adjust the wheel bearings, raise front of tractor and remove cap (36). Remove cotter pin from adjusting nut (35) and tighten nut until bearings are fully seated, then back-off nut until bearings have zero end play and wheel rotates with a very slight drag.

Renewal of bearings, cups and retainer is obvious after an examination of the unit and reference to Fig. FO10.

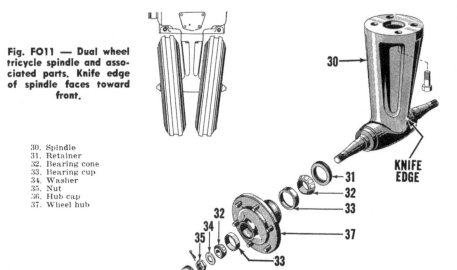

Fig. FO11 — Dual wheel tricycle spindle and associated parts. Knife edge of spindle faces toward front.

30. Spindle
31. Retainer
32. Bearing cone
33. Bearing cup
34. Washer
35. Nut
36. Hub cap
37. Wheel hub

Dual Wheel Models

2. The dual wheel spindle assembly attaches to the power steering motor shaft and must be installed with the knife edge toward front as shown in Fig. FO11. When installing the spindle, tighten the four retaining cap screws to 150-180 ft.-lbs. torque.

To adjust the wheel bearings, raise front of tractor and remove cap (36). Remove cotter pin from adjusting nut (35) and tighten nut until bearings are fully seated, then back-off nut until bearings have zero end play and wheels rotate with a very slight drag.

Renewal of bearings, cups and retainers is obvious after an examination of the unit and reference to Fig. FO11.

AXLE TYPE

Adjustable Axle Models

3. **SPINDLE BUSHINGS.** The spindle bushings (17 & 19—Fig. FO12 or FO12A) can be removed with a cape chisel after removing steering arms (3) and withdrawing spindles. (1). Bushings are pre-sized and should be installed flush with outer ends of bore using a suitable piloted driver. Tighten the steering arm retaining cap screws to 40-55 ft.-lbs. torque on tractors prior to serial number 21962 having $\frac{7}{16}$-inch cap screws, or to 65-90 ft.-lbs. torque on later tractors having ½-inch cap screws.

4. **TIE-RODS AND TOE-IN.** The tie-rod ends are the non-adjustable type and correction of faulty units is accomplished by renewal of the units.

To adjust the toe-in, turn wheel to the straight ahead position, loosen clamps on both tie-rod tubes and rotate each tube an equal amount to obtain a toe-in of ¼-inch. Tie-rods should be within ⅛-inch equal length when adjustment is complete. Tighten clamps securely.

5. **AXLE PIVOT PINS.** The axle pivot pins (9 & 15—Fig. FO12) of wide adjustable front axles have no bushings and are retained in supports (12 and 13) by cap screws.

The stay rod and rear support bracket of the all purpose front axle are equipped with renewable bushings (24 and 25—Fig. FO12A) and the pivot pins are retained in supports (12 and 13) by cap screws. The bushings are presized and should require no final sizing if carefully installed.

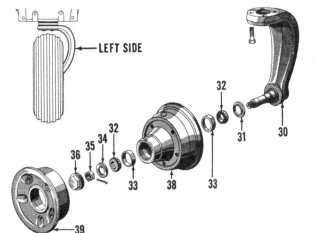

Fig. FO10—Single wheel tricycle spindle and associated parts.

30. Spindle
31. Retainer
32. Bearing cone
33. Bearing cup
34. Washer
35. Nut
36. Hub cap
38. Wheel disc
39. Wheel flange

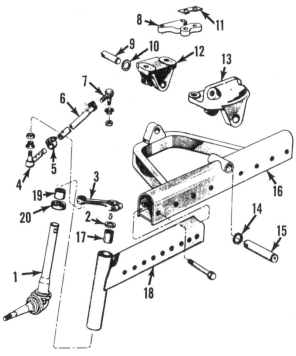

Fig. FO12 — Exploded view of the wide adjustable front axle and associated parts.

1. Spindle
2. Dust seal
3. Steering arm
4. Tie-rod end
5. Clamp
6. Adjusting tube
7. Tie-rod end
8. Center steering arm
9. Pivot pin, rear
10. Spacer
11. Bolt anchor
12. Support, rear
13. Support, front
14. Spacer
15. Pivot pin, front
16. Center main member
17. Bushing
18. Axle extension
19. Bushing
20. Bearing

To renew pivot pins, first disconnect the center steering arm (8) from power steering motor shaft, then remove the retaining cap screws. Raise front of tractor enough to take weight off pins, then using a suitable drift, drive pins forward out of supports. Do not lose spacers as pivot pins are removed.

When reassembling, tighten the center steering arm retaining cap screws to 150-180 ft.-lbs. torque.

6. AXLE MAIN MEMBER. To remove the axle main member, first remove pivot pins as outlined in paragraph 5. Raise front of tractor until supports will clear axle assembly and roll assembly forward. Any further disassembly will be obvious.

NOTE: Depending on the work required, some mechanics prefer to remove the axle extensions and lower the center main member, using a rolling floor jack.

7. AXLE EXTENSIONS. The axle extensions are retained in the center main member by two bolts. To renew an axle extension, first remove the steering arm (3) and withdraw spindles (1) from extensions. Remove the two bolts and pull extension from center main member.

Bushings (17 and 19) are final sized and should be installed flush with outer ends of bore using a suitable piloted driver. Tighten the steering arm retaining cap screws to 40-55 ft.-lbs. torque on tractors prior to serial number 21962 having $\frac{7}{16}$-inch cap screws, or to 65-90 ft.-lbs. torque on later tractors having ½-inch cap screws.

FRONT SUPPORT

8. REMOVE AND REINSTALL. The front support will not normally require any service; however, should it become necessary to remove or renew a front support, proceed as follows: Remove hood front side panels, radiator grilles, top hood and splash pan. Relieve the hydraulic system pressure by cycling the system, then disconnect lines from unload valve and drain hydraulic system. Disconnect all lines from hydraulic reservoir, then unbolt and remove hydraulic reservoir.

Disconnect pressure line from accumulator, disconnect mounting brackets and lift out accumulator assembly.

Remove steering gear and motor assembly as outlined in paragraph 16, disconnect head light wires, then unbolt front hood from hood rails and front support and lift same from tractor with a hoist. On adjustable axle tractors, remove front axle assembly by unbolting front and rear axle supports. Attach hoist to front support, remove retaining cap screws and pull same from side rails.

Upper and lower bearing cups and seals can be driven from front support, if necessary. When renewing the lower shaft seal proceed as follows: Place steering gear and motor assembly in the front support and support unit approximately ⅛-inch above its mounting pads. Use a 3½-inch O. D. seal driver and bump seal into its bore until rubber gasket of seal contacts steering motor base. Weight of steering gear and motor assembly will compress the seal gasket and position seal assembly when installed as outlined in paragraph 16.

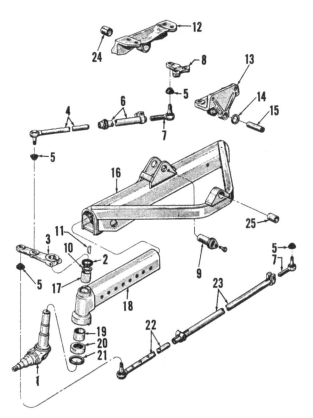

Fig. FO12A — Exploded view of the all purpose front axle and associated parts.

1. Spindle
2. Dust seal
3. Steering arm, R. H.
4. Steering arm connecting rod
5. Dust covers
6. Sleeve & clamp assy.
7. Rod ends
8. Center steering arm
9. Pivot pin, rear
10. Washer
11. Woodruff key
12. Support, rear
13. Support, front
14. Spacer
15. Pivot pin, front
16. Center main member
17. Bushing
18. Axle extension
19. Bushing
20. Bearing
21. Spacer
22. Connecting rod
23. Sleeve & clamp assy.
24. Bushing
25. Bushing

POWER STEERING

Fig. FO13 — Power steering control valve, steering gear and steering motor unit removed from tractor. Sector shaft is shown inserted for illustrative purposes.

Power steering is standard equipment on all models. The main components of the system are a gear driven roller vane type pump and a unit in which is combined a control valve, a recirculating ball nut type steering gear and a rotary vane type steering motor. Refer to Fig. FO13 for a view of the combined control valve, steering gear and steering motor unit.

SYSTEM OPERATING PRESSURE

10. A pressure test of the power steering circuit will disclose whether the pump or some other unit in the system is malfunctioning.

To make such a test, remove right front grille and side panel and radiator lower baffle, connect a pressure test gage assembly (N1100-PSA, or equivalent) capable of reading at least 2000 psi pressure in series with the pump pressure line as shown in Fig. FO14. Run engine until oil is warmed to operating temperature. With engine running at approximately 1000 rpm, turn steering wheel to full left or right, hold in this position and observe the gage pressure which should be 1100-1350 psi. Maximum working pressure of power steering pump is 1350 psi and if gage pressure approximates this reading, pump and relief valve can be considered satisfactory. If pressure is not as specified, renew the combined flow control and relief valve (4—Fig. FO15) and retest. If pressure is still considerably lower than 1100 psi, overhaul the pump as outlined in paragraph 12. If pump and relief valve are satisfactory, and the system still does not operate properly, refer to subsequent paragraphs for checking and overhauling the remainder of the system components.

POWER STEERING PUMP

11. **REMOVE AND REINSTALL.** Relieve all accumulator pressure

LUBRICATION AND BLEEDING

9. The power steering system can be considered self-bleeding; however, whenever service has been performed on any of the component parts, be sure all connections are made and tightened; then fill hydraulic reservoir with 3½ gallons of Ford M-2C-41 oil, or its equivalent. Start engine and cycle system several times to purge any air which may be present. Place hydraulic lift in lowered position, retract remote cylinders and relieve system pressure; then recheck reservoir level and refill as necessary.

Fig. FO14—Method of installing pressure gage in series with the power steering pump pressure line.

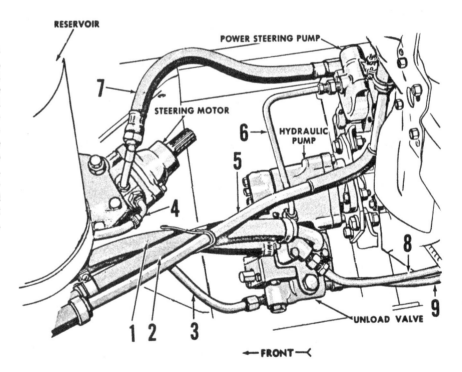

Fig. FO14A — View showing arrangement of hydraulic lines, tubes and related parts located at front of tractor. Early model tractor is shown. Later tractors differ in that the power steering by-pass line (6) is now a flexible line that is attached to a "tee" in the unload valve. The power steering return line (4) is now also attached to the "tee" instead of the reservoir.

1. Supply line	4. Power steering return line	7. Power steering pressure line
2. Return line	5. Power steering pump supply line	8. Brake return line
3. Accumulator pressure line	6. By-pass line	9. Brake pressure line

1. Plug
2. "O" ring
3. Spring
4. Flow control valve
5. Relief valve
6. Cover
7. Drive pin
9. Shaft
10. "O" ring
11. Rotor
12. Roller
13. Cam ring
14. Dowel pin
15. Inlet hose
16. Housing
17. "O" ring
18. Gasket
19. Oil seal
20. Snap ring
21. Bearing
22. Flange
23. Spacer
24. Driven gear
25. Washer
26. Nut

Fig. FO15 — Exploded view of power steering pump. Flow control valve (items 4 and 5) is serviced as a complete unit only.

either by actuating brakes or cycling hydraulic lift. Remove the hood front side panel, grill, baffle and splash pan, then disconnect the pump supply line (5—FO14A) at pump and drain hydraulic reservoir. Disconnect the pump to control valve pressure line (7) at control valve. Remove the by-pass line (6) between pump and unload valve. Unbolt and remove pump from timing gear cover.

Note: Removal of lower inside retaining nut will be simplified if a curved manifold wrench is used.

Reinstall by reversing the removal procedure and fill and bleed system as outlined in paragraph 9.

12. **OVERHAUL.** Disconnect the pump to control valve pressure line from pump, then mount pump in a vise with drive shaft end up. Unstake retaining nut (26—Fig. FO15) and remove nut, lock washer (25), gear (24) and spacer (23). Remove the two Allen screws and separate pump flange (22) from pump. Remove gasket (18), "O" ring (17), snap ring (20) and bearing (21) from flange. Remove the cover to housing cap screws and remove pump from vise;

then, separate pump and remove the two "O" rings. Remove the shaft (9), rotor (11), rollers (12), and cam ring (13), cam locating pin (14) and oil seal (19) from pump housing. Remove plug (1), spring (3) and flow control valve (4) from pump cover.

Note: The combined flow control and relief valve is not a serviceable item and if relief valve (5) or flow control valve is faulty, the complete unit must be renewed.

Wash all parts except the sealed roller bearing (21) in a suitable solvent and inspect as follows: Check bushings in housing and cover and if excessively worn, renew housing and/or cover as bushings are not available separately. Inspect cam ring, rotor and rollers for excessive wear. Renew rollers if they are scored or out-of-round. Check the length of rollers and thickness of rotor and if rollers are more than 0.0008 shorter than rotor thickness, renew rollers. Place cam ring and rotor in pump housing. Place a straight edge across machined surface of housing and use a feeler gage to measure between straight edge and rotor as shown in Fig. FO16. If clearance between cam ring and rotor exceeds 0.0016, renew rotor. A power steering pump repair kit which contains the cam ring, rotor and rollers is available for service.

When reassembling use new "O" rings and coat shaft and double lip of seal with Lubriplate, or its equivalent. Use a seal protector or shim stock when installing shaft in housing to avoid damage to seal. Install rotor so roller slots are positioned, and will travel, as indicated in Fig. FO17. Complete reassembly by reversing disassembly procedure. Flow control valve is installed with extension toward inside. Torque housing to cover cap

screws to 25-30 ft.-lbs., the flow control valve plug to 30-35 ft.-lbs. and the driven gear retaining nut to 40-50 ft.-lbs.

CONTROL VALVE

13. **R&R AND OVERHAUL.** Remove right front side panel, grille, baffle and splash pan. Remove all accumulator pressure either by actuating brakes or cycling hydraulic lift system. Disconnect supply line from power steering pump and drain hydraulic reservoir. Note: Reservoir can also be drained by removing check valve from "T" fitting on right side of hydraulic lift housing and running engine. Stop engine as soon as the oil flow stops. Disconnect steering shaft brackets from side rail and remove center and front steering shafts. Disconnect return line and pressure line from top of control valve. Disconnect steering motor lines from lower right side of control valve, then loosen the connectors at steering motor and rotate lines out of the way. Place a scribe line across control valve assembly to aid in reassembly, then remove the control valve cover. Unstake retaining nut, hold steering shaft from turning, then remove and discard nut. Pull control valve assembly and thrust bearings from steering shaft. See Fig. FO18.

Note that embossed letters "RT" and "PR" are toward aft end of steering shaft. Also note the groove machined in the I. D. of the valve spool prior to removing valve (spool) from control body. Be sure to reinstall valve spool in same position.

Wash all parts in a suitable solvent and inspect bearings and retainers for wear or damage. Inspect valve (spool)

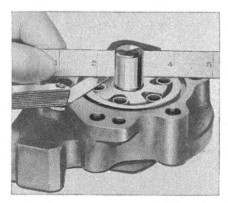

Fig. FO16 — Maximum clearance between rotor and cam ring is 0.0016 when measured as shown.

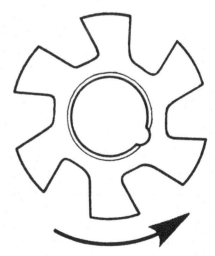

Fig. FO17—Install rotor in pump so it will turn in the direction indicated by arrow.

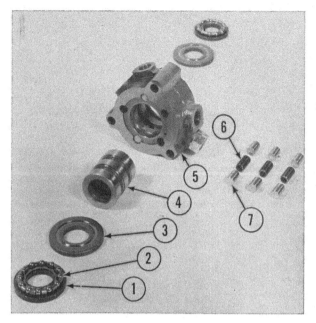

Fig. FO18—Disassembled view of power steering valve showing component parts. Note that two of the centering spring assemblies are still in body.

1. Retainer, small
2. Bearing
3. Retainer, large
4. Valve (spool)
5. Body
6. Centering spring
7. Plunger

and body for scoring, galling or other damage. If valve (4) and/or body (5) is damaged, both must be renewed as they are available only as a matched set. Inspect plungers (7) for nicks or scoring and the centering springs (6) for signs of distortion or fractures. Seal can be removed from cover after removing snap ring. Seal is installed in cover with lip facing toward spool. Lubricate lip of seal with Lubriplate, or equivalent, prior to installation.

14. Reassemble the control valve as follows: Install a small retainer (1), then a bearing (2) on steering shaft. Install valve (spool), centering springs and plungers in valve body and retain in position with the large retainers (3). Install assembly on steering shaft, install bearing and remaining small

retainer, then install a new nut and tighten only enough to hold parts in position. Use a cover which has the center portion cut out (enough to allow adjustment of nut) and install cover on valve. Rotate steering shaft in a clockwise direction and while holding steering shaft in this position, tighten nut finger tight (9-12 in.-lbs. torque). Stake nut in this position. Remove the modified valve cover and reinstall the original. Tighten cover cap screws to 15-20 ft.-lbs. torque.

Balance of reassembly is evident. Fill reservoir, if necessary, and bleed system as outlined in paragraph 9.

STEERING GEAR AND MOTOR

15. **ADJUSTMENT.** One cam adjustment, shown in Fig. FO19, is provided to adjust the steering gear assembly. Adjustment should be made with the front wheels in straight ahead position and with tractor weight off of front wheels.

To adjust, loosen pivot stud nut and the cap screw which retains the adjuster sleeve (cam) as well as the three other cap screws which retain steering gear housing to steering motor.

Note: These cap screws can be identified by the cap screw heads which are thicker than those of the top cover retaining cap screws.

With cap screws loosened slightly, rotate adjuster sleeve to move ball nut toward sector gear; that is, the rear of steering gear housing should pivot toward right. After ball nut and sector gear have butted together, back-off adjuster sleeve only enough to relieve any tension which might be

present (zero backlash). Tighten the cap screws and pivot stud nut to 40-50 ft.-lbs. torque. The front wheels should then turn from side to side without any binding condition as they are turned past straight ahead position.

Note: In some cases, the adjuster sleeve may turn completely around without obtaining zero backlash of ball nut and sector gear. When this occurs, additional adjustment can be obtained by using a pry bar on left side of steering gear housing and moving the housing toward right. Tighten cap screws while holding housing with pry bar.

16. **R&R STEERING GEAR.** Removal of steering gear requires the removal of the control valve, steering gear and power steering motor as a unit. To remove the steering gear, proceed as follows: Relieve accumulator pressure by actuating brakes or cycling hydraulic lift system. Remove hood, front side panels and both lower baffles. Turn wheels to straight ahead position. Unbolt steering shaft supports from right side rail and remove center and front steering shafts. Disconnect supply line at power steering pump and drain hydraulic reservoir, then remove line. Remove the accumulator to unload valve pressure line. Disconnect the hydraulic lift return line, the unload valve return line and the inlet lines from hydraulic reservoir and pull these to the left side of tractor so they will not cause interference during steering gear removal. Disconnect return line from control valve. Loosen adjuster sleeve cap screw and other cap screws retaining gear housing to steering motor and turn adjuster sleeves so that ball nut clears sector gear.

On tricycle type tractors, unbolt and remove wheels and spindle from steering motor shaft. On wide adjustable axle type tractors, disconnect tie-rod from right steering arm, disconnect center steering arm from steering motor shaft and move assembly rearward to clear steering motor shaft. On all purpose axle type tractors, disconnect center steering arm from steering motor shaft and move center steering arm and connecting rod (drag link) forward. Pry off dust cap at top of steering motor shaft, remove cotter key and nut, then remove shaft from bottom side of front support. Remove the bolts which retain steering gear and motor in front support and remove unit from right side of tractor. See Fig. FO13.

Fig. FO19—View showing location of steering gear adjuster sleeve (S) and sleeve retaining cap screw (C). Refer to text for adjustment procedure.

CAUTION: Be careful when removing steering gear and motor unit not to damage radiator.

Reinstall the steering gear and steering motor unit as follows: Position unit in front support. Install right hand mounting bolt downward and left hand mounting bolt upward. Install washers and nuts and tighten nuts to 190-210 ft.-lbs. torque. Turn steering gear to its mid-point of travel and align the master splines of sector gear and steering motor rotor and vane assembly. Align master spline of sector shaft with master splines of steering motor and steering gear and install sector shaft from bottom of front support. Install bearing and nut at top of sector shaft and tighten nut to a torque of 0-5 ft.-lbs. Install cotter pin and dust cap. Reconnect steering shaft and hydraulic lines. Reinstall wheels and spindle (tricycle type) or center steering arm and tie-rod (adjustable axle models), then fill and bleed hydraulic system as outlined in paragraph 9 and readjust steering gear as outlined in paragraph 15.

17. OVERHAUL STEERING GEAR. With complete unit removed as outlined in paragraph 16, place scribe line across control valve to aid in reassembly and remove control valve cover. Unstake nut and remove nut, thrust bearing assemblies and control valve from steering shaft. Discard control valve nut. Remove cap screws

and adjuster sleeve from steering gear top cover, then remove cover and lift out sector gear. Lift steering gear assembly from steering motor. See Fig. FO20.

Turn the ball nut so ball guide clamp is on top side, then remove clamp and ball guides along with the balls they contain. Turn ball nut until guide holes are on bottom side, rotate steering shaft and remove remaining balls.

Note: A total of 66 balls are used in the ball nut and ball nut guide circuits.

Rotate steering shaft out of ball nut and remove ball nut and steering shaft from steering gear housing. Bearings can be removed from housing, if necessary, by using a puller and slide hammer arrangement.

Wash all parts and check teeth of ball nut and sector gear for excessive wear or other damage. Check splines of sector shaft and inner diameter of sector gear for wear, chipping or other damage. Check balls and grooves of ball nut and steering shaft for wear, scoring or other damage. Renew parts as necessary. If new steering shaft needle bearings are installed, press on end which has the identification marks.

Reassemble unit as follows: Coat grooves of steering shaft and ball nut with grease, insert steering shaft through outer (rear) bearing and into ball nut. Use a rod, or other suitable tool as shown in Fig. FO21, and insert 25 balls in each of the two ball guide openings. Coat the ball guide halves with grease and install eight balls in each ball guide half. Place other

halves of ball guides over the halves containing the balls, then insert guides into ball nut and install clamp. Use a new gasket and place steering gear assembly on steering motor. Install sector gear so middle tooth meshes with center tooth space of ball nut rack. Place ½-pound of Ford M-4738 grease (high melting point), or equivalent, in housing and be sure ball nut, sector gear and steering shaft are thoroughly coated. Use new gasket and install steering gear top cover, however, tighten the retaining cap screws and nut only finger tight at this time.

Note: The cover and housing retaining cap screws are left loose at this time to aid in centering seals during installation of sector shaft and to allow backlash adjustment of steering gear unit after installation of unit in tractor.

Reinstall control valve as outlined in paragraph 14, then install unit in tractor as outlined in paragraph 16. Adjust steering gear as outlined in paragraph 15, then fill and bleed system as outlined in paragraph 9.

18. R&R STEERING MOTOR. Steering motor and steering gear are removed and reinstalled as a unit. Refer to paragraph 16 for procedure.

19. OVERHAUL STEERING MOTOR. With steering gear and motor unit removed from tractor, remove the steering gear and control valve assembly from steering motor housing. To aid in reassembly, scribe a line across steering motor assembly. Remove the retaining cap screws from base of steering motor and remove base and cover from steering motor housing. See Fig. FO22. Mark both hub and housing so hub can be rein-

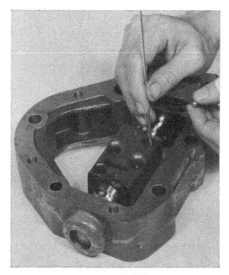

Fig. FO20 — View showing steering gear and power steering motor with top cover of steering gear removed. Sector shaft is shown installed for illustrative purposes.

Fig. FO21—When inserting balls in ball nut, work on same side and install balls as shown.

Fig. FO22 — Power steering motor with cover removed. Place correlation marks on hub and housing before removing.

stalled in the same position, then push hub from housing. Discard all "O" rings and the hub and vane seals. If desired, stops can be removed from housing; however, do not remove vanes from hub as "Loctite" was used on the Allen head screws during assembly and removal of screws could result in damage to screws and/or hub.

Check hub and vanes for scoring, burrs or excessive wear. Check splines in hub for chipping, wear or other damage. Inspect bearings in base and cover and if new bearings are installed, press on ends which carry identification markings.

When reassembling, place new seals on vanes and stops and note that the legs of stop seals are $\frac{1}{16}$-inch longer than those of the vane seals. Lubricate all parts prior to assembly, and with the previously affixed marks aligned, install hub and vanes in hous-

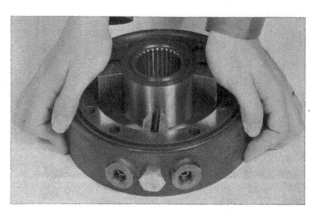

Fig. FO23 — Install hub and vanes as shown. Use care not to damage the square seals.

ing as shown in Fig. FO23.

Note: If stops were removed, tighten their retaining cap screws only after unit is assembled and be sure cap screw seals are installed. Torque value for these cap screws is 90-100 ft.-lbs.

Place new "O" rings in hub and housing, align scribe mark, then install base and cover and torque cap

screws to 55-60 ft.-lbs. If necessary, the stop retaining cap screws can now be tightened to 90-100 ft.-lbs. torque.

Reinstall steering gear and control valve assembly on steering motor, then reinstall complete unit in tractor as outlined in paragraph 16. Adjust steering gear as outlined in paragraph 15, then fill and bleed system as outlined in paragraph 9.

ENGINE AND COMPONENTS

The basic design of the gasoline and diesel engines used in the series 6000 and Commander 6000 tractors is similar and many of the component parts are interchangeable. The non-diesel engine has a bore of 3.62 inches, a stroke of 3.60 inches and a displacement of 223 cubic inches. The diesel engine has a bore of 3.62 inches, a stroke of 3.90 inches and a displacement of 242 cubic inches.

R&R ENGINE ASSEMBLY
All Models

25. Remove precleaner, muffler cover (series 6000), muffler, hood, rear hood, front side panels, rear side panels, radiator baffles and splash pan. Actuate brakes or hydraulic lift and relieve the accumulator pressure. Disconnect inlet and return lines from unload valve and drain hydraulic system. Drain cooling system. Disconnect battery cables and remove battery. Remove the hood rear support. Disconnect headlight wires. Disconnect wire from fuel gage sending unit. Shut off fuel and remove fuel supply line. On diesel models, disconnect leak-off line from cylinder head. On all models, remove fuel tank hold down bolts and lift off fuel tank. Remove the air cleaner to manifold (diesel) or carburetor (gasoline) tube. Disconnect transmission cooling line and filter line clips from right hood mounting rail. Disconnect transmission filter bracket. Remove hood front

support, loosen radiator upper support brackets, then disconnect hood rails and remove hood rails and air cleaner as a unit. Disconnect radiator hoses and transmission cooling lines from radiator and remove radiator. Disconnect wiring from generator, manifold heater, starter, coolant temperature sending unit and oil pressure sending unit and pull harness from positioning clips. Disconnect Proof-Meter cable. On diesel models, disconnect fuel shut-off wire, then disconnect both ends of injection pump control rod and move rod rearward as far as possible. On gasoline models, disconnect choke cable, loosen travel limit clips, then disconnect both ends of throttle rod and move rearward as far as possible. On all models, remove starter and manifold heat shield. Disconnect the two steering shaft supports from right side rail and remove center and front steering shafts. Disconnect power steering pump pressure line at control valve. Remove power steering pump by-pass line and supply line. Disconnect the brake valve return line, brake valve pressure line and accumulator line from unload valve, then disconnect brake lines from cylinder block. Attach a Ford N600AB engine lifting fixture, or equivalent, to left side of engine, then attach hoist to fixture and take weight of the engine. Unbolt the front cross member and remove the cross member, hydraulic

pump and unload valve as an assembly. Remove both right and left side transmission to side rail spacer blocks and the engine fire wall. Unbolt engine from transmission, move engine forward to clear transmission input shaft, then lift engine from tractor.

R&R CYLINDER HEAD
Diesel Models

26. Remove muffler cover (series 6000) muffler, hood and fuel tank. Drain cooling system and disconnect upper radiator hose from thermostat housing. Unbolt fan support bracket and remove fan and bracket. Remove breather tube from tappet cover and remove tappet cover from cylinder head. Disconnect fuel filter from cylinder head. Disconnect excess fuel line from the tee connection and injectors and remove excess fuel line. Disconnect injector pressure lines from injectors.

Note: To prevent the entrance of dirt or other foreign material, cap off all openings as diesel lines are disconnected.

Remove the rocker arm shaft oil inlet and outlet tubes, then remove the rocker arms and shaft assembly, push rods and exhaust valve caps. Disconnect air cleaner tube and manifold heater wires from manifold, then remove manifold and heat shield. Disconnect coolant temperature sending wire from sending unit at right rear of cylinder head, then unbolt and remove cylinder head.

Reinstall by reversing removal procedure. Tighten cylinder head cap screws in the sequence shown in Fig. FO24 and to a torque of 95-100 ft.-lbs (dry threads).

Non-Diesel Models

27. Remove muffler cover (series 6000), muffler, hood and fuel tank. Drain cooling system and disconnect upper radiator hose from thermostat housing. Unbolt fan support bracket and remove fan and bracket. Remove breather tube from tappet cover and tappet cover from cylinder head. Remove the rocker arm shaft oil inlet and outlet tubes, then remove the rocker arms and shaft assembly, push rods and exhaust valve caps. Shut off fuel and disconnect choke control, throttle rod and fuel line from carburetor. Unbolt and remove manifold, carburetor and heat shield. Disconnect coolant temperature sending wire from sending unit at right rear of cylinder head, then unbolt and remove cylinder head.

Reinstall by reversing the removal procedure. Tighten cylinder head cap screws in the sequence shown in Fig. FO24 and to a torque of 95-105 ft.-lbs. (lubricated)

VALVES AND SEATS
All Models

28. Exhaust valves of both non-diesel and diesel engines are equipped with free type rotators as shown in Fig. FO25 and seat on renewable seat inserts which are a shrink fit in the cylinder head. Valve seat inserts are available in standard size only for all engines except the Commander 6000 LP-Gas engines which also has a 0.020 oversize seat available.

All valves of non-diesel engines are fitted with umbrella type stem seals which should be renewed each time valves are serviced and renewal procedure for these seals is obvious.

Intake valves of diesel engines are fitted with a boot type stem seal and seals should be renewed each time valves are serviced. Installation procedure is as follows: Install valve in guide and place plastic sleeve (furnished) over end of valve stem. If sleeve extends more than 1/16-inch below lower groove of valve stem,

cut off the excess length of sleeve. Lubricate sleeve, and while holding valve head in position, push stem seal assembly over sleeve and on to valve stem. NOTE: Be sure to push on the inner plastic (white) ring and not the rubber sleeve of seal. Continue to push seal down on valve stem until rubber sleeve is over valve guide. Using installation tool or two screw drivers, push on wire ring until seal is tight against top of valve guide. Remove plastic sleeve from valve stem, install valve spring, retainer and keepers. Compress spring only enough to install keepers. Compressing spring too much could result in damage to seal.

All valves, except Commander 6000 LP-Gas engine exhaust valves, have a face angle of 44 degrees and a seat angle of 45 degrees. The Commander 6000 LP-Gas engines have exhaust valves with a face and seat angle of 30 degrees and the cylinder heads of these engines are now identified by a "P" stamped near the number 1 spark plug hole as shown in Fig. FO24A. Desired seat width is 0.060-0.090 for the intake; 0.090-0.110 for the exhaust. Total seat run-out should not exceed 0.0015.

On diesel models, exhaust valve stem diameter is 0.3405-0.3415 and intake valve stem diameter is 0.3415-0.3425. On non-diesel models, exhaust valve stem diameter is 0.3398-0.3410 and intake valve stem diameter is 0.3415-0.3423. Diesel model valves are available in oversizes of 0.003 and 0.015. Non-diesel model valves are available in oversizes of 0.003, 0.015 and 0.030.

Recommended operating clearance of valves in guides for diesel models is 0.0005-0.0025 for the intake and 0.0015-0.0035 for the exhaust. For non-diesel models, operating clearance is 0.001-0.0035 for the intake valves and 0.0028-0.0042 for the exhaust valves.

VALVE GUIDES AND SPRINGS
All Models

29. All valve guides are integral with cylinder head and are not renewable. Valve guides can be reamed to provide installation of valves with oversize stems. Refer to paragraph 28 for valve stem dimensions and operating clearances.

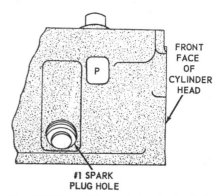

Fig. FO24A — The cylinder head of the Commander 6000 series LPG engines are now identified with the letter "P" as shown. These cylinder heads have a 30° exhaust valve seat.

Intake and exhaust valve springs are interchangeable within the same engine but not between non-diesel and diesel engines. Gasoline engine valve springs have a free length of 2.09 inches and should exert 71-79 pounds pressure when compressed to 1.39 inches. LP-Gas engine valve springs have a free length of 1.940 inches and should exert 31-39 pounds pressure when compressed to 1.809 inches. Diesel engine valve springs have a free length of 2.12 inches and should exert 54-62 pounds when compressed to a length of 1.82; or 124-140 pounds when compressed to a length of 1.505 inches. Renew any spring which is at least 7 pounds below specifications or has the ends more than $\frac{1}{16}$-inch out-of-square.

When reinstalling springs on early model diesel engines, be sure spacer washer is installed between spring and cylinder head on exhaust valves. Dampener coils (closed end) of springs are next to cylinder head.

EXHAUST VALVE ROTATORS
All Models

30. Refer to Fig. FO25. The free-type rotators will not function unless

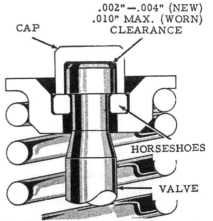

Fig. FO25 — Sectional view of free valve (release type) exhaust valve rotator. Note required end clearance between cap and end of valve stem.

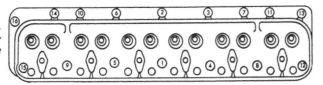

Fig. FO24 — Tighten cylinder head cap screws for all models in the sequence shown.

there is a measurable clearance (0.002-0.004) or end gap between the end of the exhaust valve stem and the inside floor of the cap when the open end of the cap just contacts the spring keeper or horseshoe as shown. Desired end gaps should be checked and if necessary adjusted each time the valves are reseated.

One of the simpler methods of checking is as follows: From a strip of 0.010 flat shim stock, cut a $\frac{3}{16}$-inch diameter disc. Lay this disc, which must be flat, on inside floor of rotator cap and install valve lock or keeper on valve stem. Now, while simultaneously pressing downward on valve lock and upward on rotator cap, measure with a feeler gage the gap between cap and valve lock. If gap measures anywhere between 0.006 and 0.008, it is within desired limits. If gap is less than 0.006, grind or lap open end face of cap; if more than 0.008, grind end of valve stem, however do not remove more than 0.010 from end of valve stem.

VALVE TAPPETS

All Models

31. Intake and exhaust valve tappet gap should be set to 0.015 when engine is at operating temperature.

The 0.4989-0.4995 diameter mushroom type tappets are available in standard size only and operate directly in the cylinder block bores with a clearance of 0.0005-0.0021. To remove the tappets, it is first necessary to remove the camshaft as outlined in paragraph 36.

ROCKER ARMS

All Models

32. Remove hood and fuel tank. Remove tappet cover. Remove oil inlet and outlet tubes from rocker arm supports. If grommet is missing from lower end of rear (inlet) oil tube, it can usually be "fished" out with a wire; if not, remove engine side plate and extract the grommet. Remove the cap screws which retain rocker arm supports to cylinder head and lift off rocker arms and shaft assembly.

All of the rocker arms are identical and interchangeable in the same type engine. The 0.780 - 0.781 diameter rocker arm shaft should have a clearance of 0.002-0.004 in the rocker arms. Renew rocker arms and/or shaft if clearance exceeds 0.007.

If I. D. of rocker arm shaft requires cleaning, drill a hole in one end plug and use a long rod inserted through hole to bump out plug on opposite end. The remaining plug can now be bumped out.

When reassembling rocker arm assembly, refer to Fig. FO27 to see the location of the different size spacers which are used on the diesel models. Wide spacers are 0.165 thick; narrow spacers are 0.0085 thick.

When reassembling rocker arm shaft to cylinder head, tighten the rocker arm support cap screws to a torque of 45-55 ft.-lbs. When installing the rocker arm oil lines, install new grommet on lower end of the rear oil line and be sure grommet is seated in the counter-bore in the cylinder block. Adjust the intake and exhaust tappet gap to 0.015 when engine is at operating temperature.

TIMING GEAR COVER

All Models

NOTE: Early engines are fitted with a vibration dampener which can be removed from the crankshaft with little or no difficulty. Late engines have vibration dampeners which are a press fit on the crankshaft. To re-move the vibration dampener from late engines, use two cap screws threaded into dampener and OTC puller number 518, or its equivalent. When reinstalling dampener, align index marks of dampener and crankshaft and using a ⅝ x 1¾-inch cap screw and washer, force dampener about half way on splines, then remove the 1¾-inch cap screw, install the original retaining cap screw (⅝ x 1¼-inch) and tighten cap screw to a torque of 45-55 ft.-lbs.

33. To remove the timing gear cover, first remove the hood front side panels, lower baffles and lower splash plate, then drain radiator and remove lower radiator hose. Loosen fan bracket and generator adjusting strap and remove fan belt and both generator belts. Exhaust accumulator pressure by actuating brakes or cycling the hydraulic lift. Disconnect power steering pump pressure line at control valve, remove pump by-pass line, then unbolt and remove power steering pump. Disconnect lines from hydraulic system unload valve, then unbolt and remove the engine front support, hydraulic pump and unload valve as a unit. Remove crankshaft pulley (vibration dampener). On non-diesel models, disconnect linkage from governor arm. On all models, unbolt timing gear cover from engine and oil pan, separate oil pan gasket from cover and remove cover.

Crankshaft front oil seal can be renewed at this time. For information on non-diesel governor and related parts, refer to paragraph 69.

Reinstall by reversing removal procedure and tighten cover retaining cap screws to 9-12 ft.-lbs. torque. BE SURE to mate timing (punch) marks of crankshaft pulley and crankshaft and torque retaining cap screw to 45-55 ft.-lbs.

TIMING GEARS

All Models

34. CAMSHAFT GEAR. To remove the camshaft gear, first remove the timing gear cover as outlined in paragraph 33, then prior to removing camshaft gear, check camshaft end play by measuring between gear hub and thrust plate. Normal camshaft end play is 0.003 - 0.007 and if the clearance exceeds 0.007, renew thrust plate during reassembly.

Remove snap ring which retains camshaft gear to camshaft, then using a suitable puller, pull gear from shaft.

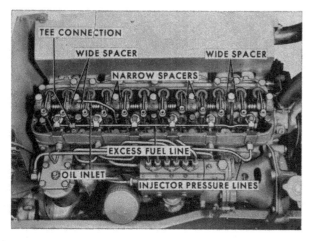

Fig. FO27 — When reassembling rocker arm assembly on diesel engines be sure spacers are located as shown.

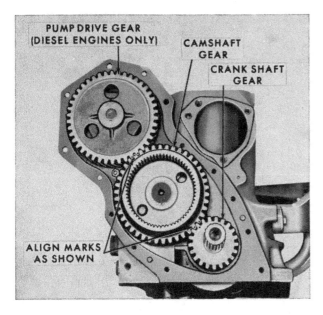

Fig. FO28—Timing gears are properly installed when punch marks are aligned as shown.

Recommended timing gear backlash is 0.007-0.013 with a maximum allowable limit of 0.018. Excessive backlash is corrected by renewal of timing gears. When reassembling, mesh the punch marked tooth space of camshaft gear with punch marked tooth of crankshaft gear as shown in Fig. FO28.

Heat gear in hot oil prior to installation and if necessary, remove fuel pump, or fuel pump hole cover and buck-up camshaft while installing camshaft gear.

35. CRANKSHAFT GEAR. To remove the crankshaft gear, first remove timing gear cover as outlined in paragraph 33. On non-diesel engines, slide governor assembly from crankshaft. On diesel engines, remove the spacer and oil slinger. Use a puller which will engage rear face of gear and pull gear from crankshaft.

To install gear, use a suitable pusher as shown in Fig. FO29. In the absence of this pusher, use a washer and a piece of pipe in conjunction with a cap screw threaded into end of crankshaft. Mate punch marked tooth of crankshaft gear with punch marked tooth space of camshaft gear as shown in Fig. FO28.

CAMSHAFT AND BEARINGS

All Models

36. CAMSHAFT. To remove the camshaft first remove engine from tractor as outlined in paragraph 25. Remove timing gear cover, rocker arm cover, rocker arms and shaft assembly and push rods. On non-diesel models, remove distributor. On diesel models, remove the injection pump, fuel pump and the oil pump drive

unit. On all models, invert engine so tappets will fall away from camshaft. Check clearance between camshaft gear hub and thrust plate prior to removal of camshaft. Normal clearance (end play) is 0.003-0.007 and if clearance exceeds 0.007, renew thrust plate during reassembly. Work through holes in camshaft gear and unbolt thrust plate from cylinder block and carefully pull camshaft and gear from cylinder block. Remove snap ring from forward end of camshaft, press shaft from gear and remove thrust plate.

Check the camshaft against the following values:

Camshaft journal
 diameter 1.9255-1.9265
Camshaft journal diameter,
 min. allowable 1.9245
Camshaft journal, max. out-
 of-round 0.001
Camshaft end play.....0.003 -0.007
Camshaft bushing I.D. ..1.9275-1.9285

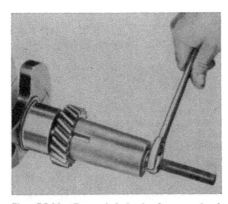

Fig. FO29—Tapped hole in front end of crankshaft permits use of a pusher to install crankshaft gear.

Camshaft operating
 clearance ....0.001 -0.003
Camshaft operating clearance,
 max. allowable 0.005

Reassemble by reversing the disassembly procedure and align the camshaft gear timing marks as follows: The double punch marked tooth space with single punch marked tooth of crankshaft gear. On diesel engines, also align the single punch marked tooth with the double punched tooth space of the injection pump gear. See Fig. FO28.

37. CAMSHAFT BEARINGS. To renew the camshaft bearings, it will be necessary to remove the camshaft as outlined in paragraph 36 as well as the engine oil pan, flywheel and rear rear end plate. Drill a ½-inch hole in expansion plug at rear of rear camshaft bushing and remove plug using Ford tool 7600E, or equivalent. Use Ford tool N6261, or equivalent, and drive bushings from their bores.

New bushings are pre-sized and will require no reaming after installation, if carefully installed. Number three bushing is not interchangeable with the other bushings.

When installing bushings, be sure to align oil holes in bushing with oil holes in cylinder block. Front bushing is installed with front face 0.005-0.020 below front surface of bearing bore. Rear bearing is installed with front of bushing flush with front of bushing bore to allow installation of expansion plug. Use sealant on edge of expansion plug during installation. Intermediate bushings are installed with front side flush with front of bushing bores.

Check camshaft and bushings against the following values:

Camshaft journal
 diameter 1.9255-1.9265
Camshaft journal diameter,
 min. allowable 1.9245
Camshaft journal, max. out-
 of-round 0.001
Camshaft end play......0.003 -0.007
Camshaft bushing I.D....1.9275-1.9285
Camshaft operating
 clearance ....0.001 -0.003
Camshaft operating clearance,
 max. allowable 0.005

Camshaft bushings are also available in 0.015 undersize to accommodate worn or reground camshafts.

Reinstall camshaft by reversing the removal procedure and tighten the thrust plate cap screws to 12-16 ft.-lbs. torque. On non-diesel engines, retime ignition as outlined in paragraph 82.

On diesel engines, retime injection pump as outlined in paragraph 63. Valve clearance for all models is 0.015 hot.

CONNECTING ROD AND PISTON UNITS

All Models

38. Connecting rod and piston units are removed from above after removing cylinder head and oil pan. Be sure top ridge is removed from cylinder bores before attempting to remove assemblies.

Connecting rod and bearing caps are numbered to correspond to their respective bores and numbers are on the side opposite camshaft. Also note the oil squirt hole on camshaft side of connecting rod.

When reassembling, make certain oil squirt hole is on camshaft side of engine (rod and cap numbers opposite) and dimple or word "FRONT" in top of piston is toward front of engine. Tighten the connecting rod nuts or bolts to a torque of 45-50 ft.-lbs. Torque value is for lubricated threads on non-diesel engines and dry threads for diesel engines.

PISTON RINGS

All Models

39. Pistons are cam ground and non-diesel engine pistons are fitted with two compression rings and one oil ring while diesel engine pistons are fitted with three compression rings and one oil ring. Ring sets are available in standard size as well as oversizes of 0.020, 0.030 and 0.040.

Note: When renewing rings, refer to Fig. FO29A.

All rings should have an end gap of 0.010-0.020. Ring side clearance should be as follows:

Non-Diesel
Top compression ring... 0.002 -0.0035
Second compression ring 0.002 -0.0035
Oil control ring....... 0.0015-0.003
Diesel
Top compression ring... 0.004 -0.0055
Second compression ring 0.002 -0.0035
Third compression ring. 0.002 -0.0035
Oil control ring....... 0.0015-0.003

PISTONS AND BORES

All Models

40. The cam ground pistons operate directly in the block bores and are available in standard size as well as oversizes of 0.001, 0.020, 0.030, and 0.040 for non-diesel engines; and 0.002, 0.020, 0.030 and 0.040 oversizes for diesel engines. With pistons removed

from engine, measure cylinder bores both lengthwise and crosswise of engine at top and bottom of piston travel. If taper exceeds a maximum of 0.008, or out-of-round exceeds a maximum of 0.003, rebore cylinder to next larger size.

To fit pistons in bores, refer to the following table and proceed as follows: Attach a ½-inch feeler ribbon of the proper thickness to a spring scale. Invert piston and position feeler ribbon at 90 degrees from piston pin hole. Insert piston and feeler ribbon into cylinder bore until piston is about ½-inch below top of cylinder block. Keep piston pin hole parallel with crankshaft. Now withdraw feeler ribbon by pulling on spring scale and note reading on scale. If reading is not within limits shown in table, try another piston or hone cylinder bore to obtain proper fit.

Note: Before fitting new piston and rings in a used bore, remove any high polish or glaze by passing a hone through the cylinder bore a few times. Hone only enough to rough up surface.

New Piston In New Bore
  (Non- Diesel)
Gage thickness 0.0015
Spring scale pull..........5-10 lbs.

New Piston In Used Bore
  (Non- Diesel)
Gage thickness 0.002
Spring scale pull..........5-10 lbs.

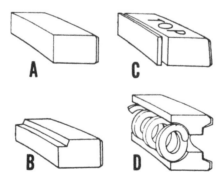

Fig. FO29A—Cross-sectional views of piston rings. Ring "A" is used in production for top compression ring in both diesel and non-diesel engines, but is not included in service ring sets. Ring "B" is used in production for 2nd and 3rd compression rings in diesel engines and for 2nd compression ring in non-diesel engines; in service ring sets, ring "B" is used for top and 2nd compression ring in diesel engines and for top compression ring in non-diesel engines. Ring "C" is not used in production; in service ring sets, ring "C" is used for 3rd compression ring in diesel engines and for 2nd compression ring in non-diesel engines. Ring "D" is used for oil control ring in both production and service for both diesel and non-diesel engines.

Used Piston In Used Bore
  (Non- Diesel)
Gage thickness 0.003
Spring scale pull...........5-10 lbs.
New Piston In New Bore
  (Diesel)
Gage thickness 0.005
Spring scale pull...........5-10 lbs.
New Piston In Used Bore
  (Diesel)
Gage thickness 0.006
Spring scale pull...........5-10 lbs.
Used Piston In Used Bore
  (Diesel)
Gage thickness 0.007
Spring scale pull...........5-10 lbs.

NOTE: Semi-finished cylinder sleeves (liners) are available for engines which are beyond rebore limits. To install sleeves in the cylinder block, bore the cylinder block to the following values:

Cylinder bore diameter (Includes
  final honing) 3.7585-3.7595
Sleeve flange counterbore
  diameter ....3.906 -3.910
Sleeve flange counterbore
  depth ....0.177 -0.181

After sleeve is pressed into place, the sleeve must be honed to provide an inside diameter of 3.6244-3.6254.

PISTON PINS

All Models

41. The floating type piston pins are retained in the piston pin bosses by snap rings and are available in standard size only. Non-diesel piston pins have an outside diameter of 0.9120-0.9123 while diesel piston pins have an outside diameter of 1.1240-1.1243. Piston pin should have a clearance of 0.0001-0.0003, or a light thumb push fit, in connecting rod bushing and pistons. Replacement bushings are of the unsplit type without an oil supply hole. Bushings must be final sized and the oil hole (0.187) drilled after installation. When assembling, oil squirt hole in rod must face toward camshaft side of engine and dimple or word "FRONT" in top face of piston must face toward front of engine.

CONNECTING RODS AND BEARINGS

All Models

42. Connecting rod bearings are of the non-adjustable, slip-in type, renewable from below after removing the oil pan and connecting rod bearing caps. When installing new bearings, make certain that tang on bearings engage the milled slot in connecting rod and bearing cap. Cylinder numbers on rod and cap must be in register and opposite to camshaft side of engine.

Connecting rod bearing inserts are available in standard size as well as undersizes of 0.001, 0.002, 0.003, 0.010, 0.020, 0.030 and 0.040. The Ford Motor Company states that diesel engine crankshafts **must not be reground**; hence, only standard size and .001, 0.002 and 0.003 undersize bearings should be used for servicing diesel engines.

Check the crankshaft connecting rod journals and the rod bearing inserts against the values which follow and bear in mind that the regrind information applies only to non-diesel crankshafts.

Crankpin diameter 2.2980-2.2988
  Regrind if
    out-of-round 0.001
  Regrind if
    tapered 0.001
Bearing running
  clearance 0.0006-0.0025
Renew bearing if
  clearance exceeds 0.0032
Rod side play.........0.003 -0.009

When reassembling, tighten the connecting rod nuts (non-diesel) or bolts (diesel) to a torque of 45-50 ft.-lbs. Torque value is for lubricated threads on non-diesel engines and dry threads for diesel engines.

CRANKSHAFT AND MAIN BEARINGS

All Models

43. Crankshaft is supported in four main bearings of the non-adjustable slip-in type, renewable from below after removing oil pan, oil pump lines and main bearing caps.

Main bearing inserts are available in standard size as well as undersizes of 0.002, 0.010, 0.020, 0.030 and 0.040. The Ford Motor Company states that diesel engine crankshafts **must not be reground**; hence only the standard size, 0.002 and 0.003 undersize bearings should be used in servicing diesel engines.

Check the crankshaft main journals and the main bearing inserts against the values which follow and bear in mind that the regrind information applies only to non-diesel crankshafts. (Also see NOTE following table of values.)

Crankpin diameter 2.2980-2.2988
  Regrind if
    out-of-round 0.001
  Regrind if
    tapered 0.001
Main journal diameter...2.4963-2.4971
  Regrind if
    out-of-round 0.001
  Regrind if
    tapered 0.001

Main bearing running
  clearance 0.0019-0.004
Crankshaft end play....0.004 -0.008
Bolt torque (lubricated)
  ½-inch bearing
    bolts ....95-105 ft.-lbs.
  9/16-inch bearing
    bolts ....105-115 ft.-lbs.

NOTE: Be extremely careful when servicing crankshaft and main bearings of the early models of series 6000 tractors. Some early versions of both diesel and non-diesel engines were made with ½-inch main bearing bolts, and in addition, the non-diesel engine may have a crankshaft with a main bearing journal diameter of 2.4976-2.4984. If a non-diesel engine with this size crankshaft is encountered, fit main bearings as follows: Numbers 1, 2 and 3, 0.0025-0.005; number 4, 0.001-0.0029.

43A. When installing crankshaft, proceed as follows: Install crankshaft along with numbers 1, 2 and 4 main bearings and tighten main bearing bolts to specified torque. Position number 3 main bearing cap and tighten bearing bolts finger tight, being sure that cap is seated. Bump the crankshaft forward; then to rear to align thrust surfaces of bearing liners and while holding bearing cap in this position, tighten bearing bolts to specified torque. Check crankshaft end play and if end play exceeds 0.008, renew thrust bearing. If end play is less than 0.004, check thrust surfaces for nicks, burrs or foreign material. If thrust surfaces appear satisfactory, repeat the installation procedure already given as incorrect bearing

Fig. FO30 — Vertical seals for rear main bearing cap. Refer to text for installation information.

alignment is the likely cause of insufficient crankshaft end play.

44. When installing the crankshaft rear oil seal, proceed as follows: Install upper half of seal in cylinder block with lip toward front of engine. Place lower half of seal on crankshaft and rotate entire seal so as to move the parting lines of seal assembly out of alignment with parting surface of rear main bearing cap. About ⅛-turn, or 45 degrees is sufficient. Install rear main bearing assembly. Dip the two vertical seals (Fig. FO30) in oil and immediately bump into position until firmly seated. Ends of vertical seals should extend approximately 3/64-inch beyond oil pan mounting surface to provide a good seal.

NOTE: Do not oil the vertical seals until they are ready to be installed. Seals begin to swell immediately after oil is applied.

CRANKSHAFT OIL SEALS

All Models

45. The crankshaft front oil seal is located in the timing gear cover and can be renewed after removing timing gear cover as outlined in paragraph 33.

The two-piece lip type crankshaft rear oil seal is located in a groove in the cylinder block and rear main bearing cap. Seal can be renewed after removing oil pan and following the procedure given in paragraph 44.

FLYWHEEL

All Models

46. The flywheel can be removed after separating transmission from engine and removing the torque limiting clutch. Holes of flywheel bolt circle are unequally spaced and flywheel can be installed in one position only.

Starter ring gear can be removed and reinstalled on flywheel by heating same uniformly to about 360 degrees F.

OIL PUMP AND RELIEF VALVE

All Models

47. The rotor type oil pump is mounted to underside of cylinder block. Pump is driven by an intermediate shaft which is pinned to the distributor drive gear on non-diesel models, or the oil pump drive gear assembly on diesel models.

To remove the oil pump, first drain and remove oil pan as outlined in paragraph 49, then unbolt and remove oil pump from cylinder block.

With pump removed, refer to Fig. FO31, unbolt and remove oil inlet tube and screen assembly (10). Remove pump cover (1) and withdraw rotor and shaft assembly (2) from pump body (3). Unstake plug (8), then remove plug, relief valve spring (7) and relief valve plunger (6).

Clean and inspect parts as follows: Check pump cover and body for visible wear, scoring, cracks or other damage and renew as necessary. Place rotor and shaft assembly in pump body and position inner and outer rotor identification marks as shown in Fig. FO32; then, turn shaft and measure clearance at points shown. Clearance should not exceed 0.008. With rotor assembly still in pump body, place a straight edge across rotor assembly and pump body. Use a feeler gage and measure between straight edge and the inner and out rotors as shown in Fig. FO33. If this clearance exceeds 0.004, renew the rotor assembly and/or the pump body. Check the relief valve spring tension which should be 8.9-9.7 lbs. @ 1.53 inches.

Reassemble pump by reversing disassembly procedure. Be sure rotor assembly is positioned as shown in Fig. FO32. Stake plug (8—Fig. FO31) securely. Mount pump to cylinder block and install oil pan.

NOTE: When ordering repair parts for the oil pump, be sure to note tractor serial number. Early pumps had a rotor shaft diameter of 0.4935 which was later increased to 0.5655.

OIL PUMP DRIVE ASSEMBLY

Diesel Models

48. An oil pump drive assembly, shown in Fig. FO34, is used on diesel engine tractors to drive the oil pump. This unit is mounted in the same well that accepts the distributor on non-diesel engine tractors. Removal of the unit is obvious after removing the injection pump. Refer to paragraph 64 for information concerning removal of injection pump.

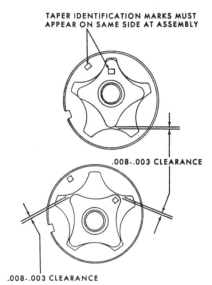

TAPER IDENTIFICATION MARKS MUST APPEAR ON SAME SIDE AT ASSEMBLY

.008-.003 CLEARANCE

.008-.003 CLEARANCE

Fig. FO32 — Install rotor assembly in body and check clearances as shown.

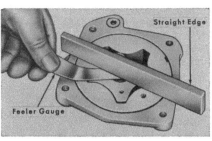

Straight Edge

Feeler Gauge

Fig. FO33—Use straight edge when checking rotor end clearance. Clearance should not exceed 0.004.

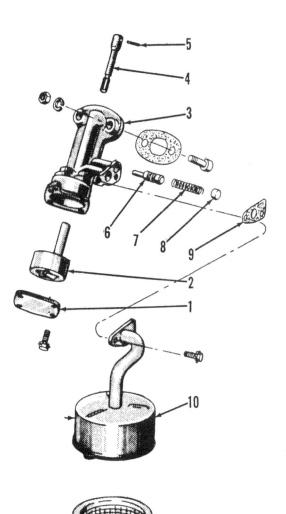

Fig. FO31 — Exploded view of oil pump and relief valve assembly.

1. Cover
2. Shaft & rotor assembly
3. Body
4. Drive shaft
5. Groove pin
6. Relief valve plunger
7. Spring
8. Plug
9. Gasket
10. Inlet tube
11. Screen

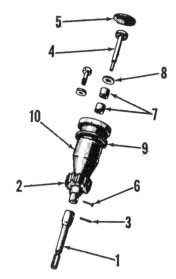

Fig. FO34 — Oil pump drive assembly used in diesel engine tractors.

1. Intermediate shaft
2. Gear
3. Pin
4. Drive shaft
5. Expansion plug
6. Rivet
7. Bushings
8. Thrust washer
9. Gasket
10. Housing

Individual parts are catalogued separately and are available for service.

OIL PAN

All Models

49. Oil pan is a heavy cast part and removal of same requires the use of a rolling floor jack to support and lower the oil pan after unbolting.

To remove oil pan, first drain oil, then place a rolling floor jack under pan to provide support. Support front of transmission, then unbolt side rail spacers from transmission and side rails and bump lower ends of spacers

slightly rearward. Unbolt rear of pan from transmission housing and front of pan from side rails. Remove cap screws holding pan to cylinder block and timing gear cover and lower pan with floor jack.

CARBURETOR

Gasoline Models

50. The gasoline engines are fitted with a Zenith carburetor. Two adjustments are provided; an idle speed stop screw to regulate the engine low idle rpm and a needle to regulate the idle fuel mixture.

Initial setting of idle mixture screw is ⅞-1 turn open and clockwise rotation of the screw richens the mixture.

Adjust the idle speed stop screw to obtain an engine low idle speed of 650-700 rpm. Float level setting is 1 5/32-inch, measured from gasket surface to farthest edge of float.

Removal and overhaul of the carburetor is obvious after an examination of the unit and reference to Fig. FO-34A.

Additional Zenith information on carburetor follows:

Idle jet	C55-22-13
Idle adjusting needle	C46-54
Main jet	C52-7-32
Venturi	B38-76-28
Float needle and seat	C81-68-62
Float	C85-115
Gasket set	C181-339
Repair kit	K12902

LP – GAS SYSTEM

SYSTEM ADJUSTMENTS

Models So Equipped

50A. Initial adjustments of the LP-Gas carburetor are 2½ turns open for the idle fuel screw and 3½ turns open for the main fuel adjustment screw. See Fig. FO34B for location of adjusting screws.

Start engine and bring to operating temperature. Set throttle stop screw to obtain an engine low speed rpm of 650-700 rpm. Turn idle fuel adjustment screw in or out until engine runs smoothly. Stop the engine and disconnect any three spark plug wires from the plugs. Restart engine, place throttle in wide open position and open main fuel adjustment screw until highest engine rpm is obtained. After reinstalling plug wires, check idle fuel adjustment and readjust if necessary; then reset throttle stop screw to obtain a low idle speed of 650-700 rpm.

FUEL TANK AND LINES
Models So Equipped

50B. **SERVICING.** The filler valve and vapor return valve are located on

lower right hand side of fuel tank as shown in Fig. FO34C. The relief valve and overflow valve are located on top side of fuel tank and the vapor withdrawal and liquid withdrawal valve are located at top rear of fuel tank. Valves are serviced as complete assemblies only. Before renewal of any valve is attempted, be sure tractor is in a well ventilated area and allow engine to run until fuel is exhausted, then open overflow valve to allow any remaining pressure to escape.

Fuel gage assembly consists of a dial face unit which can be renewed at any time, and a float unit which can only be renewed if the fuel tank is completely empty.

The safety relief valve is preset at the factory to protect the tank against excessive pressures and under no circumstances should an attempt be made to adjust this valve.

U-L regulations in most states prohibit any welding or repair on LP-Gas containers and the tank must be renewed rather than repaired if damaged.

Fuel lines can be safely renewed at any time without emptying tank if

Fig. FO34A — Exploded view of Zenith carburetor typical of those used on Ford 6000 tractors.

1. Gasket	14. Venturi
2. Throttle plate	15. Float
3. Throttle shaft	16. Choke bracket
4. Packing retainer	17. Nozzle air vent jet
5. Shaft seal	18. Packing retainer
6. Spring	19. Packing
7. Idle mixture needle	20. Main metering jet
8. Throttle body	21. Bowl
9. Idle jet	22. Main nozzle
10. Needle and seat assy.	23. Plug
11. Gasket	24. Drain plug
12. Float shaft	25. Plug
13. Spring	26. Choke plate

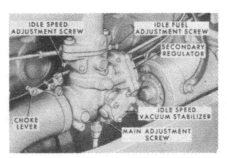

Fig. FO34B — View of LP-Gas carburetor installed. Note points of adjustment and refer to text for adjustment procedure.

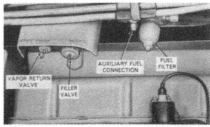

Fig. FO34C — Filler valve, vapor return valve and filter are located as shown. Auxiliary fuel connection is for use with a portable LP-Gas bottle.

liquid and vapor withdrawal valves are closed and the engine allowed to run until it stops from lack of fuel.

50C. **FUEL FILTER.** The cartridge type fuel filter is located at lower right front of fuel tank. (See Fig. FO-34C). To clean the disc-type filter, close both fuel tank withdrawal valves and run the engine until the fuel is exhausted from the system. Remove filter body from filter head. Remove element from filter body, clean element in a suitable solvent and dry thoroughly before installing. If compressed air is used during cleaning operation, be sure to direct air stream through length of filter element. Air pressure directed on side of element can render it ineffective (See Fig. FO34D).

VAPORIZER

Models So Equipped

50D. **OPERATION AND R&R.** The LP-Gas vaporizer serves two purposes in the system. First, it vaporizes the liquid fuel as it is received from the fuel tank; and second, it acts as a primary regulator to reduce and control the pressure of the vapor prior to entering the carburetor.

The vaporizer is mounted in the water outlet elbow at the front of the cylinder head. Before removing vaporizer or disconnecting any lines, close both tank withdrawal valves and allow engine to run until all fuel is exhausted from vaporizer, fuel lines and carburetor. Turn off the ignition switch after engine stops. Drain cooling system. Disconnect the vaporizer to carburetor connection and tank to vaporizer hose. Remove the vaporizer retaining cap screw and withdraw vaporizer from water outlet elbow. Refer to Fig. FO34E for an exploded view of vaporizer. Test and/or overhaul the unit as outlined in paragraph 50E.

50E. **TEST AND OVERHAUL.** The removed vaporizer assembly can be tested for external or internal leaks without disassembly, and an almost complete diagnosis of vaporizer con-

dition made. To test the unit, proceed as follows:

Connect the vaporizer inlet to a source of compressed air and completely immerse the unit in a water tank. External leaks will show up as air bubbles. Note especially the areas around vaporizer coil and mounting plate (1-Fig. FO34E) and around diaphragm cover (16). Air bubbles emerging from vent hole (V) in top of diaphragm cover indicate a leaking diaphragm.

To check the vaporizer inlet valve and seat, install a low-pressure test gage in the vaporizer outlet port, and connect inlet to air pressure. The gage reading should be 9-11 psi and hold steady. If pressure continues to rise, a leaking fuel valve or valve seat "O" ring is indicated and vaporizer should be overhauled.

To disassemble the vaporizer, remove four alternate screws from diaphragm cover (16) and install aligning studs (Zenith Tool Part No. C161-195). Apply thumb pressure to top of diaphragm cover and remove the remaining diaphragm screws, diaphragm cover (16), spacer (15) and springs (13 and 14). To renew any part of the fuel valve assembly, remove the fuel valve assembly, remove the fuel valve seat (8) using a suitable socket wrench. To renew "O" rings (2 or 3), remove the four screws retaining coil and plate (1) to body, and withdraw the plate.

When reassembling, install fuel valve (6) with long stem toward diaphragm. Make sure all screw holes are aligned in gasket (10), baffle (11), diaphragm (12) and cover (16). Tighten the retaining screws evenly, leaving aligning studs installed until cover is tight. Recheck for leaks after assembly by immersing in water or using a soap solution.

CARBURETOR

Models So Equipped

50F. **OPERATION.** The Zenth pressure regulating carburetor serves

both as a secondary regulator and as a carburetor. See Fig. FO34F for a cross sectional view of a typical pressure regulating carburetor. The idle speed vacuum stabilizer diaphragm assembly is not shown.

The fuel valve seat (2) is adjustable so that position of diaphragm lever (9) can be varied with relation to the diaphragms. The seat is locked in position by means of lock screw (3) and nylon plug (4) as shown in inset.

The two diaphragms (10 and 11) control the pressure and flow of incoming fuel to maintain the proper fuel-air mixture.

50G. **R&R AND OVERHAUL.** To remove the carburetor, first close both withdrawal valves and allow engine to run until fuel is exhausted from regulator, lines and carburetor. Turn off ignition switch, disconnect choke and throttle linkage and the fuel inlet line. Unbolt and remove the carburetor assembly.

Remove the six screws securing the diaphragm cover, spacer and the two diaphragms, and remove the diaphragms. Remove the hex plug and spring from idle diaphragm housing, remove cover and diaphragm, then remove the diaphragm housing from side of carburetor. Lever axle, lever and valve disc can now be removed.

Remove the fuel inlet fitting from bottom of carburetor and remove fuel

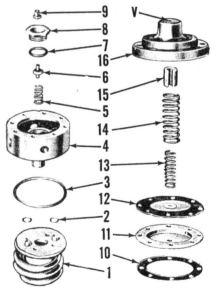

Fig. FO34E — Exploded view of vaporizer assembly.

1. Coil & plate	10. Gasket
2. "O" ring	11. Baffle
3. "O" ring	12. Diaphragm
4. Body	13. Inner spring
5. Valve spring	14. Outer spring
6. Valve	15. Spacer
7. "O" ring	16. Cover
8. Valve seat	V. Vent hole
9. Follower	

DIRECT AIR STREAM THROUGH LENGTH OF FILTER ELEMENT

FILTER ELEMENT

FILTER COVER

Fig. FO34D — View of filter element with filter cover removed. Note method of using compressed air if used during cleaning. Also refer to text.

valve seat (2—Fig. FO34F). NOTE: The valve should be held off its seat by applying light pressure to dia-phragm lever (9). Be sure locking screw (3) is loosened.

Remove the diaphragm lever shaft plug from side of carburetor body, and remove shaft, lever and valve assembly. Remove the main and idle needle valves and the main jet (24) using the special tool (Zenith C161-193).

Clean all metal parts in a suitable solvent and examine for wear or damage. Examine diaphragms for cracks, pin holes or deterioration and renew as necessary. Reassemble by reversing the disassembly procedure, using new gaskets and seals. When installing the fuel valve seat (2), hold valve open by applying pressure to valve lever (9). Use Step 2 of Zenith Tool C161-194; or adjust so that lever (9) is 1/16-inch to rear of rear surface of carburetor body as shown at (A—Fig. FO34G). Tighten the locking screw (3—Fig. FO34F) when adjustment has been obtained. When reinstalling the diaphragms, spacer, cover and gasket, make sure the parts are arranged so that air passages are open.

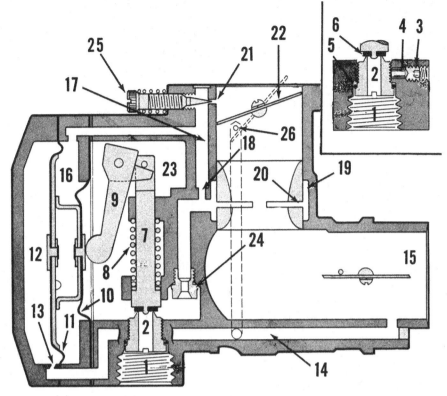

Fig. FO34F—Cross-sectional view of pressure regulating carburetor showing main components. Inset shows fuel valve seat locking arrangement.

1. Fuel inlet	8. Fuel valve spring	14. Air passage	20. Venturi
2. Fuel valve seat	9. Diaphragm lever	15. Air intake	21. Idle needle seat
3. Locking screw	10. Inner diaphragm	16. Inner diaphragm	22. Throttle fly
4. Locking plug	11. Outer diaphragm	chamber	23. Pressure chamber
5. "O" ring	12. Outer diaphragm	17. Idle fuel passage	24. Main jet
6. Sealing disc	chamber	18. Idle orifice	25. Idle needle
7. Fuel valve	13. Air passage orifice	19. Annulus	26. Economizer orifice

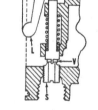

Fig. FO34G—Schematic view of regulating valve assembly used on Zenith Pressure Regulating carburetors.

A. Adjustment
L. Lever
S. Valve seat
V. Valve

DIESEL FUEL SYSTEM

The diesel fuel system consists of four basic components: fuel filter, fuel supply pump, injection pump and injection nozzles. When servicing any unit associated with the fuel system, the maintenance of absolute cleanliness is of utmost importance. Of equal importance is the avoidance of nicks or burrs on any of the working parts.

QUICK CHECKS

Diesel Models

51. If the engine does not run properly and the fuel system is suspected as the source of trouble refer to the accompanying trouble-shooting chart and locate points which require further checking. Many of the chart items are self-explanatory; however, if the difficulty points to the fuel filters, injection nozzles and/or injection pump refer to the appropriate sections which follow:

FILTERS AND BLEEDING
Diesel Models

52. **MAINTENANCE.** The fuel filter

	Lack of Fuel	Engine Surging or Rough	Cylinders Uneven	Engine Smokes or Knocks	Injection Pump Does Not Shut Off	Engine Dies at Low Speed	Loss of Power
Defective Speed Control Linkage	★				★		★
Air in Fuel System	★			★		★	
Clogged Filter	★					★	★
Fuel Lines Leaking or Clogged	★	★	★			★	★
Friction in Injection Pump		★					★
Inferior or Contaminated Fuel		★		★		★	★
Faulty Injection Pump Timing		★		★		★	★
Defective Nozzle or Injector		★	★	★			★
Faulty Governor and/or Linkage Adjustment		★		★			★
Faulty Primary Pump	★						★
Faulty Distribution of Fuel	★	★	★	★			★
Injection Pump Not Turning	★						

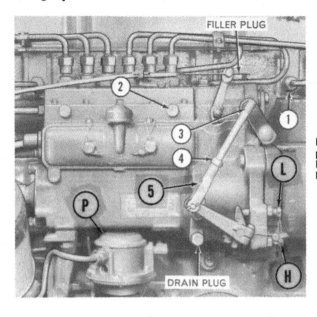

Fig. FO35 — Injection pump bleed screw (2) is located as shown. Rear bleed screw need not be used.

is fitted with a renewable type element which should be renewed every 200 hours of operation.

To renew the element, shut off fuel and unscrew filter through bolt until filter case and element are freed from filter cover. Remove and discard old element and clean interior of filter case. DO NOT use any rags which will leave lint. Install new element in filter case; then, using a new gasket, install assembly to filter cover. Open fuel shut-off valve; then, open filter bleed screw and crank engine. When air-free fuel flows from bleed screw, close bleed screw.

53. **BLEEDING SYSTEM.** To bleed the complete fuel system, proceed as follows: Open fuel shut-off valve and the bleed screw on top of filter. Crank engine and when air-free fuel flows from bleed screw, tighten bleed screw.

Open the forward one of the two bleed screws (2—Fig. FO35) in injection pump fuel gallery and crank engine. When air-free fuel flows from bleed screw, tighten the bleed screw.

Loosen injector lines at injectors and crank engine. When the fuel escaping at the connections is free of air bubbles, tighten the injector line connections.

INJECTION NOZZLES
Diesel Models

54. Diesel engines are fitted with Simms model NL141 injector nozzles. These injectors have four orifices each. Orifices are 0.011 in diameter with a spray angle of 150 degrees. Nozzle opening pressure is 2700-2800 psi (2750 desired).

WARNING: Fuel leaves the injector nozzle with sufficient force to penetrate the skin. When testing a nozzle, keep your person clear of the nozzle spray.

55. **TESTING AND LOCATING A FAULTY NOZZLE.** If engine does not run properly and a faulty injection nozzle is indicated, such a nozzle can

be located as follows: With engine running, loosen the high pressure line fitting on each nozzle holder in turn, thereby allowing fuel to escape at the union rather than enter the cylinder. As in checking spark plugs in a spark ignition engine, the faulty unit is the one which least affects the running of the engine when its line is loosened.

56. **NOZZLE TESTS.** A complete job of testing and adjusting the nozzle requires the use of a special tester, such as that shown in Fig. FO36. The nozzle should be tested for opening pressure, spray pattern, seat leakage and leak-back.

Operate the tester until oil flows and attach nozzle and holder assembly. Close the tester valve and apply a few quick strokes to the tester handle. If undue pressure is required to operate the lever, the nozzle valve is plugged and should be serviced as in paragraph 62.

57. **OPENING PRESSURE.** While operating the tester handle, observe the gage pressure at which the spray occurs. This gage pressure should be 2700-2800 psi. If pressure is not as specified, remove the cap nut (Fig. FO37) and turn adjusting nut as required. If opening pressure cannot be brought to within the specified limits (2700-2800 psi), overhaul nozzle as outlined in paragraph 62.

58. **SPRAY PATTERN.** Operate the tester handle and observe the spray pattern. All four (4) sprays must be similar and spaced approximately 90 degrees to each other in a horizontal plane. Each spray must be well atomized and should spread into a two inch cone at a six inch distance from injector. If spray pattern is not as outlined, overhaul the nozzle as outlined in paragraph 62.

Fig. FO36—To completely test an injection nozzle requires the use of a nozzle tester such as the one shown.

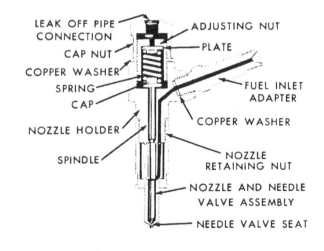

Fig. FO37 — Sectional view of the four orifice Simms injector.

59. SEAT LEAKAGE. Wipe nozzle tip dry with clean blotting paper, then operate tester handle and bring pressure to 150 psi **below** nozzle opening pressure and hold this pressure for one minute. Apply a clean piece of blotting paper to nozzle tip as shown in Fig. FO38. The fuel oil stain should not exceed ½-inch in diameter. If above conditions are not met, overhaul injector as outlined in paragraph 62.

60. NOZZLE LEAK-BACK. Operate tester handle and bring gage pressure to above 2200 psi and note the amount of time it takes for the pressure to drop from 2200 psi to 1500 psi. This time should be between six and forty-five seconds.

If the elapsed time is not as specified, nozzle should be overhauled as outlined in paragraph 62.

Note: A leaking tester connector, check valve or pressure gage will show up in this test as excessively fast leak-back. If, in testing a number of injectors, all fail to pass this test, the tester rather than the injectors should be suspected.

61. REMOVE AND REINSTALL. Remove valve rocker cover and gasket. Disconnect the excess fuel line at the top of each injector; then, remove line by unscrewing the hex fitting at the "tee" connection in the line to the cylinder head. Valve push rods interfere with the removal of the injectors and can be repositioned as follows: Turn engine so that both push rods of cylinder from which injector to be removed are free to turn. Place a ½-inch box end wrench over head of adjusting screw and lift the rocker arm enough to release the valve push rod which can then be moved sideways away from injector. Lift rocker arms and reposition push rods before turning engine to next position.

Carefully clean all dirt and foreign material from the pressure line connections at the injectors and disconnect the lines. Loosen lines at injection pump so they can be moved out of the way. Cap-off lines and the injector inlets. Injectors can now be pulled by using Ford special tool No. DDN17098, or its equivalent. CAUTION: Be sure that legs of tool bear on flange of cylinder head and not on the rubber seal. Remove injector seat washer by using a bent wire. If washers are stuck in bore, the following method of removal can be used. Grind the blade of a screw driver to a taper that will permit it to protrude not more than ⅛-inch through the I. D. of seat washer. Place blade of screw driver through I. D. of stuck washer and rap the handle sharply so blade will bite into washer. Washer can now be twisted free.

Always use new injector seat washers when reinstalling injectors. Make sure that injector bores in cylinder head are absolutely clean and that washers are flat in bore. To make certain, use a light and mirror if necessary. Install each injector by positioning the nozzle end in its bore in cylinder head and guiding the injector rubber seal into its slot. Work each injector downward, using hands only. DO NOT use a hammer. Align injector and install mounting bolts. Torque each bolt evenly to 15 ft.-lbs. Reposition the valve push rods to their proper place. Reinstall the excess fuel line by reversing the removal procedure. Flush pressure lines and injectors prior to connecting the lines by cranking engine with starter and directing the fuel discharge from pressure lines into injector inlets. After all lines and injectors have been flushed, connect lines and tighten; then, back-off two turns. Set hand throttle to low idle position and crank engine with starter. When fuel spits out at injector connections, tighten same. Start engine and bring to operating temperature, then set valve tappet gap to 0.015. Reinstall valve cover and gasket. Use a non-hardening sealant at corners of grommets where they contact the cylinder head and rocker cover gasket.

62. OVERHAUL. Unless complete and proper equipment is available (Kent-Moore J8537, or equivalent) do not attempt to overhaul diesel nozzles.

Refer to Fig. FO37 and proceed as follows: Secure injector holding fixture in a vise and mount injector in fixture. NEVER clamp the injector body in a vise. Remove the cap nut, back-off adjusting nut, then lift off the upper spring disc, injector spring and spindle. Remove the nozzle retaining nut using Kent-Moore tool J8537-14, or equivalent, and remove the nozzle and valve. Nozzles and valves are a lapped fit and must never be interchanged. Place all parts in clean fuel oil or calibrating fluid as they are disassembled. Clean injector assembly exterior as follows: Soften hard carbon deposits formed in the spray holes and needle tip by soaking in a suitable carbon solvent, then use a soft wire brush to remove carbon from the needle and nozzle exterior. Rinse the nozzle and needle immediately after cleaning to prevent the carbon solvent from corroding the highly finished surfaces. Clean the pressure chamber of the nozzle with a 0.043 reamer as shown in Fig. FO39. Clean the spray holes in the nozzle with a 0.010 diameter wire probe held in a pin vise as shown in Fig. FO40. To prevent breakage, the wire probe should protrude from pin vise only

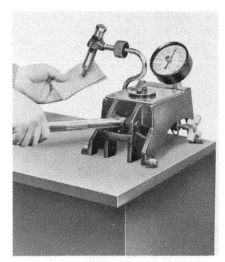

Fig. FO38 — With nozzle pressure brought to within 150 psi of opening pressure and held for one minute, the fuel stain should not exceed a ½-inch diameter when clean blotting paper is applied to nozzle tip.

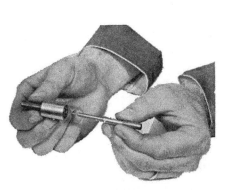

Fig. FO39 — Clean the pressure chamber of the nozzle with a 0.043 reamer as shown.

Fig. FO40 — Orifices being cleaned with a wire probe in a pin vise. Refer to text.

far enough to pass through spray holes. Rotate pin vise without applying undue pressure.

The valve seats are cleaned by inserting the small end of a valve seat scraper into the nozzle and rotating. Then reverse the scraper and clean the upper chamfer with the large end. See Fig. FO41. The annular groove in top of the nozzle and the pressure chamber are cleaned by using (rotating) the pressure chamber tool as shown in Fig. FO42.

With the above cleaning accomplished, back flush nozzle and needle by installing the reverse flushing adapter on the injector tester and positioning the nozzle and valve in adapter, tip end first. Secure with the knurled nut and rotate the needle in the nozzle while flushing to make sure it is free. After nozzle is back flushed, the seat can be polished by using a small amount of tallow on the end of a polishing stick and rotating the nozzle as shown in Fig. FO43.

If the leak-back test time was greater than 45 seconds (paragraph 60), or if the valve is sticking slightly, correction can be made by remating the needle and nozzle assembly. This is accomplished by using a polishing compound (Kent-Moore No. J8537-28)

consisting of tallow and a small amount of very fine lapping compound and proceeding as follows: Hold needle in a chuck and polish same using a piece of felt coated with a very small amount of the above mentioned special compound; or, place the nozzle in the chuck of a drill having a maximum speed of not more than 450 rpm; then, apply a small amount of special compound on the needle valve and insert same in nozzle body. Turn nozzle to lap and be sure to hold needle up off the pressure chamber shoulder during operation to avoid damage to needle. See Fig. FO44. Care should be taken to see that lapping compound does not damage the needle seat. Back flush and clean assembly.

Before assembly, rinse all parts in clean fuel oil or calibrating fluid and install while still wet. The injector inlet adapter normally does not need to be removed. However, if adapter is removed, use a new copper washer when reinstalling. Position the nozzle and needle valve on injector body and make sure dowel pins in body are correctly located in nozzle as shown in Fig. FO45. Install the nozzle retaining nut and torque to 50 ft-lbs. Note: Place injectors in holding fixture to torque nut. Install the spindle, spring, upper spring disc and spring adjusting nut. Tighten the adjusting nut until pressure from spring is felt. Connect the injector to the nozzle tester and adjust opening pressure to 2700-2800 psi. Use a new copper gasket and install cap nut. Recheck nozzle opening pressure to see that it has not changed.

Retest the injector as outlined in paragraphs 57 through 60, and if the injector fails to pass the tests, renew the nozzle and needle.

NOTE: If injectors are to be stored, it is recommended that they be cleaned in calibrating fluid prior to storage. Fuel oil tends

to separate and allow the lapped surfaces to score. Storage periods of more than thirty days may result in the necessity of disassembling and cleaning injectors in order to obtain satisfactory performance.

INJECTION PUMP

Diesel Models

The fuel injection pump is a self-contained unit which includes the engine governor and components for metering and delivering fuel to the injectors. Pump is mounted on right side of engine and is driven from engine camshaft gear. Other than renewing the unit as a complete assembly, the only service requirements are timing and the adjustment of the engine high and low idle speed.

63. **TIMING.** The diesel engine is timed to 26 degrees BTDC. To time the injection pump remove starting motor on early models, or remove access plate in engine rear cover plate on later models (Fig. FO45A), turn the crankshaft in the direction of normal rotation until the number one pis-

Fig. FO44 — Needle and nozzle assembly can be remated by using an electric drill and polishing compound as shown. Refer to text.

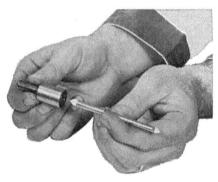

Fig. FO41 — Valve seat and upper chamfer are cleaned by using scraper as shown. Refer to text.

Fig. FO42—Clean annular groove in top of nozzle by using pressure chamber tool as shown.

Fig. FO43 — Nozzle seat can be polished by using a small amount of tallow on a polishing stick and rotating nozzle as shown.

Fig. FO45 — When positioning nozzle and needle on body be sure that body dowel pins are correctly aligned in nozzle.

ton is coming up on compression stroke and continue turning crankshaft until the proper timing mark (26 degrees BTDC) on the crankshaft vibration dampener aligns with pointer on timing gear cover. With crankshaft positioned as indicated, the timing mark on the injection pump drive flange should be aligned with the pointer in the injection pump front cover as shown in Fig. FO46. If the drive flange timing mark and pointer are not aligned, loosen the two coupling cap screws (C) and rotate drive flange to align timing mark and pointer. Tighten cap screws.

64. R&R INJECTION PUMP. To remove the injection pump, first shut off fuel, then disconnect fuel supply line from injection pump and filter and remove line. Disconnect injector pressure lines from injection pump.

Note: To prevent the entrance of dirt or other foreign material, cap off all openings as lines are removed.

Disconnect throttle control rod, fuel shut-off control and Proof-Meter cable from injection pump. Remove cap screw retaining injection pump bracket to engine, then unbolt and remove injection pump and bracket from pump mounting adapter (bell), then remove bracket from pump.

When reinstalling, it will be necessary to mate doweled bolt which protrudes from pump drive coupling member with hole in drive coupling flange. Tighten injection pump mounting bolts prior to tightening the bracket to engine mounting bolt.

Bleed system as outlined in paragraph 53 and if necessary, adjust engine high and low idle speeds as outlined in paragraph 65.

65. GOVERNOR ADJUSTMENT. To adjust the governor, start engine and bring to operating temperature. Check linkage for proper travel and operation, then place hand throttle lever in low idle position and check engine speed which should be 800-850 rpm. If engine low idle rpm is not as specified, loosen jam nut and rotate screw (L—Fig. FO47) as required to obtain correct engine low idle speed. Tighten jam nut.

With engine low idle rpm adjusted, place hand throttle lever in high idle position and check the engine high idle speed which should be 2475-2525 rpm for series 6000 tractors or 2645-2695 for Commander 6000 tractors. If engine high idle rpm is not as specified, loosen jam nut and rotate cap screw (H) as required to obtain cor-

rect engine high idle speed. Tighten jam nut.

66. THROTTLE LINKAGE. Should it be that the injection pump governor lever will not contact both stop screws, adjust rod (3—Fig. FO47) as follows: Place hand throttle lever in low idle position and disconnect rod

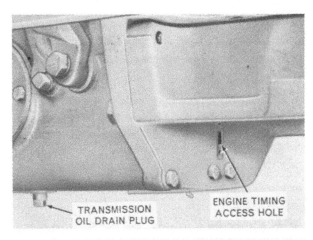

Fig. FO45A — An access hole is located as shown on late tactors. Engine can be turned by using a screw driven or small bar inserted through hole.

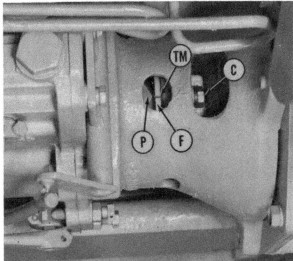

Fig. FO46 — Pointer and drive flange timing mark will align as shown when injection pump is properly timed. Refer to text.

C. Cap screw
F. Flange
P. Pointer
TM. Timing mark

Fig. FO47 — View showing injection pump linkage and the points of adjustment.

1. No. 1 injector
2. Bleed screw
3. Control rod
4. Jam nut
5. Clevis
H. High idle adjusting screw
L. Low idle adjusting screw
P. Fuel supply pump

(3) from arm of upper shaft. Loosen jam nut (4) and rotate rod in or out of clevis until upper end of rod will just fit on to upper shaft lever pivot stud when injection pump governor lever is held in the low idle position.

67. INJECTION PUMP DRIVE GEAR. Injection pump is driven by a gear which meshes with, and is timed to, the engine camshaft gear. To remove the injection pump drive gear assembly, first remove injection pump as outlined in paragraph 64, then loosen clamp bolt of drive flange and remove flange assembly from drive shaft and bearing assembly. Remove timing gear cover as outlined in paragraph 33. Work through holes in injection pump drive gear and remove drive assembly from engine front plate. Use a pusher which will bear only on outer race of the shaft and bearing assembly and press the shaft and bearing and gear from drive gear housing. Press gear from shaft and bearing assembly. Remove dust seal from drive gear housing.

Reinstall by reversing removal procedure and be sure timing marks are aligned as shown in Fig. FO28. Time injection pump as outlined in paragraph 63.

68. **FUEL SUPPLY PUMP.** The fuel supply pump (P—Fig. FO47) is a serviceable item and if pump is defective, removal and overhaul are obvious after an examination of the unit and reference to Fig. FO47A. Pump cover (3) and valves (1) are not available separately. Removal of pump will require bleeding of the system as outlined in paragraph 53.

GOVERNOR
(Non-Diesel)

69. Non-diesel model tractors are fitted with a centrifugal flyweight type governor mounted on forward end of crankshaft. Drive is imparted to the governor by clamping action of the crankshaft pulley. Refer to Fig. FO48 for an exploded view showing governor parts and their relative positions.

70. **ADJUSTMENT.** Before attempting to adjust governor, be sure all linkage operates freely and no binding exists. If necessary, adjust linkage as follows: Disconnect governor arm to carburetor rod at carburetor; hold both governor arm and carburetor throttle shaft in wide open (high idle) position and adjust rod to exactly fit this distance. Move hand throttle control lever to low idle position, start engine and bring to operating temperature. Adjust engine low idle speed to 650-700 rpm. Turn idle mixture adjusting screw in

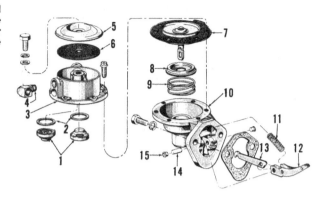

Fig. FO47A — Exploded view of the diesel engine fuel supply pump. Cover (3) and valves (1) are not serviced separately.

1. Valves
2. Gaskets
3. Pump cover
4. Elbow
5. Cover plate
6. Pulsator chamber diaphragm
7. Diaphragm assembly
8. Spring seat
9. Diaphragm spring
10. Pump body
11. Rocker arm spring
12. Rocker arm
13. Link
14. Rocker arm pin
15. Plugs

(clockwise) until engine starts to roll from a too rich mixture; then, backout screw until engine runs smoothly. Recheck low idle rpm and readjust if necessary. With hand control lever in low idle position, check to see that idle speed adjusting screw seats firmly against stop on carburetor without governor control rod being buckled. Adjust control rod length if necessary, to prevent any buckling. Advance hand control lever to obtain a high idle speed of 2575-2625 rpm for series 6000, or 2675-2725 rpm for Commander 6000 tractors, then slide clamp on governor control rod against battery support and tighten it at this point.

If throttle tends to creep toward closed position, correct by tightening the friction disc assembly located on inner end of hand control lever shaft.

71. **R&R AND OVERHAUL.** To remove the governor assembly, first remove the timing gear cover as outlined in paragraph 33, then slide governor assembly from crankshaft.

To disassemble governor, remove snap ring (1—Fig. FO48) and separate parts of weight assembly. Drive roll pin from governor fork and remove fork and control shaft from timing gear cover. Control shaft seal retainer, seal and bearing can now be driven from timing gear cover.

Note: If necessary, the crankshaft front oil seal can also be renewed at this time.

Inspect all parts for nicks, burrs, undue wear or other damage. Renew parts as necessary. Inspect timing gear cover for nicks, burrs or cracks. Machined surfaces of timing gear cover can be smoothed with a fine stone if nicked or burred.

Reassemble by reversing the disassembly procedure and adjust governor as outlined in paragraph 70.

COOLING
SYSTEM

RADIATOR

All Models

72. **REMOVE AND REINSTALL.** To remove radiator, first drain cooling system, then remove front side panels and hood. Either disconnect rear side of upper baffle, or completely remove it, whichever is desired. Disconnect upper hose from radiator. Disconnect the rubber bumpers (anti-rattlers) and stabilizer brackets from hood support rails. Remove fan and fan bracket assembly.

Note: It is not imperative that fan and fan bracket be removed; however, radiator removal will be eased if assembly is removed, especially if upper baffle was not completely removed.

Loosen lower radiator hose clamps, disconnect hose from radiator and slide same rearward on to water pump. Disconnect transmission cooler lines from radiator bottom tank and plug lines. Unbolt radiator from supports and lift radiator up and out of hood support rails.

WATER PUMP

All Models

73. **REMOVE AND REINSTALL.** Drain cooling system. Loosen radiator lower hose clamps and either remove lower hose or slide it rearward on to water pump. Loosen generator and

Fig. FO48 — Exploded view of governor assembly used on non-diesel engines.

1. Snap ring
2. Fork base
3. Thrust bearing
4. Race
5. Sleeve and driver
6. Spacer
7. Weight assembly

remove belts from generator. Loosen fan bracket and remove fan belt and generator belts from water pump pulley. Unbolt water pump from cylinder block and remove from left side of tractor.

Note: If difficulty is encountered with lower radiator hose removal, remove generator to provide working room.

Reinstall by reversing the removal procedure and adjust the fan belt and generator belt tension until a 15 pound pull midway between pulleys will deflect the fan belt $\frac{7}{16}$-½-inch and the generator belts ¼-$\frac{5}{16}$-inch. Both generator belts should be nearly equal in tension.

74. OVERHAUL. Remove water pump as outlined in paragraph 73. Refer to Fig. FO49 and proceed as follows: Support pulley (7) and press shaft from pulley. Remove snap ring (8); then, pressing from impeller end, push shaft and bearing and slinger (9) from impeller (13) and body (11). Seal (12) can now be removed from body and slinger from shaft and bearing assembly.

Clean and inspect all parts and renew as necessary. Shaft and bearing assembly is available as a unit only. Slinger is available separately but it included with a new shaft and bearing assembly.

When reassembling and a new slinger is being used, install same with flange toward bearing and press on shaft until flange of slinger is 0.82 from aft end of bearing. Press new seal into housing until seal flange bottoms. Press shaft and bearing into front of housing until snap ring grooves align and install retaining snap ring. Support front of shaft and

press on impeller until there is 0.010-0.020 clearance between impeller vanes and housing. Support aft end of shaft and press pulley on front end of shaft.

FAN AND FAN BRACKET
All Models

75. R&R AND OVERHAUL. Fan and fan bracket can be removed as an assembly and the procedure for doing so is obvious.

With assembly removed, unbolt and remove fan from pulley. Remove retaining snap ring (4—Fig. FO49) and press shaft and bearing (3) and pulley (2) from bracket (5). Use a piece of pipe to support pulley, and press shaft and bearing from pulley.

Fan shaft and bearing (3) is available as a unit only.

Reassemble by reversing the disassembly procedure. Install unit on tractor and adjust fan belt so a 15 pound pull midway between pulleys will deflect belt $\frac{7}{16}$-½-inch.

THERMOSTAT
All Models

76. REMOVE AND REINSTALL. Drain cooling system and loosen fan belt adjustment. Unbolt water outlet housing from cylinder head and pull outlet housing and fan assembly away from cylinder head far enough to remove thermostat.

Non-diesel engine thermostat starts to open at 157-162 degrees F. and is fully open at 180-184 degrees F. Diesel engine thermostat starts to open at 175-180 degrees F. and is fully open at 198-202 degrees F.

After reinstalling thermostat, adjust fan belt tension until a 15 pound

pull midway between pulleys will deflect belt $\frac{7}{16}$-½-inch. Radiator cap pressure valve opens at 6.5-7.5 psi.

IGNITION AND ELECTRICAL SYSTEM

GENERATOR AND REGULATOR
All Models

77. A Ford designed, shunt wound, two-pole type generator is used. Generator output is controlled by a two-unit type regulator. Ford part numbers assigned to these units are C3NF-10002-A for the generator and CONF-10505-A for the regulator.

Specifications are as follows:

Generator

Armature part No.....C3NF-10005-A
Engine rpm for output test......1500
Watts 375
Brush spring tension.......20-28 oz.
Min. brush length............½-inch
Generator output (max)....25 amps.
Field part No...............2900619
Field resistance, ohms...8.0 @ 70°F.

Regulator

Cut-out volatge @ 75°F.
  Opening 12.2
  Closing 12.2-13.3
Voltage limiter, volts.......14.6-15.4
               @ 4-6 amps.
Reverse current at min.
  of 12 volts...............8 amps.
             @ 12.2 volts

STARTING MOTOR
Non-Diesel Models

78. Non-diesel engines are fitted with a Ford designed, 12-volt starting motor with a centrifically actuated drive mechanism. The Ford part number assigned to this unit in CONF-11002-A, C3NF-11002-B or C3NF-11002-E.

Specifications for all units are as follows:

Max. amperes, no load..85@12 volts
Ampere load cranking
  warm engine ....175@9 volts
Min. eng. cranking speed, rpm...150
No load rpm, max..............5200

Diesel Models

79. Diesel engines are fitted with a Delco-Remy, model 1113138, 12-volt starting motor which has a lever actuated drive mechanism operated by an integral solenoid. Ford part number assigned to this unit is CONN-11001-A.

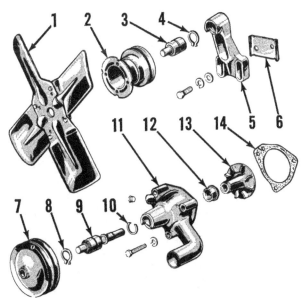

Fig. FO49 — Exploded view of water pump and the fan and bracket assembly.

1. Fan
2. Pulley
3. Bearing and shaft
4. Snap ring
5. Fan bracket
6. Guide
7. Pulley
8. Snap ring
9. Bearing, shaft and slinger
10. Snap ring
11. Housing
12. Seal
13. Impeller
14. Gasket

Specifications are as follows:
Brush tension (min.) .... 80 oz.
No Load Test
  Volts 11.5
  Amps. (min.)) 57*
  Amps. (max.) 70*
  Rpm (min.) 5000
  Rpm (max.) 7400
Lock Test
  Amps. 500
  Volts 3.4
  Torque (min.) .... 22 ft.-lbs.
    *Includes solenoid

IGNITION SYSTEM

Non-Diesel Models

80. **SPARK PLUGS.** Spark plugs are the 18mm size. Electrode gap is 0.028-0.032 for gasoline engines, or 0.018-0.022 for LP-Gas engines. When installing spark plugs, tighten same to 26-30 ft.-lbs. torque. Ford Motor Company recommends that spark plugs be renewed after every 300 hours of operation.

81. **DISTRIBUTOR.** The distributor is mounted on right side of engine and is driven by a gear which meshes with a gear on the engine camshaft. An intermediate shaft is pinned to the distributor driven gear and is used to drive the engine oil pump.

Refer to paragraph 82 for information on timing the distributor.

Removal and overhaul of distributor is conventional and is obvious upon an examination of the unit; however, the following points should be observed while servicing the distributor. Renew distributor bushing and/or shaft when clearance exceeds 0.0045. Install new bushing in same position as original bushing and ream inside diameter of bushing to 0.500-0.5005. When installing distributor driven gear, align holes in shaft and gear and before installing pin, check shaft end play which should be 0.005-0.008. Add or subtract shims between gear and base (housing) as necessary, to obtain the specified shaft end play.

Note: In some cases, replacement shafts may not have a hole. In this case, adjust shaft end play; then, drill shaft to accept pin.

Distributor specifications are as follows:
Ford part No......... CONN-12127-E
Point gap 0.024-0.026
Breaker arm spring
  tension .... 17-20 oz.
Rotation .... Clockwise
Condenser capacity..... 0.21-0.25 mfd
Dwell, degrees 28-32

Advance data—degrees at rpm:
  Start advance .... 0@ 275
  Intermdiate advance ...... 2@ 400
  Maximum advance ...... 12@1200
Advance data given above is in distributor degrees and rpm. To convert these values to engine degrees and rpm, double the listed vales.

82. **IGNITION TIMING.** Procedure for setting both static and running timing will be given. If possible, always time distributor with a timing light (running timing).

83. To set static timing, be sure breaker points are set at 0.024-0.026, then turn engine in direction of normal rotation until number one piston is coming up on compression stroke, and continue to turn engine until the 4 degree BTDC mark on vibration dampener is aligned with pointer on timing gear cover. Turn on ignition switch, hold wire of number one spark plug about $\frac{3}{16}$-inch away from spark plug, or cylinder block, and rotate distributor assembly counter-

clockwise until spark occurs. Tighten distributor mounting bolts.

Note: If distributor requires more than 10 or 12 degrees of rotation, remove mounting bolts, lift distributor from well far enough to turn shaft, then turn shaft clockwise enough to engage next gear tooth. Always approach the number one cylinder firing position in a counter-clockwise direction.

84. To check running timing, place a white mark on the 24 degrees BTDC mark on the vibration dampener. Connect timing light to number one spark plug wire, then start engine and run at 2400 rpm. Direct timing light on vibration dampener and rotate distributor assembly to align the previously marked 24 degree BTDC timing mark with pointer on timing gear cover. Tighten distributor mounting bolts.

Refer to distributor specifications in paragraph 81 for intermediate advance data to be used in checking distributor advance mechanism.

TRANSMISSION

The "Select-O-Speed" transmission is installed on all Series 6000 and Commander 6000 tractors. The transmission is available in three options (models) as follows: No pto; two speed engine driven pto; or two speed engine driven pto, plus ground driven pto (deluxe). All descriptions in this manual will refer to the deluxe transmission with special references to any service differences for other models.

OPERATION

All Models

90. The transmission is basically an arrangement of three planetary gear systems coupled hydraulically, by three bands and four multiple disc clutches, through a hydraulic control panel, giving a selection of 10 forward and two reverse speeds selected and engaged by movement of a hand selecter lever. Operation of a foot clutch is not necessary in changing from one gear to another or in starting or stopping the tractor. A foot operated feathering valve is provided for interrupting the gear train in case of emergency or for close maneuvering such as the hitching and unhitching of implements.

Refer to Fig. FO55 for a cross-sectional view of the deluxe type transmission showing the location and re-

lationship of the various parts. Planetary systems are designated, front to rear as "A", "B" and "C". Bands and clutches are designated as 1, 2 and 3 from front to rear. Servos 1, 2 and 3 are the activating piston and spring assemblies for the clutch band of the corresponding number. The roller vane type hydraulic pump is mounted on, and driven by, the transmission input shaft.

No attempt will be made to follow the power flow through the transmission in the various gear ratios; however, a brief explanation of the function of certain assemblies will provide the informantion necessary for a better understanding of the problems encountered.

Bands No. 2 and 3 are mechanically engaged by spring action and hydraulically released when the servo is activated by the movement of the selecter lever. Band No. 1 and the four multiple disc clutches are hydraulically engaged and mechanically released. Thus, when hydraulic pressure is cut off from all of the clutches and servos, band 2 and band 3 will be applied, and band 1 and the C1, C2 and C3 clutches will be released. The transmission will then be in the "Park" position with the output shaft locked from turning and the input shaft free to rotate. This condition will prevail if:

(a) The selector lever is moved to the park position with the engine running.

(b) The engine is stopped with the selector lever in any position.

(c) In the event of the complete hydraulic power failure from any cause.

When the transmission neutral position is obtained by movement of the selector lever to N position, band 2 is hydraulically released, band 3 is mechanically applied, none of the clutches, except the direct drive clutch, are activated and both the input and output shafts are free to rotate. When neutral position is obtained by means of the foot operated feathering valve, either band 2 or band 3 is hydraulically released depending on the speed position of the selector lever, band 1 and the C1, C2 and C3 clutches are not activated and both the input and output shafts are free to rotate.

Planetary system "A" is the transmission input unit and delivers power to the remainder of the transmission at engine speed or overdrive, depending on the action of the No. 1 band and direct drive clutch. Power enters the "A" planetary unit at the planet carrier and is transmitted to the transmission assembly through the ring gear. In the direct drive position, band No. 1 is released and the sun gear is prevented from rotating faster than the carrier, by the action of the direct drive clutch, which locks the carrier and the sun gear together causing the "A" planetary system to rotate as a unit at engine speed. When band No. 1 is applied the direct drive clutch is released, the sun gear is held stationary in the transmission case and the ring gear drives the remainder of the transmission at overdrive ratio.

The five basic forward and one basic reverse speeds are obtained by connecting the "B" and "C" planetary systems in various ways by means of the hydraulically disengaged bands 2 and 3 and the three hydraulically engaged clutches.

91. **HYDRAULIC CIRCUITS.** The control valve assembly separates the hydraulic system into four separate circuits. Normal pump flow is from hydraulic pump to the oil distributor where it is distributed to the direct drive clutch and the control valve and through the system relief valve.

The system relief valves is set to maintain a pressure of 170-190 psi in the transmission main circuits, then to the lubrication gallery which is ported to lubricate the transmission main assemblies. A relief valve is located in the lubrication gallery and adjusted to maintain a minimum pressure of 11¾ psi at the lubrication indicator light sending switch. Pump flow not required for transmission operation or lubrication is exhausted through this valve back to the reservoir. A hydraulic pressure switch, located in the oil distributor assembly, is ported to the lubrication gallery and connected to the lubrication indicator lamp on the instrument panel. The transmission and pto regulating valves, set at 145-155 psi, act as relief valves at this lower pressure when the pto or transmission feathering valves are activated. These valves

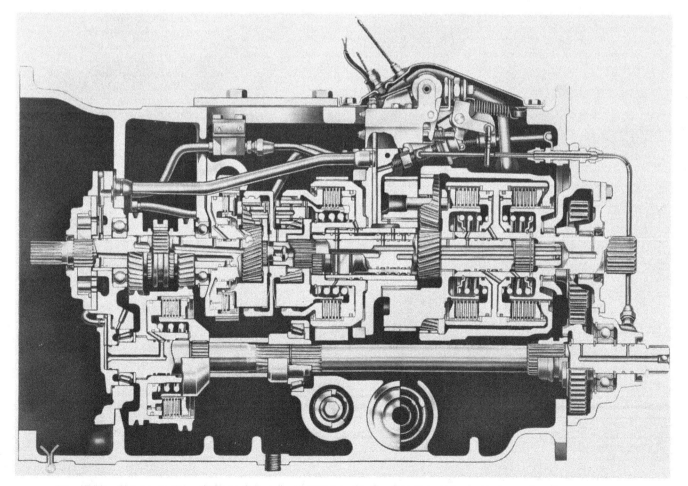

Fig. FO55—Cut-away view of Select-O-Speed transmission showing location and arrangement of main components.

maintain a pressure of 145-155 psi at the two servos controlling the hydraulically released bands 2 and 3 and prevent the transmission from going into park when either of the feathering valves are moved.

The first movement of the inching pedal opens the secondary transmission circuit to the sump, dropping the pressure in that circuit to relieve any action imposed on the three clutches. The transmission regulating valve maintains high pressure in the primary circuit and on servos 1, 2 and 3. Further movement of the inching pedal blocks the line leading to the secondary circuit and the system relief valve again comes into action. Depressing the inching pedal stops the power flow in the transmission by cutting off the pressure to the three clutches and the transmission will go into neutral.

Movement of the pto feathering valve will have a similar effect on system pressure but will not interrupt transmission power flow.

92. CONTROL POSITIONS. The following table gives the clutch and band units activated by hydraulic pressure to obtain each of the 12 operating and two stationary power combinations. It will be noted that R-2 is the overdrive condition of R-1, and that forward speeds 3, 4, 7, 8 and 10 are the overdrive conditions of speeds 1, 2, 5, 6 and 9.

Selector Position	Hydraulic Action C-Clutch B-Band
P	Direct drive clutch engaged.
R2	C1 engaged, B2 disengaged, B1 engaged.
R1	C1 engaged, B2 disengaged, direct drive clutch engaged.
N	B2 disengaged, direct drive clutch engaged.
1	C3 engaged, B2 disengaged, direct drive clutch engaged.
2	C2 engaged, B2 disengaged, direct drive clutch engaged.
3	C3 engaged, B2 disengaged, B1 engaged.
4	C2 engaged, B2 disengaged, B1 engaged.
5	C3 engaged, B3 disengaged, direct drive clutch engaged.
6	C2 engaged, B3 disengaged, direct drive clutch engaged.
7	C3 engaged, B3 disengaged, B1 engaged.
8	C2 engaged, B3 disengaged, B1 engaged.
9	C1 engaged, C2 engaged, B2 disengaged, B3 disengaged, direct drive clutch engaged.
10	C1 engaged, C2 engaged, B1 engaged, B2 disengaged, B3 disengaged.

The following table lists the three bands and the four clutches and indicates the speed positions in which they are activated:

Unit	Speed Positions
Clutch 1 (engaged)	R1, R2, 9, 10
Clutch 2 (engaged)	2, 4, 6, 8, 9, 10
Clutch 3 (engaged)	1, 3, 5, 7
Direct drive clutch (engaged)	R1, 1, 2, 5, 6, 9
Band 1 (engaged)	R2, 3, 4, 7, 8, 10
Band 2 (disengaged)	R1, R2, 1, 2, 3, 4, 9, 10
Band 3 (disengaged)	5, 6, 7, 8, 9, 10

SYSTEM ADJUSTMENTS

All Models

Malfunctions of the "Select-O-Speed" transmission, for which service is required, could be from a number of causes; the most common of which, will be maladjustment of one or more units of the transmission correctable by a complete unit adjustment. The first step in correcting troubles, therefore, would be a complete operational adjustment on the three transmission bands, the four regulating valves and the selector assembly. As part of the adjustments are made with the engine running and the selector lever in an operational position, the first step is to disengage the traction coupling as follows:

93. TRACTION COUPLING. All tractors incorporate a traction coupling sleeve which can be shifted to disengage the transmission output shaft from the differential pinion. The shift lever is located on the left side of the rear axle center housing (Fig. FO56). Note: Traction coupling must be disconnected if it is necessary to tow tractor. It may be necessary to rock tractor slightly to disengage or re-engage traction coupling.

93A. FLUID LEVEL CHECK. Before an attempt is made to start or service the tractor, first check the fluid level. To check, remove the pipe plug located on the right side of the transmission housing. Fluid should be at the plug with tractor standing level. If fluid is low, fill to plug level, through plug opening, with Ford Hydraulic Fluid, Specification M2C-41.

94. FEATHERING PEDAL ADJUSTMENT. Be sure all linkage is free and lubricated, then adjust the feathering (inching) pedal assembly as follows: Check and adjust, if necessary, the engine low idle and high idle speeds; then, as a matter of safety, disengage the traction coupling. Remove the left rear hood panel and check the kick-down support bracket which should be parallel with top edge of frame side rail.

With engine stopped, push feathering pedal down until roller on feath-

GEAR RATIO	DIRECT DRIVE * CLUTCH	BAND SERVO			CLUTCH PACK		
		B_1*	B_2**	B_3**	C_1*	C_2*	C_3*
Park (P)	A		A	A			
R2		A			A	A	
R1	A			A	A		
Neutral (N)	A			A			
1st	A			A			A
2nd	A			A		A	
3rd		A		A			A
4th		A		A		A	
5th	A		A				A
6th	A		A			A	
7th		A	A				A
8th		A	A			A	
9th	A				A	A	
10th		A			A	A	

* Pressure Applied

**Spring-Applied

The above chart indicates by the capital letter "A" the application of bands and clutches for each gear ratio.

ering lever just starts to move rearward after passing over crown (high point) of pedal cam. At this point, turn the adjusting bolt in the PTO kick-down arm until the radius portion (rear) of the kick-down arm contacts the lower cam surface of the feathering pedal. This adjustment causes the feathering pedal to work against the kick-down arm spring and provides "feel" for the pedal.

Start engine and run at low idle speed, then push down on the feathering pedal and observe the transmission lube warning light as it comes on. Continue to slowly push feathering pedal down and when the lube warning light goes out, the pedal should have an additional ⅛-inch travel before pedal contacts the kickdown lever. Adjust the length of the rod between feathering pedal and transmission, if necessary to obtain this measurement.

Now operate engine at high idle rpm and depress feathering pedal until roller of feathering lever is on the high point of the feathering pedal cam. At this point, engine speed should be 1100-1150 rpm. If engine speed is not as specified, adjust the rod between feathering lever and the lower throttle rod to obtain this engine rpm.

With engine still at high idle rpm, completely depress the feathering pedal and check the engine speed which should be 2150-2230 rpm. If engine speed is not as specified, adjust stop screw in feathering pedal as required.

95. **PRESSURE CHECK.** As stated in paragraph 91, the transmission hydraulic system is divided into four separate but inter-related systems as follows:

a. Direct Transmission System, serving bands 1, 2 and 3, direct drive clutch, and the other three systems through their respective regulating valves.

b. Indirect Transmission System, serving three clutches, and connected to the direct transmission circuit by the 145-155 psi transmission regulating valve.

c. Indirect PTO System, serving the pto clutch and connected to the direct transmission circuit by the 145-155 psi power take-off regulating valve.

d. Transmission Lubrication System, with lubrication passages to the three multiple disc clutches and the shaft bearings. The lubrication circuit is connected to the direct transmission circuit by the 170-190 psi system relief valve, and protected by its own 43 psi lubrication pressure relief valve.

Note: Lubrication pressure is checked at the lubrication light sender switch, where pressure should be a minimum of 11¾ psi; difference in pressure is due to normal leakage in lubrication circuit.

Provisions are made for installation of a pressure gage in servos 1, 2 and 3 as illustrated in Fig. FO57 and Fig. FO58. Gages in all three locations may be used; however, all pressure valve adjustments and a complete systems diagnosis can normally be made by the installation of a gage in No. 2 servo.

To check the pressure, disengage the traction coupling, bring the transmission fluid up to operating temperature and install a 0 to 300 psi gage in servo No. 2. Shift the selector lever into neutral position and operate the tractor at 800 rpm. The pressure gage should read 170-190 psi. A higher or lower reading will indicate that the system relief pressure is incorrect and the system regulating valve will need to be reset. If 170-190 psi system pressure is not obtained with transmission operating in neutral position, stop the engine, move the pressure gage to servo 3 location, restart the engine, shift to 5th position and compare reading with one previously made. A higher reading would indicate leakage at servo 2, and servo 3 and 5th speed position should be used for remainder of test. Partially depress inching pedal until lube light

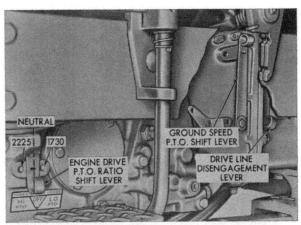

Fig. FO56—View showing traction disconnect lever and the two pto shift levers.

Fig. FO57—Right side of transmission showing Band 2 adjusting screw and Servo 1 cover.

B2. Band 2 adjusting screw
S1. Servo No. 1

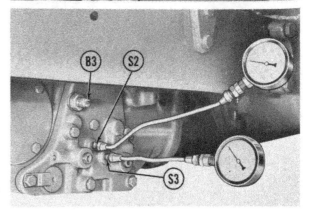

Fig. FO58 — Left side of transmission showing interlock plate with gages installed in Servos 2 & 3.

B3. Band 3 adjusting screw.
S2. Servo No. 2.
S3. Servo No. 3.

flashes on and note reading on pressure gage while lube light is burning. Pressure reading should drop to 145-155 psi. A reading higher or lower than 145-155 psi will indicate that the transmission regulating valve setting is incorrect and will need to be reset.

Release inching pedal and move pto control handle to the halfway position. System pressure reading should drop to 145-155 psi and lubrication indicator lamp should flash on. A higher or lower reading will indicate that pto regulating valve setting is incorrect and will need to be reset.

If lubrication indicator lamp reacted normally in above checks; that is, flashing on when pressure dropped below 170-190 psi and going off when ppressure returned to the 170-190 psi reading, it is safe to assume that the lubrication relief valve is functioning properly. If the indicator lamp should remain lighted it would indicate a

faulty pressure switch, grounded wire or inoperative relief valve. If indicator lamp should fail to light, it would indicate a broken wire or loose connection, faulty pressure switch, or burned out bulb. These items should be checked and corrected and the system again tested before attempting to adjust the lubrication pressure valve. Note: In some instances, broken or leaking main shaft seals will cause the lube pressure to be maintained due to leakage from the main to lube system.

If above conditions are not met, refer to paragraph 96 for information regarding regulating (control) valves adjustment.

96. REGULATING (CONTROL) VALVES ADJUSTMENT. To adjust the regulating (control) valves, it is first necessary to remove the transmission top rear cover as outlined in

paragraph 105, then proceed as follows:

NOTE: Some differences may exist in the manner in which filter and cooling lines exit from the transmission. For those early series 6000 tractors having lines attached to the heat exchanger which protrudes through transmission top cover, proceed as outlined in paragraph 96A. For those tractors which have filter and cooling lines attached to right side of transmission, proceed as outlined in paragraph 96B.

96A. Remove clevis pin from top end of inching pedal rod and remove inching pedal return spring (Fig. FO59). Unbolt heat exchanger manifold from control valve and remove heat exchanger and discharge tube. Unbolt starter safety switch bracket and remove switch and bracket. Remove the remaining two mounting cap screws at left side of valve and lift out control valve. Note: Do not remove the two cap screws nearest cam as they retain control valve upper body to lower body. It may be necessary to bump or pry control valve loose from gasket seal. Reinstall control valve by reversing removal procedure. Tighten cap screws to a torque of 6-8 ft.-lbs.

96B. Remove clevis pin from top end of inching pedal rod and remove inching pedal return spring (Fig. FO-59). Unbolt starter safety switch bracket and remove switch and bracket. Unbolt oil outlet line manifold from control valve and remove the two remaining valve mounting cap screws at left side of valve. Note: Do not remove the two cap screws nearest cam as they retain control valve upper body to lower body. It may be necessary to bump or pry the valve loose from mounting gasket seal. Pivot the oil outlet line up with the control valve assembly and remove valve from transmission. Remove oil outlet line from transmission case.

To reinstall valve, proceed as follows: Disconnect upper cooling line from side of transmission. Insert oil outlet line, without "O" ring, through hole in transmission case, install new "O" ring on outer end of line and push line back into proper place. Pivot the line and control valve assembly into place as a unit and complete remainder of installation by reversing removal procedure. Tighten cap screws to a torque of 6-8 ft.-lbs.

96C. With control valve removed, remove the two screw retainers (Fig. FO69) on front of upper control body, then reinstall control valve on tractor. Ground out the starter safety switch

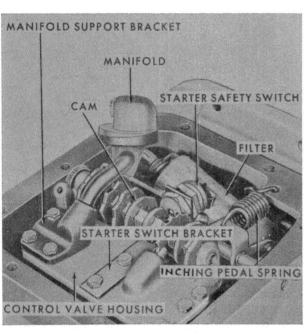

MANIFOLD SUPPORT BRACKET

MANIFOLD

STARTER SAFETY SWITCH

CAM

FILTER

STARTER SWITCH BRACKET

INCHING PEDAL SPRING

CONTROL VALVE HOUSING

Fig. FO59—View showing control valve installed. View shown is for early production tractors which have cooler lines attached to heat exchanger manifold. Late production tractors have cooler lines attached to right side of transmission housing.

OFF-SET SCREW DRIVER

CAM ASSEMBLY

VALVE MOUNTING BOLTS

Fig. FO60—When adjusting regulating valves, rotate cam assembly three positions in direction shown by arrow. Unit shown is early model transmission; however, series 6000 transmissions are similar.

and restart tractor. Rotate the control cam assembly three positions in the direction of rotation shown in Fig. FO60 and adjust the regulating valves by turning the adjusting screws in or out as required. Fig. FO61 illustrates the location of the four regulating valves. The long adjusting screw retainer makes a good screwdriver for adjusting the valves.

If it is impossible by means of system regulating valve adjustment, to obtain the 170-190 psi system operating pressure, first check the lubrication indicator lamp with the engine running. If the light is off, stop the engine and turn the ignition switch back to the "On" position. If the light then comes on it would indicate that the lubrication system is functioning and the trouble lies in the system relief valve and the valve must be removed and serviced as outlined in paragraph 109. If the lubrication indicator lamp remains on and the 170-190 psi pressure cannot be obtained, it would indicate that the pump is at fault or a leak exists in the system. Start the engine and examine the transmission for evidence of leaks or oil turbulence which would indicate a leak. If none are found, the pump will need to be removed and serviced as outlined in paragraph 112. If leaks are found, the transmission must be disassembled and the leaks repaired before proceeding further.

When proper adjustment has been made on the regulating valves, remove the control valve and reinstall the adjusting screw retainers. If adjustment of the regulating screw slots is necessary for retainer installation, turn the adjusting screws inward to increase the pressure until proper alignment is obtained. Reinstall the control valve. Complete balance of reassembly.

97. **BAND ADJUSTMENTS.** Band adjustments are all made with the traction coupling disengaged and when adjusting Bands 2 and 3, have the engine running at 800 rpm. Engine must be stopped at adjust Band 1.

Band No. 1: Hold adjusting screw (Fig. FO62) stationary with screw driver and back-off lock nut two full turns.

Install blade screw driver socket on a torque wrench and tighten adjusting screw to a torque of 5-10 ft.-lbs. as illustrated in Fig. FO63. (Note: Check to make sure that lock nut did not seat on transmission housing. If it did, back-off lock nut and recheck.) Back-off adjusting screw exactly one full turn. Start engine, move speed selector to 3rd gear position, hold adjusting screw stationary and tighten lock nut to a torque of 20-25 ft.-lbs.

Band No. 2: Place transmission in neutral with engine running at 800 rpm; then, hold adjusting screw stationary while backing off lock nut two full turns. With a torque wrench, tighten adjusting screw to a torque of 5-10 ft.-lbs. while observing pre-

cautions outlined for Band No. 1. Back-off adjusting screw exactly ¾-turn, move speed selector to Park position, hold screw stationary and tighten lock nut to a torque of 20-25 ft.-lbs. Band 2 adjusting screw is located on the interlock plate on the left side of the transmission case as shown in Fig. FO64.

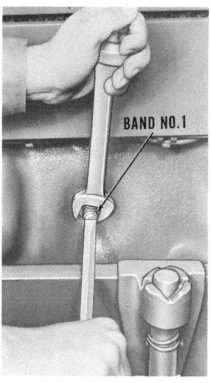

Fig. FO62. Loosening lock nut on Band 1 adjusting screw.

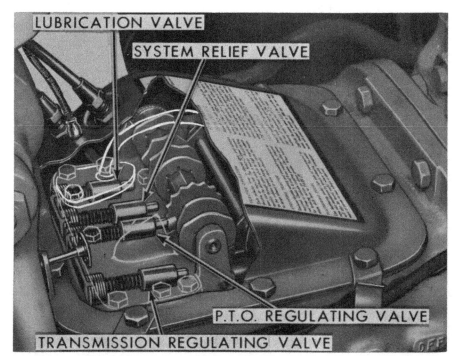

Fig. FO61 — Regulating valves are located in upper half of control valve as shown. Unit shown is early model transmission; however, series 6000 transmissions are similar.

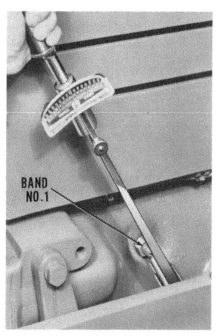

Fig. FO63—Use torque wrench as shown when making band adjustments. Refer to text.

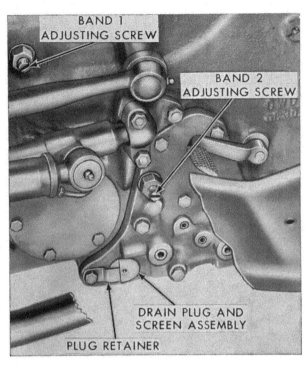

Fig. FO64 — Band 1 and Band 2 adjusting screws are located on left side of transmission as shown on this view of early type transmission.

Band No. 3: With engine running at 800 rpm, shift selector lever to fifth speed position, hold adjusting screw and back-off lock nut at least two full turns. Tighten the adjusting screw to a torque of 5-10 ft.-lbs. as shown in Fig. FO63, then back-off screw exactly ¾-turn, move speed selector to Park position, and tighten lock nut to a torque of 20-25 ft.-lbs.

NOTE: It may not be possible to reach the 5-10 ft.-lb. torque value on Band 3 adjusting screw without killing the engine; if so, tighten adjusting screw until engine begins to pull down, then back-off screw ¾-turn.

98. SELECTOR ADJUSTMENT. For positive identification of speed selections, the individual speed indications on the dial should always be positioned directly under the pointers. Adjustment for wear or misalignment can be made as follows:

Remove the selector shaft cover, move the selector lever to the park position and loosen the shaft nut. Move the dial to proper alignment and retighten nut. See Fig. FO65 for an exploded view of the selector assembly.

TROUBLE SHOOTING

All Models

99. OPERATIONAL CHECK. If the system adjustments outlined in the previous section fail to correct transmission malfunctions, the next step in trouble diagnosis would be an operational check. To perform this check, the traction coupling must first be engaged.

Start engine and set the engine speed at 800 rpm. Put the transmission in neutral by depressing the inching pedal, shift the selector lever into each speed position in turn and gradually release the inching pedal. Note the reaction in each speed position for later reference to the diagnosis guide. When the inching pedal is released, one of five conditions will prevail.

(1) The tractor will operate in an incorrect speed ratio.
(2) The tractor will go to neutral.
(3) The tractor will go to park.
(4) The tractor will lock up (stall engine).
(5) The transmission will operate properly in that control position.

A diagnosis guide in paragraph 102

Fig. FO65 — Exploded view showing component parts of speed selector assembly.

outlines the probable causes of malfunction if indicated abnormal patterns are encountered in the operational check.

100. TORQUE LIMITING CLUTCH. A torque limiting clutch is installed in the engine flywheel and functions as an overload clutch to prevent damage to the engine or transmission in the event a transmission lockup occurs. Slippage of the torque limiting clutch under normal loads will cause an interruption or lowering of transmission pump flow and system pressure and a lockup may occur. A defective torque limiting clutch is to be suspected if the transmission malfunction exists only under extremely heavy loads, especially in the higher speed ranges and the transmission operates properly when shifted to a lower speed range.

To check the torque limiting clutch, start the engine and bring the tractor up to operating temperature. Shift the selector lever into 10th position, increase the engine speed to 1500 rpm and firmly apply both brakes. If the tractor forward motion can be halted without stalling the engine the torque limiting clutch will need to be renewed as outlined in paragraph 111.

101. PRESSURE CHECKS. Leakage in any of the clutches or servos will only occur when that unit is activated. To check the various units for leakage, first disengage traction coupling, insert a 0-300 psi pressure gage (or gages) in servos 1, 2 and 3 and proceed as follows:

Operate engine at 800 rpm, move selector through all gear positions and note gear positions where low pressure readings occur. Completely de-

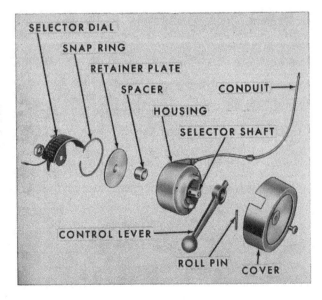

TROUBLE SHOOTING CHART

POSITION OR GEAR	DIRECT CIRCUIT								INDIRECT CIRCUIT					
	DIRECT DRIVE CLUTCH		BAND						CLUTCH PACK					
	a	r	a B1* r		a B2** r		a B3** r		a C1* r		a C2* r		a C3* r	
Park (P) Inoperative	P	**A**	L-P	P	P **A**	N	P **A**	N	L	P	L	P	L	P
R2 Inoperative	L	R2	R2 **A**	N	L R2	R2 **A**	N R2	**A**	N	L	R2	L	R2	R2
R1 Inoperative	R1	**A**	N L-R2	R1	L R1	R1 **A**	N R1	**A**	N	L	R1	L	R1	R1
Neutral (N) Inoperative	N	**A**	N L-N	N	P N	N **A**	N	R1	N	2	N	1	N	N
1st Inoperative	1	**A**	N L-3	1	L 1	1 **A**	N	L	1	L	1	1 **A**	N	
2nd Inoperative	2	**A**	N L-4	2	L 2	2 **A**	N	L	2	2 **A**	N	L	2	
3rd Inoperative	L	3	3 **A**	N	L 3	3 **A**	N	L	3	L	3	3 **A**	N	
4th Inoperative	L	4	4 **A**	N	L 4	4 **A**	N	L	4	**A**	N	L	4	
5th Inoperative	5	**A**	N L-7	5	5 **A**	N	L	5	L	5	5 **A**	N		
6th Inoperative	6	**A**	N L-8	6	6 **A**	N	L	6	L	6	6 **A**	N	L	6
7th Inoperative	L	7	7 **A**	N 7	**A**	N	L	7	L	7	7 **A**	N		
8th Inoperative	L	8	8 **A**	N 8	**A**	N	L	8	L	8	8 **A**	N	L	8
9th Inoperative	9	**A**	N L-10	9	L 9	L	9	9 **A**	N 9	**A**	N	9	9	
10th Inoperative	L	10	10 **A**	N	L 10	L	10	10 **A**	N 10	**A**	N	10	10	

A, applied for that position.
L, indicates lock-up.
a, applied and does not release.
r, released and does not apply.

* pressure applied.
** spring applied.
▨ hydraulic pressure on.

press the inching pedal in the gear positions which had the low pressure readings and note the lubrication warning light and the pressure readings at servos 1, 2 and 3. If the pressure readings increase and the warning light goes out when the inching pedal is completely depressed, leakage is occuring in the indirect circuit which activates Clutches 1, 2 and 3. If pressure readings remain low, leakage is occurring in the direct circuit which activates Bands 1, 2 and 3.

To check for leakage in the power take-off clutch circuit, move selector lever to any position giving 170-190 psi reading on any of the Servo 1, 2 or 3 gages and pull PTO control handle out to the engaged position. If gage reading decreases, leakage is occuring in the power take-off circuit or clutch unit.

With faulty circuit located, proceed to the Diagnosis guide in paragraph 102.

102. **DIAGNOSIS GUIDE.** Complete failure of one of the clutch or band units will cause a definite pattern of failure in certain speed positions. The following, lists these patterns and the units involved.

Pattern: Operates in reverse only.

Cause: Broken mainshaft; B carrier spline damaged; clutch 2 or clutch 3 housing damaged or broken.

Pattern: Goes to neutral in R1, N, 1, 2, 5, 6 and 9. Cause: Direct drive clutch does not engage.

Pattern: Goes to neutral in 3, 4, 7, 8, 10 and R2. Cause: Band 1 does not engage.

Pattern: Goes to neutral in P, 5, 6 7 and 8. Cause: Band 2 does not engage.

Pattern: Goes to neutral in P, 1, 2, 3, 4, R1 and R2. Cause: Band 3 does not engage.

Pattern: Goes to neutral in 9, 10, R1 and R2. Cause: Clutch 1 does not engage.

Pattern: Goes to neutral in 2, 4, 6, 8, 9 and 10. Cause: Clutch 2 does not engage.

Pattern: Goes to neutral in 1, 3, 5 and 7. Cause: Clutch 3 does not engage.

Pattern: Locks up in 3, 4, 7, 8, 10 and R2. Cause: Direct drive clutch does not release.

Pattern: Locks up in P, R1, N, 1, 2, 5, 6 and 9. Cause: Band 1 does not release.

Pattern: Locks up in 1, 2, 3, 4, 9, 10, R1 and R2 and goes to P in neu-

tral. Cause: Band 2 does not release.

Pattern: Locks up in 5, 6, 7, 8, 9 and 10. Cause: Band 3 does not release.

Pattern: Locks up in P, 1, 2, 3, 4, 5, 6, 7 and 8 and goes to R1 in neutral. Cause: Clutch 1 does not release.

Pattern: Locks up in P, 1, 3, 5, 7, R1 and R2 and goes to 2 in neutral. Cause: Clutch 2 does not release.

Pattern: Locks up in P, 2, 4, 6, 8, R1 and R2 and goes to 1 in neutral. Cause: Clutch 3 does not release.

Pattern: Operates only in P, N, R2, 3, 4, 7, 8 and 10. Band 1 does not release due to stuck valve.

The accompanying Trouble Shooting Chart indicates, by the shaded areas, which elements have hydraulic pressure applied and the bold faced capital letters "A" indicate what units are applied in normal operation. The lower case "a" and "r" at the top of each column indicates whether the unit involved remains applied or released during actual operation of the transmission. The letters and numbers under the "a" and "r" heading indicate the reaction which occurs in each gear position when unit involved is applied or released, and these are what determines the shift patterns previously given.

As an example: Assume the pressure check (paragraph 101) indicated low pressure in the indirect circuit and the gear positions affected were 2, 4, 6, 8, 9 and 10. Since the indirect circuit is involved, the shaded areas of the chart will show Clutch 2 is the unit involved since this is the only unit activated by the indirect circuit in these gear positions.

TRANSMISSION OVERHAUL

All Models

Access to the control valve assembly is obtained by removal of the transmission top rear cover. Access to the hydraulic pump, torque limiting clutch, input shaft and pto system is obtained by splitting the tractor between the transmission housing and engine assembly. All other transmission components may be removed and reinstalled from the rear of the transmission case after detaching the transmission housing from the rear axle center housing. The first step in transmission overhaul is to determine, as nearly as possible, which components require service and remove only the parts necessary to obtain access to those components.

Note: The "Select-O-Speed" transmission is a hydraulic unit and merits the same standards of care and cleanliness accorded any hydraulic or diesel unit. Disassembly or service should only be attempted in a clean, dust free shop and the removed assemblies stored and serviced only where good housekeeping is observed.

103. DRAINING THE TRANSMISSION. To drain the transmission on tractors prior to 1963 production, loosen the cap screw holding the drain plug retainer on the right side of transmission case. Insert a ¼-20 bolt into drain plug threads and withdraw plug sufficiently to expose the "O" ring. Rotate the plug 180 degrees and withdraw the plug until the drain hole is exposed. Effective with 1963 production, a ½-inch drain (pipe) plug was incorporated into bottom side of transmission housing. Use this plug to drain transmission before removing plug and strainer assembly.

Clean strainer each time transmission oil is drained.

104. R&R TOP FRONT COVER. To remove transmission top front cover, first remove rear side panels. Disconnect starter solenoid from battery tray and disconnect throttle linkage. Disconnect battery cables and remove battery, battery tray and transmission top front cover.

105. R&R TOP REAR COVER. To remove the transmission top rear cover, remove rear side panels, hood and rear hood panel. Disconnect starter solenoid from battery tray and disconnect throttle linkage. Disconnect battery cables and remove battery and battery tray. Pull the pto control handle all the way out and loosen cable conduit nut at transmission cover. Loosen cable conduit nut at control handle end; then, loosen lock nut which retains control handle to instrument panel. Rotate pto cable clockwise until cable disengages from pto connector in the transmission.

Place selector handle in "Park" position and disconnect cable conduit at transmission. Move selector lever to the 10th speed position and while flexing cable, disconnect the upper and lower cables. Turn lower cable clockwise to disengage it from wheel of control valve. Note: If selector position indicator shows correct alignment, measure the distance that lower cable protrudes from cover and use this measurement when reassembling. Disconnect starter safety switch and oil pressure switch wires at transmission connectors. Remove floor plate and the remaining two bolts which retain hood support to hood support brackets. Disconnect muffler bracket from hood rail. Raise instru-

ment panel pedestal enough to allow placing a 4x4-inch block between fuel tank and engine valve rocker arm cover. Pivot bottom of instrument panel pedestal upward toward rear and support with a wood block. If so equipped, disconnect coolant lines from heat exchanger, then remove the rubber heat exchanger seal. Remove cover retaining cap screws, lift cover enough to disconnect starter safety switch and oil pressure switch wires, then lift off cover.

NOTE: Prior to installing top rear cover, obtain the latest cover gasket which is a steel gasket with a ⅛-inch bead that follows the bolt circle and encircles each bolt hole. Gasket is installed with bead next to transmission case and must be used only once as bead is collapsed during assembly.

To reinstall transmission top rear cover, be sure control valve cam is in "Park" position, with valves in the out position and the cam followers setting even. Thread the lower selector cable into the pto cable connector to act as a guide and lower cover into position temporarily. Check to see that bolt holes align and cable operates freely. Reposition heat exchanger, if necessary, to allow bolt hole alignment. Raise rear of cover enough to permit connecting the starter safety switch and oil pressure switch wires. Reposition cover and install cover cap screws but do not tighten at this time. Remove the lower selector cable from

the pto connector and start the pto control cable into the pto connector. Thread pto cable into connector by turning control handle counter-clockwise until the lower end of conduit contacts the cover connection. Be sure tab of instrument panel is in groove in threads of control handle assembly and that control handle is in a vertical position. Secure control handle to instrument panel and tighten upper and lower conduit nuts. Check control for freedom of operation. Thread the lower selector cable into wheel of control valve and check for freedom of movement. If tractor is so equipped, the conduit connector in cover can be shifted, if necessary, by tapping carefully using a hammer and punch. Tighten cover cap screws and recheck cable operation. Turn cable in or out as required to obtain a length of 2¾ inches, or the previously measured distance, from top of conduit connection to upper side of notch in the cable connector as shown in Fig. FO66. Be sure control valve is still in "Park" position and speed selector handle is in 10th speed position. Connect the two selector cables and move the selector handle to "Park" position. This will bring the conduit into contact with connection on cover and conduit nut can be tightened. Connect starter safety switch and oil pressure switch wires.

Complete reassembly of tractor by reversing the disassembly procedure.

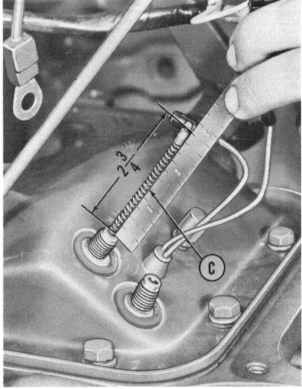

Fig. FO66—Correct lower selector cable (C) installation dimension is 2¾ inches. Refer to text.

106. **SPLIT (CENTER HOUSING FROM TRANSMISSION.)** To separate center housing and transmission, proceed as follows: Disengage traction coupling, loosen set screw on top side of traction coupler shaft and pull shifter assembly outward. Acutate hydraulic lift to relieve pressure in system, then drain hydraulic system. Drain transmission if it is to be serviced. Disconnect brake lines from center housing, loosen opposite ends at brake valve and swing lines out of the way. Disconnect the two rear hydraulic lines from the three-way connectors. Disconnect the two hydraulic lines which are across top of center housing from the three-way connectors and hold in a raised position with a piece of wire. Position a rolling floor jack under drawbar hanger and support frame side rails. Unbolt frame side rails from center housing and center housing from transmission. Disconnect tail light wire and roll the center housing and real axle assembly away from transmission.

107. **SPLIT (TRANSMISSION FROM ENGINE).** To separate (split) transmission from engine proceed as follows: Remove hood rear side panels, hood and rear hood panel. Disconnect starter solenoid from battery tray. Disconnect throttle linkage, then disconnect battery cables and remove battery and battery tray. Pull pto handle out and disconnect conduit at both ends; then loosen lock nut which retains pto control handle to instrument panel. Turn pto cable clockwise until it disengages from connector in

transmission. Place gear selector handle in "Park" position and disconnect cable conduit at transmission top rear cover. Move selector to 10th speed position and while flexing cable conduit, detach upper selector cable from lower selector cable. Disconnect starter safety switch and oil pressure switch wires. Disconnect wiring and remove starting motor. Remove floor plate and the two remaining instrument panel support attaching bolts. Disconnect muffler bracket from left hood rail. Raise instrument panel enough to install a 4x4-inch block between fuel tank and engine valve rocker cover, then pivot instrument panel upward and support same with a suitable wood block. Disconnect the transmission cooler lines at transmission (or cover, if so equipped). Remove clevis pin from upper end of inching pedal rod, then unbolt inching pedal supports and remove the inching pedal assembly. Actuate hydraulic system to relieve hydraulic pressure and drain system. Disconnect the two hydraulic lines, which run across center housing, from the three-way connectors. Disconnect the two rear hydraulic lines from the three-way connectors. Disconnect brake lines from brake valve and remove brake valve from frame side rail. Remove the left and right hand spacer blocks from between transmission and side rails and unbolt rear of side rails from center housing. Support side rails at a point ahead of the spacer blocks location. Place a hydraulic jack under engine oil pan. Place a rolling floor jack under the

transmission assembly. Disengage traction coupling, unbolt the transmission from engine and roll the transmission and center frame assembly away from tractor.

When joining transmission to engine, apply lubricant (Ford M1C-43) to splines of input shaft and torque limiting clutch. Be sure to clean splines thoroughly.

108. **TRANSMISSION R&R.** To completely remove the transmission for servicing, first separate tractor as outlined in paragraph 107. Support both transmission and center housing, then unbolt and remove transmission from center housing.

109. **CONTROL VALVE R&R AND OVERHAUL.** To remove the control valve refer to paragraph 96.

The same standards of care and cleanliness necessary in the disassembly and overhaul of any hydraulic or diesel components should be observed in the dissassembly of the control valve. Place the control valve on lint free paper wipers and if necessary, affix scribe marks on upper and lower valve bodies to insure correct alignment during reassembly. Remove the two bolts joining the upper and lower bodies; then, separate the two halves. Remove the six spool valves and the two feathering valves, together with their springs and retainers, from the lower valve body. The six spool valves are interchangeable but the two feathering valves are not. The pto feathering valve can be identified by the tapered land on inner end of valve. Fig. FO67 shows the lower valve body

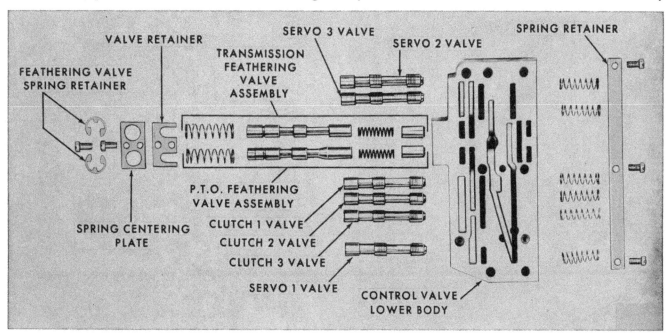

Fig. FO67 — Exploded view of the control valve lower housing showing control (clutch) valves, feathering valves, servo valves, springs and retainers.

with its component parts. The lower valve body and any of the component valves, springs and retainers are serviced separately.

Remove the adjusting screw retainers, the four adjusting screws, and the regulating valves and springs from the upper control valve body. The four pressure regulating valves are identical, and the three springs controlling the system relief valve and the transmission and pto regulating valves are identical but the lubrication valve spring is different. Remove the two snap rings from the cam follower shaft and slide the shaft out of the cam followers and spacers. The six cam followers are identical but three different lengths of spacers are used to properly align the cam followers with the six spool valves cams, and they should be identified prior to removal. Using a fully threaded ¼-20 x 2 inch socket head cap screw as a forcing screw, remove cable wheel trunnion and wheel from right end of camshaft. Note: Do not

use a common steel or rethreaded bolt for this purpose; such a bolt will ordinarily twist off in the trunnion. Remove trunnion from left end of camshaft in similar manner; the socket and flat washer shown in Fig. FO68 are not required. Remove the camshaft assembly and the detent assembly and spring from the upper control valve body. Fig. FO69 shows the upper control valve body with its component parts.

Clean all parts thoroughly in a suitable solvent and examine for wear, scoring or broken or bent parts. Check all valves for free movement in their bores. The table at the end of this paragraph lists the sizes and specifications of the control valve parts.

Before reassembly, examine the mating surfaces of the two valve bodies for burrs, distortion or evidence of oil leakage. The two surfaces must be absolutely flat within 0.0003 over the entire surface. Also check to see that the steel balls which seal

oil passages are securely in position. If balls are loose or missing, renew body.

To assemble, reverse the disassembly procedure. Press cable wheel into camshaft with 0.015 clearance between wheel and valve body as shown in Fig. FO70. Position the two valve halves together and tighten the retaining bolts enough to prevent movement of the two halves, rotate the camshaft to the "Park" position (all spool valves in the out position and all cam followers setting even) and align the scribe lines; or if a new housing is being installed, turn the valve over and align the two halves by measuring the clearance between the two outer spool lands and the side of the discharge port with a feeler gage as shown in Fig. FO71. This clearance should be equal for both spools and measure 0.047-0.067 (0.057 desired). To adjust, loosen the retaining bolts and shift the two halves until alignment is obtained, tighten the valve body bolts equally to 5-8 ft.-lbs. torque. Note: Excessive tightening may distort the bodies, causing the valves to stick.

Insert regulating valve adjusting screws as follows to obtain pressure setting which should be approximately correct: Transmission and pto regulating screws flush with housing; system pressure relief valve screw flush with housing plus one turn; lubrication pressure valve adjusting screw all the way down.

When control valve has been assembled, use new gasket, reinstall on

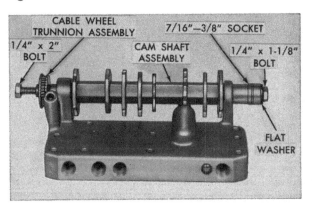

Fig. FO68—Remove camshaft trunnions using method shown for removing cable wheel trunnion assembly; socket shown is not required. Refer to text.

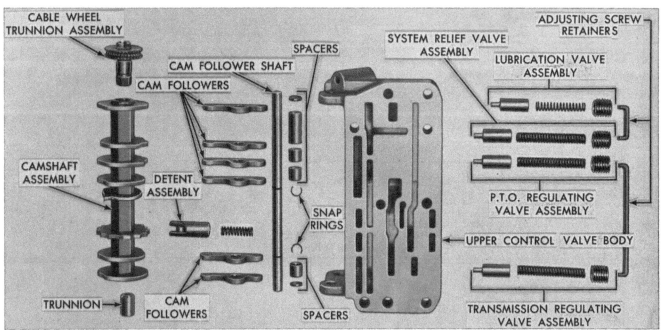

Fig. FO69 — Exploded view of control valve upper housing showing the four regulating valves, springs and retainers and the camshaft and cam followers.

transmission and adjust pressure as outlined in paragraph 96. Remove valve, reinstall adjusting screw retainers, then install valve and tighten the six mounting bolts equally to a torque of 5-8 ft.-lbs.

Specifications for the inspection of the control valve parts are as follows:
Valve spool diameter....0.3738-0.3742
Valve bore diameter....0.3751-0.3758
Left trunnion diameter..0.4350-0.4355
Right trunnion diameter...0.437-0.443

Valve Spring Free Lengths

Feathering valves

Plunger springs 1.220

Return springs 1.300

Detent ....0.070 approx.

Spool valves return.....1.240 approx.

System relief 2.06

Regulating valves 2.06

Lubrication valve 1.47

Valve Spring Test

Feathering valves

Plunger springs

........0.78 in.@ 1.9 - 2 lbs.

Return springs

........0.86 in.@ 7.85- 8.85 lbs.

Detent ....0.58 in.@17 -20 lbs.

Spool valves return

........0.45 in.@ 3.10- 3.50 lbs.

System relief.1.66 in.@15.5 -17 lbs.

Regulating valves

........1.66 in.@15.5 -17 lbs.

Lubrication valve

........1.03 in.@ 3.76- 4.24 lbs.

110. **DIRECT DRIVE CLUTCH VALVE.** Direct drive clutch valve can be removed after the transmission top front cover is removed by unbolting valve from valve support. Valve support can be removed as follows, after removing front adapter plate as outlined in paragraph 124. Pull direct drive clutch supply line forward to disconnect it from valve support. Disconnect the direct drive clutch inlet line and the No. 1 servo line from rear of support and remove the fittings. Support can now be unscrewed from transmission case.

Disassemble direct drive clutch valve by removing the two screws, plate, stop plate, spring and valve from housing.

Clean all parts and inspect valve and valve bore for scoring, nicks, burrs or other damage. Valve should slide freely in its bore. Inspect spring for distortion or possible fractures. Check that all fluid passages are clear. See Fig. FO86 for an exploded

view of the direct drive clutch valve.

Renew "O" rings on direct drive clutch supply line during reassembly.

111. **TORQUE LIMITING CLUTCH.** To remove the torque limiting clutch, first split the tractor between the transmission and engine assembly as outlined in paragraph 107 and unbolt and remove the clutch assembly from the flywheel.

Examine the clutch housing for cracks or damage and the flywheel and pressure plate friction faces for wear or scoring. Renew clutch disc or facings if they are glazed, excessively worn or oil soaked. Total thickness of a new disc with facings is 0.333-0.347. Renew the Belleville washer if it is discolored, cracked or if the dish measures less than 0.178-0.182 when laying on a flat surface. Thickness of new Belleville washer is 0.125. Renew clutch disc and/or transmission input shaft if splines are excessively worn.

In reassembly, tighten the mounting bolts evenly to a torque of 25-30 ft.-lbs. Dish of Belleville washer is toward front of tractor and hub of clutch disc is toward rear of tractor. Hub of clutch will serve as a pilot.

Fig. FO72 shows a partially assembled view of the torque limiting clutch parts.

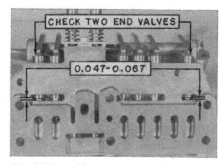

Fig. FO71 — If necessary, use feeler gages to align control valve housings. Refer to text.

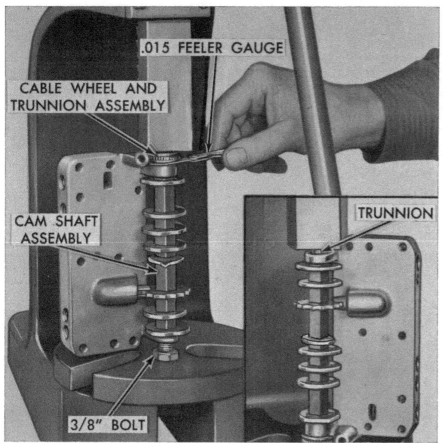

Fig. FO70 — When assembling camshaft, support left end of shaft with ⅜-inch bolt as shown and press cable wheel and trunnion assembly in until there is 0.015 clearance between wheel and body. Then, screw ¼-20 puller bolt into cable wheel trunnion until it bottoms and press right trunnion in as shown in inset.

Fig. FO72 — Partially disassembled torque limiting clutch. Clutch is mounted in flywheel and serves as a safety coupler.

112. **TRANSMISSION PUMP**. To remove the transmission pump first split the tractor between the transmission and engine as outlined in paragraph 107 and unbolt and remove pump.

The 10 gpm roller vane type hydraulic pump can only be renewed as an assembly, however, the shaft seal and pump cover "O" ring are serviced separately. The seal can be renewed from the outside without disassembling pump. Use OTC step plate 630-5, or its equivalent, to drive seal into pump housing. To renew the "O" ring, remove the two screws retaining the back cover to the pump body, remove the cover, and lift out the "O" ring with a pointed tool. Renew the pump if any of the parts are damaged, scored or excessively worn. Always renew the mounting gasket when reinstalling the pump. Tighten pump mounting bolts evenly to a torque of 15-18 ft.-lbs.

113. **TRANSMISSION SHAFT END PLAY**. Transmission shaft end play is controlled by bronze thrust washers placed between each of the rotating members. The position and relative size of the thrust washers are shown in Fig. FO73. Cumulative wear of all the thrust washers will show up as end play in the transmission. Variation in total length of the components is compensated for during assembly by providing two selective fit washers (See Fig. FO73) to hold the end play within specified limits. Cumulative wear on all of the washers will require that this end play be checked and corrected at each overhaul, and renewal of the selective fit washers with one of a greater thickness will provide a correction for wear. End play should be checked before disassembly of the transmission components, as outlined in paragraphs 114 and 115. The overhaul procedures outlined in this manual are based on the supposition that

the two selective fit washers are the only ones which will need to be renewed; however, if any of the other thrust washers in the transmission must be renewed, the additional thickness of the new thrust washers must be taken into account when determining the thickness of the selective fit washers to be installed. Also, if any transmission component, such as a clutch unit or planetary carrier, is renewed, end play may be affected.

114. **TRANSMISSION FRONT END PLAY**. To check front end play, remove the transmission front top cover as outlined in paragraph 104. Back-off the No. 2 band adjusting screw until No. 2 band is completely free from "B" ring gear. Using a small pry bar, pry the "A" ring gear toward the rear and check the clearance between the "A" ring gear and "B" carrier as follows: Clamp a dial indicator to transmission case, make contact between "A" ring gear and dial indica-

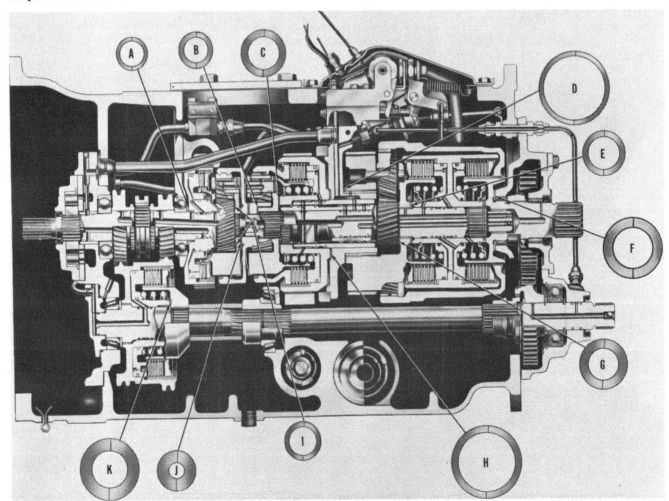

Fig. FO73—Location and relative size of transmission thrust washers. The two selective fit thrust washers are available in various thicknesses and increase in thickness in increments of 0.010. Refer to text.

A. "A" sun gear	D. Distributor to "C" carrier	G. "C" sun gear to carrier	I. "A" sun gear to carrier
B. "B" sun gear	E. "C" carrier to housing	H. Clutch 1 to distributor (selective)	J. "A" ring to carrier
C. Clutch 1	F. Rear thrust (selective)		K. PTO bevel gear

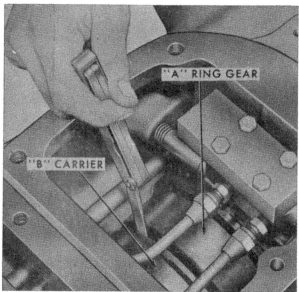

Fig. FO74—View showing point of measurement which can be used for checking transmission front end play on early model transmissions. Use dial indicator for late model transmissions as outlined in paragraph 114.

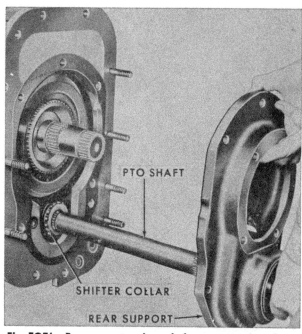

Fig. FO76—Rear support and pto shaft are removed as shown.

tor and zero the dial indicator. Now pry "A" ring gear forward as far as it will go and note reading on dial indicator. End play should be 0.005-0.015. If end play exceeds 0.015, renew the Clutch 1 to distributor thrust washer with one of the correct thickness to give the proper clearance. Selective washers are available in thicknesses ranging from 0.062-0.122 in increments of 0.010. Refer also to Fig. F074.

115. TRANSMISSION REAR END PLAY. After splitting the tractor as outlined in paragraph 106 and before removing any of the transmission components, back-off the Band 3 adjusting screw to completely release the band. Mount a dial indicator on the rear of the transmission so that the contact button rests on the output shaft. Push the output shaft forward and zero the indicator; then pull the shaft rearward and measure the end play. Rear transmission end play should be 0.005-0.015. Renew the rear (output shaft to rear support) thrust washer during transmission reassembly with one of sufficient additional thickness to provide the correct end play. Thrust washer is available in thicknesses of 0.062-0.102 in increments of 0.010.

116. REAR SUPPORT AND PTO REAR SHAFT. To remove the rear support and rear pto shaft, first split tractor between transmission and rear axle center housing. Remove transmission top covers and check transmission end play as outlined in paragraphs 114 and 115 and record values for use during assembly.

On models with ground speed pto, disconnect and remove the rear (external) oil tube (33 — FO75). Disconnect front oil tube (35) from connector (34), then remove connector. Remove gear cover (3), then remove snap ring and slide the ground speed gear (8) from output shaft. On all

models, unbolt rear support from transmission case. Note: On models with no ground speed pto the top center bolt extends through the transmission case and is fitted with a nut and washer. Withdraw the rear support and pto rear shaft as a unit as shown in Fig. FO76. Tapped holes are

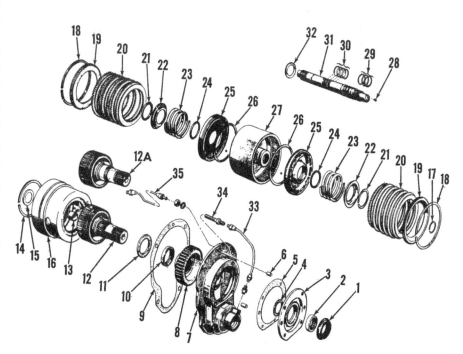

Fig. FO75 — Exploded view of output shaft, rear support, clutches 2 and 3 and mainshaft.

1. Oil seal	10. Bearing	17. Snap ring	27. Clutch 2 & 3
2. Bearing	11. Thrust washer	18. Snap ring	housing
3. Retainer	(selective)	19. Pressure plate	28. Cup plug
4. Snap ring	12. Output shaft (deluxe)	20. Clutch plates	29. Mainshaft seals
5. Gasket	12A. Output shaft (No pto	21. Snap ring	30. Mainshaft seals
6. Dowel	or 2-speed)	22. Spring seat	31. Mainshaft
7. Rear support	13. Bushing	23. Piston spring	32. Thrust washer
8. Ground speed drive	14. Snap ring	24. "O" ring	33. Oil tube
gear	15. Thrust washer	25. Piston	34. Connector
9. Gasket	16. "C" ring gear	26. Piston seal	35. Oil tube

Fig. FO77—"C" ring gear and output shaft can be removed after rear support is off.

provided for jack screws to aid in removal of rear support.

NOTE: Beginning with 1963 production tractors, the rear support (and front plate) are fitted to the transmission case and the assembly line bored. Consequently, none of the three units are available separately.

Refer to Fig. FO75 for an exploded view of output shaft, rear support, "C" ring gear, Nos. 2 and 3 clutches and the main shaft.

On models with ground speed pto, remove oil seal and bearing snap ring from rear of pto shaft. Remove snap ring next to gear at inner side of gear, then press shaft and gear from rear support and slide gear from pto shaft. Examine the output shaft seal (1), bearings (2 and 10) and renew as necessary. Examine output shaft and all gear teeth for wear or damage and renew if necessary. If transmission rear end play was excessive, measure thrust washer (11) and renew with one of proper thickness to obtain the 0.005-0.015 transmission end play.

To reassemble, reverse disassembly procedure. Torque rear support bolts to 35-40 ft.-lbs.

117. "C" RING GEAR AND OUTPUT SHAFT. To remove the "C" ring gear and output shaft, first remove

rear support as outlined in paragraph 116. Remove the pto ground speed shifter collar and position the pto ground speed shift lever to rear position; then, remove the "C" ring gear and output shaft as a unit as shown in Fig. FO77.

Disassembly procedure is evident if renewal of parts is required. Use caution not to lose or damage the selective fit thrust washer on output shaft.

Fig. FO79—Removal of "C" sun gear shaft. Seals are cast iron rings.

Fig. FO80—To remove clutch piston, compress spring with a suitable straddle mounted tool and remove snap ring.

118. CLUTCHES 2 AND 3, "C" CARRIER, MAINSHAFT AND "C" SUN GEAR SHAFT. After removal of "C" ring gear and output shaft as outlined in paragraph 117, the No. 2 and 3 clutch assemblies, "C" carrier and mainshaft assembly, can be withdrawn as a unit as shown in Fig. FO78, after completely loosening the No. 3 band adjustment.

Note: Mainshaft oil seals (29 & 30 —Fig. FO75) are cast iron rings. To avoid breakage, keep assembly together if service on the units is not indicated.

Remove the "C" sun gear shaft by carefully withdrawing from distributor body, being careful not to damage cast iron sealing rings. See Fig. FO79.

If service on Clutches 2 or 3 is indicated, slide "C" carrier off mainshaft, remove snap ring (17—Fig. FO75) from rear of mainshaft and carefully withdraw mainshaft from clutch housing.

To disassemble the clutch, break weld on ends of snap ring (18), remove snap ring, pressure plate (19), bronze clutch plates and steel clutch plates from clutch housing. Using Nuday tool N-775, or other suitable straddle mounted fixture as shown in Fig. FO80, place clutch housing in a press and depress spring seat (22—Fig. FO75) sufficiently to expand and remove snap ring (21). Carefully guide spring seat while releasing the press to prevent the seat from entering the snap ring groove and remove return spring from housing. Place an

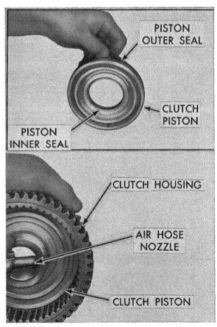

Fig. FO81—Use compressed air to force clutch piston from housing.

Fig. FO78 — Removal of mainshaft, "C" carrier and clutches 2 and 3.

air hose nozzle in the piston port of the clutch hub and force the piston out of the clutch housing as shown in Fig. FO81. Renew piston seals whenever piston is removed and renew housing or piston if scored or excessively worn.

Note: All clutch plates are flat except the steel clutch plates in the pto clutch. Check plates with a straight edge and renew any clutch plates (except pto steel plates) which are warped or distorted. Check the coning of the steel pto clutch plates with a feeler gage as shown in Fig. FO82. Renew any plates that have less than 0.015 cone, or that are warped or otherwise distorted.

During reassembly, install piston in the housing being careful not to damage the new piston seals. Coating the seals with vaseline or Lubriplate before installing the piston will assist in assembly. Reinstall piston return spring by reversing the disassembly procedure. On all except No. 3 clutch, install steel and bronze clutch plates alternately, starting with a steel plate next to the piston. When installing plates in the No. 3 clutch, proceed as follows. Install thin bronze plate with smooth side next to clutch piston. Then install steel plates and the five thick bronze plates alternately with a steel plate next to the thin bronze plate. Install the second thin bronze plate with grooved side towards the last steel plate and install pressure plate against the smooth side of the thin bronze plate. In the No. 3 clutch, steel plates have internal splines and bronze plates have external splines; in all other clutches, steel plates have external splines and bronze plates

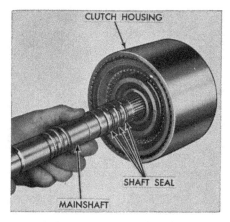

Fig. FO84—Installing mainshaft into clutch housing. Mainshaft serves as an oil distributor line for the three clutches.

have internal splines. Also, note that pressure plate for clutch 3 is thinner than for clutch 2 and is identified by the number 3 etched on back surface.

NOTE: After installation of snap ring (18—Fig. FO75), braze ends using a brass rod. This prevents the possibility of ring contracting and coming out of its groove.

Examine gear teeth and wearing surfaces of all components for wear or damage and renew if indicated. Bushings and planetary gear assemblies of "C" carrier are not available separately.

To reassemble the components in the housing, renew, and center, the three rear mainshaft seals on the shaft and insert shaft into clutch hub as shown in Fig. FO84. Lock into place with the snap ring. Note: The cast iron sealing rings throughout the transmission expand in their grooves to seal against the mating surface of their

housing. The entering edge of the housing bore is always chamfered to allow the rings to contract and enter. To avoid breakage of rings during reassembly, always align the ring ends at the top of the shaft as shown in Fig. FO84 and insert the shaft with a slight jiggling motion while applying light entering pressure. Stand hub on end as shown in Fig. FO85, install thrust washer (32—Fig. FO75) and carefully slide the "C" carrier into place over mainshaft. Make sure rear splines on the "C" carrier enter and bottom in the discs in Clutch 2, then install "C" sun gear thrust washer (13—Fig. FO 86) over mainshaft. Renew and center the four remaining mainshaft seals and the six seals on the "C" sun gear shaft. Coat the sun gear shaft seals with vaseline to hold them in place and slide the "C" sun gear into place over the transmission mainshaft. Reinstall "C" carrier front thrust washer (14—Fig. FO86) over distributor body hub and carefully reinstall assembly in housing, turning assembly slightly to make sure the front splines on mainshaft and "C" sun gear engage the internal splines in "B" carrier and Clutch 1 hub. Reinstall "C" ring gear and output shaft, install and reposition ground speed pto collar, install rear support and oil line and tighten retaining bolts to a torque of 35-40 ft.-lbs.

119. **DISTRIBUTOR, CLUTCH NO. 1 AND "B" PLANETARY.** The distributor, Clutch No. 1 and the "B" planetary system can be removed after removing the mainshaft and "C" planetary system as outlined in paragraph 118.

Fig. FO82 — Check coning of pto clutch steel plates with a feeler gage. Renew any plates which are coned less than 0.015.

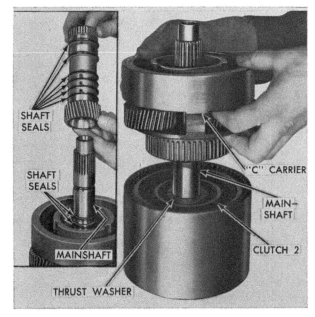

Fig. FO85—Installing "C" carrier and "C" sun gear shaft on mainshaft.

Before removal of the distributor, Band No. 3, Servo 3 actuating pin and lever, Servo 3 adjusting screw, the interlock cover and the two servo hydraulic fluid tubes must be removed as follows:

To remove Band No. 3, manually compress band while holding adjusting strut and actuating strut and remove band and the two struts. See Fig. FO88.

To remove Servo 3 actuating pin and lever, first remove Servo 2 and 3 retaining nuts and flat washers, retainer bolt and pin retainer from the right side of the transmission housing as shown in Fig. FO89 and remove the actuating pin and lever from the inside as an assembly. Remove Servo 3 adjusting screw by threading it out the inside of the housing.

Remove Band 2 adjusting screw lock nut and washer and completely loosen the adjusting screw. Remove the eight interlock cover attaching bolts and slide the cover off evenly to avoid damage to the interlock valve on Servo 3 piston (Fig. FO90). Loosen, but do not remove, the four distributor retaining bolts, and remove the two servo hydraulic fuild tubes by pulling with a twisting motion as shown in the inset of Fig. FO90.

Disconnect the pto hydraulic fluid line fitting from the distributor. Move the ground speed pto shifter fork toward the rear and pivot it up out of the way, then remove the distributor, Clutch No. 1, "B" ring gear and "B" carrier from the housing as an assembly as shown in Fig. FO91.

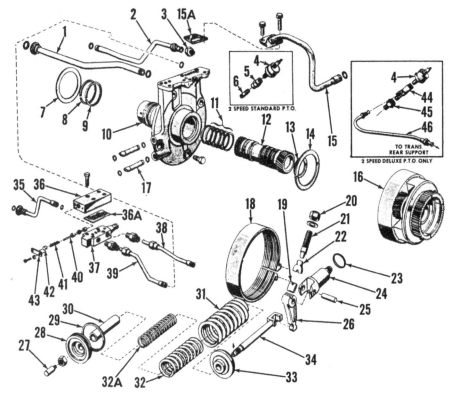

Fig. FO86—Exploded view of distributor, "C" carrier, No. 3 band assembly, direct drive shuttle valve and related parts.

1. Pump pressure line	25. Pin	36. Valve body (direct
2. Pto pressure line	26. No. 3 lever	drive clutch)
3. Connector	13. Thrust washer	27. Interlock valve	36A. Gasket
4. Pressure switch	14. Thrust washer	28. Servo piston	37. Valve support
5. Adapter	15. Outlet tube	29. "O" ring	38. No. 1 servo tube
6. Nipple	15A. Gasket	30. Sleeve	39. Valve inlet tube
7. Thrust washer	16. "C" carrier	31. Outer spring	40. Valve
(selective)	17. Servo oil tubes	32. Center spring	41. Spring
8. Seal ring	18. No. 3 band	32A. Inner spring	42. Seat
9. Seal ring	19. Actuating strut	33. Retainer	43. Plate
10. Oil distributor	21. Adjusting screw	34. Piston rod	44. Tee
11. Sealing rings	22. Adjusting strut	35. Direct drive clutch	45. Adapter
12. "C" sun gear shaft	23. "O" ring	oil tube	46. PTO oil tube
	24. Actuating pin		

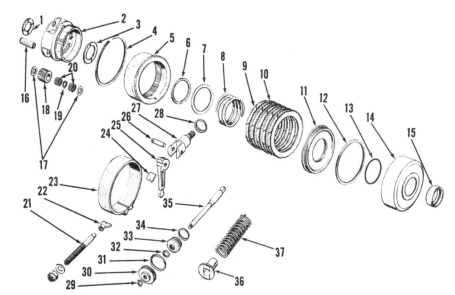

Fig. FO87—Exploded view of "B" carrier, No. 1 clutch, No. 2 band and related parts. Bushing (15) and planetary gear assembly (items 16 through 20) are not available separately.

1. Thrust washer
2. "B" carrier
3. Thrust washer
4. Snap ring
5. "B" ring gear
6. Snap ring
7. Spring seat
8. Clutch spring
9. Bronze plate
10. Steel plate
11. Piston
12. Piston seal
13. "O" ring
14. No. 1 clutch housing
15. Bushing
16. Pinion shaft
17. Washer
18. Pinion
19. Spacer
20. Needle bearings
21. Adjusting screw
22. Adjusting strut
23. No. 2 band
24. Actuating strut
25. Actuating lever
26. Pin
27. Actuating pin
28. "O" ring
29. Snap ring
30. Servo piston
31. "O" ring
32. "O" rng
33. Guide
34. "O" ring
35. Piston rod
36. Retainer
37. Servo spring

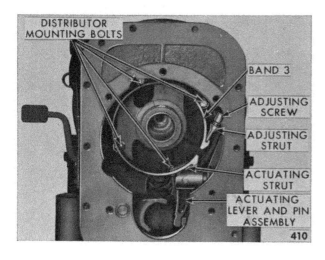

Fig. FO88—Band 3 shown in position. Adjusting and actuating struts are secured only by band, actuating lever and adjusting screw.

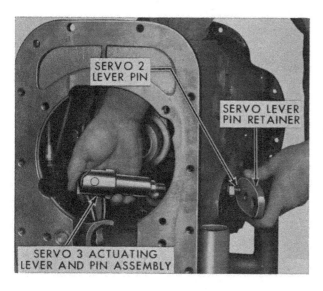

Fig. FO89 — Removing servo 3 actuating pin and lever assembly. Pin retainer prevents pins from rotating in housing.

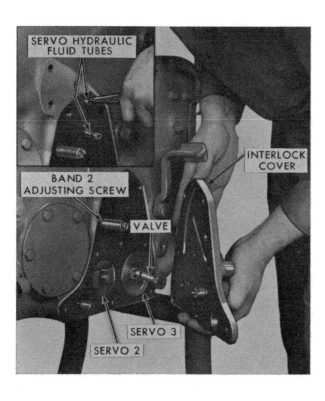

Fig. FO90—Remove interlock cover to pull the servo oil tubes.

Planetary "B" ring gear also serves as Clutch No. 1 pressure plate. After removal of "B" ring gear, service procedures for Clutch No. 1 are identical to that of Clutches 2 and 3. Clutch service is covered in paragraph 118.

Inspect the upper face of the distributor for burrs, scratches and flatness and check the oil passages for obstruction. Inspect bore for sealing ring wear. Inspect "B" and "C" bushing contact surfaces as well as thrust washer contact surface.

Fig. FO86 shows an exploded view of distributor, "C" carrier, No. 3 band and related parts. "B" carrier, No. 1 clutch, No. 2 band and related parts are shown in Fig. FO87.

To reassemble, place distributor on bench rear end down as shown in Fig. FO92 and place selective fit washer (7—Fig. FO86) over distributor. If transmission front end play was incorrect when measured before disassembly as outlined in paragraph 114, measure thrust washer and renew with one of proper thickness. Renew and center the distributor shaft seals (8 and 9—Fig. FO86), lock rear seal to hold it in place and carefully compress front seal in its groove with two screw drivers as shown in Fig. FO92 while lowering "B" carrier into position.

Note: An alternate and better method is to install an interlocking rear seal in both the front and rear grooves, eliminating the need of compressing the seal with screw drivers. The seals are identical except for the interlocking feature.

Reinstall assembly in transmission housing, being careful to align the pto and main supply hydraulic fluid tubes. Install but do not tighten the four distributor mounting bolts and reinstall the two servo fluid tubes from the left side of the housing.

Fig. FO91—Distributor, clutch No. 1, "B" ring gear and "B" carrier shown being removed.

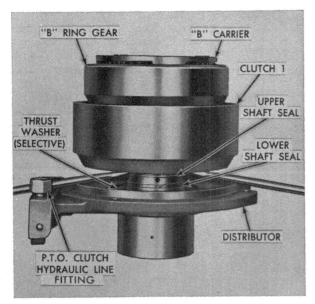

Fig. FO92—Compress distributor lower seal with screwdrivers when installing the clutch No. 1 and "B" carrier assembly.

Tighten distributor mounting bolts to a torque of 20-25 ft.-lbs., connect and tighten the pto fluid line fitting. Reinstall the control valve assembly.

Reinstall interlock cover, using a new gasket and tighten the attaching bolts to a torque of 35-40 ft.-lbs. Reinstall Band 3 adjusting screw and Servo 3 actuating pin and lever. Reinstall actuating pin retainer, bolt and the two actuating pin flat washers and nuts. Tighten nuts to a torque of 100-120 ft.-lbs. Reinstall remainder of transmission components as outlined in paragraph 118.

120. "A" RING GEAR, "A" PLANETARY AND DIRECT DRIVE CLUTCH. After distributor, clutch No. 1 and "B" planetary have been removed as outlined in paragraph 119, the "A" ring gear and "B" sun gear, "A" planetary and direct drive clutch assembly can be removed as shown in Fig. FO93. Also see Fig. FO94.

To disassemble and overhaul the direct drive clutch assembly, proceed as follows: Remove the large snap ring from inside diameter of clutch housing, then lift out the pressure plate and the four clutch plates. Place spring compressor tool (Nuday No. N-488) shown in Fig. FO95 inside clutch housing and reinstall the large snap ring, making sure that ends of snap ring are not positioned over slot of compression tool.

NOTE: Notice that compressor tool has three stepped, slotted studs spaced equidistantly around its outer edge. The steps are cut in increments of 0.006 and are used in compressing the Belleville washer to allow removal of the spiral type snap ring.

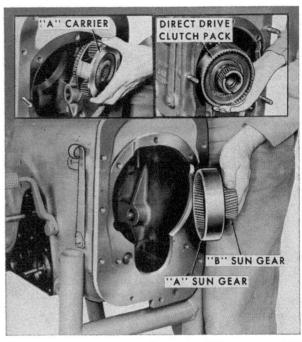

Fig. FO93—Removing "A" ring gear, "B" sun gear, "A" carrier (planetary) and direct drive clutch from transmission case.

Use lever (N-488-3) and screw driver as shown in Fig. FO96 and depress compressor until lowest (bottom) step of each stud can be engaged under the snap ring. Repeat this operation for the remaining two steps of the three studs and after the highest (top) steps of the studs are

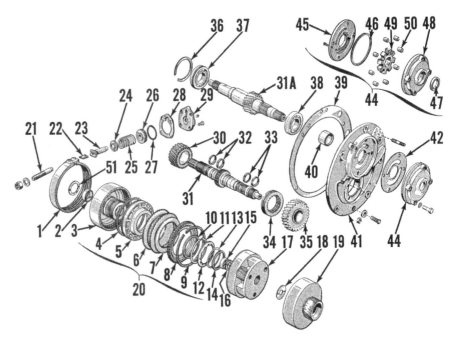

Fig. FO94—Exploded view of pump, input shaft, "A" carrier, No. 1 band assembly and the direct drive clutch.

1. No. 1 band	13. Pilot ring	25. Spring	37. Bearing
2. Bushing	14. Snap ring (spiral)	26. Piston	38. Bearing
3. Clutch housing	15. Bushing	27. "O" ring	39. Gasket
4. Outer seal	16. Thrust washer	28. Gasket	40. Sleeve
5. Piston	17. "A" carrier	29. Cover	41. Adapter plate
6. Clutch plate	18. Thrust washer	30. Pto drive gear	42. Gasket
(ext. spline)	19. "A" ring & "B"	31. Input shaft	44. Oil pump
7. Clutch plate	sun gear	(2-speed pto)	45. Back plate
(int. spline)	20. Direct drive clutch	31A. Input shaft (no pto)	46. "O" ring
8. Pressure plate	21. Adjusting screw	32. Front seals	47. Oil seal
9. Snap ring	22. Strut	33. Rear seals	48. Pump body
10. "O" ring	23. Piston rod	34. Pto coupling	49. Pump rotor
11. Pilot ring	24. Guide	35. Pto drive gear	50. Pump roller
12. Belleville washer		36. Snap ring	51. Thrust washer

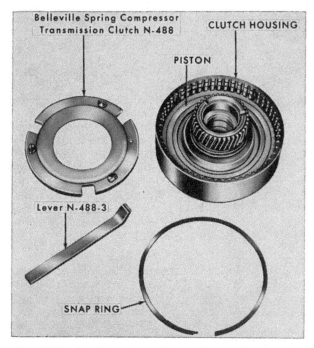

Fig. FO95—Special tool needed to disassemble the direct
drive clutch.

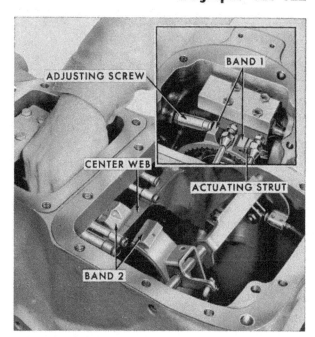

Fig. FO97—Band No. 2 shown ready to be removed. Band
No. 2 is removed by rotating around web in transmission case.

engaged under the large snap ring,
the spiral snap ring can be removed
from hub of clutch housing. Remove
piston from clutch housing by applying compressed air to oil port of
clutch housing.

121. **PLANET PINIONS, SHAFTS,
BEARINGS AND CARRIERS.** The
planetary carriers are not a serviceable component and if damaged, the
complete planetary carrier must be
renewed as parts are not catalogued
separately. The planet pinions should
have 0.008-0.025 end play in the carrier and are installed with thrust
washers on each side of gear.

122. **BANDS 1 AND 2, SERVOS
AND INTERLOCK PLATE.** After removal of "B" sun gear and "A"
planetary and direct drive clutch as
outlined in paragraph 120, remove
Band 1 by loosening the adjusting

screw, compressing the band and removing it from the case. To remove
Band 2, compress the band and remove the two struts, then remove the
band by rotating it around the center
web of the transmission case as shown

in Fig. FO97. Remove Servo 2 actuating lever and pin by withdrawing
from the inside of the case. Remove
the four Servo 1 cover attaching bolts
and remove the cover and servo as
shown in Fig. FO98.

Fig. FO98—Servos Nos. 1,
2 and 3 being removed.

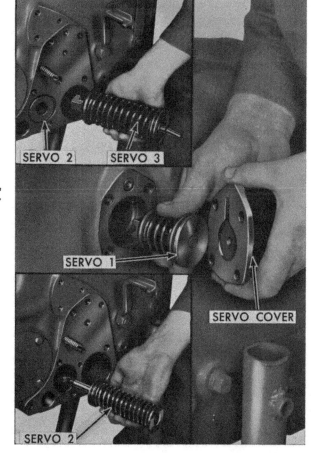

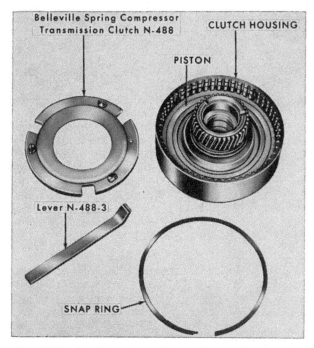

Fig. FO96 — Compress clutch Belleville
washer using method shown. Refer to text.

Servo seals can be renewed without disassembling the transmission if care is used not to dislodge the loose band struts. To renew the servo seals, proceed as follows:

SERVO: 1: Refer to Fig. FO98 and remove the four cover attaching bolts and Servo 1 cover. Back-off the lock nut on Band 1 adjusting screw while holding the adjusting screw stationary; then, carefully tighten the adjusting screw until the servo piston is forced from its bore just enough to expose the "O" ring seal. Apply slight pressure on the servo piston while renewing the seal to avoid dislodging the strut. Back-off the adjusting screw while applying pressure on the servo piston, reinstall the cover and tighten the retaining cap screws to 20-25 ft.-lbs. torque. Readjust the band as outlined in paragraph 97.

NOTE: Except for "O" ring, servo 1 is serviced as a unit only. Spring should test 75 lbs. when compressed to a length of 1.44 inches.

SERVO 2: Drain the transmission as outlined in paragraph 103, completely loosen Band 2 and Band 3 adjusting screws and remove the interlock cover. Remove Servo 2 and 3 actuating pin nuts and flat washers, and retainer bolt and retainer. Unbolt and remove the right pto side opening cover for access during reassembly. Tighten Band 2 adjusting screw until servo spring is forced from transmission housing enough that it can be grasped firmly by one hand, rotate spring either way 180 degrees to disengage notch in servo piston rod from actuating lever and

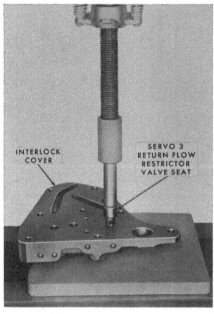

Fig. FO100—Press Servo No. 3 return flow restrictor valve into interlock cover with the opening in a vertical position as shown.

carefully withdraw servo assembly from housing. Servo guide (33—Fig. FO87) will usually be removed with servo assembly; if, not, extract guide with a suitable hooked tool and renew "O" ring seals on guide and servo piston. Servo spring has a free length of 7.22 inches and should test 375-423 lbs. when compressed to a length of 4.36 inches. Reinstall servo guide over piston rod and insert the assembly in the transmission housing with the notch forward. Feeling

through the remove pto side cover, be sure the notched end of the servo piston is in line with the lower end of actuating lever (25—Fig. FO87), then, rotate outer end of actuating pin with a wrench until the milled flat is vertical and the actuating pin retainer can be reinstalled. Tighten acuating pin retaining nuts to a torque of 100-120 ft.-lbs. Care must be taken not to move actuating lever inward to dislodge the actuating strut while removing and installing the servo. Reassemble the remainder of the transmission parts, refill the transmission and adjust the bands as outlined in paragraph 97.

SERVO 3: Drain the transmission as outlined in paragraph 103, completely loosen Band 2 and 3 adjusting screws and remove interlock cover. Grasp the interlock valve attached to the end of Servo 3 and withdraw the servo to expose the sealing "O" ring. If sufficient slack is not available to expose the seal, loosen the adjusting screw while withdrawing the servo. Renew the "O" ring while being careful not to force the piston back in the housing bore. Tighten the adjusting screw to draw the servo piston into its bore, then reinstall the interlock plate and the remainder of the parts. Fill the transmission and adjust the bands as outlined in paragraph 97.

If renewal of the piston rod, piston, springs or retainers is necessary, disassembly of the servo will be required. Servos are spring loaded and

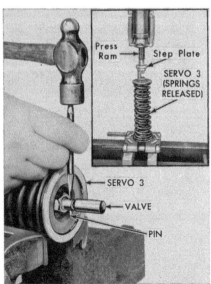

Fig. FO99 — Remove servo valve prior to disassembling Servo 3 as shown in inset.

Fig. FO101—Transmission control linkage on deluxe transmission. Refer to text for adjustment.

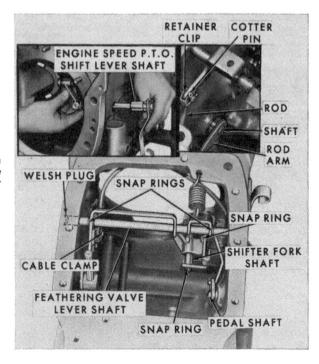

must be compressed in a suitable press before removal of the retaining nut as shown in the inset (Fig. FO99). On Servo 3, remove interlocking valve as shown in Fig. FO99 before releasing spring tension.

Spring specifications are as follows:

Inner spring
    Free length ..6.44 in.
    Test ....4.13 in.@136-148 lbs.

Middle spring
    Free length ..7.22 in.
    Test ....4.36 in.@375-423 lbs.

Outer spring
    Free length ....9.90 in.
    Test ....4.88 in.@467-515 lbs.

Renew the bands if they are damaged or worn until most of the band adjustment has been used.

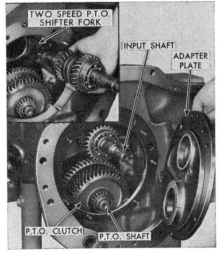

Fig. FO102 — Removing the adapter plate and input shaft.

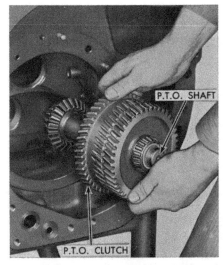

Fig. FO103—Removing the pto clutch and shaft assembly.

Servo 3 interlock valve seat can be renewed by removing the plug, spring and ball from the outside of the interlock plate and driving the seat out. Press the new seat in with a suitable press with the cross opening in a vertical position as shown in Fig. FO100. Be sure not to lose the straight pin in Servo 3 interlock valve seat which acts to unseat the check valve ball.

123. TRANSMISSION CASE LINKAGE. To remove the inching pedal linkage, drive out the roll pin securing pedal adapter to shaft and slide the adapter from shaft. Remove cotter pin and the upper retaining clip from control rod and remove rod and pedal shaft as an assembly. "O" ring seal on pedal shaft can be renewed at this time. To remove the feathering valve lever shaft, remove the right hand Welch plug and the two retaining snap rings and withdraw the shaft from the housing. On the ground speed pto (deluxe) transmission, it will be necessary to first loosen the interlock cable clamp. To remove the pto interlock mechanism and the

ground speed shift lever and fork, remove the snap rings from the lever shaft and fork shaft and withdraw the shafts as shown in Fig. FO101. The "O" ring on the ground speed pto shift lever shaft can be renewed at this time.

After reassembly of the transmission, the ground speed pto interlock mechanism will have to be adjusted as outlined in paragraph 125.

124. INPUT SHAFT, FRONT PLATE AND PTO SYSTEM. To remove the input shaft or pto assembly, first remove the transmission assembly as outlined in paragraph 108, and the major components as outlined in paragraphs 113 through 120, then proceed as follows: Remove the transmission front plate retaining bolts and using jack screws in the two tapped holes provided in the front plate, remove the front plate as shown in Fig. FO102. Main hydraulic pressure line, pto clutch line, direct drive clutch line or their seals can be renewed at this time. Remove the input shaft by lifting it from the housing.

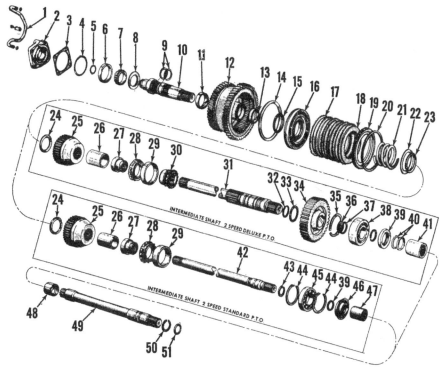

Fig. FO104—Exploded view of pto assembly.

1. Pto pressure tube	14. Sealing ring, outer	25. Side shaft drive gear
2. Bearing retainer	15. Sealing ring, inner	26. Spacer
3. Gasket	16. Clutch piston	27. Adapter
4. "O" ring	17. Clutch plate	28. Bearing cone
5. Oil seal	(ext. spline)	29. Bearing cup
6. Bearing cup	18. Clutch plate	30. Coupler
7. Bearing cone	(int. spline)	31. Intermediate shaft
8. Thrust washer	19. Pressure plate	32. Retaining ring
9. Sealing rings	20. Snap ring	33. Washer
10. Front shaft	21. Clutch spring	34. Ground drive gear
11. Bushing	22. Retaining washer	35. Retaining ring
12. Gear & clutch	23. Snap ring	36. Snap ring
housing	24. Thrust washer	37. Thrust washer
13. Bushing		

38. Bearing
39. Oil seal
40. Seal
41. Coupling
42. Rear shaft
43. Snap ring
44. Snap ring
45. Bearing
46. Oil seal
47. Coupling
48. Bushing
49. Rear shaft
50. Snap ring
51. "O" ring

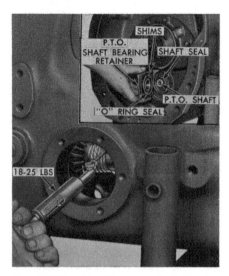

Fig. FO105—Checking pto shaft bearing preload. Inset shows location of adjusting shims.

NOTE: Beginning with 1963 production tractors, the front plate (and rear support) are fitted to the transmission case and the assembly line bored. Consequently none of the three units are available separately.

After removal of the input shaft, the pto assembly may be raised slightly and removed from the front of the housing as shown in Fig. FO-103. Refer to Fig. FO104 for an exploded view showing components of pto assembly.

Service on the pto clutch is identical to that on the three transmission clutches and procedures are outlined in paragraph 118. However, bear in mind that the pto steel clutch plates are coned 0.015-0.020. Renew any plate which is coned less than 0.015. Concave sides of steel plates are toward pressure plate.

125. PTO ADJUSTMENTS. Pto end play is adjusted by removal of shims at the pto front bearing retainer, which adjusts the preload of the tapered roller bearings. Rolling torque of the pto shaft when bearing preload is properly adjusted should be 15-20 in-lbs. To check the adjustment, remove the right hand side cover as shown in Fig. FO105 and wrap a piece of cord around the shaft spacer. The reading on the pull scale when the shaft is rolling should be 18-25 lbs. to obtain the required 15-20 in-lbs. preload.

On the transmission with ground speed pto, the pto ground drive interlock assembly must be properly adjusted to prevent the ground speed and engine speed pto from being en-

gaged at the same time. To adjust, proceed as follows:

Remove the transmission top rear cover as outlined in paragraph 105, shift the pto feathering valve into the disengaged position and the ground speed pto lever into the engage position. Loosen the set screw on the interlock clamp and pull the cable through the clamp until the pto feathering valve lever just contacts the feathering valve. Tighten the cable clamp to retain the cable in this position, then shift the ground speed pto lever rearward to the disengaged position and pull up on the pto cable connector to engage the engine pto. If the interlock is properly adjusted, the ground lever cannot now be moved forward to the engaged position.

After proper adjustment of the interlock mechanism has been made, reinstall the transmission top rear cover as outlined in paragraph 105.

126. SELECTOR ASSEMBLY. If service is required on the selector assembly, proceed as follows: Remove hood and hood rear top panel. Place selector lever in "Park" position, then remove the Phillips head screw

and cover (Fig. FO65). Remove nut from inner end of selector shaft and pull shaft and control lever from housing and remove dial. Disconnect conduit and control cable from transmission. Disconnect conduit from housing and pull conduit from cable, then unbolt and remove selector housing from instrument pedestal. Remove snap ring, retainer plate and spacer. The wheel and cable assembly can now be removed from housing.

To assemble, reverse the disassembly procedure. Lubricate wheel and cable during assembly and be sure to synchronize dial before installing nut.

126A. EXTERNAL FILTER. An externally mounted transmission oil filter is incorporated into the transmission oil cooler lines as shown in Figs. FO106A and FO106B. Renewal of the throwaway type filter is obvious after an examination of the unit.

Note that aft ends of cooling lines on late production tractors are attached to right side of transmission housing instead of being attached to heat exchanger manifold as were the cooler lines of earlier production tractors.

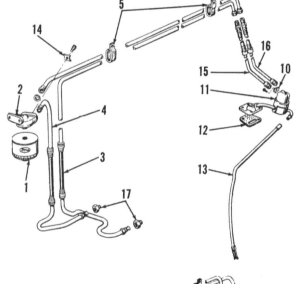

Fig. FO106A — View of transmission oil cooler lines and external filter (early production tractors).

1. Filter assembly
2. Mounting bracket
3. Inlet tube
4. Filter to cooler lines
5. Clamps
10. Clip
11. Manifold assembly
12. Gasket
13. Manifold outlet tube
14. Clip
15. Inlet tube, rear
16. Cooler outlet tube, rear

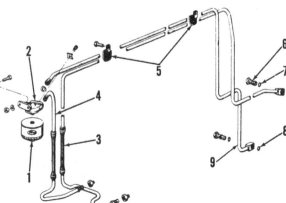

Fig. FO106B — View of transmission oil cooler lines and external filter (late production tractors).

1. Filter assembly
2. Mounting bracket
3. Inlet tube
4. Filter to cooler line
5. Clamps
6. Adapter bolt
7. "O" ring
8. "O" ring
9. Outlet tube

DIFFERENTIAL, BEVEL GEARS AND REAR AXLES

DIFFERENTIAL

All Models

127. **R&R AND OVERHAUL.** To remove the differential, first remove the left axle and housing assembly as outlined in paragraph 130, then pull differential assembly from center housing.

Note: If renewal of main drive bevel gear is required, it will be necessary to also remove the main drive bevel pinion as these are available only as a matched set.

When overhauling the differential, place correlation marks on the differential case halves so they can be reassembled in their same relative position. Remove the eight differential case bolts and separate the case halves. See Fig. FO107. Remove spider, differential pinions and thrust washers. Remove differential side gears and thrust washers. Check the differential carrier bearings and renew if required. If bearing cups are to be renewed, both axle and axle housings and differential bearing retainers must be off tractor.

Differential case and main drive bevel gear are riveted together during factory assembly and are usually renewed as a unit. However, bevel ring gear and bevel pinion are available as a matched set in a kit which also includes twelve special bolts, nuts and cotter pins for attaching bevel ring gear to differential case. Remove rivets, if old ring gear is so attached, by drilling center of rivets with a $\frac{7}{16}$-inch drill, then pressing out rivets. Note: Drill through rivet heads only, do not drill rivets completely out as this may enlarge holes. Check differential case for trueness before installing new ring gear. Position new gear on case, install the twelve special bolts, tighten the nuts to a torque of 90-110 ft.-lbs. and install cotter pins. Note: Rivets are also available for installing ring gear on differential case, but must be installed by upset riveting the rivets cold in a press. Check the assembled ring gear and case for trueness. The differential case halves are not available separately. Backlash of bevel gears is not adjustable. Align matching numbers or correlation marks when assembling differential case halves. Tighten differential case bolts to a torque of 80-100 ft.-lbs.

When installing differential in center housing, be sure that marked tooth of pinion meshes between marked teeth of ring gear.

MAIN DRIVE BEVEL GEARS

All Models

NOTE: Main drive bevel gears are not interchangeable between series 6000 and Commander 6000 tractors.

128. **BEVEL PINION.** To remove the main drive bevel pinion, split center housing from transmission as outlined in paragraph 106, then remove left axle assembly as outlined in paragraph 130 and the differential as outlined in paragraph 127. Remove coupler (17—Fig. FO107), then drive roll pin from traction disconnect lever and remove lever. Loosen set screw and remove shifter shaft and bearing block from inside of center frame. Remove snap ring (16) and coupler flange (15) from pinion shaft, then unbolt and remove bearing retainer (14), oil trough and shims (13). Pinion shaft and bearings can now be removed. Be sure to save shims (13).

If bevel pinion is renewed it will also be necessary to renew the bevel ring gear as outlined in paragraph 127. In addition to the pinion shaft, if any other pinion assembly parts are renewed it is necessary to check and adjust preload of the pinion shaft bearings as follows: Wrap a length of twine around the bevel pinion shaft and thread loose end through oil hole in top of center housing. Attach a spring scale to end of string and vary shims (13) until 5-7 lbs. (used), or 5-10 lbs. (new) pull on scale is required to rotate bevel pinion shaft. Shims are available in thicknesses of 0.002 and 0.003. Tighten bearing retainer cap screws to a torque of 80-100 ft.-lbs.

129. **BEVEL RING GEAR.** To renew the bevel ring gear follow the procedure given in paragraphs 127 and 128.

REAR AXLE SHAFT AND HOUSING

All Models

NOTE: Beginning with the Commander 6000 series tractors, the rear axle and housing assembly has been redesigned.

The planet ring gear inside diameter has been increased and an additional tooth has been added. The planetary pinions now have nineteen teeth instead of the previous twenty. The sun gear now has sixteen teeth instead of the previous twenty and is larger in diameter on the gear end. The rear axle shaft and axle housing have been changed to accept the above redesigned parts.

The Commander 6000 axle components are not interchangeable with the earlier series 6000 tractors, however, service procedures remain unchanged except that a new special tool Nuday no. SW6-65 is required to remove and install the planet ring gear.

Service of left and right rear axle shaft and housings is the same. However, since removal of the left unit is required for service on the differ-

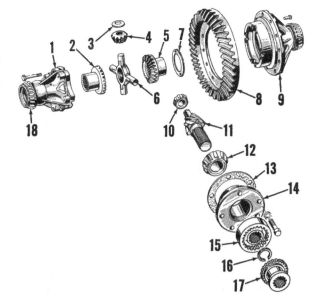

Fig. FO107 — Exploded view of differential and main drive bevel pinion.

1. Differential half, RH
2. Differential gear, RH
3. Thrust washer
4. Pinion gear
5. Differential gear, LH
6. Spider
7. Thrust washer
8. Bevel ring gear
9. Differential half, LH
10. Bearing cone
11. Drive pinion
12. Bearing cone
13. Shims
14. Bearing retainer
15. Coupler flange
16. Snap ring
17. Coupler
18. Differential bearing

ential, and other components, the following procedure will apply to the left hand unit. The same procedure may be used for the right hand unit except for removing additional hydraulic lines.

130. REMOVE AND REINSTALL.
Actuate hydraulic system to relieve system pressure. Drain center housing. Disconnect tail light wire and remove left fender. Support tractor under center housing and transmission and remove the left wheel and tire assembly. Remove floor plate and the remaining bolt which retains rear hood support to left hood support bracket. Remove pins and disconnect lower lift links. Unbolt side rail from transmission mounting bracket and axle housing Disconnect brake line and cap off openings. Attach a hoist to axle and housing assembly and unbolt housing from center frame. Pull axle and housing assembly outward and at the same time, either pry or wedge the side rail outward and remove the axle and housing assembly from tractor. Hold bearing retainer on the center frame studs during removal of axle assembly.

Note: A piece of 2 x 4 about 10 inches long, cut wedge shaped, and driven between front of transmission case and side rail will hold side rail out and facilitate reinstallation of assembly.

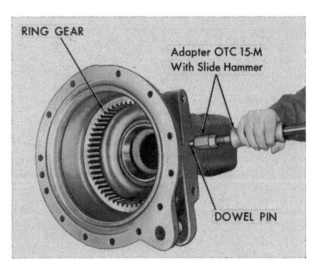

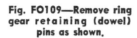
Fig. FO109—Remove ring gear retaining (dowel) pins as shown.

131. REAR AXLE SHAFT. To remove the rear axle shaft, first remove the axle and housing assembly as outlined in paragraph 130, then proceed as follows:

Remove snap ring (17—Fig. FO108), spacer (14) spring (13), shim (15) (series 6000 only), pressure plate (12), brake disc and sun gear (11 and 18). Remove retaining bolt (9) and nut (10), then remove brake housing (8). Note: The Commander 6000 brake housing (8) has been changed and shim (15) is no longer used. Remove the retainer (19) and oil seal (21), then loosen lock nut (22) and remove lock nut and adjusting nut (23) to-

gether. Remove the two-piece thread collar (24) and spacer key (29). Remove outer bearing (25) and push axle assembly from housing.

NOTE: If the two-piece thread collar (24) does not have identification markings on outer edges, affix markings on both halves before removing.

With axle removed, inspect bearings and planet gears and renew as necessary. Inner bearing can be removed from axle, using a suitable puller. If planet gears are to be renewed, press planet gear shafts from carrier toward outer end of axle (bearing side of carrier). When installing needle bearings in planet gears, they can be held in position by coating bore of gear with heavy grease. Be sure spacer (4) is between the two rows of bearings. Install planet gear, bearings and thrust washers using a dummy shaft approximately 1.725 inches in diameter and four inches long. Install planet gear shaft from bearing side of carrier. Dummy shaft will be removed as planet gear shaft is pressed in.

When installing axle in housing, start both nuts (22 and 23) on collar (24), then tighten adjusting nut (23) until axle bearings have 0-3 lbs. preload (axle turns freely by hand with no end play). Tighten lock nut (22).

NOTE: When installing the brake components, be sure hub of brake disc (11) is toward outer end of axle for series 6000 tractors and toward the center housing for the Commander 6000 series tractors.

132. AXLE PLANET RING GEAR.
The planet ring gear (7—Fig. FO108) is retained in the axle housing (26) by three retaining (dowel) pins (28). Ring gear is renewable; however, it should not be attempted unless proper equipment is available.

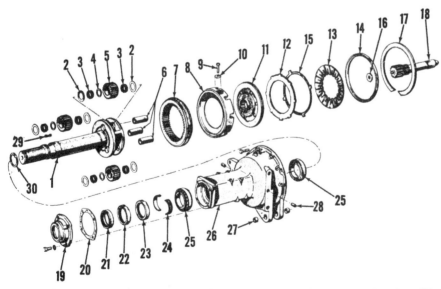

Fig. FO108—Exploded view of rear axle housing, axle assembly and brake housing. Shim (15) is not used on Commander 6000 tractors.

1. Rear axle		16. Thrust washer	23. Adjusting nut
2. Thrust washer	9. Retaining bolt	17. Snap ring	24. Lock nut collar
3. Needle bearings	10. Retaining nut	18. Sun gear (stub shaft)	25. Bearing assembly
4. Bearing spacer	11. Brake disc	19. Oil seal retainer	26. Axle housing
5. Planet gear	12. Pressure plate	20. Gasket	27. Bushing
6. Shaft	13. Spring	21. Oil seal	28. Retaining pin
7. Ring gear	14. Spacer	22. Lock nut	29. Key
8. Brake housing	15. Shim		30. Retaining ring

New ring gears require drilling and ream fitting to accept retaining pins. A removing and replacing tool (Nuday N-4075-A for series 6000, or Nuday SW6-65 for Commander 6000) is required to remove and reinstall ring gear, and in addition, a press with a 10-ton minimum capacity is required.

To renew an axle planet ring gear, proceed as follows: Remove axle assembly as outlined in paragraph 131. Use adapter OTC 15-M, or equivalent, and pull retaining (dowel) pins as shown in Fig. FO109. Install removing tool (N-4075-A) for series 6000 as follows: Place tool on ring gear so teeth of tool mesh with teeth of ring gear. Push down on tool until tool teeth extend below ring gear teeth, then look through sighting hole and turn tool until both sets of teeth are aligned. Hold tool in this position by tightening locking pin. See Fig. FO-110. Install removing tool (SW6-65) for Commander 6000 as follows: Tip tool and insert it through planet ring gear, then engage the stepped cor-

Fig. FO110A—View showing the SW6-65 removal and installation tool positioned for ring gear removal. This tool is required for service on Commander 6000 tractors.

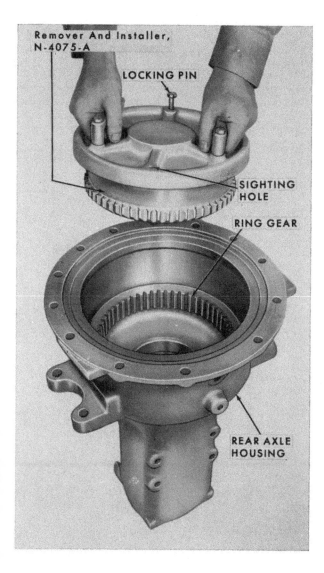

Fig. FO110 — Removing and replacing tool shown ready to be positioned. Tool shown is for series 6000 models; a different tool is required for Commander 6000 tractors. Refer to text for procedure.

ners of tool with teeth of gear and install the tool retaining bar. See Fig. FO110A. Ring gear can now be pressed from axle housing.

To install a new ring gear, attach the removing and replacing tool to the ring gear as described above and press ring gear into axle housing. Now using drill and ream kit (N-4075-AA), shown in Fig. FO111, tap the drill guide into one of the retaining pin holes until it bottoms against ring gear. Insert pilot drill through collar, spacer and into guide until drill contacts ring gear, then position collar against spacer and tighten set screw. Remove spacer and drill pilot hole until collar contacts guide (⅜-inch). Remove guide, then using secondary drill, drill to bottom of pilot hole. Use reamer to finish sizing the hole. Repeat operation on the remaining two holes and drive the retaining pins into position.

NOTE: During second drilling, BE SURE to keep drill in a vertical position and stop drill before removing in order to preclude any possibility of enlarging the hole.

133. OIL SEAL. The rear axle oil seal (21—Fig. FO108) is contained in retainer (19) and can be renewed after wheel and retainer are removed. Procedure for renewing seal is obvious.

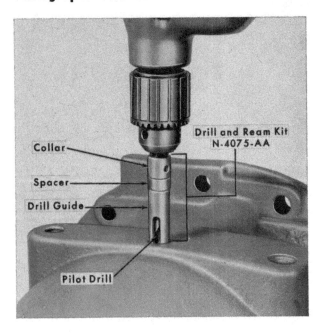

Fig. FO111 — View showing installation of drill and ream kit. Refer to text.

BRAKES

The wet type, hydraulically operated power brakes receive their pressurized operating fluid from the hydraulic system and are controlled by individual, foot operated control valves.

All Models

134. ADJUSTMENT. No provision is made for adjusting brakes. However, the brake pedal rods are fitted with a clevis so that pedal height can be equalized and simultaneous brake application accomplished.

135. R&R AND OVERHAUL BRAKES. To service brakes, first remove rear axle and housing assembly as outlined in paragraph 130. Remove differential bearing retainer from inner end of axle housing, then remove snap ring (Fig. FO112) from inner side of retainer. Place retainer, piston side down, on wood blocks as shown in Fig. FO113, apply compressed air to hydraulic port and force piston from retainer.

NOTE: On tractors prior to serial number 15939, differential bearing retainer was fitted with a separate brake cylinder as shown in Fig. FO-112. On tractors serial number 15939 and up, the brake cylinder is an integral part of the differential bearing retainer.

On early models, the cylinder may, or may not, be removed when compressed air is applied to remove piston. If cylinder remains in the retainer, it can be removed as follows: Drill and tap three ¼-inch holes, 120

degrees apart and $\frac{9}{16}$-inch deep, in the exposed edge of the piston. Drill three matching holes in a steel plate and attach steel plate to cylinder with ¼-inch cap screws. Invert retainer and press cylinder from retainer.

Refer to Fig. FO108 and remove snap ring (17), spacer (14), spring (13), shim (15) (Series 6000 only), pressure plate (12), brake disc (11) and shaft (18). Remove retaining bolt

Fig. FO112 — Differential bearing retainer with brake piston and cylinder removed; early production unit is shown. Cylinder is not removable on late production models.

Fig. FO113 — Remove brake piston with compressed air as shown.

(9) and nut (10), then pull brake housing (8) from axle housing.

Clean all parts in a suitable solvent and inspect. Renew parts showing undue wear or damage. Linings on brake disc (11) are bonded and are not available separately; however, a new disc and lining assembly is available which can be riveted to hub.

Use all new "O" rings and back-up washers and reassemble components by reversing the disassembly procedure. Torque retaining bolt (9) to 40-50 ft.-lbs. Shim (15) (series 6000 only) is installed with beveled edge toward spring (13).

NOTE: Beginning with the Commander 6000 series tractors, the brake housing (8) has been redesigned and shim (15) is no longer used. This also requires that the brake disc be installed with hub toward center frame rather than toward outer end of axle as it is on the series 6000 tractors.

After brake assembly is installed in axle housing (series 6000 only) be sure clearance between spacer (14) and spring (13) is within limits of 0.028-0.033. If clearance exceeds 0.033, install additional shim, or shims (15). Reinstall axle and housing assembly on center frame.

136. BRAKE CONTROL VALVE. To remove the brake control valve, first actuate hydraulic system or brakes to relieve pressure, remove operators platform, then disconnect hy-

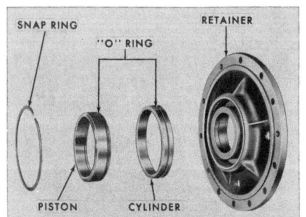

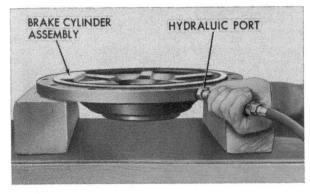

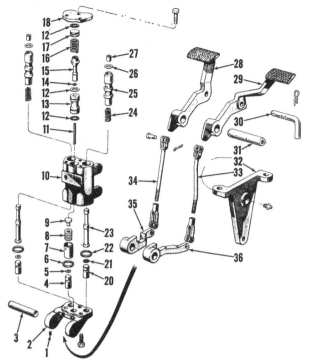

1. Set screw
2. Head
3. Rocker arm shaft
4. Valve plunger
5. "O" ring
6. Seal
7. Spacer
8. Spring
9. Spring seat
10. Valve body
11. Regulating valve plunger
12. "O" ring
13. Bushing
14. "O" ring
15. Regulating valve
16. Spring
17. Plug
18. Cover
20. Valve plunger
21. "O" ring
22. Seal
23. Control valve
24. Spring
25. Control valve bushing
26. Seal
27. Stop
28. Brake pedal, LH
29. Brake pedal, RH
30. Brake lock
31. Pedal shaft
32. Bracket
33. Brake rod, RH
34. Brake rod, LH
35. Rocker arm, LH
36. Rocker arm, RH

the following diameters:

Yellow 0.4374-0.4376
Green 0.4376-0.4378
Blue 0.4378-0.4380

When reassembling, use all new "O" rings and coat all internal parts with oil.

Note: A brake valve repair kit is available which contains all of the springs, "O" rings and seals.

Commander 6000 series tractors are equipped with a parking brake assembly which allows either the left or both pedals to be held in applied position. Any service required will be obvious.

A kit is also available for installation on earlier tractors and instructions are supplied with the kit.

POWER TAKE-OFF

The pto reduction unit, shown exploded in Fig. FO116, mounts on rear of center housing and is driven by an intermediate shaft which is coupled to the transmission pto shaft. Both 540 and 1000 rpm operation can be obtained by shifting the reduction (cluster) gear shaft either forward or rearward and inserting the correct output shaft.

draulic lines and immediately cap off lines to retain fluid. Disconnect brake rods from valve levers, then unbolt and remove control valve from side rail.

To disassemble brake valve, refer to Fig. FO114 and proceed as follows: Remove valve body head, then remove parts as shown. BE SURE to keep brake valves identified with their original bores as brake valves and their bushings in the housing are matched units. To remove the brake valve bushings, remove cover from end of housing, then remove bushings by pressing from brake valve end.

Clean all parts in a suitable solvent and inspect all valves and bushings for scoring, nicks, burrs or other damage. Regulating valve is available as a unit only. Check all springs for fractures, distortion or other damage. Spring specifications are as follows:

Spring   **Inches @ lbs.**

Brake valve return spring
.... 1.62 @ 38- 42

Regulating valve
return spring ...... 0.56 @ 18- 22

Regulating valve
plunger spring ..... 1.06 @ 275-325

Reassembly is the reverse of disassembly; however, the following must be considered. Brake valve bushings must have a 0.0002-0.0006 interference fit in housing and if new bushings are installed, the bushings are available with the following outside diameters:

White 0.8669-0.8671
Blue 0.8671-0.8673
Yellow 0.8673-0.8675
Green 0.8675-0.8677
Orange 0.8677-0.8679

If new bushings are installed, select a brake valve which will be a close fit in bushing yet will not bind at any point. Brake valves are available with

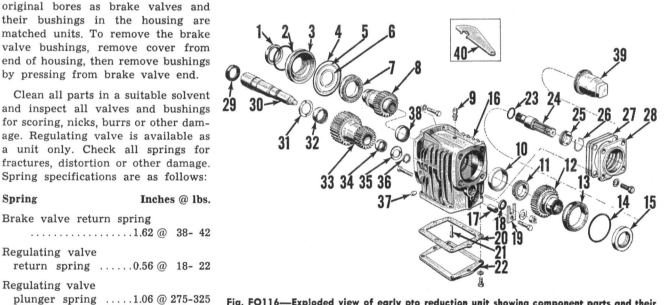

Fig. FO116—Exploded view of early pto reduction unit showing component parts and their relative positions. Refer to Fig. FO116A.

1. Oil seal
2. "O" ring
3. Front bearing retainer
4. "O" ring
5. Snap ring
6. Snap ring
7. Bearing assy.
8. Input gear
9. Vent
10. Bearing cup
11. Bearing cone
12. Output gear
13. Bearing cup & cone
14. "O" ring
15. Oil seal
16. Housing
17. Locator
18. Seal
19. Retainer
20. Gasket
21. Interlock pin
22. Plate
23. "O" ring
24. Output shaft
25. Shaft nut
26. Lock ring
27. Shims
28. Rear bearing retainer
29. Oil seal
30. Cluster gear shaft
31. Thrust washer
32. Bearing
33. Cluster gear
34. Bearing
35. Thrust washer
36. Snap ring
37. Plug
38. Bearing
39. Cap

Beginning with tractor serial number 61003, the pto reduction unit was redesigned and strengthened and the new unit can be easily identified by not having the cooling fins as does the early unit. See Fig. FO116A. A parts kit is available to modify the early pto reduction unit and while service procedures remain basically the same, the parts and installation information given in the following paragraph MUST be observed.

The late pto reduction unit can be installed on any tractor prior to serial number 61003 without any modification providing tractor is equipped with only a 540 rpm pto. However, if tractor is equipped with the 540 and 1000 rpm pto reduction unit the blind hole in center housing must be enlongated to accomodate the reduction unit countershaft. To rework the blind hole, proceed as follows: Clean the machined area directly below the blind hole and apply a layer of chalk or crayon to the machined surface. Use a scribe, or pencil, and mark a contour line as indicated by the dotted semi-circle shown in Fig. FO116B. Use a center punch, start at vertical center line and locate the eleven hole locations as shown. Seal the bore above the blind hole, then using a $\frac{3}{16}$-inch drill, drill holes 1 through 5 to a depth of 1.250 inches. See Fig. FO116C. Use a 5/32-inch drill and drill hole number 6. Use a chisel and break out all excess material, then use a high speed grinder to finish the hole to the desired diameter and depth.

If the parts kit is being used to update the early pto reduction unit, the following information MUST be observed.

If tractor has a serial number of 10749, or later, the field kit can be installed by removing the countershaft and replacing the existing parts

with the parts contained in the kit. If tractor has a serial number prior to 10749, and pto reduction unit has an output gear with Ford part number CONN-A737-B, it must be replaced with an output gear, Ford part number CONN-A737-C, or damage to reduction unit assembly will result.

NOTE: For positive identification of the cluster gear and output gear the following procedure is recommmended. Remove pto reduction unit from tractor, then remove countershaft assembly from housing. Take the cluster gear that was removed (original) from unit and set it on a flat surface so small gear (21 teeth) end is on top side. Now take cluster gear from kit (new) and place it on top side of original cluster gear with small gear (21 teeth) on bottom. Use a straight edge and align teeth on one side of the small gears of cluster gears. Move straight edge to opposite side of gear teeth and note whether the two small gears are the same diameter. If the small gears of cluster gears are the same diameter, the output gear need not be changed if it is in otherwise satisfactory condition. If the small gear of the new cluster gear has a larger diameter (0.025-0.035), it will then be necessary to remove the old output gear and install a new one of the correct diameter (Ford part number CONN-A737-C).

For removal and overhaul information refer to paragraph 137.

NOTE: Lubrication recommendation for the late pto and the early pto units that have been modified has been changed. Recommended lubricant is now Ford Specification M-4864-A (SAE 80 extreme pressure) instead of M2C41. DO NOT mix these

lubricants as they are not compatible.

For information on the transmission pto shaft, refer to the transmission section. For information on the pto reduction unit, refer to the following paragraphs.

All Models

137. **R&R AND OVERHAUL.** To remove the pto reduction unit, first remove the drawbar, then unbolt drawbar hanger plate from pto and remove drawbar hanger assembly. Leave the output shaft bearing retainer in place until unit is removed. If pto unit is to be disassembled, make a provision for catching oil, then remove bottom plate (22—Fig. FO116). Either support unit with a rolling floor jack, or attach a hoist, then unbolt and remove unit from center frame.

With unit removed and bottom plate off, disassembly is as follows: If intermediate shaft remained with pto unit, pull same from input gear. Remove nut, washer, locator (17) and retainer (19) from aft end of cluster gear shaft (30). Tap cluster gear shaft forward until snap ring (36) can be lifted from its groove and moved rearward on shaft. Remove cluster gear shaft from front of housing and the cluster gear (33), snap ring (36) and thrust washers (31 and 35) from bottom of housing. If necessary, bearings (32 and 34) can be removed from cluster gear by using a suitable puller. Lift interlock (stop) pin (21) from web of housing. Remove "O" ring (2) from pilot of front bearing retainer (3), then using two screw drivers in the "O" ring groove, pry retainer from housing. Pull input shaft (8) and bearing (7) from housing. Remove snap ring (6) and bump input

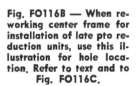

Fig. FO116B — When reworking center frame for installation of late pto reduction units, use this illustration for hole location. Refer to text and to Fig. FO116C.

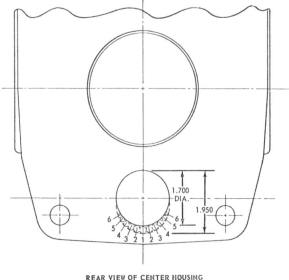

REAR VIEW OF CENTER HOUSING

FO116A — View showing late pto reduction unit. Note absence of cooling fins on housing as shown in Fig. FO116.

HYDRAULIC LIFT SYSTEM

shaft from bearing. If necessary, remove snap ring (5) from outer bearing race. Seal (1) can be driven from retainer. Remove cap (39), then remove rear bearing retainer (28) and shims (27). Disengage and remove locking ring wire (26) and using special shaft nut wrench which is furnished with tractor, remove shaft nut (25) and pull output shaft (24) from output gear (12). Remove "O" ring (23) from I.D. of output gear. Output gear (12), bearings and rear bearing cup can now be bumped from housing. Bearings (11 and 13) can be removed from output gear by using suitable pullers. Use Ford tool NCA-851, or its equivalent, to remove bearing (38) from housing. Use a suitable puller to remove bearing cup (10) from housing. If necessary, remove "O" ring (14) and oil seal (15) from rear bearing retainer.

When reassembling, install output gear and shaft assembly first and check output gear bearing preload as follows: Either use shorter cap screws, or spacers on original cap screws, to secure rear bearing retainer to housing. Wrap a length of cord around output shaft, attach a spring scale, and check the amount of pull required to rotate output shaft. Bearing preload is correct when a scale reading of 14-22 lbs. is required to keep output shaft in motion. If necessary, vary the shims (27) between retainer and housing to obtain this reading. Shims are available in thicknesses of 0.002, 0.005 and 0.013. Complete reassembly by reversing disassembly procedure and fill to level of filler hole with correct lubricant.

The hydraulic lift system incorporates both automatic draft control and implement position control. The power steering system, the braking system and the hydraulic lift system all share a common reservoir which is located on left side of tractor in front of the radiator. The hydraulic lift and brake systems also have an accumulator included in their circuits. The accumulator, located on right side of tractor in front of radiator, is charged with dry nitrogen gas and in addition to storing the hydraulic fluid under pressure, also tends to provide smoother action and quicker response. The basic pressure control of the hydraulic lift system is provided by an unload valve bolted to the axial piston type pump. The hydraulic pump is driven from the front end of the engine crankshaft. The power steering pump is mounted on the engine timing gear cover, is driven by the engine camshaft and has its own relief valve. A schematic view of the hydraulic lift, brake and power steering circuits is shown in Fig. FO117.

TROUBLE SHOOTING

All Models

139. To trouble shoot and/or isolate malfunctions of the hydraulic lift system, proceed as follows: Cycle hydraulic system to relieve system pressure. Obtain a pressure gage capable of registering at least 3000 psi and, depending upon how the tractor is equipped, install gage as follows: If tractor is equipped with a hydraulic lift and a remote control valve, install gage in remote control valve as shown in Fig. FO118. If tractor has a hydraulic lift and no remote control valve, install gage in hydraulic lift pressure line as shown in Fig. FO118A. If tractor has no hydraulic lift or remote control valve, install gage at the "tee" behind the brake valve assembly as shown in inset of Fig. FO118A.

The following procedure is based on the assumption that the tractor is equipped with hydraulic lift and remote control valve. For those models not so equipped, ignore those paragraphs not applicable.

140. Check accumulator gas precharge pressure as follows: With hydraulic pressure gage installed as outlined in paragraph 139 and with all hydraulic pressure relieved (gage reading zero), engage engine starter while closely watching the hydraulic gage needle. The gage needle will "jump" to the accumulator gas precharge pressure. Note: Do not start engine as hydraulic pump would pump oil into the accumulator resulting in a reading of accumulator system hy-

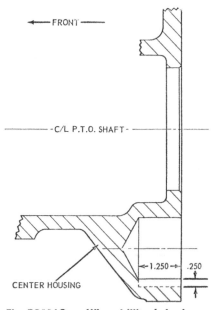

Fig. FO116C — When drilling holes in center frame, drill them to depth shown. Also refer to Fig. FO116B.

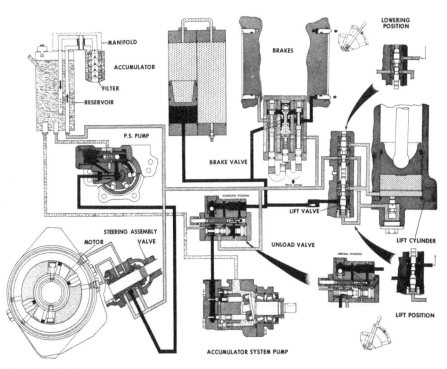

Fig. FO117—Schematic view of hydraulic lift system and power steering system. Note that power steering circuit is not included in the accumulator circuit.

draulic pressure rather than gas pre-charge pressure. If gas pre-charge pressure is not as indicated in table accompanying paragraph 146, re-charge accumulator as outlined in paragraph 146.

141. With accumulator fully charged, start engine and cycle hydraulic system several times to purge any air which might be present in system, then observe the gage at the remote control valve. Gage should read 2200-2300 psi. If pressure is not as speci-fied, remove lower splash plate then remove acorn nut from unload valve and turn adjusting screw in or out as required as shown in Fig. FO119. Operate brake pedals rapidly and observe when pressure stops dropping. Gage should read 1900-1950 psi. Leave engine running and ob-serve gage. Pressure should not drop more than 300 psi in two minutes from the 2200-2300 psi reading.

If the above conditions are met, the system, except for brakes, can be con-sidered satisfactory. However, if above conditions are not met, isolate the faulty assemblies as outlined in the following paragraphs.

142. Shut off engine, cycle hydrau-lic system to relieve system pressure, then remove remote control valve and reinstall gage in remote control valve pressure line. Plug remote control valve return line. Restart engine and observe gage. If system builds to relief pressure (2200-2300) and does not drop more than 300 psi in two min-utes, the remote control valve is faulty and should be serviced as outlined in paragraph 152.

143. If, after testing, the remote control valve is satisfactory, shut off engine, cycle hydraulic system to re-lieve system pressure, then quickly remove the cap, spring and check valve (See Fig. FO118B) from check valve at lower right of hydraulic lift unit, insert ½-inch steel ball and a ½-inch diameter brass rod $1\frac{5}{16}$-inch long and reinstall check valve cap. Restart engine and observe gage. If system builds to relief pressure and does not drop more than 300 psi in two minutes, the lift unit is faulty and should be serviced as outlined in paragraph 154.

144. If hydraulic lift unit is satis-factory, continue to run engine and depress each brake pedal independ-ently while observing pressure gage. If pressure drops rapidly with either brake pedal depressed, the respective brake cylinder is leaking and should be serviced as outlined in paragraph

135. Now shut off engine and cycle hydraulic system to relieve system pressure, then remove brake pressure line adapter bolt and reinstall a bolt which has the drilled passage plugged. Restart engine and observe gage. If system builds to relief pressure and

does not drop more than 300 psi in two minutes, the brake valve is faulty and should be serviced as outlined in paragraph 136.

145. With the remote control valve, hydraulic lift and brakes isolated, re-peat the system holding test as out-

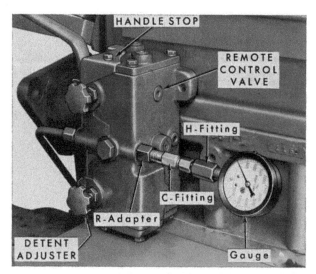
Fig. FO118—To check hy-draulic system pressure on tractors with remote con-trol valve install pressure gage as shown.

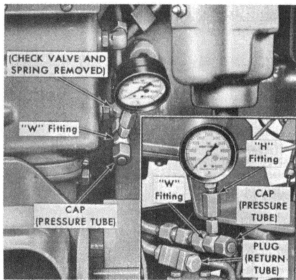
Fig. FO118A — To check hydraulic system pressure on tractors with no remote control valve, install pres-sure gage in hydraulic lift pressure line as shown. On tractors without remote control valve or hydraulic lift, use "tee" behind brake valve as shown in inset.

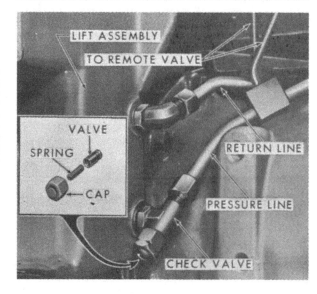
Fig. FO118B—View show-ing location of hydraulic lift check valve at lower right side of hydraulic lift housing. To isolate hy-draulic lift when trouble-shooting, remove the check valve and spring (inset) and insert a ½-inch steel ball and a ½ x 15/16 inch brass rod.

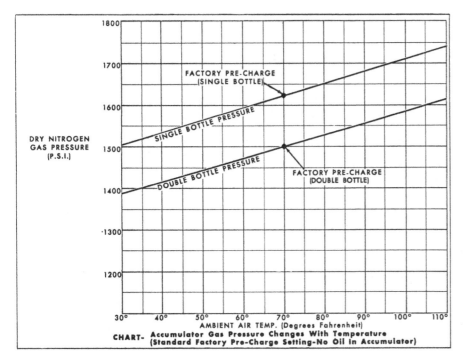

DRY NITROGEN GAS PRESSURE (P.S.I.)

FACTORY PRE-CHARGE (SINGLE BOTTLE)

SINGLE BOTTLE PRESSURE

DOUBLE BOTTLE PRESSURE

FACTORY PRE-CHARGE (DOUBLE BOTTLE)

AMBIENT AIR TEMP. (Degrees Fahrenheit)

CHART— Accumulator Gas Pressure Changes With Temperature (Standard Factory Pre-Charge Setting-No Oil in Accumulator)

Use the above chart to determine the amount of pre-charge of accumulator at existing temperatures. Note that pressure varies with temperature.

lined in paragraph 141 and if pressure drops rapidly from system relief pressure, the unload valve (or pump) is faulty and should be serviced as outlined in paragraph 151.

NOTE: Because of construction, the unload valve and hydraulic pump cannot be successfully isolated from each other. However, pump malfunctions can generally be determined by the system pressure not reaching specifications even though relief valve is adjusted, or by the hydraulic system pressure raising or lowering as the engine speed is changed. In addition, the hydraulic pump in question could also be checked by removing it from tractor and installing it on a different tractor with a hydraulic system known to be satisfactory. Pump should build and maintain the system pressure when running at 1250 rpm or less.

NOTE: With the components of the hydraulic system isolated, pressure will be trapped in the system with no way to relieve it. To relieve the trapped pressure, loosen the hydraulic lift check valve cap about one turn to allow the check valve ball to unseat. The lift control lever can then be actuated and the pressure relieved. Then, remove the ½-inch steel ball and brass rod from check valve and reinstall check valve, spring and cap. Remove the plugged bolt from brake pressure line and reinstall drilled bolt with new "O" rings.

ACCUMULATOR

All Models

146. **CHARGING.** The accumulator is pre-charged with dry nitrogen gas which may be a water dry type or an oil dry type. Either type can be used and the Ford N-925-A Nitrogen Pre-Charge Kit contains both left hand thread (water dry nitrogen) and right hand thread (oil dry nitrogen) adapters for use on the supply tank.

To charge the accumulator, proceed as follows: Cycle hydraulic lift (or brakes) until all pressure in hydraulic system is relieved, then remove hood front side panels, hood and radiator air baffles. Be sure valve stem of kit valve is turned completely in (off), then attach kit valve to accumulator gas valve which is on top side of cylinder on single accumu-

lator and on bottom side of gas cylinder on double accumulators. Attach kit hose to supply tank, then open the kit valve. Refer to the accompaning table to determine the proper charging pressure at the existing temperature, then slowly open the supply tank valve and charge accumulator to the pre-determined pressure.

Close the supply tank valve and observe the gage for about thirty seconds to make sure charge pressure remains steady, then close the kit valve and remove kit assembly from accumulator.

NOTE: With all hydraulic pressure relieved from system, accumulator gas pressure should never be less than 1200 psi nor more than 1775 psi. As seen in the table the accumulators are factory charged to 1625 psi for single accumulators, or to 1500 psi for double accumulators at 70 degrees Farenheit.

146A. **TESTING.** If accumulator external gas leakage is suspected, test accumulator as follows: Charge the accumulator with dry nitrogen gas, if necessary, then using a soap solution, coat gas valve, gas valve base and gas cylinder end plug. If no leakage is noted, gas valve assembly and gas end plug can be considered satisfactory.

147. **R&R AND OVERHAUL.** To remove and service the accumulator, first actuate the hydraulic lift, or brakes, to relieve system pressure. Remove hood front side panels, hood and radiator air baffles. Disconnect hydraulic reservoir lines from unload valve and drain reservoir. Disconnect hydraulic pressure line from bottom of accumulator, then unbolt accumulator cylinders from bottom support bracket and upper mounting strap from hood side rail. Remove accumulator from tractor.

Fig. FO119 — To adjust hydraulic lift system pressure, remove acorn nut and turn adjusting screw in or out as required.

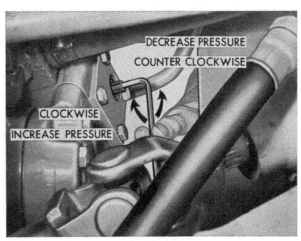

DECREASE PRESSURE
COUNTER CLOCKWISE

CLOCKWISE
INCREASE PRESSURE

With accumulator removed from tractor, open gas valve slightly and allow the nitrogen gas to escape. Remove gas valve and the "U" shaped pipe from top of accumulator. Remove mounting strap and separate cylinders. Remove end plugs from cylinders, then using a long clean stick or a hammer handle, push piston from end of oil cylinder which has threads on outside diameter. See Fig. FO120.

NOTE: Utility models use a single cylinder accumulator and scribe marks should be affixed on end plugs prior to removal from cylinder.

Discard all "O" rings and inspect bore of oil cylinder for scoring or burrs. Small defects can be corrected by using crocus cloth. If necessary, renew the seal on piston.

When reassembling, refer to Fig. FO120A for view showing piston seal arrangement. If piston has two seal grooves, install seals with lips opposed and toward ends of piston. If piston has three seal grooves, disregard the two end grooves and install the "T" seal in center groove as shown in Fig. FO 120A. Be sure chamfered ends of the back-up rings overlap each other. Coat piston seal with Lubriplate, or equivalent, and install piston in oil cylinder from end of cylinder having the external threads and with closed end toward oil pressure end of cylinder. Install new "O" rings on end

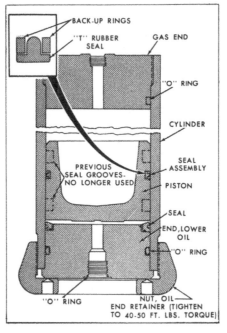

Fig. FO120A—View showing seals that have been used. Refer to text for types used on two and three groove pistons.

plugs and coat same with Lubriplate, or equivalent. Install end plugs in their original positions and screw plugs having external threads in cylinders until they protrude 0.000-0.020 from cylinders and the affixed scribe marks (utility) align. Tighten oil end retainer nut (14—Fig. FO120) to 40-50 ft.-lbs. torque. Use new gasket, install gas valve and torque same to 50-60 ft.-lbs. Mount cylinders on bottom support and install cap screws. Install mounting strap. Use new "O" rings and install the "U" shaped tube on top end of cylinders. Be sure larger "O" rings are between fittings and end plugs. Pre-charge accumulator with dry nitrogen gas using Pre-Charge Kit N-925-A as outlined in paragraph 146. Connect hydraulic pressure line to oil cylinder end plug and the reservoir lines to unload valve. Fill reservoir, start engine and cycle the hydraulic lift system until all air is purged from system.

While normal accumulator gas pressure is 1500 psi at 70° F. for double accumulator; or 1625 psi at 70° F. for single accumulators, both ambient air temperature and oil temperature will affect pressure gage readings. When charging accumulators, bear in mind that for every 10 degrees change in temperature, gas pressure will change approximate 25 psi. For temperatures above 70° F., add to gage reading; for temperatures below 70° F., deduct from gage reading. Refer to the gas pre-charge table.

Fig. FO120 — Exploded view of dual cylinder accumulator assembly. Utility models are equipped with single cylinder accumulator.

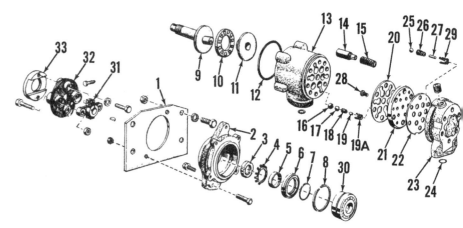

Fig. FO121—Exploded view of hydraulic pump showing component parts and their relative positions. Latest pumps have guides for valves which are not shown.

1. Adapter bolt
2. Tube assy.
3. End plug
4. "O" ring
5. Mounting strap
6. Cylinder
7. Piston seal
8. Piston
9. Seal
10. End plug
11. Gasket
12. "O" rings
14. Nut
15. Gas valve
16. Bracket
17. "O" ring
18. Connector
19. Ferrule
20. Nut

1. Mounting bracket
2. Shaft housing
3. Lock nut
4. Lock washer
5. Seal sleeve
6. Seal
7. "O" ring
8. Snap ring
9. Drive shaft
10. Thrust bearing
11. Wobble plate
12. "O" ring
13. Pump body
14. Piston
15. Piston spring
16. Valve seat
17. Steel ball
18. Spring
19. Pin
19A. Ball cage
20. Gasket
21. Valve plate
22. Gasket
23. Pump cover
24. "O" ring
25. Steel ball
26. Spring
27. Pin
28. Screw
29. Ball cage
30. Bearing assy.
31. Hub
32. Coupling
33. Spacer

HYDRAULIC PUMP

All Models

148. **REMOVE AND REINSTALL.** To remove the hydraulic pump, first cycle the hydraulic lift system to relieve system pressure, then remove the hood front side panels and the bottom splash plate and air baffle. Disconnect reservoir lines from unload valve and drain hydraulic reservoir. Remove the cap screws which retain unload valve to hydraulic pump, then remove the two self-locking nuts from coupling to drive hub studs. Unbolt and remove pump and drive coupling from mounting bracket.

149. **OVERHAUL.** With pump removed, remove drive hub and key, place pump in a vise with drive end up, then unbolt and remove drive shaft housing (2—Fig. FO121) and drive shaft (9). Lift out thrust bearing (10) and wobble plate (11). Discard "O" ring (12). Lift out the six pistons (14) and springs (15). Invert pump body in vise, unbolt cover (23) from pump body and remove inlet valve springs (18) and ball cages (19A) from pins of valve retaining plate (21). Remove pump body from vise, invert same and remove the six steel inlet valve balls. If necessary, valve seats (16) can be removed from pump body by using Nuday Seat Remover tool No. NCA-600-G. Remove screw (28) and gasket (20). Insert a tapered punch through the large cap screw hole in pump cover and carefully bump out valve plate (21). Remove the six outlet valve balls (25), springs (26) and ball cages (29).

Straighten tab of lock washer (4) and remove lock nut (3) and lock washer. Remove seal sleeve (5). Remove shaft (9), along with the inner bearing cone and spacer, from shaft housing. Remove seal (6), "O" ring and the outer bearing cone from shaft housing. Bump outer bearing cup from shaft housing, then remove the large flange spacer. Bump inner bearing cup from shaft housing. Do not remove snap ring (8) unless necessary as damage to housing could result.

Wash all parts in a suitable solvent and inspect. Renew any springs which show signs of distortion or possible fractures. Inspect valves, valve seats, pistons and piston bores for wear, scoring or galling and renew as necessary. Any faulty drive shaft bearing parts will require renewal of the complete bearing assembly comprising the bearing cones, bearing cups and the large and small spacers.

Coat all parts with Ford M-2C-41 oil, or equivalent, and reassemble pump by reversing the disassembly procedure. Be sure to align scribe marks on valve plate (21) and pump cover (23) when installing valve plate. If new valve seats (16) are being installed, use Nuday Seat Driver tool No. NCA-600-EA and be sure the chamfered ball seat is toward ball. Tighten cover and shaft housing cap screws to a torque of 35-45 ft.-lbs.

UNLOAD VALVE

All Models

150. **REMOVE AND REINSTALL.** To remove the unload valve, first cycle hydraulic lift system to relieve system pressure, then remove hood front side panels and lower splash plate and air baffle. Remove the two reservoir hoses from unload valve and drain reservoir. Remove air cleaner tube. Disconnect power steering pump bypass line, the two pressure lines and return line from unload valve, then unbolt and remove unload valve from hydraulic pump.

Note: If trouble shooting indicates leakage in the unload valve, it may be due to a faulty check valve seat (25—Fig. FO122) or ball (26). These parts can be removed and inspected without removing the unload valve and it is recommended that this be done before removing unload valve.

151. **OVERHAUL.** With unload valve removed, unbolt manifold (14—Fig. FO122) and discard "O" rings (10 and 11). Remove acorn nut (1), loosen jam nut (3) and back-off adjusting screw (4) until spring pressure is relieved. Remove front end plate (5), spring guides (6), spring (7), front ball (29) and "O" ring seal (8).

Note: Valve pin (32) will probably also come out at this time. If not, it can be removed from aft side.

Remove rear end plate (23), then remove "O" ring (36), spring (35), retainer (34), spring (33), rear ball (29) and if not already done, the valve pin (32). Depress valve stop (21) and quickly release and allow spring pressure to pop valve stop from bore. Remove spring (20). Work from front end of valve body and push shuttle valve (18) and bushing (16) out rear of body, then pull shuttle valve from bushing. Use a small brass rod inserted through front of valve body and bump unload valve seat (31) out rear of body. Remove retainer (37), spring (27) and ball (26). Use a hooked brass rod to remove seat (25) and "O" ring (24).

NOTE: Spacer (39) is a press fit in shuttle valve bore and generally will not require removal.

Discard all "O" rings, wash all parts in a suitable solvent and inspect. Check all springs for fractures and/or

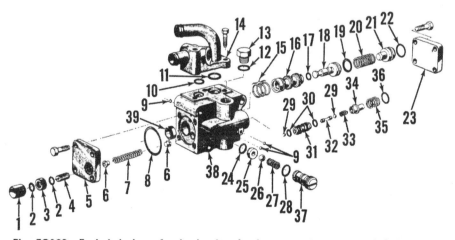

Fig. FO122—Exploded view of unload valve showing component parts and their relative positions.

1. Acorn nut	11. "O" ring	20. Spring	30. "O" ring
2. Seal	12. "O" ring	21. Stop	31. Bushing
3. Jam nut	13. Plug	22. "O" ring	32. Valve pin
4. Adjusting screw	14. Manifold	23. Rear end plate	33. Spring
5. Front end plate	15. "O" rings	24. "O" ring	34. Retainer
6. Guide	16. Bushing	25. Seat	35. Spring
7. Spring	17. "O" ring	26. Ball	36. "O" ring
8. "O" ring	18. Shuttle valve	27. Spring	37. Retainer
9. Steel ball	19. "O" ring	28. "O" ring	38. Valve body
10. "O" ring		29. Steel ball	39. Spacer

distortion and check free lengths against the values which follow:

Spring   **Load length**

Shuttle valve spring

.......1.12 in. @ 4.6 - 5.0 lbs.

Pressure adjusting

spring ....1.80 in. @ 37.4 -42.4 lbs.

Unload valve spring

.......0.27 in. @ 0.16- 0.24 lbs.

Unload valve retainer

spring ....0.56 in. @ 45 -55 lbs.

Check valve spring

.......0.50 in. @ 0.95- 1.05 lbs.

Inspect shuttle valve and bushing for galling or wear patterns. Valve should be snug fit, yet slide freely in bushing with no "O" ring installed. Shuttle valve and bushing are renewed as a set.

Inspect unload valve bushing for damaged ball seats, wear patterns, or any other damage and renew parts as necessary.

Inspect check valve ball and seat. Seat should have a sharp edge and should be renewed if it shows any appreciable wear.

Make an assembly of the unload valve bushing (31), valve pin (32) and the two balls (29) and check to be sure that valve pin is long enough to prevent both balls from seating at once. Spacer pin length is 0.656-0.658.

Reassemble unload valve as follows: Use all new "O" rings. Insert shuttle valve in bushing (seat), then install the loose shuttle valve bushing "O" ring (15) in body bore and install shuttle valve assembly. Install shuttle valve spring (20) and stop (21). Use grease and place valve pin (32) in unload bushing (31), install units in body bore and be sure bushing seats. Use grease on spring (33), install spring in retainer (34) and stick ball (29) on spring; then, carefully install assembly in body. Install spring (35) and "O" ring (36), then install rear end plate (23). Now working at opposite end, be sure inner end of adjusting screw (4) is flush with inner face of front end plate (5). Install "O" ring (8) in counterbore of body front face. Position front ball (29).

Use grease to hold guide and install assembly in body. Place front end plate in position with adjusting screw over guide (6), push down on front end plate and install retaining cap screws. Turn adjusting screw (4) in with fingers until contact with spring guide is felt then turn adjusting screw an additional two turns and tighten jam nut (3). Install check valve and tighten retainer (37) to 50-60 ft.-lbs. Install manifold (14) with hose connections toward front end plate.

Mount assembly on tractor, fill reservoir, then check and adjust system pressure, if necessary, as outlined in paragraph 141.

REMOTE CONTROL VALVE

All Models

152. **R&R AND OVERHAUL.** To remove the remote control valve, first cycle lift system to relieve system pressure, then disconnect the two lines from right side of valve and cap off lines. Disconnect and cap off remote cylinder lines, then unbolt and remove remote control valve from hydraulic lift unit.

With valve removed, remove lever stop, both end covers and discard gaskets. Use a deep, thin walled socket and remove both safety (relief) valves.

NOTE: DO NOT use a screw driver to remove and reinstall safety valves. Safety valves are preset at the factory and silver soldered to hold the adjustment.

Use Nuday tool N-875-A, or equivalent, and remove the four retainers from outer end of valves, then remove spacers, bushings and springs. Move control valve handle until valves are extended and can be grasped and removed with the fingers. Use a hooked rod to remove valve seats, "O" rings and the inner Teflon seals. Keep valves and valve seats together so they can be reassembled in the same position and with the same parts. Remove plug, spring and detent assembly. Remove both plugs (front and rear) over actuating arms and remove actuating arms. Loosen gland nuts and unscrew both flow control valves. Drive pin from collar and shaft on end of shaft opposite control handle and pull shaft from valve body. Any further disassembly required is obvious.

Discard all "O" rings, wash all parts in a suitable solvent and inspect. Valve return springs should have a free length of 1.240 and test 23.75-

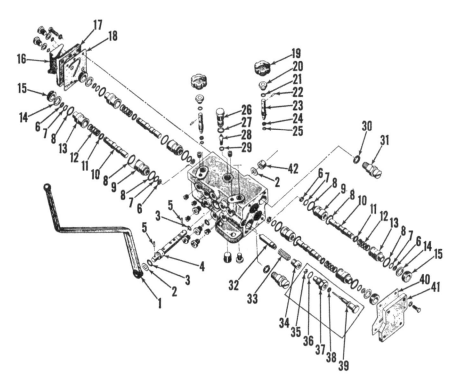

Fig. FO123—Exploded view of remote control valve showing component parts and their relative positions. Control valves, valve seats and bushings are same as those in hydraulic lift unit.

1. Control lever	12. Return spring	22. Pin	32. Detent
2. Spacer	13. Valve bushing	23. Adjusting valve	33. Detent spring
3. "O" ring	14. Spacer	24. Back-up washer	34. Plug
4. Actuating shaft	15. Retaining screw	25. "O" ring	35. Ring
5. Groove pin	16. Lever stop	26. Plug	36. Seal
6. Glide ring (Teflon)	17. Cover	27. "O" ring	37. Plug
7. "O" ring	18. Gasket	28. Arm	38. Seal
8. "O" ring	19. Knob	29. "O" ring	39. Screw
9. Valve seat	20. Gland nut	30. Gasket	40. Gasket
10. Valve	21. "O" ring	31. Safety valve	41. Cover
11. Washer			42. Hub

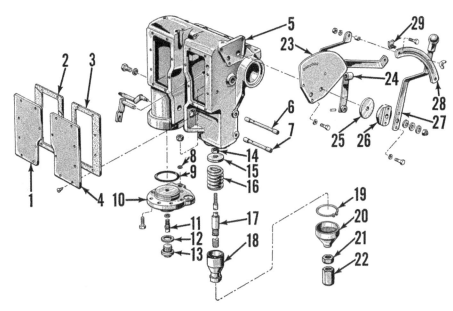

Fig. FO124—Exploded view of hydraulic lift housing, quadrant and related parts. Series 6000 unit is shown, however Commander 6000 series unit is similar.

1. Housing cover, LH	11. Safety relief valve	20. Boot
2. Gasket	12. Gasket	21. Lock nut
3. Gasket	13. Plug	22. Adjusting nut
4. Housing cover, RH	14. Nut	23. Quadrant
5. Housing	15. Belleville washer	24. Multi-Trol lever
6. Outlet tube	(15 used)	25. Friction disc
7. Inlet tube	16. Draft control spring	26. Hub
8. "O" ring	17. Plunger rod	27. Control lever
9. "O" ring	18. Spring seat	28. Lever retainer
10. Piston cover	19. Clamp	29. Stop assy.

26.25 lbs. when compressed to a length of 0.70 inches. Detent spring should have a free length of 1.495-1.505 and test 16.8-18.4 lbs. when compressed to a length of 0.70 inches. Inspect all valves, seats and bushings for cracks, scoring, galling or excessive wear and if found to be defective, renew complete valve assembly as valves and seats are not serviced separately. Safety valves can be bench tested and should pop at approximately 2450-2550 psi. Do not attempt to adjust safety valves. If safety valves are defective, renew them. Check detent for defects and operation. If detent is not held firmly on its seat by spring pressure, renew detent. Inspect flow control valves for excessive grooving, or other damage, and renew parts as necessary.

Use all new "O" rings, seals and gaskets and reassemble by reversing the disassembly procedure.

NOTE: When operating (extending) a single acting cylinder with the remote control valve, the control handle must be held in the actuating position as there is no return oil available to operate the detent assembly.

In cases where the load is not sufficient to overcome restriction of return oil through the lines and detent assembly, it may be necessary to in-

stall the adjustable detent assembly (items 35 through 39) shown in inset of Fig. FO123.

HYDRAULIC LIFT UNIT

All Models

153. **REMOVE AND REINSTALL.** To remove the hydraulic lift unit, cycle the hydraulic lift system to relieve system pressure, then disconnect the pressure and return lines from right side of hydraulic lift unit and plug each line as it is disconnected, to retain fluid. Disconnect lift links from rockshaft lift arms. Tilt seat forward, then unbolt and remove seat and tool box as a unit. Remove the three attaching cap screws from rear center and the two cap screws from the bottom front of the lift unit. Unbolt remote control valve from lift unit. Attach a chain and hoist to lift unit, then remove the three attaching cap screws from top front of lift unit and remove unit from tractor. Be careful not to lose the wear plate located between lift link hanger arm and draft control plunger.

Reinstall by reversing the removal procedure. Be sure to reinstall wear plate between hanger arm and draft control plunger.

154. **OVERHAUL.** Remove hydrau-

lic lift unit from tractor as outlined in paragraph 153. Note: Service on hydraulic lift unit will be simplified if rockshaft lift arm is bolted to bench to hold unit in a convenient position. Remove left hand housing cover (1— Fig. FO124) and the piston cover (10) assembly. Push piston out of cylinder by using a hammer handle or other similar object. Remove plug (13), then using a long, thin walled socket, remove safety relief valve (11) from piston cover.

NOTE: DO NOT use a screw driver to remove safety relief valve. Safety relief valve is preset at the factory to relieve at 2450-2550 psi and then silver soldered to hold the adjustment. Valve can be bench tested and if found to be faulty, renew same.

Use Nuday special tool N-875-A, or equivalent, and remove the valve stack retainer (36—Fig. FO125), then push control lever forward and remove spacer (35), "O" ring (34), bushing (33), return spring (32), valve (28), valve seat (27) and "O" rings (26). The "O" rings and Teflon glide rings can now be removed from I.D. of valve seat and bushing.

Note: A hooked brass rod about 12 inches long will aid in removing valve parts.

Remove pivot bolt from control lever (Fig. FO126) and pull lever from center spacer. Loosen center spacer retaining set screw and remove center spacer. Remove "O" ring and push down on exposed shoulder of upper bushing and remove upper valve parts in the reverse order to those of the bottom valve.

155. Remove right hand housing cover (4—Fig. FO124) and disconnect spring (39—Fig. FO125). Disconnect connecting rod (38) from actuating lever (41) and actuating lever from control lever shaft (43), then remove actuating lever and roller arm (42) as a unit. Remove clamp (19—Fig. FO-124) and plunger boot (20). Remove lock nut (14) from top of plunger rod (17), then unscrew plunger rod from clevis (15—Fig. FO125). Disconnect roller guide (14) from clevis and remove clevis, then disconnect and remove roller guide from position control lever (12). Remove pin from stop lever (17), then remove connecting rod, control valve lever, stop lever and stop valve lever rod from housing as a unit.

Any further disassembly of linkage, or quadrant, is obvious.

156. Remove cap screws which retain lift arms (2) to lift (rock) shaft (11) and remove arms and "O" rings

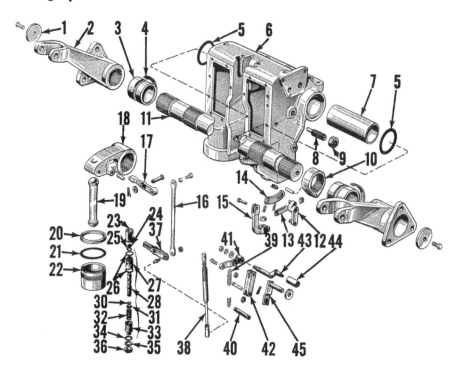

1. Retainer washer
2. Lift arm
3. Bushing
4. Felt seal
5. "O" ring
6. Housing
7. Spacer
8. Position control lever shaft
9. Lock nut
10. Position control arm
11. Rockshaft
12. Position control lever
13. Position control link
14. Guide
15. Clevis
16. Shut-off rod
17. Stop lever
18. Ram arm
19. Connecting rod
20. Back-up washer
21. "O" ring
22. Piston
23. Center spacer
24. Teflon glide ring
25. "O" ring
26. "O" ring
27. Valve seat
28. Valve
30. "O" ring
31. Teflon glide ring
32. Valve spring
33. Bushing
34. "O" ring
35. Spacer
36. Retainer screw
37. Valve lever
38. Valve connecting rod
39. Actuating lever spring
40. Anchor pin
41. Actuating lever
42. Roller arm
43. Actuating lever shaft
44. Bushing
45. Multi-Trol lever shaft

Fig. FO125—Exploded view of hydraulic lift cylinder, control valve, rockshaft and control linkage. Series 6000 unit is shown, however, Commander 6000 series is similar. Also see Fig. FO125A.

(5). Bump lift shaft from left to right out of housing and lift spacer (7), ram arm (18) and position control arm (10) from housing. Remove pin and separate connecting rod (19) from ram arm. Bushings (3) and seals (4) can now be removed, if necessary.

157. Inspect interior of housing for signs of inlet or outlet tubes leaking. If tubes require renewal, proceed as follows: Cut damaged tube about two inches from piston side of housing and pry end of tube out as shown in Fig. FO127. Grasp tube ends with pliers, or vise grips, and work pieces out of housing.

When installing new tubes, apply two drops of "Loctite" to inner end of tubes before driving them in place using tube replacer Nuday No. NA-856 as shown in Fig. FO128. Allow "Loctite" to set 10-12 hours before operating hydraulic lift system.

158. With lift unit completely disassembled, clean all parts in a suitable solvent and inspect. Control valve return springs (32—Fig. FO125) should have a free length of 1.240 inches and should test 23.75-26.25 lbs. when compressed to a length of 0.70 inches. Draft control spring free length is 3.400 inches. Linkage connecting rod (38) internal spring has a free length of 6.350 inches and should require approximately 60 pounds pull on a spring scale to start rod moving out of housing. Inspect all rods and levers for fractures, bends or distortion. Inspect control valves, seats and bushings for scoring, chipping, grooves or undue wear. Renew parts as necessary. Valves and valve seats are not serviced separately and must be renewed in sets. A service kit is available which includes valve, valve seat, seals and "O" rings.

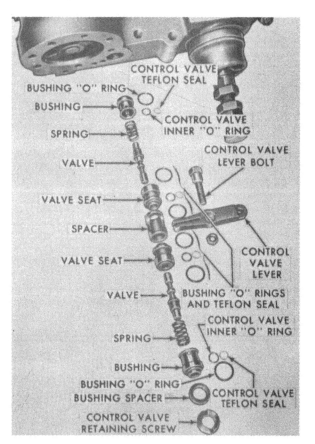

CONTROL VALVE TEFLON SEAL
BUSHING "O" RING
BUSHING
CONTROL VALVE INNER "O" RING
SPRING
CONTROL VALVE LEVER BOLT
VALVE
CONTROL VALVE LEVER
VALVE SEAT
SPACER
VALVE SEAT
CONTROL VALVE LEVER
VALVE
BUSHING "O" RINGS AND TEFLON SEAL
CONTROL VALVE INNER "O" RING
SPRING
BUSHING
BUSHING "O" RING
BUSHING SPACER
CONTROL VALVE TEFLON SEAL
CONTROL VALVE RETAINING SCREW

Fig. FO125A — Exploded view of hydraulic lift and lowering valves as they are removed from hydraulic lift housing. Refer to Fig. FO125 for exploded view showing lift cylinder, rockshaft and control linkage.

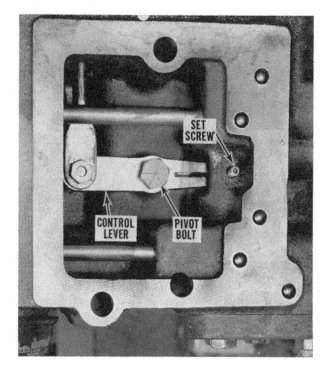

Fig. FO126—Front side of lift housing showing location of control lever, pivot bolt and the set screw which retains control valve center spacer.

Fig. FO127 — View showing hydraulic tube being removed from lift housing. Refer to text.

Fig. FO128—Use Nuday tool NA-856 to install tubes in lift housing. Refer to text.

159. To reassemble, refer to Fig. FO129 for a schematic view showing relationship of operating parts and proceed as follows: Use all new "O" rings, seals and gaskets and coat all parts with oil prior to assembly.

NOTE: Prior to starting assembly, install "O" ring in I.D. of control valve bushing and seat, then collapse (fold in a heart shape) and install the Teflon glide ring in groove over the "O" ring. Use a small, smooth brass rod to work glide ring into position. Install valve into bushing and seat, and work valve back and forth to assure proper fit of seal. See Fig. FO129A.

Turn lift housing upside down and install upper "O" ring of upper control- valve stack in position. Use a piece of brass wire, or rod, about 12 inches long to properly position "O" ring. Install upper bushing and be sure it is seated in end of bore. Install valve seat "O" ring and position same with the brass rod, then install valve spring. Install valve, valve seat and

"O" ring in bore. Place center spacer over a heavy brass rod, then compress valve parts with rod, position "O" ring and align center spacer and set screw hole with retaining set screw and tighten set screw. Use caution during this operation not to damage the "O" ring positioned between valve seat and center spacer.

Install the lower valve stack upper "O" ring and position same using the 12-inch brass rod. Install valve, valve seat and "O" ring in bore. Position "O" ring with the brass rod. Install valve spring, and lower bushing. Place "O" ring, spacer and retainer over a heavy brass rod, then compress valve, position "O" ring, spacer and retainer and tighten retainer enough to engage a few threads. Loosen center spacer set screw and using Nuday tool N-875-A, or equivalent, tighten the retainer to 35-45 ft.-lbs. torque. After tightening, retainer exposed surface must be below machined surface of housing. Retighten center spacer set screw.

160. Use grease pencil and mark sides of ram arm and position control arm adjacent to the blind spline. Position ram arm, position control arm and spacer in housing, then slide rockshaft into housing from right to left while aligning the previously affixed reference marks. Install the two "O" ring seals, lift arms, retaining washers and the self-locking cap screws. Tighten cap screws to not more than 20 ft.-lbs. torque. Screw shaft (8— Fig. FO125) into position control lever (12) if it was previously removed.

Note: Rockshaft should pivot freely. Recheck rockshaft rotation after tractor is completely assembled. Lift arms should lower by gravity alone after being raised to the "Up" position by hand. Loosen both cap screws slightly, if necessary.

Attach piston connecting rod to ram arm. Place "O" ring and back-up washer on piston with smooth side of

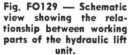

Fig. FO129 — Schematic view showing the relationship between working parts of the hydraulic lift unit.

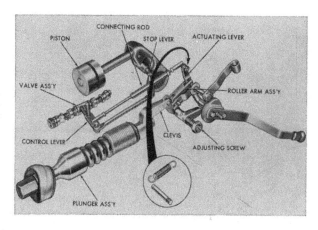

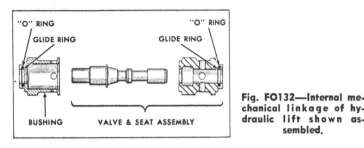

Fig. FO129A — Install "O" rings and glide rings in valve seat and bushing as shown. Also refer to text and to Fig. FO125A.

Fig. FO132—Internal mechanical linkage of hydraulic lift shown assembled.

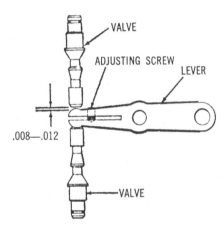

Fig. FO130 — Control valve lever should have 0.008-0.012 clearance between control valves. Refer to text.

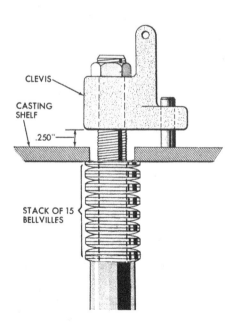

Fig. FO131—Adjust plunger rod to obtain the 0.250 clevis to housing clearance. Refer to text.

back-up washer next to "O" ring and "O" ring on pressure (closed) side of piston. Install piston in cylinder bore (open end first) and bump into place with a hammer handle. Install "O" rings on piston cover and secure cover to housing. Install safety relief valve in piston cover using a long, thin walled socket, then with a new gasket on valve plug, install plug.

161. Adjust the control lever fingers until there is a total clearance of 0.008-0.012 between end of control lever and the control valves as shown in Fig. FO130. Assemble stop lever, stop lever rod, control lever and linkage connecting rod, then install assembly in housing. Torque both control lever bolt and lock nut to 50-60 ft.-lbs. torque.

Place spring seat, draft control spring and the 15 Belleville washers on plunger rod, then insert plunger rod assembly in housing and screw rod end into the clevis. Note: Place first (bottom) Belleville washer on rod with concave (cup) side up, second washer on rod with cup side down, third washer on rod with cup side up, etc., until all 15 washers are on rod. The last (top) washer will then have the cup side up against housing web. Push up on plunger rod until the top of the Belleville washer stack contacts bottom of housing web and the shoulder of the plunger rod contacts the bottom of the Belleville washer stack. Check to see that spring is not contacting the Belleville washers (back-off spring seat if necessary), then screw plunger rod into clevis until there is 0.250 clearance between bottom of clevis and top of housing web as shown in Fig. FO131. Install and tighten lock nut to 85-95 ft.-lbs. torque to hold this adjustment. Now push plunger rod down and turn spring seat clockwise until

it contacts draft control spring (finger tight), then install and tighten lock nut to 135-145 ft.-lbs. to hold this adjustment. Install boot and clamp, then install, but do not tighten, lock nut and plunger rod adjusting nut.

Attach roller guide to position control lever and clevis. Attach roller arm assembly to actuating lever. Attach actuating lever to connecting rod assembly and control lever shaft and be sure that pin of the roller arm assembly is in the slot of the "Multi-Trol" control lever. Install actuating lever spring with open side of bottom hook toward front of tractor. Do not install rear covers until unit has been adjusted as outlined in paragraph 163. Reinstall unit as outlined in paragraph 153.

Note: It is recommended that the system be tested for leaks with a high pressure hand pump and a 3000 psi gage connected to the high pressure inlet of lift housing prior to installing lift housing on tractor.

With hydraulic lift unit assembled, install unit on tractor by reversing the removal procedure. Adjust the draft and position control linkage as outlined in paragraph 163.

163. ADJUSTMENT. Check for correct draft control adjustment as follows: Start engine to pressurize system. Place the Multi-Trol lever in the forward notch on quadrant (notch marked "D" on late model tractors). When moving the control lever slowly towards top and rear of quadrant, the lift arms should start to raise when the control lever is 1½-2 inches from lever stop at top and rear of quadrant. If not, loosen the two cap screws that retain control lever to hub and move the lever on hub to proper position. If slotted holes in the control lever does not provide adequate adjustment, remove the right hand cover

from rear face of hydraulic lift housing, loosen locknut on top end of connecting rod (38—Fig. FO125) and adjust length of rod so that lift arms start to raise when control lever retaining cap screws are tightened in center of slotted holes in lever and lever is positioned 2 inches from top rear stop on quadrant.

To check for correct position control adjustment, be sure that draft control is properly adjusted as outlined in preceding paragraph and proceed as follows: Move the Multi-Trol lever to rear notch on quadrant (notch marked "P" on late model tractors). Move control lever to bottom front stop on quadrant and start engine to pressurize system. Slowly move control lever up and to rear of quadrant; the lift arms should start to raise when the lever is 1½-2 inches away from bottom (front) stop. If not, loosen the locknut (9—Fig. FO125) and turn screw (eccentric) (8) in small increments until lift arms start to raise when control lever is positioned 2 inches from bottom stop on quadrant. Hold the adjusting screw in this position and tighten locknut. As the adjustment is controlled by an eccentric on inner end of screw, screw cannot be more than ½-turn out of adjustment. Note: On late model tractors, there is a punch mark on the face of the adjusting screw (8—Fig. FO125). This punch mark should always be forward of the center of the adjusting screw.

If considerable position control adjustment (more than ¼-turn) was required, recheck draft control adjustment; then, recheck position control adjustment.

NOTE: When system is not pressurized, spring in connecting rod (38—Fig. FO125) will pull control lever back down if it is

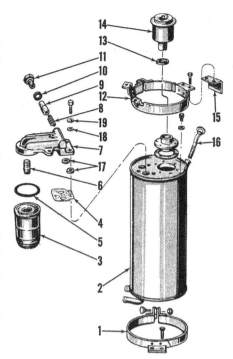

Fig. FO133—Exploded view of the hydraulic system reservoir and filter assembly.

1. Mounting strap	10. Gasket
2. Reservoir	11. Plug
3. Filter	12. Mounting strap
4. Manifold gasket	13. Gasket
5. "O" ring	14. Filler cap
6. Nipple	15. Bracket
7. Manifold	16. Indicator
8. Pressure relief	17. Seals
spring	18. Seal
9. Plunger	19. Washer

raised to top of quadrant and lift arms are in lowered position. Control lever should maintain any position on quadrant when system is operating (pressurized). If not, adjust lever tension with nut and spring washers on outer end of lever shaft. If difficulty is encountered in adjusting tension, be sure that the friction disc (25—Fig. FO124) is in good condition and that spring washers are placed as follows: Cup of inner washer should be towards control lever and cups of outer spring washers should be facing each other.

164. With hydraulic lift unit installed and adjusted, manually push lower links forward, then turn the nut (22—Fig. FO124) on end of draft control plunger downward against wear plate on lift link hanger (rocker) arm until there is no free play of hanger. Then, tighten the lock nut (21) to a torque of 135-145 ft.-lbs. to hold this adjustment.

RESERVOIR AND MANIFOLD
All Models

165. **R&R AND OVERHAUL.** To remove the reservoir and manifold assembly, first actuate lift system to relieve system pressure, then remove hood front side panels, hood and radiator air baffles. Disconnect the two reservoir hoses from unload valve and drain reservoir. Disconnect power steering return line at "tee" on bottom of reservoir (early models) then disconnect the three remaining lines. Disconnect upper bracket from hood rail, then remove lower clamp bolt and lift reservoir from tractor.

166. With reservoir removed, unscrew filter element. Unbolt and remove manifold from reservoir and discard gasket. Remove plug, gasket, bypass valve and spring from manifold. Clean all parts, except filter element, in a suitable solvent and inspect. Valve should be free of scoring, chipping or other damage. Check the valve spring for fractures and/or distortion. Spring free length should be 2.18-2.38.

Reassemble by reversing disassembly precedure and install new filter element, if necessary. Install reservoir assembly on tractor and fill with 3½ gallons of Ford M-2C-41 hydraulic fluid, or its equivalent.

CAUTION: Do not attempt to use an engine or transmission oil filter on the hydraulic system.

NOTES

FORD

Models ■ 1000 ■ 1600

Previously contained in I&T Shop Service Manual No. FO-36

SHOP MANUAL

FORD

SERIES 1000-1600

Tractor serial number and engine
serial number are stamped on vehicle
identification plate located on left side
of clutch housing.

INDEX (By Starting Paragraph)

2

CONDENSED SERVICE DATA

(NOTE: ALL MEASUREMENTS ARE METRIC)

GENERAL

Engine Make	Own
No. of Cylinders	2
Bore, mm	90
Stroke, mm	100
Displacement, cc	1272
Compression Ratio	21:1
Pistons Remove From	Above
Main Bearings, Number of	2
Cylinder Sleeves	None
Forward Speeds	9
Reverse Speeds	3

TUNE-UP

Compression, Gage at Cranking Speed	2690 kPa
Valve Tappet Gap, I&E, Cold	0.3 mm
Engine Low Idle RPM	750-850
Engine High Idle RPM	2650-2700
Engine RPM at Rated Load	2500
Engine RPM for 540 PTO RPM	1955
Injection Timing	21° BTDC
Battery Terminal Grounded	Negative

SIZES—CLEARANCES—CAPACITIES

Crankshaft Journal Diameter	67.95-67.97 mm
Crankpin Diameter	59.95-59.97 mm
Camshaft Journal Diameter:	
Front	20.00-20.01 mm
Rear	19.99-20.00 mm
Piston Pin Diameter	32 mm
Valve Stem Diameter, Intake	8 mm
Valve Stem Diameter, Exhaust	8 mm
Main Bearing Running Clearance	0.04-0.010 mm
Rod Bearing Running Clearance	0.06-0.13 mm
Crankshaft End Play	0.1-0.45 mm
Piston Skirt Clearance	0.12-0.18 mm
Cooling System, Quarts	5.3 Liters
Crankcase, Quarts,	
Without Filter Change	4.7 Liters
With Filter Change	5.0 Liters
Hydraulic System	10 Liters
Transmission and Final Drive	19.8 Liters
Fuel Tank	22 Liters

FRONT AXLE AND STEERING SYSTEM

FRONT AXLE

1. Front axle is adjustable type shown in Fig. 1. The axle pivots on shaft (2) which rides in support bushings (16). Spindles (29) are supported by bushings (22) and thrust bearings (26). All bushings and bearings are renewable.

Desired pivot shaft end play is 0.03-0.008 mm and should not exceed 0.15 mm. Thrust washers (18) are available in varying thicknesses to adjust pivot shaft end play. Desired radial clearance between pivot shaft and bushings is 0.1-0.2 mm and should not exceed 1.0 mm.

TIE RODS AND TOE-IN

2. Tie rod and drag link ends are automotive type. Adjust toe-in to 0-12.7 mm by shortening or lengthening tie rod. Steering drag link can be adjusted if necessary, to permit a full turn in either direction.

STEERING KNUCKLES

3. To remove steering knuckle, support front axle and remove wheel. Detach tie rod from right steering arm (2—Fig. 2) or tie rod and drag link from left steering arm (10). Remove steering arm bolts and steering arm. Remove key and shims and withdraw spindle from front axle.

Spindle bushings are pre-sized and can be renewed after spindle is withdrawn. Upper and lower bushings are

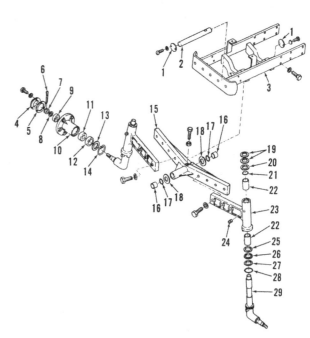

Fig. 1—Exploded view of adjustable front axle assembly.

1. Cover
2. Pivot shaft
3. Front support
4. Hub cap
5. Gasket
6. Cotter pin
7. Nut
8. Washer
9. Roller bearing
10. Hub
11. Roller bearing
12. Oil seal
13. Washer
14. Snap ring
15. Axle center member
16. Bushings
17. "O" rings
18. Thrust washers
19. Shims
20. Spacer (washer)
21. "O" ring
22. Bushings
23. Axle extension
24. Grease fitting
25. Bearing race
26. Thrust bearing
27. Thrust washer
28. "O" ring
29. Spindle

interchangeable. Install sufficient shims (19—Fig. 1) to fill gap between steering arm and spacer (20).

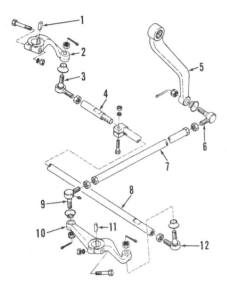

Fig. 2—Exploded view of steering linkage.

1. Key	7. Drag link
2. Steering arm, R.H.	8. Tie rod
3. Tie rod end	9. Drag link end
4. Tie rod extension	10. Steering arm, L.H.
5. Pitman arm	11. Key
6. Drag link end	12. Tie rod end

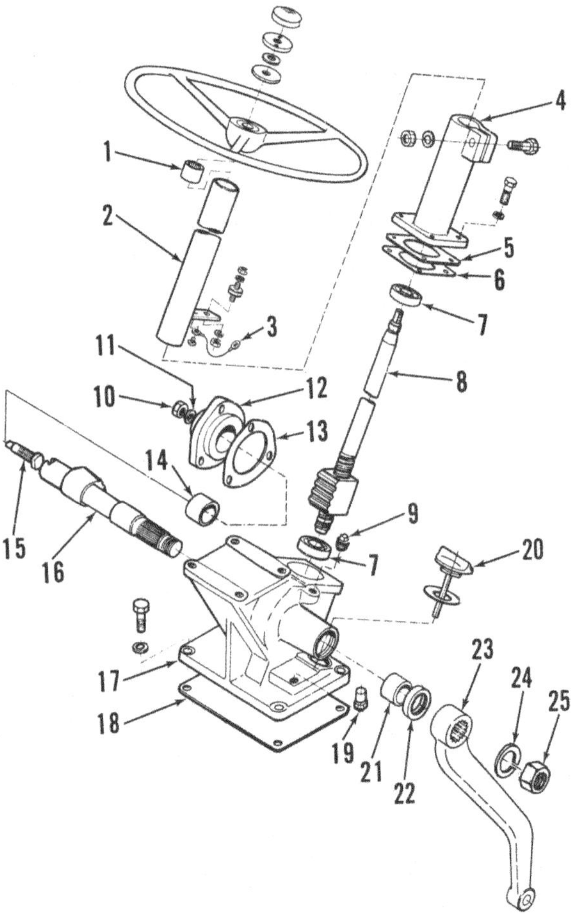

Fig. 3—Exploded view of steering gear.

1. Bearing
2. Steering column
3. Ground wire
4. Steering housing
5. Gasket
6. Shims
7. Bearings
8. Steering shaft & gear
9. Fill plug
10. Nut
11. Lockwasher
12. Side cover
13. Gasket
14. Bushing
15. Adjusting screw
16. Cross shaft
17. Body
18. Gasket
19. Breather
20. Dipstick
21. Bushing
22. Oil seal
23. Pitman arm
24. Lockwasher
25. Nut

FRONT SUPPORT

4. To remove front support, remove hood, drain cooling system, disconnect radiator hose and remove radiator. Detach drag link from pitman arm and place a support under clutch housing. Unscrew twelve screws securing front support to sides of engine, and roll front axle assembly away from tractor. Separate front support from front axle.

STEERING GEAR

5. All models are equipped with recirculating ball nut steering gear shown in Fig. 3. To remove steering gear, remove pitman arm, steering wheel and instrument panel. Unbolt and remove steering gear.

Remove capscrews securing right side cover (12) and remove cover along with pitman shaft and gear (16). Remove steering shaft housing (4) being careful not to lose shims (6). Remove steering shaft and gear (8). Inspect components and renew any which are damaged or excessively worn. Shaft and gear unit (8) is available as an assembly only.

Shims (6) are used to adjust bearing preload. Install steering shaft and gear (8), bearings (7), original shims (6), gasket (5) and steering housing (4). Tighten steering housing (4) screws

Fig. 4—Turn screw (15) to adjust steering gear backlash.

and measure force necessary to turn steering shaft. Add or remove shims (6) as needed so 1-3 kg/cm is required to turn steering shaft.

Reassemble and install steering gear by reversing disassembly procedure. Fill housing with SAE EP90 oil through plug (9) hole until oil reaches bottom of plug hole.

Steering shaft free play should be 20-35 mm measured at outer circumference of steering wheel. To adjust free play, loosen locknut and adjusting screw (15—Fig. 4). Turn screw clockwise to decrease free play or counterclockwise to increase free play.

ENGINE AND COMPONENTS

R&R ENGINE WITH CLUTCH

6. To remove engine, remove battery and disconnect all wires interfering with engine removal. Remove hood and disconnect throttle rod. Disconnect fuel tank lines, loosen clamps and remove fuel tank. Remove front support and axle assembly as outlined in paragraph 4. Disconnect hydraulic pump lines and remove muffler and air cleaner. Attach a hoist to engine, unscrew clutch housing bolts and remove engine. Install by reversing removal procedure.

CYLINDER HEAD

7. To remove cylinder head, remove hood panel and drain cooling system. Detach upper coolant hose, air inlet hose and exhaust pipe from head. Remove and cap injector lines. Remove rocker arm assemblies and push rods. Unscrew six cylinder head retaining nuts and lift off cylinder head assembly.

NOTE: One cylinder head retaining nut is located in air inlet passage.

Cylinder heat warpage should not exceed 0.12 mm. Reverse removal procedure to install cylinder head. Note tightening sequence of cylinder head retaining nuts in Fig. 5. Later 1600 models are equipped with a long stud (L—Fig. 5) and spacer (S) in air inlet passage to prevent loosening of retaining nut. Tighten cylinder head nuts on all models to 147-151 Nm. Be sure rocker arm support holes are aligned with roll pins in cylinder head.

VALVES AND SEATS

8. Valve seat and valve face angles are 45 degrees for intake and exhaust valves. All valves seat directly in head. Recommended valve seat width is 1.2-1.5 mm. Maximum allowable valve seat depth is 2.0 mm measured from cylinder head surface to valve head surface. Minimum allowable valve margin is 1.0 mm. Valve stem diameter is 8 mm for all valves.

VALVE GUIDES

9. Maximum allowable intake valve-to-guide clearance is 0.2 mm while maximum allowable exhaust valve-to-guide clearance is 0.25 mm. Replacement valve guides are not available from manufacturer.

VALVE SPRINGS AND SEALS

10. Identical springs are used on all valves. Valve spring free length should be 48 mm. Valve spring tension should be 12-14 kg at 36 mm. Identical valve seals are used on all valves.

ROCKER ARMS

11. Intake and exhaust rocker arms are interchangeable. Desired clearance between rocker arm and rocker shaft is 0.04 mm while maximum allowable clearance is 0.2 mm. Renew rocker shaft if diameter is less than 13.55 mm.

Rocker arms, shafts and valves are lubricated by oil directed through push rods from valve tappets. When installing rocker shaft support stand on cylinder head, be sure hole in support stand is aligned with roll pin in cylinder head.

VALVE CLEARANCE ADJUSTMENT

11A. Valve clearance for all valves should be 0.3 mm. Set valve clearance with engine cold.

R&R TIMING GEAR COVER

12. To remove timing gear cover, tilt hood forward, drain cooling system and remove radiator. Drain engine oil. Remove fan, fan belt, crankshaft pulley and pulley key. Disconnect oil lines to hydraulic oil pump. Unbolt and remove timing gear cover. Crankshaft front seal (8—Fig. 6) can now be renewed.

After installing crankshaft pulley, tighten pulley bolt to 40-48 Nm.

TIMING GEARS

13. Before removing any gears in the timing gear train, rocker arms should be removed to avoid possible damage to piston or valve train if either the camshaft or crankshaft should be turned independently of the other.

The timing gear train consists of the crankshaft gear, camshaft gear, injection pump drive gear, oil pump gear and an idler gears which drives the camshaft and injection pump gears. Refer to paragraph 28 for injection pump drive gear service. Tachometer drive gears are located behind camshaft gear.

14. **IDLER GEAR AND SHAFT.** To remove idler gear and shaft, unscrew two retaining screws which secure assembly to front of cylinder block and loosen oil tube banjo bolts. Renew shaft and/or gear if bushing to shaft clearance is excessive or bushing is scored. Bushing (3—Fig. 6) and gear (2) are available as a unit assembly only. Shaft oil hole must be free of dirt or foreign material. Inspect gear and renew if damaged or excessively worn.

Note that oil tubes leading to idler shaft direct oil to main oil gallery and must be properly attached to idler gear

shaft (4—Fig. 6). Do not over-tighten banjo bolts. Timing marks between idler gear and camshaft gear and idler gear and crankshaft gear must be aligned when installing idler gear. Tighten idler gear retaining screws to 18-19 Nm.

Backlash between idler gear and camshaft gear or between idler gear and crankshaft gear should be 0.1-0.3 mm. Renew gears which exceed backlash requirements.

15. **CAMSHAFT AND TACHOMETER GEARS.** To remove camshaft gear, unscrew gear retaining nut. Tachometer drive gears are now accessible and may be removed.

When reinstalling camshaft gear, align timing marks on idler gear and camshaft gear. Tighten camshaft gear retaining nut to 147-156 Nm. Backlash between idler gear and camshaft gear should be 0.1-0.3 mm.

16. **CRANKSHAFT GEAR.** The crankshaft is equipped with two keys to retain crankshaft pulley and crankshaft gear. Use a suitable puller to pull gear off crankshaft. Inspect gear, key and crankshaft.

To reinstall gear, fully seat key in crankshaft keyway and install gear with timing mark out. Backlash between idler gear and crankshaft gear should be 0.1-0.3 mm.

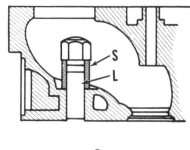

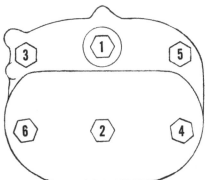

Fig. 5—Use sequence shown above when tightening cylinder head nuts. Long stud (L) and spacer (S) in air inlet passage are used on later 1600 models.

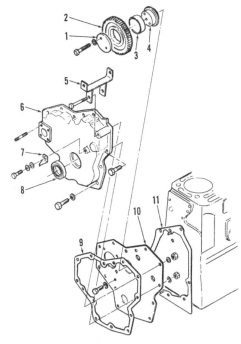

Fig. 6—Exploded view of timing gear cover and idler gear.

1. Retainer plate
2. Idler gear
3. Bushing
4. Idler shaft
5. Alternator bracket
6. Timing gear cover
7. Timing pointer
8. Oil seal
9. Gasket
10. Inner plate
11. Gasket

CONNECTING ROD AND PISTON UNITS

17. Connecting rod and piston units are removed from above after removing the cylinder head and oil pan. Be sure to remove top ridge from cylinder bores before attempting to withdraw assemblies. Connecting rods should be marked so assembly can be reinstalled in original cylinder.

Assemble connecting rod and piston so "F" mark on rod will be towards engine front and slot (S—Fig. 7) in piston crown will be towards intake side of engine. Tighten connecting rod screws to 69-73 Nm.

PISTON RINGS

18. Pistons are fitted with three compression rings and one oil control ring. Top compression ring is chrome plated while the two lower compression rings are cast iron. Oil control ring is a one-piece slotted two rail type. All rings have a straight face. Install rings so "R" mark is towards piston crown.

Piston ring end gap should be 0.3-0.5 mm. Side clearance should be 0.3-0.9 mm for compression rings and 0.3-0.7 mm for oil ring.

Piston ring sets are available in oversizes of 0.5 mm and 1.0 mm.

PISTONS AND CYLINDERS

19. Engine is equipped with full skirt type pistons which are available in oversizes of 0.5 mm and 1.0 mm. Cylinders are not sleeved.

Piston-to-cylinder clearance should be 0.12-0.18 mm. Maximum allowable cylinder taper is 0.2 mm. Cylinder must not be out-of-round more than 0.2 mm.

PISTON PINS

20. Piston pin floats in connecting rod but is a light press fit in piston. Heat piston to 100° C. to remove piston pin.

Piston pin diameter is 32 mm. Renew pin if worn more than 0.05 mm. Desired pin-to-rod clearance is 0.01 mm while maximum allowable clearance is 0.05 mm.

Refer to paragraph 17 when assembling piston and connecting rod.

CONNECTING RODS AND BEARINGS

21. Connecting rod bearings are available in standard size only. Connecting rod is fractured at big end and rod and cap must mesh perfectly during assembly. Tighten connecting rod screws to 69-73 Nm. Refer to paragraph 17 when assembling piston and connecting rod. Connecting rod bearing clearance is 0.06-0.13 mm. Desired connecting rod side play is 0.1-0.3 mm and should not exceed 0.7 mm.

CRANKSHAFT AND MAIN BEARINGS

22. Crankshaft is supported by two sleeve type main bearings; one in front cylinder block wall and one in flywheel cover. Both bearings are pressed into place. Crankshaft thrust is directed against thrust washers (12 and 17—Fig. 8) pinned to flywheel cover and front cylinder block wall.

To remove crankshaft, remove engine as outlined in paragraph 6. Refer to appropriate paragraphs and remove clutch assembly, alternator, starter cylinder head, water pump, timing gears, injection pump, and front plate. Unscrew flywheel nut and remove flywheel. Note that it may be necessary to tap end of crankshaft to free

flywheel from crankshaft taper. Remove oil pan and oil pump. Remove piston and rod assemblies. Unscrew capscrews securing flywheel cover. Remove flywheel cover while supporting crankshaft as rear end of crankshaft will be unsupported after flywheel cover removal. Withdraw crankshaft from cylinder block.

Inspect bearings, thrust washers and crankshaft. Crankshaft main bearing journal diameter is 67.95-67.97 mm. Main bearing clearance is 0.04-0.10 mm. Standard size main bearings only are available. Crankshaft main bearing journal out-of-round or taper should not exceed 0.05 mm. Crankshaft rod journal diameter is 59.95-59.97 mm. Connecting rod bearing-to-rod journal clearance is 0.06-0.13 mm. Standard size only connecting rod bearings are available. Be sure oil holes are aligned when installing main bearings.

Crankshaft end play should be 0.1-0.45 mm while maximum allowable end play is 0.7 mm. Thrust washers (12 and 17—Fig. 8) control end play. Install thrust washers so oil grooves are adjacent to crankshaft. Be sure thrust washer locating pin is 0.5 mm below thrust bearing face.

Care should be used not to damage crankshaft main bearings and thrust washers when installing crankshaft. Be sure camshaft bearing "O" ring (9) is installed and coated with a sealant. Tighten flywheel cover screws in a diagonal pattern to 44-48 Nm.

CRANKSHAFT REAR OIL SEAL

23. Crankshaft rear oil seal is located in flywheel cover (21—Fig. 8). For

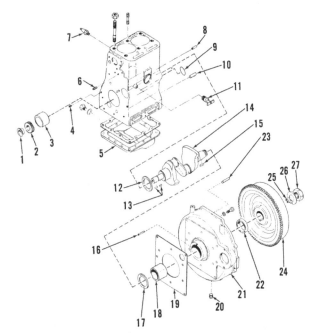

Fig. 8—Exploded view of cylinder block and crankshaft assembly.

1. Oil slinger
2. Gear
3. Bearing
4. Plug
5. Oil pan
6. Dowel pin (2)
7. Oil pressure sensor
8. Roll pin
9. "O" ring
10. Dowel pin (2)
11. Drain valve
12. Thrust washer
13. Keys
14. Crankshaft
15. Key
16. Roll pin
17. Thrust washer
18. Bearing
19. Gasket
20. Plug
21. Flywheel cover
22. Oil seal
23. Dowel pin (2)
24. Flywheel
25. Roller bearing
26. Tab washer
27. Nut

Fig. 7—Piston crown slot (S) must be towards intake side and (F) on rod must be towards engine front.

access to rear oil seal, remove flywheel cover as outlined in paragraph 22.

CAMSHAFT, BEARINGS AND TAPPETS

24. Camshaft rides in two ball bearings located in front and rear cylinder block walls. Refer to Fig. 9 for an exploded view of camshaft and tachometer drive assemblies. To remove camshaft, tappets and bearings, first remove crankshaft as outlined in paragraph 22. Remove "O" ring (9—Fig. 8) adjacent to rear camshaft bearing and withdraw bearing and camshaft. Tappets and front camshaft bearing can now be removed.

Desired camshaft lobe height is 32.05 mm while minimum allowable lobe height is 31.65 mm. Maximum allowable camshaft runout measured at center of camshaft with ends supported in V-blocks is 0.2 mm.

Be sure "O" ring (9—Fig. 8) is installed when assembling engine. Sealant should be used around "O" ring to prevent possible oil leakage.

OIL PUMP

25. Engine oil pump is a gear type located in front cylinder block wall. Pump is driven by crankshaft timing gear.

Pressurized oil from pump is directed by internal passages to relief valve adjacent to pump, to full flow oil filter, to front main bearing, then to banjo bolt and external line above pump. Oil flows through external line to idler gear shaft and from idler gear shaft by external line to banjo bolt above camshaft gear. Oil then flows through internal passages to rear main bearings and tappets. Oil is forced through tappets and up push rods to lubricate valve train.

To remove oil pump, drain engine oil and remove oil pan. Remove timing gear cover as outlined in paragraph 12. Detach oil pickup from oil pump body and remove oil pump drive gear (8—Fig. 10). Unbolt oil pump cover (10) and remove oil pump. Pump may be

disassembled after removing pump cover-to-body screws.

Clearance between oil pump body face and gear faces should be 0.07-0.14 mm with a maximum allowable clearance of 0.25 mm. Clearance between

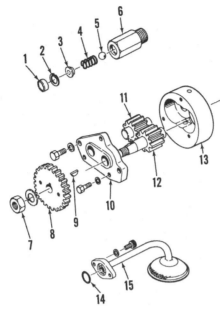

Fig. 10—Exploded view of oil pump and relief valve.

1. Cap	
2. Snap ring	9. Key
3. Spring seat	10. Oil pump cover
4. Spring	11. Driven gear
5. Ball	12. Drive gear
6. Relief valve body	13. Oil pump body
7. Nut	14. "O" ring
8. Gear	15. Oil pick-up

Fig. 11—View of fuel filter and air bleed screw (B).

gear teeth and pump should be 0.01-0.03 mm. Tighten oil pump drive gear nut to 30-34 Nm.

RELIEF VALVE

26. Relief valve for engine lubricating system is located adjacent to oil pump. Remove timing gear cover as outlined in paragraph 12 for access to relief valve. Refer to Fig. 10 for exploded view of relief valve. Later models have a cap (1) in end of valve housing. Early and late relief valve components are interchangeable.

Relief valve should open when oil pressure exceeds 245-392 kPa.

FUEL SYSTEM

All models are equipped with a diesel fuel injection system. When servicing any unit associated with the fuel system, absolute cleanliness must be maintained. Of equal importance is the avoidance of nicks or burrs on any working parts.

Probably the most important precaution that service personnel can impart to owners of diesel powered tractors is to urge them to use an approved fuel that is absolutely clean and free from foreign material. Extra precaution should be taken to make certain that no water enters the fuel storage tanks.

FUEL FILTERS AND LINES

27. Fuel filter and valve assembly is located on fuel tank bottom. Fuel filter element should be removed and cleaned after every 100 hours of operation and renewed after every 200 hours.

27A. **BLEEDING.** To air-bleed fuel system, open fuel valve and fuel filter bleed screw (B—Fig. 11). Allow fuel to flow until air bubbles disappear and fuel is clear. Close fuel filter bleed screw while fuel is flowing; open injection pump bleed screw (B—Fig. 12)

Fig. 12—View showing location of injection pump bleed screw (B).

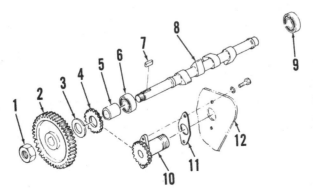

Fig. 9—Exploded view of camshaft assembly.

1. Nut
2. Gear
3. Washer
4. Tachometer gear
5. Spacer
6. Bearing
7. Key
8. Camshaft
9. Bearing
10. Tachometer drive assy.
11. Gasket
12. Inner plate

and allow fuel to flow until air bubbles disappear, then close bleed screw. Loosen high pressure fuel lines at injectors, open throttle and crank engine until fuel escapes from loosened fuel line connections. Tighten fuel line compression nuts.

INJECTION PUMP DRIVE GEAR

28. To remove injection pump drive gear, remove timing cover as outlined in paragraph 12. Note alignment marks (A—Fig. 14) on drive gear (P) and coupling (C) and remove pump drive gear from coupling.

Be sure marks (A) are aligned when installing gear on coupling. If a new gear is installed, refer to paragraph 29 for injection pump timing. Tighten gear retaining screws to 40-49 Nm.

INJECTION PUMP

29. **PUMP TIMING.** Fuel injection should occur at 21° BTDC. Timing gear cover must be removed as outlined in

paragraph 12 to check or adjust injection timing.

If original pump drive gear (1—Fig. 13) and coupling (2) are installed, pump timing is correct if marks (M—Fig. 14) on pump drive gear (P) and idler gear (I) are aligned and marks (A) on pump coupling (C) and pump drive gear (P) are aligned. See Fig. 4.

If pump drive gear or coupling has been renewed, spill timing procedure must be used. Remove injection pump delivery valve (8—Fig. 15) for number one cylinder and install a spill tube. Position pump drive gear (1—Fig. 13) on coupling (2) but do not install screws. Align marks (M—Fig. 14) on pump drive gear (P) and idler gear (I). Rotate injection pump drive shaft so drive shaft mark (D) is nearest elongated hole (E). Install injection pump gear screws finger tight. Rotate injec-

tion pump coupling until fuel just ceases to flow and tighten pump gear screws to 39-48 Nm. Mark pump gear and coupling for future reference as shown at (A).

30. **REMOVE AND REINSTALL.** To remove injection pump, disconnect fuel lines and throttle rod and remove timing cover as outlined in paragraph 12. Unscrew coupling (2—Fig. 13) nut and remove coupling and pump gear as a unit. Unscrew pump mounting nuts and remove pump.

To install pump, reverse removal procedure. Tighten pump mounting nuts to 44-48 Nm. Tighten coupling nut to 39-48 Nm. Refer to previous paragraph for pump timing procedure. Bleed fuel lines and pump as outlined in paragraph 27A.

THROTTLE LINKAGE ADJUSTMENT

31. To check and adjust governed speed, start and run engine until normal temperature is reached. Back out high speed stop screw (W—Fig. 16) and detach foot throttle rod from bell crank (B). Adjust high idle speed screw (S—Fig. 16A) to obtain 2650-2700 rpm high idle speed. Turn high speed stop screw (W—Fig. 16) in so screw contacts throttle arm stop when hand throttle is in full throttle position. If stop screw (W) will not contact stop when screwed in or if screw prevents full throttle

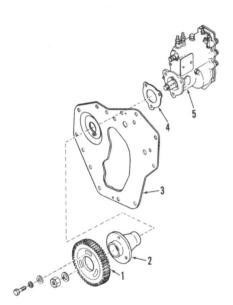

Fig. 13—Exploded view of injection pump gear drive.

1. Gear
2. Coupling
3. Inner plate
4. Gasket
5. Injection pump

Fig. 15—Exploded view of fuel delivery valve. Remove No. 1 fuel delivery valve to spill time engine.

1. Holder retainers
2. Lockwasher
3. Screw
4. Delivery valve holder
5. Spring
6. Gasket
7. Delivery valve
8. Plunger

Fig. 16A—Adjust high idle speed screw (S) to obtain 2650-2700 rpm high idle speed.

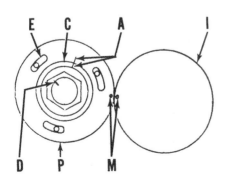

Fig. 14—Diagram of injection pump gear and coupling timing marks. See text.

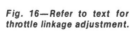

Fig. 16—Refer to text for throttle linkage adjustment.

opening, turn throttle rod turnbuckle (T) to shorten or lengthen throttle rod. Turn low idle speed stop (I—Fig. 17) to obtain low idle speed of 750-850 rpm. Attach foot throttle rod to bellcrank (B—Fig. 16) and adjust turnbuckle (FT) so foot throttle does not affect low or high speed settings.

INJECTION NOZZLES

CAUTION: Fuel leaves injection nozzles with sufficient force to penetrate skin. Care must be used when testing injection nozzles.

32. **TESTING AND LOCATING A FAULTY NOZZLE.** If engine is missing and fuel system is suspected as being the cause of trouble, system can be checked by loosening one and then the other injector line connection while engine is running at slow idle speed. If engine operation is not materially affected when injector line is loosened, that cylinder is missing. Remove and test (or install a new or reconditioned unit) as outlined in appropriate following paragraphs.

33. **REMOVE AND REINSTALL.** Before removing an injector, carefully clean all dirt and other foreign material from lines, injector and cylinder head. Detach injector leak-off line and disconnect injector line from pump and injector. Cap all lines and openings. Unscrew injector from cylinder head.

Fig. 17—Low idle speed should be 750-850 rpm and is adjusted by turning screw (I).

When installing injector, make sure machined seating surface in cylinder head is clean and free from carbon buildup. Use a new copper washer underneath injector nozzle and tighten injector to 59-68 Nm.

34. **TESTING.** Testing and adjusting injector requires special test equipment. Only clean, approved testing oil should be used in tester tank. Nozzle should be tested for opening pressure and leakage.

Before conducting test, operate tester lever until fuel flows, then attach injector to tester. Close valve to tester gage and pump tester lever a few quick strokes to be sure nozzle valve is operating and not stuck thereby requiring overhaul.

35. OPENING PRESSURE. Open tester valve and operate tester lever slowly while observing gage reading. Opening pressure should be 11760 kPa.

Opening pressure is adjusted by adding or deleting shims (7—Fig. 18) which are 0.1 mm thick. Installation or removal of one shims will raise or lower opening pressure approximately 980 kPa.

36. LEAKAGE. To check injector for leakage, operate tester until pressure reading is 9800 kPa. If external leakage is evident, especially at nozzle tip, injector must be disassembled and overhauled.

37. SPRAY PATTERN. Spray pattern should be well atomized and slightly conical, emerging in a straight axis from nozzle tip. If pattern is wet, ragged or intermittent, nozzle must be overhauled or renewed.

38. **OVERHAUL.** Hard or sharp tools, emery cloth, grinding compound or other than approved solvents or lapping compounds must never be used. An approved nozzle cleaning kit is available through a number of specialized sources.

Wipe all dirt and loose carbon from exterior of nozzle and holder assembly. Refer to Fig. 18 for an exploded view of injector.

Secure nozzle nut (2) in a soft jawed vise or holding fixture and remove nozzle body (3). Place all parts in clean calibrating oil or diesel fuel as they are removed. Be sure parts are not mixed if more than one injector is being serviced.

Clean exterior surfaces with a brass wire brush and soak in an approved carbon solvent, if necessary, to loosen hard carbon deposits. Rinse parts in clean diesel fuel or calibrating oil immediately after cleaning to neutralize solvent and prevent etching polished surfaces.

Clean nozzle spray hole from inside using a pointed hardwood stick or wood splinter as shown in Fig. 19. Scrape carbon from pressure chamber using hooked scraper as shown in Fig. 20. Clean valve seat using brass scraper as shown in Fig. 21, then polish seat using

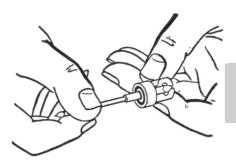

Fig. 19—Use a pointed hardwood stick to clean spray hole as shown.

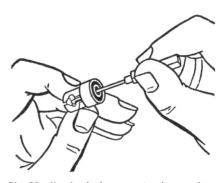

Fig. 20—Use hooked scraper to clean carbon from pressure chamber.

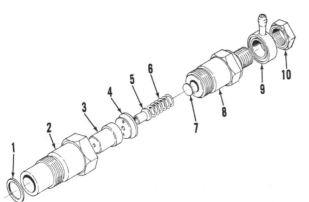

Fig. 18—Exploded view of fuel injector unit.

1. Gasket
2. Nut
3. Nozzle body
4. Nozzle plate
5. Pressure pin
6. Spring
7. Shim
8. Injector body
9. Fitting
10. Nut

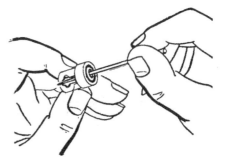

Fig. 21—Clean valve seat using brass scraper as shown.

wood polishing stick and mutton tallow as in Fig. 22.

Back flush nozzle using reverse flusher adapter. Reclean all parts by rinsing thoroughly in clean diesel fuel or calibrating oil and assemble while parts are immersed in cleaning fluid. Make sure adjusting shim pack is intact.

GLOW PLUGS

39. Each cylinder is equipped with a glow plug located on lower right side of cylinder head. Glow plugs are wired in parallel and controlled by key switch. To remove glow plugs, disconnect wiring, detach strap connecting glow plugs and unscrew glow plugs.

COOLING SYSTEM

RADIATOR PRESSURE CAP AND THERMOSTAT

40. A 90 kPa (13 psi) radiator pressure cap is used. The thermostat is located under the coolant outlet elbow on front of cylinder head. Standard thermostat opening temperature is 71° C. Thermostat should be fully opened at 85° C.

RADIATOR

41. To remove radiator, tilt hood forward and drain cooling system. Detach coolant hoses, unbolt lower mounts and support bracket and remove radiator.

Cooling system capacity is 5.3 liters.

WATER PUMP

42. To remove water pump, remove radiator as outlined in paragraph 41 and detach fan from water pump. Loosen alternator retaining bolts and remove fan belt. Unbolt and remove water pump.

Using a suitable puller, pull pump pulley (2—Fig. 23) off shaft (3). Remove rear plate (10) and bearing retaining screw (4). Press shaft and bearing (3), seal (8) and impeller (9) out of housing (6).

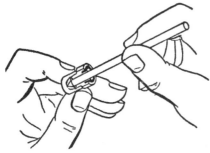

Fig. 22—Polish seat using polishing stick and mutton tallow.

To reassemble water pump, press pump pulley (2) on shaft (3). Press shaft and bearing (3) into housing (6) so bearing groove is aligned with bearing

retaining screw (4) hole. Coat housing side of seal (8) with sealer and install seal in housing. Lubricate impeller side of seal (8) with oil and press impeller

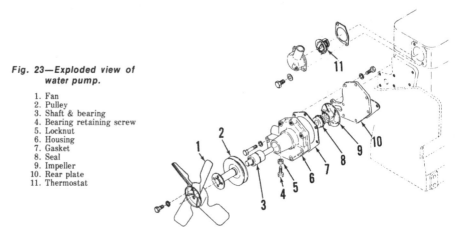

Fig. 23—Exploded view of water pump.

1. Fan
2. Pulley
3. Shaft & bearing
4. Bearing retaining screw
5. Locknut
6. Housing
7. Gasket
8. Seal
9. Impeller
10. Rear plate
11. Thermostat

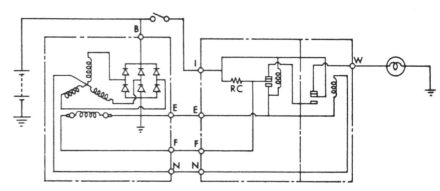

Fig. 24—Alternator and voltage regulator schematic for models prior to serial number U101421.

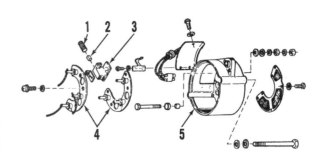

Fig. 25—Exploded view of alternator used on models prior to serial number U101421.

1. Spring
2. Brush
3. Brush holder
4. Diode assy.
5. Frame
6. Stator
7. Collar
8. Key
9. Rotor
10. Slip rings
11. Retainer
12. Bearing
13. Front frame
14. Bearing
15. Retainer

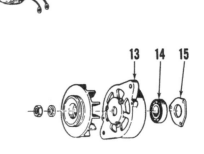

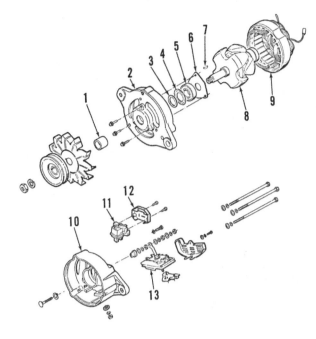

Fig. 26—Exploded view of alternator used on models after serial number U101420.

1. Spacer
2. Front frame
3. Packing
4. Retainer
5. Bearing
6. Bearing retainer
7. Key
8. Rotor
9. Stator
10. Frame
11. Brush assy.
12. Cover
13. Diode assy.

(9) on shaft (3) so clearance between impeller fins and housing is one mm. Install gasket and rear plate (10).

ELECTRICAL SYSTEM

ALTERNATOR AND REGULATOR

43. **ALTERNATOR.** Models 1000 and 1600 are equipped with a 12 volt alternator. Current output is 10 amperes for alternator number 185046020, 35 amperes for alternator number 185046070 and 20 amperes for alternator number 185046071.

CAUTION: Alternator output wire is connected directly with positive battery terminal and is "hot" at all times

battery is connected. Use caution when disconnecting or connecting terminals so output wire is not grounded.

44. **REGULATOR.** On models prior to serial number U101421, voltage regulator unit contains a voltage regulating relay and an indicator lamp control relay. Output voltage is regulated at 13.5-14.2 volts. Refer to Fig. 24 for alternator and regulator schematic.

To adjust voltage regulator on models prior to serial number U101421, refer to Fig. 27 and adjust voltage regulator relay yoke gap (Y) to 0.7 mm, core gap (C) to 1.3 mm and point gap (P) to 0.3 mm. Install regulator and reconnect all wires. Connect a voltmeter and ammeter as shown in Fig. 28. Run engine at approximately 2000 rpm and turn screw (S—Fig. 27) to obtain 13.5-14.0 volts. Ammeter should indicate 10 amperes or less after voltage adjustment.

To adjust indicator lamp relay on models prior to serial number U101421, refer to Fig. 29 and adjust yoke gap (Y) to 1.1 mm, core (C) gap to 1.2 mm and point gap (P) to 1.0-1.3 mm. Install regulator and reconnect all wires. Connect a voltmeter and ammeter as shown in Fig. 30. Start engine. Turn adjusting screw (S—Fig. 29) so that as engine rpm is increased, indicator lamp relay closes at 3-4 volts and one ampere.

Voltage regulator on models after serial number U101420 should regulate voltage between 13.8 and 14.8 volts. Closing voltage for regulator relay is 4.2-5.2 volts measured between "N" terminal of regulator and ground.

STARTER MOTOR

45. The starter motor is on left side of engine and starter solenoid is

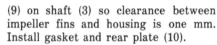

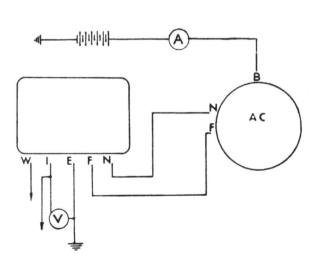

Fig. 27—Adjust voltage regulator on models prior to serial number U101421 as outlined in text.

Fig. 28—View showing connection of ammeter (A) and voltmeter (V) for regulator testing.

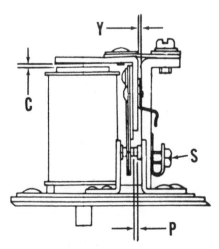

Fig. 29—Adjust indicator lamp relay as outlined in text.

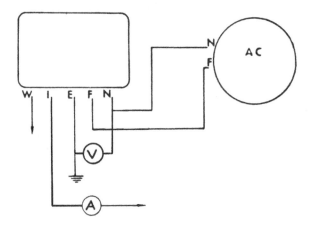

Fig. 30—Connect voltme-
ter (V) and ammeter (A) as
shown and refer to text for
indicator lamp relay adjustment.

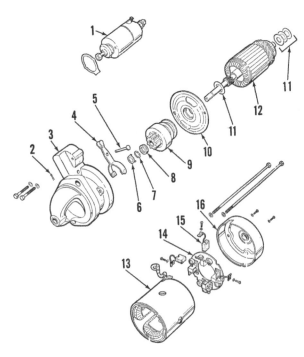

Fig. 30A—Exploded view
of starter motor.

1. Solenoid
2. Cotter pin
3. Drive housing
4. Yoke
5. Yoke pin
6. Washer
7. Wire ring
8. Pinion stop
9. Pinion assy.
10. Bushing & retainer
11. Thrust washers
12. Armature
13. Field assy.
14. Brush holder
15. Brush
16. End cover

CLUTCH

ADJUSTMENT

46. Clutch pedal free play is measured at pedal as shown in Fig. 31.
Clutch pedal free play should be 19-30
mm. To adjust free play, remove clevis
pin (P) and turn rod clevis (C).

A clutch safety switch is located on
bottom of left foot rest adjacent to
clutch pedal. Rotate safety switch
adjustable stop so switch operates
properly.

R&R CLUTCH ASSEMBLY

47. **SPLIT TRACTOR.** To split tractor for clutch removal, proceed as
follows: Remove front weights and
hood. Place supports under engine and
transmission. Disconnect battery cables
and disconnect interfering electrical
wires. Disconnect fuel line and remove
fuel tank. Remove hydraulic pump
suction and pressure lines. Detach
throttle rod and drag link. Unscrew
clutch housing screws and separate
front and rear tractor sections.

To reconnect tractor, reverse splitting procedure. Tighten clutch housing
screws to 54-68 Nm. Refer to paragraph 27A to bleed fuel line.

48. **R&R AND OVERHAUL CLUTCH.**
Split tractor as previously outlined and
alternately unscrew pressure plate
retaining screws.

Inspect clutch disc, pressure plate
and flywheel for damage and excessive
wear. Pressure plate is available as a

attached to top of starter. Refer to Fig.
30A for an exploded view of starter
motor.

Maximum no load starter current
draw is 50 amperes at 12 volts and
5500 starter rpm. Maximum current
draw with starter installed should be
250 amperes.

Fig. 32—Exploded view of
clutch housing assembly.

1. Clutch disc
2. Pressure plate
3. Bearing
4. Bearing holder
5. "O" rings
6. Sleeve carrier
7. Spring
8. Spring bracket
9. Washer
10. Plate
11. Drain plug
12. Rubber plug
13. Bushing
14. Clutch release shaft
15. Plugs
16. Clutch housing
17. Bushing
18. Fork
19. Pin
20. Dowel pin
21. Clutch shaft
22. Sleeve
23. Oil seal

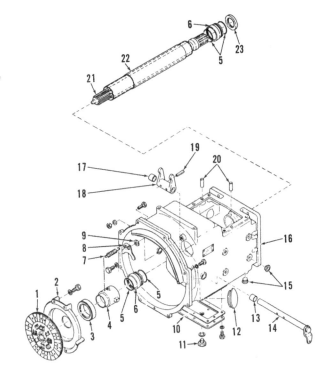

Fig. 31—To adjust clutch pedal free play,
remove clevis pin (P) and turn rod clevis (C)
on rod (R) so free play is 19-30 mm.

unit assembly only. Inspect throw-out bearing assembly.

Tighten pressure plate retaining screws to 17.6-24.5 Nm. To adjust clutch finger height, position tool number SJ-101 so outer legs contact flywheel and center leg is over clutch finger. Turn clutch finger adjusting screw so finger just touches tool center leg.

CLUTCH RELEASE SHAFT AND FORK

49. To remove clutch release shaft and fork, split tractor as outlined in paragraph 47 and remove clutch bearing (3—Fig. 32) and bearing holder (4).

Drive out roll pins (19), withdraw clutch release shaft (14) and remove fork (18). Shaft bushings (13 and 17) are renewable.

CLUTCH SHAFT

50. Clutch shaft turns in sleeve (22—Fig. 32). Sleeve (22) is held in sleeve carriers (6) and sealed by "O" rings (5).

The clutch shaft may be withdrawn after splitting tractor as outlined in paragraph 47. Rear sleeve carrier and "O" rings may be renewed after splitting tractor between clutch and center housings as outlined in paragraph 52.

TRANSMISSION

The sliding gear transmission consisting of three forward speeds and one reverse speed is combined with a range unit providing a total of nine forward and three reverse speeds.

REMOVE AND REINSTALL

51. Service on shift rails, forks and associated parts can be performed after splitting clutch housing from transmission and removing front cover and shift cover. Bevel pinion shaft may be removed without splitting tractor. All other transmission service requires transmission removal; refer to appropriate following paragraphs.

52. **TRANSMISSION SPLIT.** To split tractor between clutch and center housings, proceed as follows: Drain center housing oil and disconnect interfering wires and hydraulic lines. Hydraulic lines and openings should be plugged to prevent contamination. Detach foot throttle and clutch rods. Support center and clutch housings and remove nuts and screws securing center housing to clutch housing. Separate front and rear tractor sections.

To mate tractor sections, reverse split procedure. Tighten center housing to clutch housing screws and nuts to 54-68 Nm. Adjust clutch pedal free

travel as outlined in paragraph 46.

53. **TRANSMISSION REMOVAL.** To remove center housing, refer to paragraph 52 and split center and clutch housing. Securely support rear of tractor and remove final drive gears and differential as outlined in paragraph 56.

TRANSMISSION OVERHAUL

54. **SHIFT COVER, SHIFT FORKS AND SHIFT RAILS.** To remove shift forks and rails, split tractor as outlined in paragraph 52. Unscrew shift cover retaining screws and lift shift cover assembly off center housing. To remove shift levers, drive out roll pins (5—Fig. 33) and pivot pin (4). Remove oil baffle plate (8) and center housing front cover. Remove two spring plungers (9), six detent springs (10, 12 and 13) and four detent balls (11) from top of transmission. Remove roll pins from shifter forks and withdraw shift rails.

NOTE: An interlock pin (20) is located in housing boss between 1st-reserve and 2nd-3rd shift rails—the two center rails.

Pto rail and range rail gates (17—Fig. 34) are interchangeable. Note that detent notches on 1st-reverse rail (21) are further apart than on 2nd-3rd rail (25). Pto rail (24) is identified by detent notches located between fork and gate.

Refer to Figures 33 and 34 during installation of shift assembly. Install interlock pin (20—Fig. 33) after 2nd-3rd shift rail (25) is in position. Note that 1st-reverse rail cannot be inserted unless interlock pin meshes with notch in 2nd-3rd shift rail. Detent spring plungers (9) must be installed above detent springs (10) for pto and range shift rails. Install springs (12 and 13) above each detent ball (11) for 1st-reverse and 2nd-3rd shift rails. Roll pins (15) must not interfere with shift

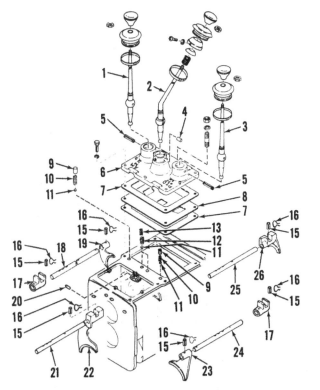

Fig. 33—Exploded view of transmission shift mechanism.

1, 2 & 3. Shift levers
4. Pivot pin
5. Roll pins
6. Cover
7. Gasket
8. Baffle plate
9. Spring plunger
10. Spring
11. Detent ball
12. Spring
13. Spring
15. Roll pin
16. Safety wire
17. Pto & range rail gates
18. Range shift rail
19. Range shift fork
20. Interlock pin
21. 1st & rev. shift rail
22. 1st & rev. shift fork
23. Pto shift fork
24. Pto shift rail
25. 2nd & 3rd shift rail
26. 2nd & 3rd shift fork

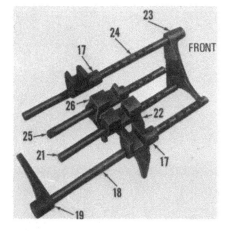

Fig. 34—View of shift rails, gates and forks. Refer to Fig. 33 for parts identification.

levers and should be secured with safety wire (16).

55. GEARS, SHAFTS AND BEARINGS. If transmission gears and shafts must be removed, refer to paragraph 53 and remove center housing. Remove shift rails and forks as outlined in paragraph 54. Refer to Fig. 35 and proceed as follows:

Unscrew nuts (18 and 20) and withdraw pinion (26) and bearing (23) out rear of center housing. Remove gears and bearing as pinion shaft is removed. Withdraw pto front shaft (29) out front of housing. Drive mainshaft (7) out front of housing and remove gears as shaft leaves housing. Remove snap rings (37 and 49) and withdraw countershaft (44) out rear of housing. Drive roll pins (1) out of reverse idler gear shaft (2), withdraw shaft and remove reverse idler gear (4) and bearing (3).

Carefully clean and inspect all parts and thoroughly clean transmission housing. Check all bearings and renew any scored, excessively worn or damaged ball bearings, tapered bearing cones or cups or needle roller bearing assemblies.

Install snap ring (10) in front web upper bore, snap rings (22) in rear web upper bore, snap ring (34) in second web lower bore and snap ring (41) in third web lower bore. Install bearing (35) against snap ring (34) and install bearing (16) in upper bore of third web with outer race snap ring towards front of housing. Install bearings (21 and 23) against snap rings (22). Install mid-range gear (47) and bearing (48) on countershaft (44). Place bearing (42) on high range gear (43) and install against snap ring (41). Insert countershaft through rear of housing and in sequence install components (40, 39, 38, 37 and 36). Install snap ring (49). Place pto gear (30) in front transmission housing compartment. Install needle bearing (8) in mainshaft (7). Install ball bearings (9), snap ring (11), reverse gear (12) and snap ring (13) on mainshaft (7). Install adapter (15) in bearing (16). Insert mainshaft through front of housing and install 2nd and 3rd gear (14) during installation. Install bearing (5) and coupler (6) with larger outside diameter of coupler towards mainshaft.

Refer to paragraph 57 if bevel pinion adjustment is required.

Install bevel pinion shaft (26), shims (24) and washer (25). Tighten inner nut (20) so 1.96-2.45 Nm force is required to turn pinion shaft with a string wrapped around pinion shaft splines.

Install lockwasher (19) and outer nut (18) then bend lockwasher tabs against nuts. Install bearing (28) on pto front shaft (29) with snap ring in outer race towards front. Install snap ring (27) on shaft. Install pto front shaft (29) through front of housing and install pto gear (30). Place concave washer (31), bearing (32) and flat washer on pto shaft so concave side of washer (31) is next to bearing (32). Insert pto shaft in countershaft and install washer (45) and bearing (46). Install reverse idler components (1, 2, 3 and 4). Install shifter assembly as outlined in paragraph 54.

DIFFERENTIAL AND BEVEL GEARS

REMOVE AND REINSTALL

56. Note when using following procedure, left axle and final drive assemblies may remain on tractor unless all components in rear center housing compartment must be removed or differential adjustment is required.

To remove differential and bevel ring gear, drain transmission oil and securely support tractor under center housing. Remove rear wheels, fenders, roll bar, if so equipped, seat and three point hitch linkage. Unscrew pto cover screws and remove pto cover, shaft and bearing holder. Remove brakes, brake drums, axle housings, final drive housings and lift cover as outlined in subsequent paragraphs.

Remove bull pinion covers (25—Fig. 36) and withdraw bull pinions (4). Lay right bull gear as far down in housing as possible. Using jackscrew holes provided, remove right bearing carrier (3) while holding differential assembly in left bearing carrier. Remove right bull pinion gear. Lift out differential assembly. Lay down left bull gear as far as possible in housing and remove left bearing carrier (20) and then bull gear.

Refer to paragraph 58 for overhaul procedure, and to paragraph 55 for removal of bevel pinion shaft. Refer to paragraph 57 for backlash and mesh position adjustment.

Reverse removal procedure to install differential.

MESH AND BACKLASH ADJUSTMENTS

57. Main drive bevel ring gear backlash should be 0.13-0.18 mm. Backlash is controlled by shims (5—Fig. 36) under bearing carrier flanges (3 and 20).

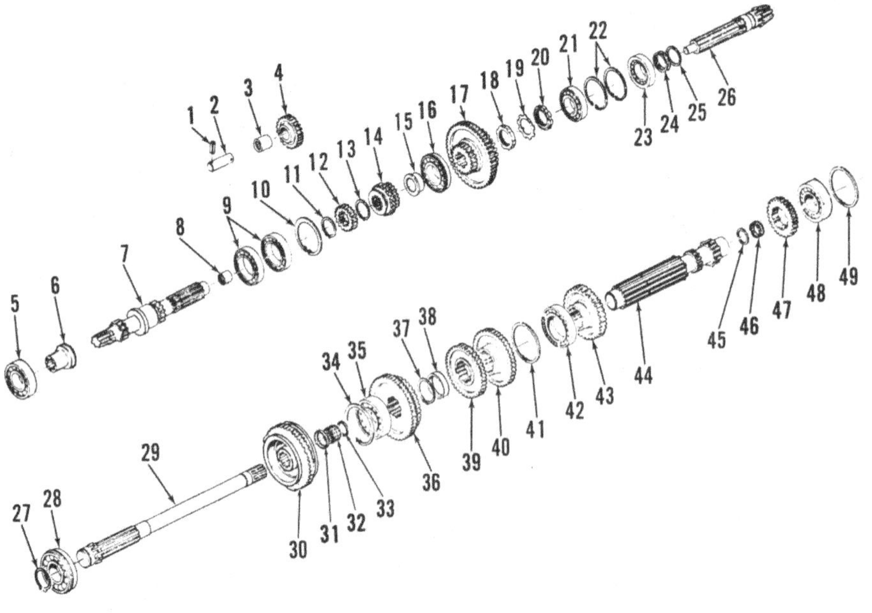

Fig. 35—Exploded view of transmission.

1. Roll pin	14. 2nd & 3rd sliding	25. Washer
2. Idler shaft	gear	26. Bevel pinion shaft
3. Bearing	15. Adapter	27. Snap ring
4. Reverse idler gear	16. Bearing	28. Bearing
5. Bearing	17. Sliding range gear	29. Pto front shaft
6. Coupler	18. Nut	30. Pto gear
7. Mainshaft	19. Lockwasher	31. Concave washer
8. Bearing	20. Nut	32. Bearing
9. Bearings	21. Bearing	33. Flat washer
10. Snap ring	22. Snap rings	34. Snap ring
11. Snap ring	23. Bearing	35. Bearing
12. Reverse gear	24. Shim	36. 1st & rev. gear
13. Snap ring		

37. Snap ring	43. High range gear
38. Spacer	44. Countershaft
39. 3rd gear	45. Washer
40. 2nd gear	46. Bearing
41. Snap ring	47. Mid-range gear
42. Bearing	48. Bearing
	49. Snap ring

Proper meshing of main drive bevel gears depends on position of bevel pinion gear as well as bevel ring gear backlash. Pinion adjustment is controlled by shims (24—Fig. 35).

If no parts are changed when reassembling differential, install original shims or assemble a trail shim pack of equal thickness. Temporarily install differential assembly and adjust backlash. Paint pressure side of bevel ring gear teeth and rotate pinion in normal direction. Check tooth contact pattern using Figs. 37, 38 and 39 as guides. Contact should center on tooth profile as shown in Fig. 37. If heavy contact area is at heel or tip of tooth as shown in Fig. 38, move bevel pinion gear toward ring gear by removing shims, then reset backlash. If heavy contact area is at base or toe of tooth as shown in Fig. 39, move bevel pinion gear away from ring gear teeth by adding shims, then reset backlash.

OVERHAUL

58. Refer to Fig. 36 for an exploded

view of differential assembly. To disassemble differential, unscrew ring gear retaining screws and remove ring gear. Unscrew differential case screws and separate differential case halves (7 and 18). Remove pinions, shafts, thrust washers and side gears from case halves.

Minimum allowable diameter of pinion shafts (15 and 16) is 17.9 mm. Minimum allowable thickness of side gear thrust washers (9) is 0.1 mm. Minimum allowable thickness of pinion thrust washers (12) is 1.1 mm.

To assemble differential, install thrust washers and side gears in differential case halves. Install long pinion shaft (15), joint (17), pinion gears and thrust washers in right differential case half (7) with notches in pinion shaft ends parallel with case mating surface. Install short pinion shafts (16), thrust washers and pinion gears in right differential case half (7) with notches in pinion shaft ends parallel with case mating surface. Join right and left differential case halves so left differ-

ential case half fits over notches in pinion shaft ends. Tighten differential case screws to 25-34 Nm. Tighten ring gear retaining screws to 49-58 Nm.

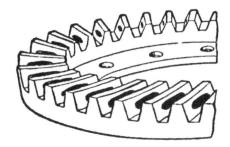

Fig. 37—Tooth contact pattern should center on tooth profile as shown.

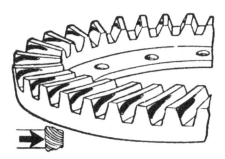

Fig. 38—If heavy contact area is at heel or tip of tooth, remove shims as shown.

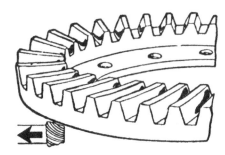

Fig. 39—If heavy contact area is at base of tooth, add shims as shown.

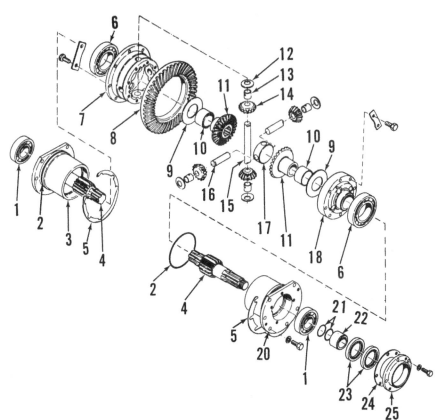

Fig. 36—Exploded view of differential and carrier assembly.

1. Bearing	7. Right differential
2. "O" ring	case
3. Right bearing	8. Bevel ring gear
carrier	9. Thrust washer
4. Bull pinion	(2 mm)
5. Shim	10. Bushing
6. Bearing	11. Side gear

12. Thrust washer	18. Left differential
13. Bushing	case
14. Pinion gear	20. Left bearing carrier
15. Long pinion shaft	21. "O" rings
16. Short pinion shaft	22. Bushing
17. Joint	23. Oil seal
	24. Gasket
	25. Retainer

DIFFERENTIAL

LOCK

OPERATION

59. The differential lock mechanism is located in transmission housing below differential. Depressing actuating pedal (1—Fig. 40) will rotate shaft and actuating pin (3). Actuating pin (3) rotates against ramp on fork (4) and as fork is forced to left, it engages coupling (8) with left bull gear. Coupling is mounted on right axle splines so that right and left axles turn together when coupling engages left bull gear.

R&R AND OVERHAUL

59A. To remove differential lock mechanism, drain transmission oil and remove seat and lift cover assemblies. Remove pin (6—Fig. 40 or 41) and washer (7). Remove actuating pin (3). Hold return spring (5) and fork (4) while withdrawing pedal and shaft (1) from transmission housing. To remove coupling (8), refer to paragraph 62 and remove right axle.

If coupling was removed, install coupling on right axle with coupling dogs towards axle end and install right axle assembly. Complete remainder of reassembly by reversing disassembly procedure.

FINAL DRIVE

REMOVE AND REINSTALL

60. To remove axle housings, drain transmission oil and support center housing. Remove seat, roll bar, if so equipped, fender, rear wheel and any interfering three-point hitch linkage. Disconnect brake linkage and remove brake drum as outlined in paragraph 64. Remove hydraulic lift cover as outlined in paragraph 78. Remove snap ring (2—Fig. 42), unbolt axle housing and separate axle housing from center housing.

If both axle assemblies have been removed, the right axle assembly should be installed first. Note that right axle end must pilot in bushing in left axle end. Be sure to install differential lock coupling on right axle with coupling dogs towards axle end. Tighten final drive housing mounting screws to 54-68 Nm.

OVERHAUL

61. **FINAL DRIVE BULL GEARS.** To remove left bull gear, remove both axle housings as outlined in paragraph 60. Remove left bull pinion cover and withdraw left bull pinion. Lay left bull gear down in housing and remove left bearing carrier (20—Fig. 36). Remove left bull gear from housing.

To remove right bull gear, remove differential as outlined in paragraph 56.

62. **AXLE SHAFT AND BEARINGS.** To remove axle shaft and bearings, remove axle housing as outlined in paragraph 60. Unscrew cover (14—Fig. 42) retaining screws.

Using a suitable press, remove axle and outer bearing assembly from inner bearing (4). Remove collar (7) and washer (8) and press bearing (9) off axle. Remove remaining components on axle.

When reassembling axle components, install seal (11) so large flange is towards bearing (9).

BRAKES

ADJUST

63. Individual shoe-type brakes should apply fully when total movement measured at pedal pad is 20-30 mm. To adjust brake, loosen locknut (L—Fig. 43) and rotate brake rod (R).

Both brakes must engage equally when pedals are depressed at same time or when both pedals are locked together.

R&R AND OVERHAUL

64. To remove either brake assembly, disconnect brake rod and remove cap screws retaining brake cover. Brake shoes and actuator will be removed with brake cover and can be examined or renewed at this time. Unscrew nut securing brake drum to bull pinion and remove brake drum.

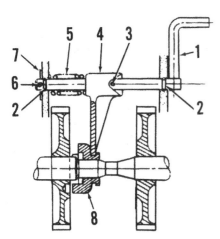

Fig. 40—Cross-sectional view of differential lock.

1. Pedal & shaft	5. Spring
2. "O" rings	6. Roll pin
3. Pin	7. Washer
4. Fork	8. Coupling

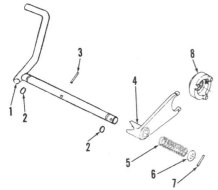

Fig. 41—Exploded view of differential lock. Refer to Fig. 40 for parts identification.

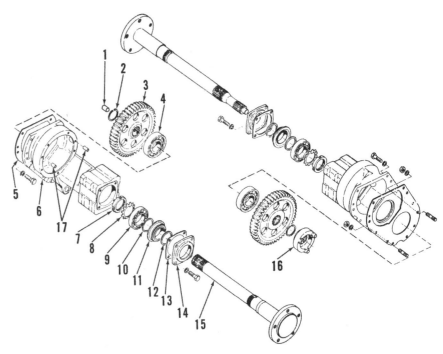

Fig. 42—Exploded view of final drive. Bushing (1) seats in left axle (15) end to support right axle end.

1. Bushing		14. Cover
2. Snap ring	6. Final drive housing	15. Left axle
3. Bull gear	7. Collar	10. Spacer	16. Differential lock
4. Bearing	8. Tab washer	11. Seal	coupling
5. Gasket	9. Bearing	12. Spacer	17. Dowel pin
		13. Gasket	

Inspect brake shoes and actuating mechanism. Renew brake drum if inner diameter is more than 170 mm.

Install brakes by reversing removal procedure. Adjust brakes as outlined in paragraph 63.

POWER TAKE-OFF

65. To remove pto shaft assembly, unscrew pto cover screws and remove cover. Withdraw pto shaft assembly. Pry bearing retainer (3—Fig. 44) from center housing. Press the two bearings off pto shaft. Remove oil seal (7) from cover (9).

To reassemble pto components, press bearings on pto shaft and install oil seal (7) in cover (9). Install gasket (2) and bearing retainer (3) in center housing. Place coupling (1) on pto shaft and insert shaft in bearing retainer being sure coupling (1) engages pto front shaft. Install pto cover (9).

HYDRAULIC SYSTEM

LUBRICATION

66. Hydraulic oil is contained in clutch housing. Hydraulic system oil capacity is 10 liters. Check oil level with dipstick (20—Fig. 3). Oil level should be between mark and dipstick end with tractor level and engine off. A drain plug is located in the plate attached to underside of clutch housing. Oil should be changed after every 600 hours of operation. A filter screen is located at end of pick-up tube.

OPERATION

67. Hydraulic system oil is contained in reservoir (S—Fig. 45) in clutch housing. Oil pump (P) mounted on front engine cover pumps oil to control valve attached to right side of lift cover. When control valve spool (V) is in neutral position, spool blocks pressurized oil from pump which forces unloading valve (U) to open and route oil into return line. When control valve spool (V—Fig. 45A) is in lowering

position, oil flows from lift cylinder (L), through port opened by spool and through flow control valve (F) after which oil returns to reservoir. Oil flow through flow control valve (F) may be varied by turning knob (K—Fig. 46). Time required to lower lift cylinder is increased when oil flow is restricted. When control valve spool (V—Fig. 45B) is in raised position, oil is routed by valve spool to rear of unloading valve (U) which closes unloading valve. Check valve (C) is forced open and pressurized oil flows to lift cylinder (L). Relief valve (R) opens when oil pressure reaches 9457-9800 kPa.

Control valve spool (V—Fig. 47) is actuated by lever (R) which is attached at its upper end to lift control cam (T) and pivots over a fulcrum carried on position control cam (M). Moving lift control lever (L) will rotate lift control cam (T) which moves actuating lever (R) against or away from control valve (V). Rockshaft is raised or lowered as control valve (V) moves in or out. As rockshaft moves, position control linkage repositions the fulcrum on position control cam (M) which acts on lever (R)

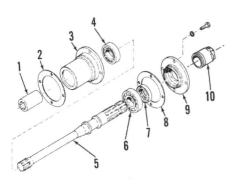

Fig. 43—Loosen locknut (L) and rotate brake rod (R) to adjust brake.

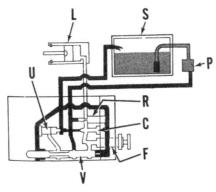

Fig. 45—Flow diagram for hydraulic lift circuit in neutral position.

C. Check valve
F. Flow control valve
L. Lift cylinder
P. Pump
R. Relief valve
S. Reservoir
U. Unloading valve
V. Control valve spool

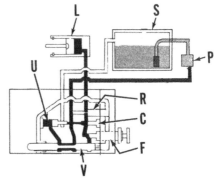

Fig. 45B—Flow diagram for hydraulic lift circuit in raise position.

Fig. 44—Exploded view of pto shaft assembly.

1. Coupling
2. Gasket
3. Bearing retainer
4. Bearing
5. Pto shaft
6. Bearing
7. Oil seal
8. Gasket
9. Cover
10. Cap

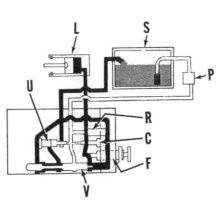

Fig. 45A—Flow diagram for hydraulic lift circuit in lower position.

Fig. 46—Turn flow control valve knob (K) to vary lowering rate.

and returns control valve (V) to neutral when rockshaft movement has been completed.

The mechanical draft control mechanism shown in Fig. 48 is used on some models. Draft control will not function if lock pin (P—Fig. 49) is inserted through holes in rocker arm and lift cover flanges. If lock pin is not installed, draft control will operate as follows: Should resistance against implement increase, implement will try to rotate around hitch pins which will force upper link (N—Fig. 48) towards tractor. Force from upper link (N) rotates rocker arm (R) forward. When rocker arm (R) rotates forward, lever (L) rotates rearward and pulls against depth link (D). Depth link (D) is connected to front of left lift arm (A). As front of lift arm (A) is pulled down by depth link (D), lift arm rockshaft rotates and both lift arms will lift the implement thereby transferring weight to rear wheels.

HYDRAULIC PRESSURE CHECK

68. System relief pressure should be 9457-9800 kPa. Pressure can be checked by installing a suitable test gage in pipe plug opening (P—Fig. 50) located in ram cylinder cap.

NOTE: Test port has British pipe threads. An adapter with 3/8-24 threads will fit with minimal leakage for testing purposes. Thread sealer should be used if a leak free fit is desired.

Tie lift arms down, run engine at 1000 rpm then move control lever to pressurize lift cylinder. Adjust relief pressure by adding or deleting relief valve shims (20—Fig. 53).

ADJUSTMENT

69. **FLOW CONTROL.** Oil flow through flow control valve (F—Fig. 45A) may be varied by turning knob

Fig. 48—View of draft control mechanism. Refer to paragraph 67 for operation.

    A. Lift arm
    D. Depth link
    K. Depth adjustment knob
    L. Lever
    N. Upper link
    R. Rocker arm

(K—Fig. 46). Lift arms will drop slower if knob is turned clockwise or drop faster if knob is turned counterclockwise.

70. **POSITION CONTROL.** Before adjusting position control, be sure relief valve operates at pressure specified in paragraph 68. Start engine and move lift control lever to full raised position. With lift arms raised to highest position, adjust position control link (L—Fig. 51) to shortest length which will NOT cause relief valve to open.

71. **DRAFT CONTROL.** Maximum implement depth on models with mechanical draft control is determined by length of depth adjustment link (D—Fig. 48). To adjust maximum implement depth, turn depth adjustment knob (K).

HYDRAULIC OIL PUMP

72. **TESTING.** Gear-type hydraulic pump is located on engine timing gear cover and driven by idler gear (2—Fig. 6). Specified pump output is 15.9 liters/min. at 9800 kPa and 2500 engine rpm.

73. **REMOVE AND REINSTALL.** To remove hydraulic oil pump, remove timing gear cover as outlined in paragraph 12. Unscrew pump drive gear

nut and remove drive gear (16—Fig. 52). Remove pump retaining nuts and separate pump from timing gear cover.

To reinstall hydraulic pump, reverse removal procedure.

74. **OVERHAUL.** Remove hydraulic pump as outlined in paragraph 73. To disassemble pump, remove end cover (1—Fig. 52) and remove components from pump housing (12). Inspect components for damage and excessive wear. Install new "O" rings and seals. Note that notches (N) on bearings (5 and 10) must be towards gears (6 and 7).

CONTROL VALVE

75. **REMOVE AND REINSTALL.** To remove control valve, detach and cap pressure line to valve. Remove

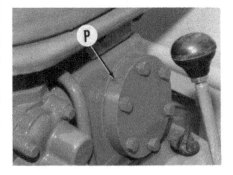

Fig. 50—Remove pipe plug (P) and install gage as outlined in paragraph 68 to check hydraulic system pressure.

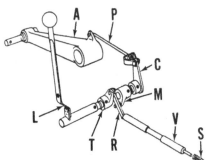

Fig. 47—Diagram of position control linkage. Refer to paragraph 67 for operation.

    A. Lift arm
    C. Position control arm
    L. Lift control lever
    M. Position control cam
    P. Position control link
    R. Actuating lever
    S. Spring
    T. Lift control cam
    V. Control valve spool

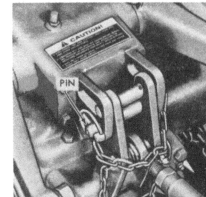

Fig. 49—Install lock pin to disable draft control linkage.

Fig. 51—Adjust position control link (L) by following procedure in paragraph 70.

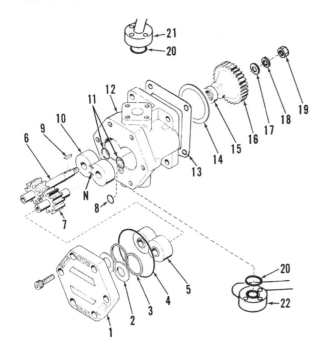

Fig. 52—Exploded view of
hydraulic oil pump. Note
notches (N) on bearings (5
and 10) must be toward
gears.

1. End cover
2. Washer
3. Seals
4. "O" ring
5. Bearing
6. Drive gear
7. Driven gear
8. "O" ring
9. Key
10. Bearing
11. "O" rings
12. Pump housing
13. Gasket
14. Aligning spacer
15. Oil seal
16. Gear
17. Washer
18. Lockwasher
19. Nut
20. "O" rings
21. Outlet oil line
22. Inlet oil line

control lever assembly and disconnect position control link (Fig. 51). Remove three screws securing control valve to lift cover and remove valve.

Use new "O" rings (28—Fig. 53) when reinstalling and reverse removal procedure. Refer to paragraph 70 if adjustment of position control link is required.

76. OVERHAUL. Refer to Fig. 53 for an exploded view of control valve. Control valve, check valve and relief valve assemblies can be removed after removing front cover (1). Control mechanism and unloading valve can be removed after removing rear cover. Flow control valve is removed from side of valve body.

Control valve spool (37) must move freely in valve bushing (17). Spool (37) and bushing (17) are available only as a matched assembly.

LIFT ARMS

77. Prior to lift arm removal, mark lift arm and rockshaft so lift arm can

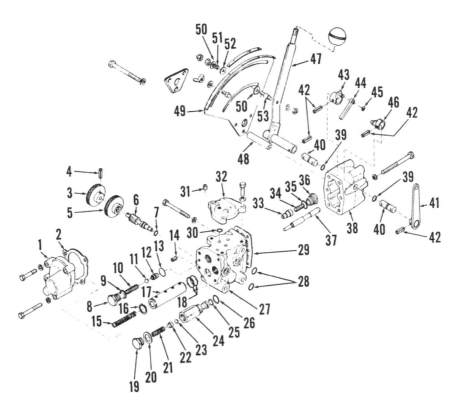

Fig. 53—Exploded view of control lever and control valve assemblies.

1. Front cover
2. Gasket
3. Flow control knob
4. Roll pin
5. Lock knob
6. Flow control valve
7. "O" ring
8. Plug
9. "O" ring
10. Spring
11. Check valve ball
12. Check valve seat
13. "O" ring
14. Set screw

15. Spring
16. Snap ring
17. Spool bushing
18. "O" rings
19. Plug
20. Shim
21. Spring
22. Relief valve seat
23. Relief valve ball
24. Relief valve body
25. "O" ring
26. "O" ring
27. Control valve body

28. "O" rings
29. Gasket
30. "O" ring
31. Plug
32. Top cover
33. Unloading valve
34. Spring
35. "O" ring
36. Plug
37. Control valve spool
38. Rear cover
39. "O" rings
40. Shafts

41. Position control arm
42. Roll pins
43. Lift control cam
44. Actuating lever
45. Snap ring
46. Position control cam
47. Lift control lever
48. Spacer
49. Quadrant
50. Washer
51. Spring
52. Washer
53. Stud

be installed in original position. To remove lift arm, disconnect linkage and remove snap ring (21 or 30—Fig. 54) on rockshaft end. When installing lift arm, be sure to align marks on lift arm and rockshaft or movement may be reduced.

LIFT COVER AND ROCKSHAFT

78. REMOVE AND REINSTALL. To remove lift cover, first remove seat assembly from cover. Disconnect oil lines from control valve and lift cover. Detach lift rods from lift arms. Unscrew screws retaining lift cover and separate lift cover assembly from center housing.

To install lift cover assembly, reverse removal procedure and tighten the retaining screws to 54-68 Nm.

79. OVERHAUL. Refer to Fig. 54 for an exploded view of lift cover assembly. If lift arms have not been removed, mark lift arms and rockshaft for future alignment and remove lift arms. Remove snap ring (16) adjacent to arm (13). Note alignment marks on arm (13) and rockshaft (17). Withdraw rockshaft (17) from lift cover and arm. Remove arm (13) and piston rod (10). To remove piston (8) and sleeve (4), remove cap (1) and extract sleeve with piston. Remove snap ring (9) and remove piston from sleeve.

Inspect components and renew any which are damaged or excessively worn. Reassemble lift cover assembly by reversing disassembly procedure. Be sure marks on rockshaft (17) and arm (13) are aligned. Tighten cap (1) retaining screws to 54-68 Nm.

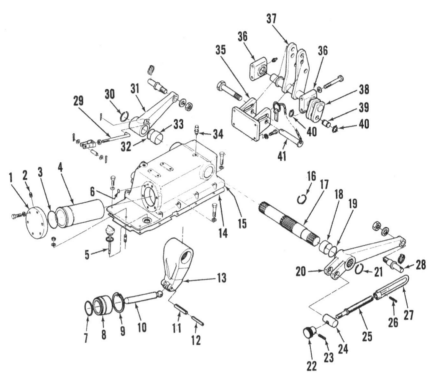

Fig. 54—Exploded view of hydraulic lift assembly.

1. Cap	12. Roll pin	22. Depth adjustment knob	32. "O" ring
2. Plug	13. Arm	23. Roll pin	33. Bushing
3. "O" ring	14. Gasket	24. Pivot pin	34. Breather
4. Sleeve	15. Lift cover	25. Depth link	35. Bracket
5. Dipstick	16. Snap ring	26. Roll pin	36. Bushing block
6. "O" ring	17. Rockshaft	27. Guide	37. Rocker arm & shaft
7. "O" ring	18. Bushing	28. Link pin	38. Lever
8. Piston	19. "O" ring	29. Position control link	39. Pivot pin
9. Snap ring	20. Left lift arm	30. Snap ring	40. Snap rings
10. Piston rod	21. Snap ring	31. Right lift arm	41. Lock pin
11. Roll pin			

FORD

Models ■8000 ■8600 ■8700 ■9000 ■9600 ■9700
■TW-10 ■TW-20 ■TW-30

Previously contained in I&T Shop Service Manual No. FO-39

FORD

SHOP MANUAL

MODELS 8000—8600—8700—9000—9600—9700—TW-10—TW-20—TW-30

These Ford tractors are equipped with a six-cylinder diesel engine. The 9000 series, TW-20 and TW-30 models engines are equipped with a turbocharger and due to increased power output, several components on these models have been strengthened in comparison to the 8000 series and TW-10 models. An eight-speed gear type transmission and disc type clutch is standard on all models. A Dual Power planetary gear assembly which provides under-drive ratios in all transmission speeds is standard on TW-30 models and available for all other models.

Identification numbers pertaining to Models 8000, 8600, 9000 and 9600 are located on a plate inside the tool box cover.

Numbers pertaining to Models 8700 and 9700 are located on a plate mounted on the underside of the radiator filler cap access door.

Numbers pertaining to Models TW-10 and TW-20 are on a plate located above the right front corner of the radiator and are accessible after removing the right front grill panel.

Numbers pertaining to Model TW-30 are on a plate located between the radiator and front fuel tank; they are accessible by sliding the right front grill panel forward.

INDEX (By Starting Paragraph)

INDEX CONT.

CONDENSED SERVICE DATA

GENERAL	8000 8600, 8700	9000 9600, 9700	TW-10	TW-20	TW-30
Torque Recommendations			See End of Shop Manual		
Engine make	Own	Own	Own	Own	Own
No. of Cylinders	6	6	6	6	6
Bore, Inches	4.4	4.4	4.4	4.4	4.4
troke, Inches	4.4	4.4	4.4	4.4	4.4
Displacement, Cubic Inches	401	401	401	401	401
Compression Ratio	16.5:1	16.5:1	16.3:1	15.6:1	15.6:1
Pistons Removed From	Above	Above	Above	Above	Above
Main Bearings, Number of	7	7	7	7	7
Cylinder Sleeves	None	None	None	None	None
Generator Make	Own	Own	......	......	......
Alternator Make			Motorola		
tarter Make	Own	Own	Own	Own	Own
Turbocharger Make			See Paragraph 134		
Injection Pump Make	Simms	Simms	Minimec	Minimec	Minimec
Forward Speeds	8	8	8	8	......
With Dual Power	16	16	16	16	16
Reverse Speeds	2	2	2	2	......
With Dual Power	4	4	4	4	4

CONDENSED SERVICE DATA CONT.

TUNE-UP	8000, 8600, 8700	9000, 9600, 9700	TW-10	TW-20, TW-30
Firing Order		1-5-3-6-2-4		
Compression, Gage Lbs. at				
Cranking Speed of 200 Rpm	380-480	380-480	300-400	275-375
Maximum Allowable Variation				
Between Cylinders, Psi	20	20	20	20
Valve Tappet Gap, Intake, Hot		0.014-0.016 inch		
Valve Tappet Gap, Exhaust,				
Hot		0.017-0.019 inch		
Engine Low Idle Rpm	700-800	700-800	700-800	700-800
Engine High Idle Rpm	2530-2580	2420-2470	2530-2580	2420-2470
Engine Rpm at Rated Load	2300	2200	2300	2300
Engine Rpm for 540 Pto Rpm	1900	*1900	1873	*1873
Engine Rpm for 1000 Pto Rpm	1935	1935	1918	1918
Injection Timing		23° BTDC		
Battery Terminal Grounded		Neg.		

*540 rpm pto shaft not available on 9000, TW-30 and some 9600 models.

SIZES—CAPACITIES—CLEARANCES [INCHES]				
Crankshaft Journal Diameter		See Paragraph 98		
Crankpin Diameter		See Paragraph 97		
Camshaft Journal Diameter		2.3895-2.3905		
Piston Pin Diameter	1.4997-1.5000	1.6246-1.6251	1.4997-1.5000	1.6246-1.6251
Valve Stem Diameter, Intake		0.3711-0.3718		
Valve Stem Diameter, Exhaust		0.3701-0.3708		
Main Bearing Running				
Clearance		0.002-0.0045		
Rod Bearing Running Clearance,				
Aluminum Bearings		0.0025-0.0046		
Copper-Lead Bearings		0.0017-0.0038		
Camshaft Bearing Running				
Clearance		0.001-0.003		
Crankshaft End Play		0.004-0.008		
Camshaft End Play		0.001-0.007		
Piston Skirt Clearance		See Paragraph 95		
Cooling System, Quarts	18	19	20	20
Crankcase, Quarts, Without				
Filter Change	10	12	18	18
With Full Flow Filter				
Change Only	12	14	20	20
With Both Filters Changed	13 (8000 only)			
Rear Axle & Hydraulic Systems,				
Quarts		See Paragraph 197		
Power Take-Off, Quarts		See Paragraph 217		
Transmission, Quarts		See Paragraph 197		
Fuel, Gallons	43	43	58	**100
Power Steering System, Quarts	4.2	4.2	3.9	3.9

**TW-20 fuel capacity is 58 gallons.

FRONT SYSTEM AND STEERING
MODELS 8000—8600—9000—9600

WIDE ADJUSTABLE FRONT AXLE

Exploded view of row crop axle is shown in Fig. 1; all purpose type is similar except center (main) member is reversed. All parts except center steering arm (13) and inner tie rod ends (12) are interchangeable between row crop and all purpose types. Refer to Figs. 2 and 3 for center steering arm and tie rod installation.

1. **R&R FRONT AXLE ASSEMBLY.** To remove either the row crop or all purpose type front axle, proceed as follows: Straighten tabs of locking plates and unbolt center steering arm (13—Fig. 1) from steering motor shaft. Support front end of tractor, unbolt front pivot pin support (21) from front support and roll axle assembly forward. Unbolt and remove rear pivot pin support (16).

Reinstall front axle by reversing

removal procedure, making sure a thrust washer (17) is placed on each pivot pin. Tighten pivot pin support and center steering arm cap screws to a torque of 180-220 ft.-lbs., then bend locking plates against steering arm cap screw heads.

2. **SPINDLE BUSHINGS.** To renew spindle bushings, support front of tractor and remove front wheels. Remove steering arm clamp bolts on models so equipped, pull arms from spindles and remove Woodruff keys (8—Fig. 1). On models equipped with spindles having splined end is shown in Fig. 1, remove retaining nut (24) and pull steering arm from spindle. Withdraw spindles downward out of axle extensions. Remove seals (7) and thrust bearing (3). Remove thrust bearing spacers (2) if worn or damaged. Drive bushings from axle extensions and install new bushings using piloted drift or driver; bushings are pre-sized and should not require reaming if carefully installed. Be sure grease holes are aligned. Pack thrust bearings with grease.

NOTE: Upper ends of early spindles with a key were 1½ inches in diameter whereas late spindles with a key are approximately 1-5/8 inches. Only the large diameter spindle is available. If renewing small spindle, a new large diameter steering arm must also be installed. Be sure correct size spindle seal is installed. Tighten steering arm clamp bolts on models equipped with keyed spindles to a torque of 135-165 ft.-lbs. On models with splined spindles, tighten retaining nuts (24—Fig. 1) to a torque of 200-250 ft.-lbs. Stake spindle threads to nut.

3. **AXLE CENTER MEMBER, PIVOT PINS AND BUSHINGS.** To remove axle center member (20—Fig. 1), support front of tractor, remove clamp bolts from tie rods and unbolt axle extensions from center member. Withdraw axle extensions with spindles and wheels from center member and tie rod sleeves. Unbolt and remove center steering arm with tie rods. Support center member and unbolt front pivot pin support. Move center member and front pivot forward until clear of rear pivot pin and lower to floor. Unbolt and remove rear pivot pin and support.

Pivot pins are integral parts of pivot pin supports; renew the pin and support assembly if pin is excessively worn. Renew bushings (18) in center member if member is otherwise serviceable; bushings are pre-sized and should not require reaming if carefully installed. Be sure grease holes are aligned and install plug (19) in bore through axle tube. Renew thrust washers (17) if worn.

Reinstall by reversing removal procedure. Tighten pivot pin support and center steering arm cap screws to a torque of 180-220 ft.-lbs., then bend locking plates against steering arm cap screw heads. Tighten tie rod clamp bolts to a torque of 25-35 ft.-lbs. and check toe-in as outlined in paragraph 4.

4. **TIE RODS AND TOE-IN.** Tie rod ends are of the non-adjustable automotive type and procedure for renewing same is evident. Tighten clamp bolts on outer tie rod ends to a torque of 25-35 ft.-lbs. and tighten jam nut at inner end of tie rods to a torque of 100-125 ft.-lbs.

Toe-in on both row crop and all purpose models should be ¼- to ½-inch. Position center steering arm at center line of tractor as shown in Fig. 2 or Fig. 3 before checking toe-in at spindle height. If toe-in is not correct, remove

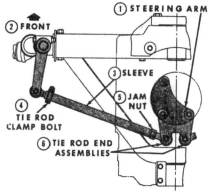

Fig. 2—When checking toe-in on row crop wide adjustable axle, be sure steering arm is centered as shown. View is from top. When installing center steering arm, be sure steering motor is centered and install arm as shown.

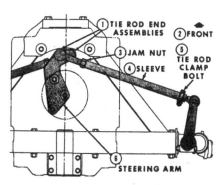

Fig. 3—View showing position of all purpose type center steering arm for checking toe-in. Note that center arm (6) and tie rod inner ends are different than those used on row crop front end. View is from top.

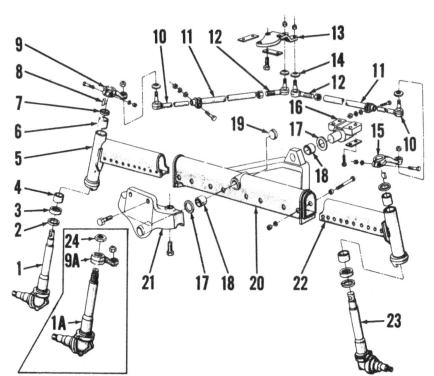

Fig. 1—Exploded view of row crop type wide adjustable front axle. All purpose type front axle is similar except that center member (20) is reversed to provide shorter wheelbase. Note that later models are equipped with spindles (1A) and steering arms are retained by a nut as shown in inset. All parts except center steering arm (13) and inner tie-rod ends (12) are interchangeable between row crop and all purpose types. Refer to Figs. 2 and 3 for center steering arm views.

1. & 1A. Spindle, R.H.		19. Plug
2. Thrust spacer	8. Woodruff key	20. Axle center member
3. Thrust bearing	9. & 9A. Steering arm	21. Front pivot pin &
4. Lower bushing	10. Tie rod ends, outer	support
5. Axle extension, R.H.	11. Tie rod tube	22. Axle extension, L.H.
6. Upper bushing	12. Tie rod ends, inner	23. Spindle, L.H.
7. Seal	13. Steering arm, center	24. Nut
	14. Dust seals	
	15. Steering arm, L.H.	
	16. Rear pivot pin &	
	support	
	17. Thrust washers	
	18. Pivot pin bushings	

clamp bolts and loosen jam nuts on both tie rods, then turn each tie rod sleeve an equal amount as necessary. Refer to preceding paragraph for tightening torques.

TRICYCLE FRONT SPINDLE

5. The dual wheel tricycle spindle is bolted directly to the power steering motor shaft; procedure for removing and installing spindle is obvious. Tighten spindle to steering motor shaft cap screws to a torque of 209-231 ft.-lbs. Spindle can be installed on steering motor shaft in one position only due to offset bolt holes.

FRONT SUPPORT (PEDESTAL)

6. To remove front support, first remove steering motor assembly as outlined in paragraph 22. Remove wide front axle assembly as outlined in paragraph 1. Unbolt and remove side plates from front support and transmission. Attach hoist to front support, then unbolt front support from engine cylinder block and oil pan; be careful not to lose shims on the two oil pan bolts and label shims for reinstallation, if same pedestal, oil pan and cylinder block are to be reinstalled.

If front support, oil pan, and/or engine cylinder block have been renewed, it will be necessary to select shim thickness for installing front support as follows: Install the three bolts and one cap screw retaining front support to cylinder block and tighten to a torque of 180-220 ft.-lbs. Install the two cap screws retaining front support to oil pan, tighten them to a torque of 180-220 ft.-lbs., then measure gap between front support and oil pan at the two bolting points using a feeler gage. Remove the two front axle support to oil pan cap screws, then reinstall with shims equal to measured gap and tighten cap screws to a torque of 270-330 ft.-lbs. Shims are available in thicknesses of 0.015, 0.018, 0.021, 0.024 and 0.027 inch.

If front support is being reinstalled with same engine cylinder block and oil pan, reinstall shims as removed on the two front support to oil pan cap screws. Tighten the front support to cylinder block bolts and cap screws to a torque of 180-220 ft.-lbs., then tighten the two front support to oil pan cap screws to a torque of 270-330 ft.-lbs.

HYDROSTATIC POWER STEERING SYSTEM

CAUTION: The maintenance of absolute cleanliness is necessary when servicing the hydrostatic power steering

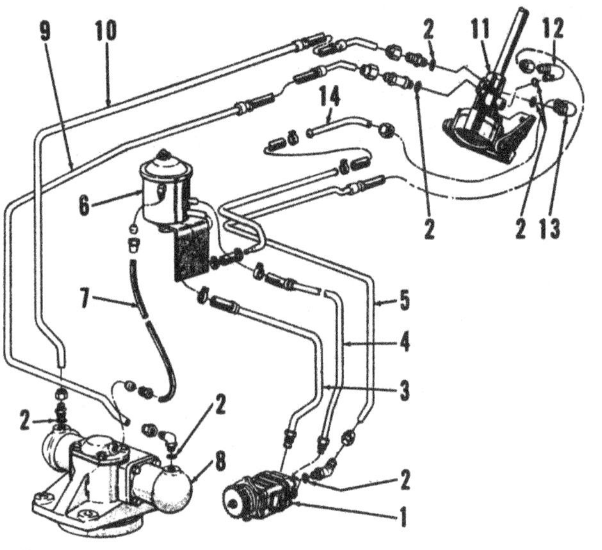

Fig. 4—View showing hydrostatic power steering system used on 8000 and 9000 models and early 8600 and 9600 models. Note that tube (9) from left end of steering motor (8) is connected to lower port on front side of Hydramotor steering unit (11) and is connected by long fitting to offset connecting nuts. Pressure tube (5) from pump is connected to left side port of Hydramotor unit by elbow (12); elbow (13) is fitted in rear port of Hydramotor unit and connects to return tube (14) to reservoir. Refer to Fig. 5 for later system.

1. Pump assy.	5. Pressure tube	9. Left cylinder tube
2. "O" rings	6. Reservoir assy.	10. Right cylinder tube
3. Pump inlet tube	7. Return tube	11. Hydramotor steering
4. By-pass tube	8. Steering motor	unit
		12. Elbow connector
		13. Elbow connector
		14. Return tube

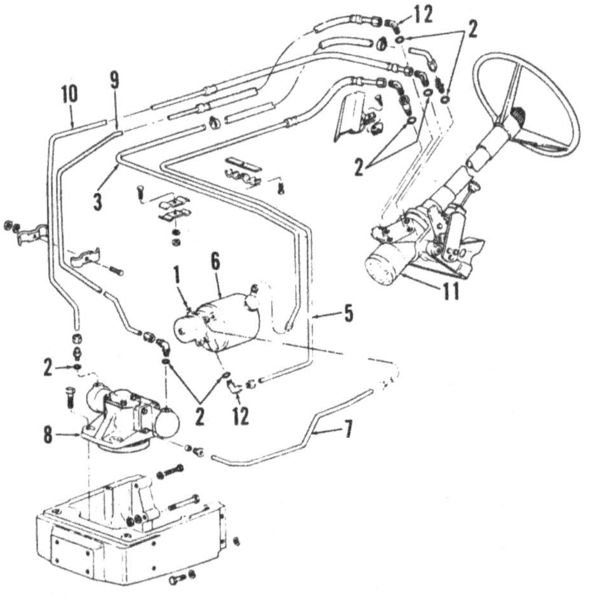

Fig. 5—8600-9600 hydrostatic power steering system with integral hydraulic pump and reservoir. Refer to Fig. 4 for identification of components. Steering motor (11) is pressurized from tube (5) to lower port. Upper port returns unused fluid through tube (3) to reservoir (6). Tubes (9) and (10) are routed to front steering motor (8) from side ports.

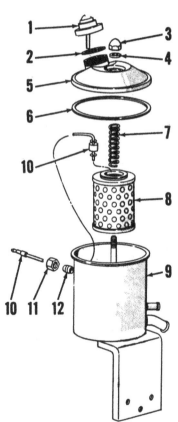

Fig. 6—Exploded view of power steering system reservoir used on 8000 and 9000 models and early 8600 and 9600 models. Oil level switch (10) is connected to warning light on instrument panel. Filter element (8) and oil in reservoir should be renewed after each 600 hours of use.

1. Dipstick & filter cap	7. Element retainer
2. Gasket	spring
3. Acorn nut	8. Filter element
4. Sealing washer	9. Reservoir
5. Cover	10. Warning light switch
6. Gasket	11. Nut
	12. Ferrule

system. Avoid use of shop towels or rags in wiping internal parts as any lint can cause malfunction of the system.

All 8000 and 9000 models and early 8600 and 9600 models were equipped with a remote power steering oil reservoir, a Saginaw Hydramotor power steering motor and a power steering pump with a flow control valve. The relief valve on pump is accessible from the outside.

Late 8600 and 9600 models were equipped with a Ross power steering motor which uses a spool to control flow. Steering pumps on these models have the reservoir as an integral part of pump. It is necessary to remove this type pump from engine to change filter or relief pressure.

7. **FLUID, BLEEDING AND SYSTEM RESERVOIR.** Recommended power steering fluid is Ford M-2C-41 oil. Maintain fluid level to full mark on dipstick. A low oil level switch assembly (10—Fig. 6) is used on models with remote fluid reservoir and is connected to a warning light on instrument panel. The light (located in Proof-Meter dial) should be on when starter switch is turned to "ON" position and go out when engine is started. If light remains on after engine is started, check for low oil level or malfunction in warning light system.

After each 600 hours of use, renew filter element. On models with remote reservoir, remove all oil from reservoir with suction gun, install new element and refill reservoir with new oil. On models with integral reservoir, remove pump from engine, drain reservoir, remove bolt (1—Fig. 8) and renew filter and "O" rings.

On all models the power steering system is self-bleeding. When any unit has been removed or disconnected, refill reservoir, start engine and cycle system by turning steering wheel from lock to lock. System is fully bled when front wheels respond directly to steering wheel movement and oil stays at level mark. Check fluid level and add oil as required to maintain full reservoir when cycling system.

8. **CHECKING SYSTEM PRESSURE.** On models with remote reservoir the power steering pump assembly incorporates a pressure relief valve and a flow control valve. System relief pressure should be 1450-1550 psi. On later models with integral pump and reservoir, pressure should be 1550-1650 psi.

To check system relief pressure, disconnect fitting and remove elbow in pressure line (5—Fig. 4 or Fig. 5) and connect a 0-2000 psi gage to pump, using

an "O" ring on fitting to pump. With the engine running, gage reading should be as stated above. On models with remote reservoir, if pressure is not as specified, remove the pressure relief valve cap (7—Fig. 7) and add or remove shims (6) as required. If adding shims under the pressure relief valve cap will not increase system pressure, clean flow control spool in pump. If pressure is still low remove and overhaul power steering pump as outlined in paragraph 11.

CAUTION: When checking system relief pressure, run engine only long enough to observe gage reading; pump may be damaged if engine is allowed to run for an excessive length of time.

On 8600 and 9600 models, with pump and reservoir as an integral unit, if pressure is not as specified, pump must be removed from engine. Drain reservoir and refer to Fig. 8. Remove reservoir (2) and filter (3). Remove relief valve body (24), shims (25) and spring (26). Shims are available in 0.010, 0.015 and 0.060 inch thicknesses. Each shim will change the pressure by the following approximate values: 0.010 inch—70 psi, 0.015 inch—105 psi, 0.060 inch—420 psi. Tighten relief valve to 30-40 ft.-lbs. torque.

9. **STEERING SYSTEM TROUBLE-SHOOTING** [Models with remote reser-

Fig. 7—Exploded view of power steering pump used with remote reservoir. Note position of flow control valve spring (4) and that small tip on valve (3) is towards spring.

1. Cap plug
2. "O" ring
3. Flow control valve
4. Spring
5. Tubing seats
6. Shims
7. Cap plug
8. "O" ring
9. Spring
10. Pressure relief valve
11. Seal ring
12. Outlet elbow
13. Cap plug
14. "O" ring
15. Rear cover
16. "O" ring
17. "O" ring
18. Rear plate
19. Inner seal ring
20. Outer seal ring
21. Bearing block
22. Drive gear & shaft
23. Driven gear & shaft
24. Pump body
25. Dowel rings (2)
26. Bearing block
27. Outer seal ring
28. Inner seal ring
29. Front cover
30. Seal
31. Snap ring
32. "O" ring
33. Drive gear
34. Tab washer
35. Nut
36. Woodruff key

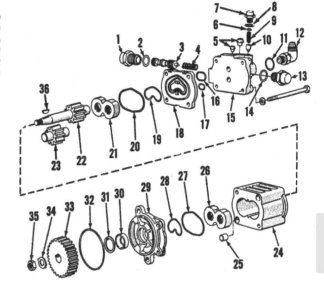

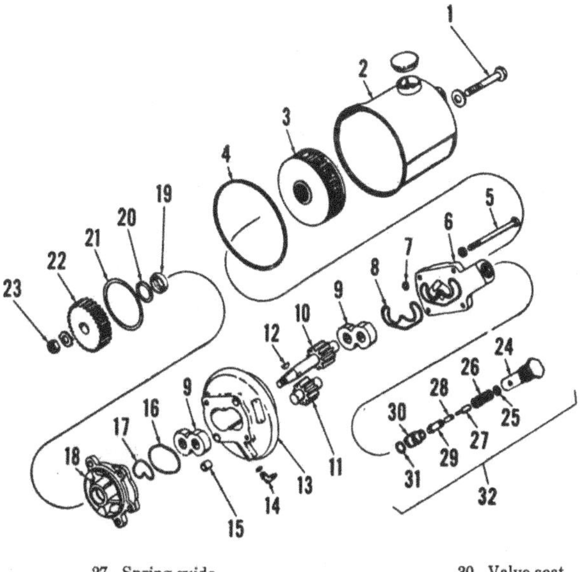

Fig. 8—Exploded view of power steering pump with integral reservoir used on late 8600 and 9600 models.

1. Bolt
2. Reservoir
3. Filter
4. Gasket
5. Through-bolt
6. Cover
7. "O" ring
8. Seal ring
9. Bearing block
10. Driven gear
11. Follow gear
12. Woodruff key
13. Body
14. Outlet elbow
15. Ring dowel
16. Seal ring
17. Seal ring
18. Flange housing
19. Oil seal
20. Snap ring
21. "O" ring
22. Drive gear
23. Shaft nut
24. Valve body
25. Shim pack
26. Spring
27. Spring guide
28. Seal
29. Valve head
30. Valve seat
31. "O" ring
32. Relief valve

voir). Refer to the following paragraphs for checking causes of steering system malfunction:

HARD STEERING. Check column bearings and bearings in Hydramotor unit; renew if rough or damaged. Check ring, rotor and vanes for wear and renew the assembly if necessary. Check for sticking control valve spool or blocking spool in Hydramotor; clean valves or renew Hydramotor parts as required.

EXCESSIVE WHEEL DRIFT. Check blocking spool spring and guide assembly and renew if spring is broken. Check for leakage past blocking valve; if excessive, renew valve body housing assembly. Check seals on steering cylinder pistons and renew pistons and/or cylinders as required.

STEERING WHEEL TURNING UN-AIDED. Check the Hydramotor unit for sticking control valve spool, broken valve spool spring, actuator shaft binding or torque shaft (inside actuator shaft) broken. Clean spool and bore or renew valve body housing assembly as required.

STEERING WHEEL SLIPPAGE. Hydramotor control valve spool scored (renew valve body housing assembly) or rotor seals leaking (renew seals).

EXCESSIVE NOISE. Hydraulic lines vibrating against tractor frame or broken control valve spool spring; insulate lines from tractor or renew valve body housing assembly if spring is broken.

ERRATIC MOVEMENT OF FRONT WHEELS. Check Hydramotor ring, rotor or vanes for scoring, wear or binding condition; renew the ring and rotor assembly if necessary.

WILL NOT STEER IN EITHER DIRECTION. The manual steer check ball between pump return and pressure passages in Hydramotor unit may not be seating. Disassemble unit and clean passage with solvent and dry with compressed air. Renew pressure plate assembly if check ball cannot be made to seat.

FRONT WHEELS JERK OR TURN WITHOUT MOVING STEERING WHEEL. Check for sticking rotor vanes, rotor springs out of place or broken, scored pressure plate, scored rotor ring, scored housing, ball check valves in pressure plate leaking, improper assembly causing gap between rotor components. Disassemble the Hydramotor

unit, carefully clean and inspect all parts and renew components as necessary.

10. **STEERING SYSTEM TROUBLESHOOTING. (Models with integral pump and reservoir).** Refer to the following paragraphs for checking causes of steering system malfunction:

HARD STEERING. Check column bearings and bearings in steering motor; renew if rough or damaged. Check rotor and stator assembly for wear or damage and renew assembly if necessary. Check for leaks from damaged valve spool; renew steering motor if spool is damaged. Check for binding at all pivot points in steering; free up and lubricate as necessary. Check for jammed valve spool; if unable to free up, renew steering motor assembly.

EXCESSIVE WHEEL DRIFT. Check for leakage past valve spool; if spool is worn or damaged, renew spool assembly.

EXCESSIVE NOISE. Hydraulic lines vibrating against tractor frame. Insulate lines from tractor.

ERRATIC WHEEL MOTION. Rotor vanes sticking or damaged. Check vanes, rotor and stator for free movement; renew if necessary.

WHEELS JERK FROM STOP-TO-STOP. Rotor vane springs jammed; check for proper seating of vane springs, renew complete assembly if damaged.

11. **R&R AND OVERHAUL PUMP. (REMOTE RESERVOIR).** Thoroughly clean pump, lines and surrounding area. Disconnect lines from pump and allow fluid to drain. Cap all openings to prevent dirt from entering pump or lines, then unbolt and remove pump assembly from engine front plate. When reinstalling pump, use new sealing "O" ring and tighten retaining bolts to a torque of 23-30 ft.-lbs. Reconnect lines, fill and bleed system as in paragraph 7.

Refer to exploded view of remote reservoir model pump in Fig. 7 and disassemble pump as follows: Scribe an assembly mark across pump covers and body. Straighten tab on washer (34) and remove nut (35). Pull drive gear (33) from pump shaft and remove key (36). Remove the four through-bolts and separate rear cover assembly (15), plate (18), body (24) and front cover (29). Remove bearing blocks (21 and 26) and gears (22 and 23) from pump as a unit. Remove caps (1, 7 and 13) from rear cover (15) and withdraw flow control valve (3), pressure relief valve (10) and related parts. Remove locating

snap ring (31) and the oil seal (30) from front cover. Clean all parts in a suitable solvent, air dry, then lightly oil all machined surfaces.

Inspect bearing blocks (21 and 26) for signs of seizure or scoring on face of journals. (When disassembling bearing block and gear unit, keep parts in relative position to facilitate reassembly). Light score marks on faces of bearing blocks can be removed by lapping bearing block on a surface plate using grade "O" emery paper and kerosene. Examine body for wear in gear running track. If track is worn deeper than 0.0025 inch on inlet side, body must be renewed. Examine pump for excessive wear or damage on journals, journal bores in bearing blocks or teeth. Runout across the gear face to gear tooth edge should not exceed 0.001 inch. If necessary, the gear journals may be lightly polished with grade "O" emery paper to remove wear marks. The gear faces may be polished by sandwiching grade "O" emery paper between gear and face of scrap bearing block, then rotating the gear. New gears are available in matched sets only. If flow control valve (3) or rear cover (15) are scored or damaged, they must be renewed as a matched set only.

When reassembling pump, install all new seals, "O" rings and sealing rings. Insert new drive shaft oil seal (30) in front plate and install locating snap ring. Install flow control valve (3), spring (4) and plugs (1 and 13) with new "O" rings (2 and 14). Install pressure relief valve (10), spring (9) and plug (7), being sure that all shims (6) are in plug and using new "O" ring (8). Assemble pump gears to bearing blocks and insert the unit into pump body. Be sure the two bolt rings (hollow dowels) are in place in pump body, then position the front cover on body. Place the rear plate (18) at rear of body and install rear cover. Tighten the four cap screws (through-bolts) to a torque of 13-17 ft.-lbs. Install the pump drive gear key, drive gear, tab washer and nut. Tighten the nut to a torque of 55-60 ft.-lbs. and bend tab of washer against flat on nut.

12. **R&R AND OVERHAUL INTEGRAL RESERVOIR PUMP.** For exploded view of parts used on models with integral pump and reservoir refer to Fig. 8. Clean pump and surrounding area and disconnect pump pressure and return lines. Remove the two cap screws securing pump to engine front cover and lift off pump and reservoir as a unit. Drain the reservoir and remove through-bolt (1), reservoir (2) and filter (3).

Relief valve cartridge (32) can now be removed if service is indicated. For access to shims (25) grasp seat (30) lightly in a protected vise and unscrew body (24). Shims (25) are available in thicknesses of 0.010, 0.015 and 0.060 inch. Starting with the removed shim pack substitute shims, thus varying total pack thickness, to adjust opening pressure. Available shims permit thickness adjustment in increments of 0.005 inch and each 0.005 inch in shim pack thickness will change opening pressure about 35 psi. If parts are renewed, the correct thickness can only be determined by trial and error, using the removed shim pack as a guide.

To disassemble the pump, bend back tab washer and remove shaft nut (23), drive gear (22) and key (12). Mark or note relative positions of flange housing (18), pump body (13) and cover (6); then remove pump through bolts (5). Keep parts in their proper relative position when disassembling pump unit. Pump gears (10 and 11) are available in a matched set only. Bearing blocks (9) are available separately but should be renewed in pairs if renewal is because of wear. Bearing blocks should also be renewed with gear set if any shaft or bore wear is evident. Examine body (13) for wear in gear running track. If track is worn deeper than 0.025 inch on inlet side, body must be renewed. Renew all "O" rings and seals.

When reassembling the pump, tighten through bolts (5) to a torque of 25 ft.-lbs. and drive gear nut (23) and relief valve body (24) to a torque of 30-40 ft.-lbs.

13. SAGINAW HYDRAMOTOR STEERING UNIT. Refer to the following paragraphs 14 through 18 for information on removal, overhaul and installation of the Saginaw Hydramotor steering unit which is used on 8000 and 9000 models and early 8600 and 9600 models. If parts are not available for repair of Hydramotor unit, a conversion kit is available to install the later Ross unit on early tractors. Refer to paragraph 9 for troubleshooting information. For the Ross unit used on 8600 and 9600 models, refer to paragraphs 19, 20 and 21.

14. R&R HYDRAMOTOR UNIT. To remove Hydramotor, first remove hood top, right and left side panels, then proceed as follows:

Remove cap (1—Fig. 9) from adjuster knob (4) and remove nut (2) and washer (3). Knob can then be removed from locking rod in shaft (11), then remove steering wheel (7) and shaft (8) as an assembly. Disconnect the four tubes from Hydramotor unit, then cap or plug all openings. Loosen both jam nuts (24) and unscrew the pivot studs (23) from support (21). Then remove steering unit from below the instrument panel.

NOTE: Remove intake manifold air tube if necessary for clearance.

To reinstall, position unit in support with tilt quadrant engaged in lock plunger and turn pivot studs in to support unit. Reconnect the four tubes and reinstall steering wheel and adjuster knob. With steering shaft shortened to full extent and steering wheel in lowered position, attach pull scale to steering wheel rim and release quadrant latch. Tighten pivot studs until a pull of 18-22 pounds will lift steering wheel from lowered position, then tighten jam nuts to a torque of 180-220 ft.-lbs. and recheck pivot stud adjustment.

NOTE: Do not attempt to position steering wheel on shaft as slippage in unit will not allow wheel to remain in any relative position to front wheel movement.

R&R STEERING COLUMN JACKET AND SHAFT ASSEMBLIES (Hydramotor Models). With Hydramotor unit removed as outlined in paragraph 14, proceed as follows:

Loosen clamp (15—Fig. 9) and pull column jacket assembly (10) from control valve housing (16). Unscrew the hex nut (14) until it nearly contacts control valve

housing. Nut was staked when assembled and will turn hard. Drive the tapered collar (13) towards until collar is loose, then turn collar until hole in collar is over locking ball hole in outer shaft (11) and shake the ball (12) out of hole. The outer shaft, tapered collar and hex nut can then be removed from the Hydramotor actuator shaft.

Reassemble the unit before reinstalling on actuator shaft as follows: Install tapered ring (13) on outer shaft (11), with large I.D. first. Install a new nut (14) just far enough to catch one or two threads of outer shaft. Engage splines of outer shaft on splines of actuator shaft. Align hole in tapered ring, hole in outer shaft and groove around the actuator shaft, then drop locking ball in hole and groove and turn tapered collar ¼-turn. Tighten the hex nut to a torque of 40-50 ft.-lbs. and stake nut into slot in outer shaft as shown in Fig. 11.

16. R&R BLOCKING SPOOL (REACTION) VALVE (Hydramotor Models). The blocking spool valve and related parts can be removed and reinstalled after the Hydramotor steering unit has been removed as outlined in paragraph 14. Refer to Fig. 12 and proceed as follows:

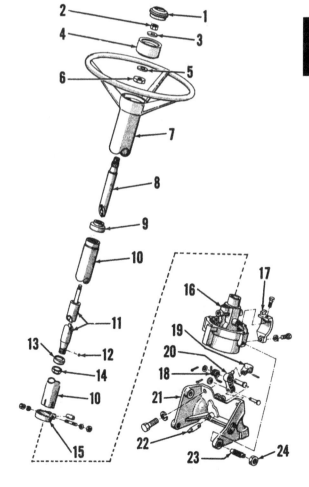

Fig. 9—View showing adjustable length steering shaft and variable position Hydramotor steering unit bracket on 8000 and 9000 models and early 8600 and 9600 models. Lock plunger (22) engages one of eight notches in quadrant (17) to hold steering wheel at desired tilt position. Adjusting knob (4) tightens tapered end of steering shaft (8) against tapered end of lower shaft to lock steering wheel at desired height.

1. Cap
2. Nut
3. Flat washer
4. Adjusting knob
5. Flat washer
6. Nut
7. Steering wheel & outer tube
8. Steering wheel shaft
9. Shaft scraper & seal
10. Steering column jacket
11. Hydramotor outer shaft
12. Steel ball (11/64-inch)
13. Taper ring
14. Nut
15. Clamp
16. Hydramotor unit
17. Tilt quadrant
18. Spring
19. Tilt knob
20. Tilt pivot
21. Support
22. Tilt lock plunger
23. Pivot studs
24. Jam nut

Remove the lockout adjuster nut (1). Plug (3) and spool valve (4) may now be removed by pushing the plug into bore against spring pressure with screwdriver, then quickly releasing the plug to allow spring to pop it out of bore. Remove plug and, if spool sticks in bore, invert the unit and tap housing (12) with soft faced mallet to jar spool out. Invert unit and allow spring (5) and spring and guide assembly (6) to drop from bore.

Spool is not serviced separately, but is available in a complete housing kit, which includes necessary parts to rebuild housing assembly.

NOTE: On some Hydramotor housings, oil leakage around blocking valve adjuster (1) may be due to mismatch of counterbore in valve spool bore and position of "O" ring (2) on plug (3). To stop oil leakage, install plug with part No. C9NN-3R675-A. Note difference in plugs shown in Fig. 14. Later models will all be equipped with the later design plug.

To reassemble, install parts in bore of housing (12—Fig. 12) as shown in exploded view, renewing the "O" ring (2) on plug (3) and tightening adjuster nut to a torque of 10-15 ft.-lbs.

NOTE: The adjuster (1) is not accessible after tractor is fully assembled; thus, the adjuster must be in the down,

or closed position when unit is being reinstalled.

17. R&R COVER RETAINING SNAP RING (Hydramotor Models). To remove snap ring (7—Fig. 10) used to retain cover (30) to housing (12) proceed as follows:

With unit removed from tractor as outlined in paragraph 14, check to see that end gap of snap ring is near hole in cover as shown in Fig. 13; if not, bump snap ring into this position with hammer and punch. Insert a pin punch into hole and drive punch inward to dislodge snap ring from groove. Hold punch under snap ring and pry ring from cover with screwdriver. Usually, the coil spring (27—Fig. 10) will push housing from cover; if not, bump cover loose by tapping around edge with mallet.

To reinstall the cover retaining snap ring, housing must be held in cover, against spring pressure. It is recommended that the unit be placed in an arbor press and the housing be pushed into cover with a sleeve as shown in Fig. 15.

CAUTION: Do not push against end of shaft (14—Fig. 10).

Place snap ring over housing before placing unit in press. Carefully apply force on housing with sleeve until flange on housing is below snap ring groove in

cover. Note that lug on housing which prevents rotation must enter slot in cover. If housing binds in cover, do not apply heavy pressure; remove unit from press and bump cover loose with mallet.

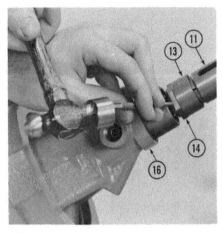

Fig. 11—After installing steering shaft on Hydramotor actuator shaft, stake hex nut to slot in steering (11) with center punch.

11. Outer shaft   14. Hex nut
13. Tapered ring   16. Housing

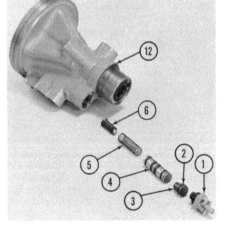

Fig. 12—Exploded view of Hydramotor housing and blocking valve components. Blocking valve can be removed without disassembly of Hydramotor.

1. Lockout
2. "O" ring
3. Plug
4. Blocking valve
5. Spring
6. Spring & guide assy.
12. Housing

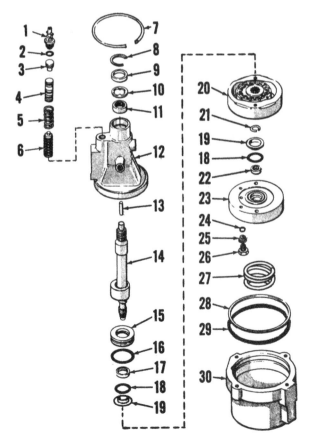

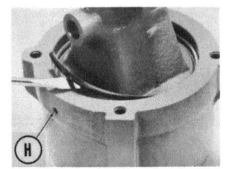

Fig. 13—To remove cover retaining snap ring, drive pin punch through hole (H) in cover to disengage snap ring from groove.

Fig. 10—Exploded view of the Saginaw Hydramotor steering unit. Also refer to Figs. 11 to 27 for photos showing disassembly and reassembly techniques.

1. Blocking valve lockout
2. "O" ring
3. Plug
4. Block valve
5. Spring
6. Spring & guide assy.
7. Snap ring
8. Snap ring
9. Dust seal
10. Oil seal
11. Needle bearing
12. Housing
13. Dowel pins (2)
14. Actuator shaft & control valve spool
15. Bearing support
16. "O" ring
17. Needle bearing
18. "O" ring
19. Rotor seal ring
20. Ring, rotor & vane assy.
21. Snap ring
22. Needle bearing
23. Pressure plate assy.
24. Check valve balls (2)
25. Check valve springs (2)
26. Retaining plugs (2)
27. Pressure plate spring
28. Back-up ring
29. "O" ring
30. Cover

When housing has been pushed far enough into cover, install snap ring in groove with end gap near hole in cover as shown in Fig. 13.

18. **OVERHAUL SAGINAW HYDRAMOTOR STEERING UNIT.** With the unit removed from tractor as outlined in paragraph 14 and the cover retaining snap ring removed as outlined in paragraph 17, proceed as follows:

Clamp flat portion of Hydramotor housing in a vise and remove cover (30—Fig. 10) by pulling upward with a twisting motion. Remove the pressure

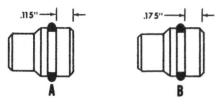

Fig. 14—Note difference in "O" ring position on above plugs. Plug (B) is Ford part No. C9NN-3R675-A and should be installed to stop oil leakage described in paragraph 16.

Fig. 15—Using sleeve and arbor press to push housing into cover to allow installation of cover retaining snap ring (7).

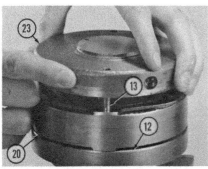

Fig 16—Lifting pressure plate (23) from dowel pins (13).

plate spring (27), then lift off the pressure plate (23) as shown in Fig. 16. Remove the dowel pins (Fig. 17), then remove snap ring (21) from shaft (14) with suitable snap ring pliers and screwdriver; discard the snap ring. Pull pump ring and rotor assembly (20) up off of shaft as shown in Fig. 18. Tap outer end

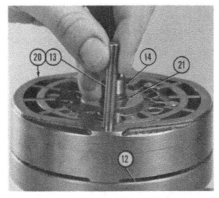

Fig. 17—Removing dowel pins (13) from motor ring and housing. Then, remove the snap ring (21) retaining rotor to actuator shaft (14).

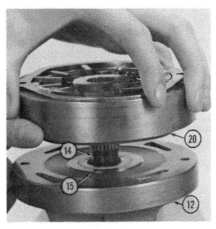

Fig. 18—Lifting the motor ring, rotor and vane assembly (20) from actuator shaft (14) and housing (12).

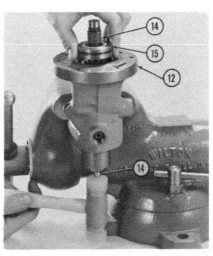

Fig. 19—Tap on outer end of actuator shaft (14) to bump bearing support (15) from the housing (12).

of shaft with soft faced mallet as shown in Fig. 19 until bearing support (15) can be removed, then carefully remove the actuator shaft assembly from housing as shown in Fig. 21.

NOTE: As the actuator shaft and control valve spool asembly is a factory balanced unit and is not serviceable except by renewing the complete

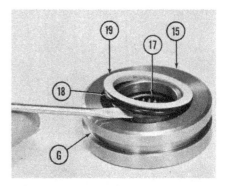

Fig. 20—Lifting the Teflon rotor seal (19) and "O" ring (18) from bearing support (15). Needle bearing (17) is serviced separately from bearing support. Groove (G) is for support sealing "O" ring (16—Fig. 10). Identical seals (18 and 19) are used in pressure plate.

Fig. 21—Removing the actuator shaft assembly from housing. Be careful not to cock control valve spool in bore of housing.

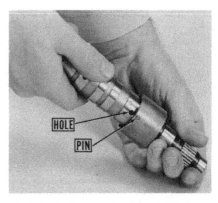

Fig. 22—Pin in actuator sleeve must be engaged in hole in end of control valve spool before actuator assembly is installed. If spool cannot be pulled out of sleeve, pin is engaged.

housing assembly, it is recommended that this unit not be disassembled.

Carefully clean and inspect the removed units. Refer to paragraph 16 for information on the blocking valve assembly. If the housing control valve bore or blocking valve bore are deeply scored or

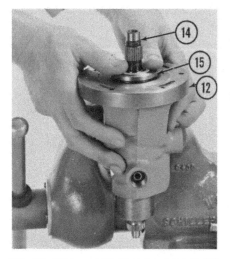

Fig. 23—When pushing bearing support (15) into housing (12), be careful not to damage the "O" ring on outside groove of support.

Fig. 25—Be sure all vane springs are engaged behind the rotor vanes. Springs can be pried into place with screwdriver as shown.

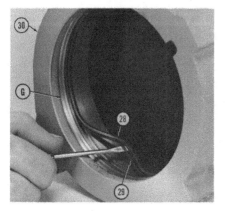

Fig. 26—"O" ring (28) and back-up ring (29) are installed in cover (30); be sure back-up ring is to outside (open side) of cover. Groove (G) is for cover retaining snap ring.

worn, or if the blocking valve spool or the actuator shaft and control valve spool assembly are damaged in any way making the unit unfit for further service, a complete new housing assembly must be installed. If these components (housing, blocking valve and actuator assembly) are serviceable, proceed with overhaul as follows:

Remove the check valve retainers (26—Fig. 10), springs (25) and check valve balls (24) from pressure plate (23) and blow passages clear with compressed air. Renew the pressure plate assembly if check valve seats or face of plate are deeply scored or damaged. Renew needle bearing (22), springs (25) and/or check valve balls (24) if damaged and pressure plate is otherwise serviceable.

NOTE: Drive or press on lettered (trademark) end of bearing cage when installing new needle bearing.

If bearing support (15) is otherwise serviceable, a new needle bearing (17) may be installed; drive or press only on lettered end of bearing cage.

Remove snap ring (8), dust seal (9) and oil seal (10) and inspect the needle bearing (11); renew needle bearing if

worn or damaged. Press only on lettered (trademark) end of bearing cage when installing new bearing. Install oil seal with lip towards inside (needle bearing), then install dust seal and retaining snap ring.

If the motor ring, rotor or vanes are worn, scored or damaged beyond further use, or if any of the vane springs are broken, renew the unit as a complete assembly (20). If unit was disassembled and is usable, reassemble as follows: Place ring on flat surface and place rotor inside ring. Insert six vanes in rotor slots which are in line with large inside diameter of ring. Make sure rounded edges of vane are facing outward. Turn rotor ¼-turn and insert remaining vanes. Hook the springs behind the vanes with a screwdriver as shown in Fig. 25, then turn the assembly over and hook springs behind the vanes on opposite side of rotor.

To reassemble the Hydramotor unit, place housing (12 – Fig. 10) (with needle bearing, seals and snap ring installed) in a vise with flat (bottom) side up. Check to be sure that pin in actuator is engaged with hole in valve spool; if spool can be pulled away from actuator as shown in Fig. 22, push the spool back into actuator and be sure pin is engaged in one

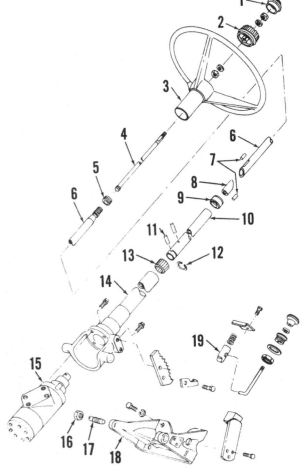

Fig. 27—Exploded view of steering column and motor used on late 8600 and 9600 models. Unit differs from Saginaw models mainly in motor (15).

1. Cap
2. Adjusting knob
3. Steering wheel & tube
4. Shaft lock rod
5. Shaft taper ring
6. Motor inner shaft
7. Retainer pin
8. Motor shaft lock
9. Upper bearing
10. Inner column jacket
11. Motor shaft pin
12. Dowel ring
13. Bearing
14. Column
15. Steering motor
16. Jamnut (2)
17. Pivot stud (2)
18. Support
19. Tilt lock plunger

of the holes in spool. Then, lubricate spool and shaft and carefully insert the assembly into bore of housing. Place bearing support (15) with outside "O" ring and needle bearing installed on shaft, and carefully push the support into housing as shown in Fig. 23. Insert rotor sealing "O" ring and rotor seal into bearing support. Place ring and rotor assembly (20—Fig. 10) on shaft and housing with chamfered outer edge of ring up (away from housing). Install a new rotor retaining snap ring (21) and insert the dowel pins through ring into housing. Using heavy grease, stick the "O" ring and rotor sealing ring (19) into pressure plate, then install the pressure plate and bearing assembly over shaft, pump ring and rotor assembly and the two dowel pins. Place the coil spring on top of pressure plate. Install new "O" ring and backup ring in second groove of cover (Fig. 26), lubricate the rings and push cover down over the pressure plate and ring. While holding the cover on the assembly, place the unit in an arbor press and insert the cover retaining snap ring as outlined in paragraph 17. Reinstall Hydramotor as outlined in paragraph 14.

19. **R&R STEERING COLUMN SHAFT AND MOTOR ASSEMBLY** (Late 8600-9600 Models). Refer to Fig.

27. Remove cap (1) from adjuster knob (2), remove nut, washer and knob. Use a suitable puller to remove steering wheel and tube assembly from inner shaft (6). Remove left and right hood panels. Disconnect the four tubes from power steering motor (15) and cap or plug all openings. Remove the heat deflector panels below steering motor to allow removal out the bottom. Loosen two jam nuts (16) and unscrew pivot studs (17) from support (18) until assembly can be brought down and out right side of tractor.

To reinstall, reverse removal procedure. Before installing pivot studs (17), lubricate studs and tighten until a pull of 18-22 lbs. will tilt assembly. Tighten jam nuts to a torque of 180-220 ft.-lbs. Reconnect four oil tubes so that all tubes are at right angles to engine when tight. Tighten steering wheel nut to 60-80 ft. lbs. torque. After knob (2) is placed on shaft, use knob to rotate locking shaft counter-clockwise until it stops, then tighten nut to 11-15 ft.-lbs. torque.

20. **R&R STEERING COLUMN JACKET AND SHAFT ASSEMBLIES** (Late 8600-9600 Models). With steering motor and shaft assembly removed as outlined in paragraph 19, proceed as follows:

Remove four bolts holding column

assembly (14—Fig. 27) to steering motor (15). Remove motor and shaft assembly from column and remove motor shaft pin (11).

Reassemble in reverse order of disassembly. Tighten four bolts from column to motor to 18-22 ft.-lbs. torque.

21. **OVERHAUL ROSS STEERING MOTOR UNIT.** Remove column and steering motor assembly as outlined in paragraph 19 and separate motor assembly from column as outlined in paragraph 20. To disassemble the removed motor, refer to Fig. 28 and proceed as follows: Install a fitting in one of the four ports in housing (27), then clamp fitting in a vise so that input shaft (18) is pointing downward. Remove cap screws and end cover (1).

NOTE: Lapped surfaces of end cover (1), commutator set (5 and 6), manifold (7), stator-rotor set (8), spacer (plate) and housing (27) must be protected from scratching, burring or any other damage as sealing of these parts depends on their finish and flatness.

Remove seal retainer (4) and seal (3), then carefully remove washer (2), commutator set (5 and 6) and manifold (7). Lift off the spacer plate (10), drive link (9) and stator-rotor set as an assembly. Separate spacer plate and drive link from stator-rotor set. If the pin in end cover (1) or cover is damaged, cover must be renewed, since pin is not serviced separately.

Remove unit from vise, then clamp fitting in vise so that input shaft is pointing upward. Place a light mark on flange of upper cover (32) and housing (27) for aid in reassembly. Unbolt upper cover from valve body, using a 5/16-inch 12-point socket, then grasp input shaft and remove input shaft, upper cover and valve spool assembly. Remove and discard seal ring (31). Slide upper cover assembly from input shaft and remove spacer (17). Remove shims (11) from cavity in upper cover or from face of thrust washer (13) and note number of shims for aid in reassembly. Remove retaining ring (36), brass washer (34) and seal (33). Retain seal ring (35) and retaining ring (36) for reassembly. Do not remove needle bearing (30) unless renewal is required.

Remove ring (12), thrust washer (13) and thrust bearing (14) from input shaft. Drive out pin in input shaft (18) and withdraw torsion bar (21) and spacer (20). Place end of valve spool on top of bench and rotate input shaft until drive ring (19) falls free, then rotate input shaft clockwise until actuator ball (24) is disengaged from helical groove in input shaft. Withdraw input shaft and remove actuator ball. Do not remove actuator

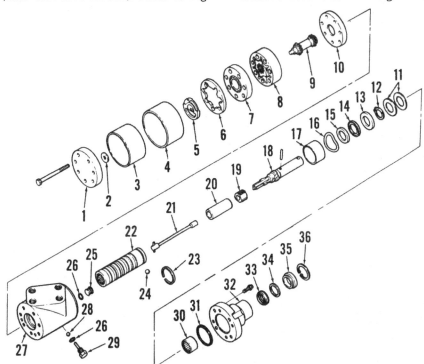

Fig. 28—Exploded view of Ross power steering motor used on late models.

1. End cover w/pin	10. Spacer (plate)	19. Drive ring	28. Recirculating ball
2. Washer	11. Shims	20. Spacer	29. Plug w/pin
3. Seal	12. Retaining snap ring	21. Torsion bar	30. Needle bearing
4. Seal retainer	13. Thrust washer	22. Valve spool	31. Seal ring
5. Commutator	14. Valve thrust bearing	23. Ball retaining spring	32. Upper cover
6. Commutator ring	15. Thrust washer	24. Actuator ball	33. Seal
7. Manifold	16. Spring washer	25. Plug	34. Brass washer
8. Stator-rotor set	17. Spacer	26. "O" ring	35. Stepped spacer
9. Drive link	18. Input shaft	27. Valve body	36. Retaining ring

ball retaining spring (23) unless renewal is required.

Remove plug (29) and recirculating ball from valve body.

Thoroughly clean all parts in a suitable solvent, visually inspect parts and renew any showing excessive wear, scoring or other damage.

If needle bearing (30) must be renewed, press same out toward flanged end of cover. Press new bearing in from flanged end of cover to the dimension shown in Fig. 29. Press only on numbered end of bearing, using a piloted mandrel.

Using a micrometer, measure thickness of the commutator ring (6—Fig. 28) and commutator (5). If commutator ring is 0.0015 inch or more thicker than commutator, renew the matched set.

Place the stator-rotor set (8) on the lapped surface of end cover (1). Make certain that vanes and vane springs are installed correctly in slots of the rotor.

NOTE: Arched back of springs must contact vanes.

Position lobe of rotor in valley of stator as shown in Fig. 30. Center opposite lobe on crown of stator, then using feeler gage, measure clearance between rotor lobe and stator. If clearance is more than 0.006 inch, renew stator-rotor assembly. Using a micrometer, measure thickness of stator and rotor. If stator is 0.002 inch or more thicker than rotor, or vanes are worn to 0.250 inch or less in length, renew the assembly. Stator, rotor and vanes are available only as an assembly.

Before reassembling, wash all parts in clean solvent and air dry. All parts, unless otherwise indicated, are installed dry. Install plug (25—Fig. 28) and new "O" ring in valve body. Install recirculating ball (28) and plug (29) with new "O" ring in valve body and tighten plug to a torque of 10-14 ft.-lbs. Clamp fitting

(installed in valve body port) in a vise so that top end of valve body is facing upward. Install thrust washer (15), thrust bearing (14), second thrust washer (13) and snap ring (12) on input shaft (18). If actuator ball retaining spring (23) was removed, install new retaining spring in spool (22). Place actuator ball (24) in its seat from the inside of valve spool (22). Insert input shaft into valve spool, engaging the helix and actuator ball with a counter-clockwise motion. Use the mid-section of torsion bar (21) as a gage between end of valve spool and thrust washer as shown in Fig. 31, to insure assembly in the neutral position of the ball on ramp. Place the assembly in a vertical position with end of input shaft resting on a bench. Insert drive ring (19—Fig. 28) into valve spool until drive ring is fully engaged on input shaft spline. Remove torsion bar gage. Install spacer (20) in

valve spool, over drive ring (19). Distance from top of spool (22) to top of spacer (20) should be 11/16 inch. Install torsion bar into valve spool. Align cross-holes in torsion bar and input shaft with a 0.120 inch pin punch and install pin into input shaft (18). Pin must be pressed into shaft until end of pin is about 1/32-inch below flush. Place spacer (17) over spool and carefully install spool assembly into valve body. Position original shims (11) on thrust washer (13) (if the original input shaft and cover are to be used), lubricate new seal ring (31), place seal ring in upper cover (32) and install upper cover assembly. Align the match marks on cover flange and valve body and install cap screws finger tight. Tighten a worm drive type hose clamp around cover flange and valve body to align the outer diameters, (as shown in Fig. 32), then tighten cap screws to a torque of 18-22 ft.-lbs.

NOTE: If either input shaft (18—Fig. 28) or upper cover (32) or both have been renewed, the following procedure for shimming must be used.

With upper cover installed (with original shims) as outlined above, invert unit in vise so that input shaft is pointing downward. Grasp input shaft, pull downward and prevent it from rotating. Engage drive link (9) splines in valve

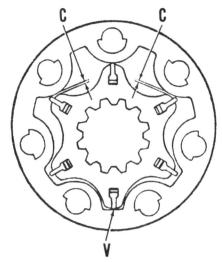

Fig. 30—With rotor positioned in stator as shown, clearance "C" must not exceed 0.006 inch. Refer to text.

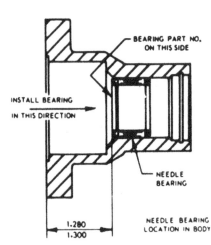

Fig. 29—When installing needle bearing in upper cover, press bearing into dimension (inches) shown.

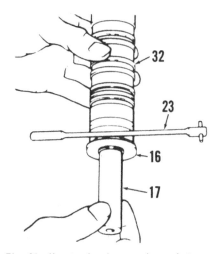

Fig. 31—Use torsion bar as shown between thrust washer and end of spool, to establish neutral position. Refer to Fig. 28 for parts identification.

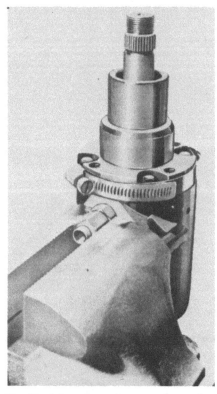

Fig. 32—A large hose clamp may be used as shown to align cover to valve body before tightening cap screws.

spool and rotate drive link until end of spool is flush with end of valve body. Remove drive link and check alignment of drive link and check alignment of drive link slot to torsion bar pin. Install drive link until its slot engages torsion bar pin. Check relationship of spool end to body end. If end of spool is protruding from body and is within 0.0025 inch of being flush with end of body, no additional shimming is required. If not within 0.0025 inch of being flush, remove cover and add or remove shims (11) as necessary. Reinstall cover and recheck spool to valve body position.

With drive link installed, place spacer plate (10) on valve body with plain side up. Install stator-rotor set over drive link splines and align cap screws holes. Make certain vanes and vane springs are properly installed. Install manifold (7) with circular slotted side up and align cap screw holes with stator, spacer and valve body. Install commutator ring (6) with slotted side up, then install commutator (5) over drive link end with counter-bore for washer (2) facing out. Make certain that link end is engaged in the smallest elongated hole in commutator. Install seal (3) and retainer (4). Apply a few drops of hydraulic fluid on commutator (5) and manifold (7). Use a small amount of grease to stick washer (2) in position over pin on end cover (1). Install end cover making sure that pin engages center hole in commutator. Align holes and install cap screws. Alternately and progressively tighten cap screws while rotating input shaft. Final tightening should be 18-22 ft.-lbs. torque.

Relocate the unit in vise so input shaft is up. Lubricate seal (33) and carefully work seal over shaft and into bore with lip toward inside. Install brass washer (34) then install stepped spacer (35) with flat side up. Install retaining ring (36) with rounded edge inward.

Remove unit from vise and remove fitting from port. Turn unit on its side with hose ports upward. Pour clean hydraulic fluid into inlet port, rotate input shaft until fluid appears at outlet port, then plug all ports.

Reinstall unit to steering column by installing motor shaft pin (11—Fig. 27). Tighten the four bolts from column to steering motor to a torque of 18-22 ft.-lbs. and reinstall assembly as outlined in paragraph 19.

22. R&R FRONT POWER STEERING MOTOR (All Models). To remove power steering motor from front support, first remove radiator and shell as an assembly as outlined in paragraph 152, then proceed as follows:

On tricycle model, support front of tractor and remove spindle assembly from steering motor shaft. On wide front

axle models, unbolt center steering arm from motor shaft. Disconnect power steering tubes from steering motor and cap or plug all openings. Remove engine oil cooler, then unbolt and remove power steering motor from pedestal.

When reinstalling, tighten steering motor retaining cap screws to a torque of 180-220 ft.-lbs. and oil cooler cap screws to 20-26 ft.-lbs. Fill and bleed the power steering system as outlined in paragraph 7.

23. OVERHAUL FRONT POWER STEERING MOTOR. With assembly removed as in paragraph 22, proceed as follows:

To renew piston seals, unbolt cylinders (20—Fig. 34) from housing (8) and withdraw cylinders from rack pistons (13). Remove cap screws (18), flat washers (17), pistons and spacer sleeves (16) from each end of rack (7). Remove Teflon ring (15) and "O" ring (14) from pistons. Check to see that the spacer sleeves (16) are not crushed; pistons

must be free (except for "O" ring tension) between end of rack and flat washer for alignment to cylinder bores. Reinstall pistons with new "O" rings (12) and tighten retaining cap screws to a torque of 14-20 ft.-lbs. Install new "O" rings (14) in piston grooves. Soften new Teflon rings (15) in hot water, install rings over the "O" rings and force the Teflon rings into groove using a piston ring compressor. Renew cylinder(s) if scored or excessively worn. Lubricate pistons and cylinder bores with power steering fluid, then install cylinders with new "O" rings (19). Tighten cylinder retaining cap screws to a torque of 30-40 ft.-lbs.

To overhaul rack, pinion gear, shaft, bearings and seal assembly, first remove cylinders and pistons as outlined in preceding paragraph, then proceed as follows: Remove cap (1), then unstake and remove nut (3). Press shaft (26) downward out of housing, remove upper bearing cone (4) from housing and remove pinion (22) and spacer (23). In-

Fig. 33—Power steering motor is accessible after removing shell and radiator assembly. Engine oil cooler is mounted on top of motor. Oil return line carries any leakage past rack pistons back to reservoir.

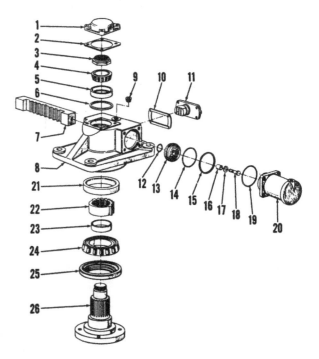

Fig. 34—Exploded view of power steering motor assembly. Motor shaft bearings are adjusted by tightening nut (3). Shims (10) control pressure of rack guide (slipper) (11) against front side of rack (7), pinion (22) and steering motor shaft (26).

1. Cap
2. Gasket
3. Nut
4. Bearing cone
5. Bearing cup
6. Retaining ring
7. Rack
8. Housing
9. Plug
10. Shims
11. Slipper guide
12. "O" ring
13. Piston
14. "O" ring
15. Teflon ring
16. Spacer
17. Flat washer
18. Cap screw
19. "O" ring
20. Cylinder
21. Bearing cup
22. Pinion
23. Spacer
24. Bearing cone
25. Seal
26. Motor shaft

spect bearing cone (24) and if necessary to renew, use a suitable puller to remove from shaft. Remove rack guide (slipper) (11) from housing, taking care not to lose shims (10), and withdraw rack from housing. Remove upper (5) and lower (21) bearing cups from housing, if necessary to renew and remove oil seal (25) from housing and discard.

Carefully clean and inspect all parts and renew any that are scored or excessively worn. Reassemble using new oil seal, gasket, sealing rings and other parts required as follows:

If they have been removed, install snap ring (6), then drive upper bearing cup (5) into housing tightly against snap ring. If removed, install new bearing cup (21) and oil seal (25) in bottom of housing. Use a suitable step plate on bearing cup and make sure it bottoms in bore. Install bearing cone (24), spacer (23) and pinion (22) on shaft, making sure that pinion to shaft timing marks are aligned as shown in Fig. 35. Place housing over spindle, then install rack so that timing marks on rack and pinion are aligned as in Fig. 35, then install guide (11—Fig. 34) with shims as removed; tighten guide retaining cap screws finger tight only. Install upper bearing cone (4) and retaining nut (3). Tighten a ¾-inch cap screw in center hole of shaft flange so that torque wrenches may be used to check effort to turn shaft. Tighten nut (3) so that a torque of 52-62 in.-lbs. is required to turn shaft with guide retaining cap screws loose, then stake nut to shaft and install cap (1) with new gasket. Install rack pistons and cylinders as previously outlined, then tighten rack guide retaining cap screws to a torque of 20-26 ft.-lbs. With rack centered, breakaway torque required to turn shaft with torque wrench should be 20-25 ft.-lbs. If breakaway torque is not within this range, decrease shim pack (10) thickness to increase guide pressure or add thickness to decrease pressure. Shims are available in thicknesses of 0.007, 0.009, 0.010, 0.012, 0.015 and 0.020 inch.

When rack guide is properly shimmed, remove plug (9) and fill housing with power steering fluid. Reinstall unit as outlined in paragraph 22.

FRONT SYSTEM AND STEERING

(MODELS 8700, 9700, TW10, TW20 AND TW30)

Refer to paragraph 44 for front wheel drive models.

28. Models 8700 and 9700 are equipped with a front axle which may be adjusted to track widths from 56 to 84 inches. Axles on Models TW-10, TW-20 and TW-30 may be adjusted from 60 to 84 inches. A short (93.2 inches) or long (109.7 inches) tractor wheelbase may be obtained by reversing front axle and replacing spindle steering arms, tie rod sleeve and anchor for power steering cylinder.

All models are equipped with a tilting steering column and hydrostatic power steering which consists of an engine driven pump with fluid reservoir, a Ross-type steering motor and a steering cylinder.

FRONT AXLE

29. **REMOVE AND REINSTALL.** To remove front axle, remove front weights and support front of tractor. Disconnect and cap steering cylinder hoses. Unbolt and remove front support bracket (21—Fig. 40), then roll front axle assembly away from tractor.

To install front axle, reverse removal

procedure. Tighten front support bracket (21) screws to 180-220 ft.-lbs. Cycle steering wheel several times in both directions to bleed steering cylinder and check fluid level in reservoir.

NOTE: Tractor wheelbase may be changed from long to short or vice versa by reversing direction of front axle center member (23—Fig. 40) and replacing steering arms (3 and 7), tie rod sleeve (5) and steering cylinder anchor (11). Steering cylinder hoses must be rerouted and lengthened or shortened.

30. **SPINDLE BUSHINGS.** Spindle bushings (13 and 16—Fig. 40) are pre-sized and can be renewed without removing axle extension. Pull old bushings and install new ones using a piloted driver to prevent damage to bushing and axle bore. Tighten spindle retaining nuts to a torque of 360-440 ft.-lbs. on all TW-30 models and on TW-10 and TW-20 models equipped with a heavy duty front axle. Tighten spindle nuts on all other models to a torque of 100-125 ft.-lbs.

31. **AXLE CENTER MEMBER, PIVOT PINS AND BUSHINGS.** To remove axle center member (23—Fig. 40), support front of tractor, remove tie rod ends and cylinder rod as necessary, then unbolt axle extensions from center member. Withdraw axle extensions with spindles and wheels from center member and tie rod sleeves. Support center member and unbolt front pivot pin support. Move center member and front pivot forward until clear of rear pivot pin and lower to floor.

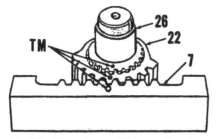

Fig. 35—When reassembling steering motor, be sure that timing marks (TM) on rack (7), pinion (22) and shaft (26) are aligned as shown.

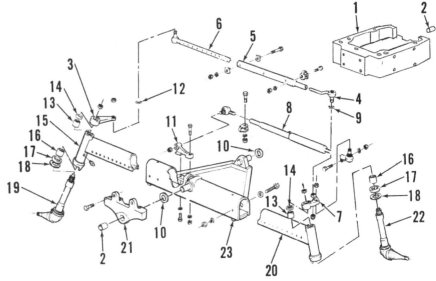

Fig. 40—Exploded view of front axle assembly used on Models 8700, 9700, TW10, TW20 and TW30.

1. Front support	13. Bushing
2. Bushing	14. Seal
3. Steering arm, R.H.	15. Axle extension, R.H.
4. Tie rod end	16. Bushing
5. Tie rod sleeve	17. Thrust bearing
6. Tie rod	18. Spacer
7. Steering arm, L.H.	19. Spindle, R.H.
8. Steering cylinder	20. Axle extension, L.H.
9. Seal	21. Support bracket
10. Thrust washer	22. Spindle, L.H.
11. Anchor	23. Front axle center member
12. Seal	

To remove pivot pins and bushings from front support (1—Fig. 40) and bracket (21), remove front axle for access. Renew front axle center member (23) if pivot pins are excessively worn or damaged. Use suitable driving tool to remove bushings. New bushings are pre-sized.

32. **TIE ROD AND TOE-IN.** Recommended toe-in is ¼ to ½-inch. To adjust toe-in, remove bolt securing sleeve (5—Fig. 40) to tie rod (6) and loosen clamp on sleeve. Turn sleeve (5) until desired toe-in is obtained. Tighten clamp, install bolt securing sleeve to tie rod and recheck toe-in measurement.

FRONT SUPPORT (PEDESTAL)

33. **REMOVE AND REINSTALL.** To remove front support casting (1—Fig. 40), remove radiator as outlined in paragraph 152. Remove front axle assembly as outlined in paragraph 29. Attach a hoist or other device to front support casting and unscrew retaining bolts and nuts.

To install front support casting, attach front support casting to engine using four upper bolts and tighten bolts to 180-220 ft.-lbs. Be sure oil pan cap screws are tight and measure gap between front support and oil pan at attaching bolt holes. Loosen the four installed cap screws, then install the two lower screws using shims equal to the measured clearance. Shims are available in thicknesses of 0.014, 0.017, 0.021, 0.024 and 0.027 inch. Tighten front support to cylinder block bolts to 180-220 ft. lbs., then tighten front support to oil pan cap screws to 270-330 ft.-lbs. Install front axle and radiator by reversing removal procedure.

POWER STEERING SYSTEM

These models are equipped with a hydrostatic power steering system consisting of a steering cylinder attached to the front axle, a Ross-type steering motor and a tilting steering gear mechanism.

34. **FLUID AND BLEEDING.** Use Ford M-2C41A fluid in this system. Fluid level should be kept at bottom of reservoir filler neck. System capacity is 7.8 US pints. Reservoir (2—Fig. 41) should be disassembled for cleaning and new filter (3) installed after each 600 hours of operation.

System is self-bleeding. Whenever power steering system has been disassembled for any reason, after reassembly, refill fluid reservoir, start engine to run at low idle rpm and operate steering full right to full left through at least five complete cycles. Replenish reservoir

fluid as needed to maintain level. System is free of trapped air when no bubbles appear, steering is firm and reservoir level remains steady at full.

35. **PRESSURE AND FLOW.** System pressure is regulated at 1950-2100 psi on TW-30 or 1550-1650 psi on all other models by the pump pressure relief valve. Rate of flow should be 4.2 gpm at 1000 engine rpm.

36. **R&R AND OVERHAUL PUMP.** For exploded view of pump refer to Fig. 41. Clean pump and surrounding area and disconnect pump pressure and return lines. Remove two cap screws securing pump to engine front cover and lift off pump and reservoir as a unit. Relief valve is now accessible. Shims (25) are available in thicknesses of 0.010, 0.015 and 0.060 inch to adjust relief valve opening pressure. A change of shim pack thickness of 0.005 inch will alter opening pressure approximately 35 psi.

To disassemble the pump, bend back tab washer and remove shaft nut (23), drive gear (22) and key (12). Mark or note relative positions of flange housing (18), pump body (13) and cover (6); then remove pump through bolts (5). Keep parts in their proper relative position when disassembling pump unit. If necessary, flow control valve components can now be removed. Pump gears (10 and 11) are available in a matched set only. Bearing blocks (9) are available separately but should be renewed in pairs if renewal is because of wear. Bearing blocks should also be renewed

with gear set if any shaft or bore wear is evident. Examine body (13) for wear in gear running track. If track is worn deeper than 0.025 inch on inlet side, body must be renewed. Renew all "O" rings and seals.

When reassembling the pump, tighten through bolts (5) to a torque of 25 ft.-lbs. and drive gear nut (23) and relief valve body (24) to a torque of 30-40 ft.-lbs.

37. **R&R STEERING MOTOR.** Remove hood and side panels and braces necessary to gain access to steering motor. Disconnect and tag hydraulic lines leading to steering motor. Hydraulic lines and motor parts should be sealed to prevent contamination. Loosen pinch bolt (4—Fig. 42) in coupler, remove steering column lower bracket screws and remove motor with bracket. Unscrew motor retaining screws and separate motor from bracket.

Reverse removal procedure to install steering motor. Refer to paragraph 34 for bleeding procedure.

38. **OVERHAUL STEERING MOTOR.** Refer to paragraph 21 and Fig. 28 for steering motor overhaul.

FRONT WHEEL DRIVE

44. **Front wheel drive is offered as an option on 8700, 9700, TW-10, TW-20 and TW-30 models. Major components are the transfer box, front axle, differential, wheel hubs, final drive planetary units**

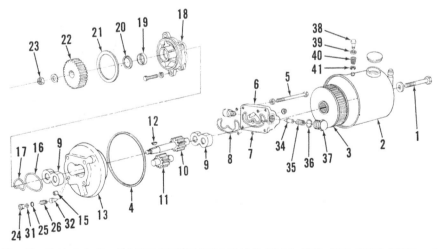

Fig. 41—Exploded view of power steering pump used in Models 8700, 9700, TW10, TW20 and TW30.

1. Bolt	10. Driven gear	20. Snap ring			
2. Reservoir	11. Follow gear	21. "O" ring	34. Flow control valve		
3. Filter	12. Woodruff key	22. Drive gear	35. Spring		
4. Gasket	13. Body	23. Nut	36. Seal		
5. Through-bolt	15. Ring dowel	24. Plug	37. Plug		
6. Cover	16. Seal ring	25. Shim	38. Vent		
7. "O" ring	17. Seal ring	26. Spring	39. Seal		
8. Seal ring	18. Flange housing	31. "O" ring	40. Spring		
9. Bearing block	19. Oil seal	32. Relief valve	41. "E" ring		

and front axle assembly. The front wheel drive transfer box bolts directly onto the transmission handbrake housing assembly and is lubricated by oil overflow from rear axle housing.

There are three axles used on six cylinder tractors equipped with front wheel drive. Axle APL 4053/CK is used on 8700, 9700 models, axle APL 3052/B is used on TW-10 and TW-20 models and axle APL 3054/BK is used on TW-30 models. The axle type number is stamped on an identification plate attached to the rear face of the axle next to the differential housing.

The differential is offset to the left of the axle center line and is driven from the transfer box by a drive shaft having a U-joint at both ends and a sliding coupling at the rear. Late 8700, 9700 models and all TW models are equipped with a limited slip differential.

At each axle end is a hub that contains a planetary reduction gear set. Each hub is held in place by a pair of king pins. Steering is controlled by a single, adjustable tie rod connected between the steering knuckles. The power steering cylinder is bolted to the steering arm on the left-hand steering knuckle.

FRONT AXLE

45. R&R FRONT AXLE AND PIVOT SUPPORT. To remove front axle (11—Fig. 47) and pivot support (6), first remove drive shaft. Remove front weights and weight bracket, then support front of tractor and remove bolt securing steering cylinder hose support to rear of pivot support (6). Disconnect steering cylinder from left hand steering arm and pivot support. Temporarily secure steering cylinder to underside of tractor. Support front axle (11), remove bolts securing pivot support (6) to front support, then roll axle assembly (11) away from tractor.

Inspect components for excessive wear or damage and renew as necessary. Install shims (4) to reduce pivot support (6) end play to 0.008-0.016 inch. To install front axle (11) and pivot support (6) assembly, reverse removal procedure. Tighten fasteners to following torques: pivot support to axle housing bolts—302 ft.-lbs.; steering cylinder to steering arm nut—217 ft.-lbs.; drive shaft flange nuts—36 ft.-lbs.

45A. TOE-IN. Toe-in is adjusted by detaching tie rod end and rotating end. Correct toe-in is 0-1/8 inch. Check for bent or excessively worn parts if toe-in is not correct.

46. WHEEL HUB AND PLANETARY CARRIER. To remove wheel hub

(25—Fig. 48) and planetary assembly (12), raise tractor and remove wheel on side requiring service. Drain oil in hub (25) and scribe alignment marks on planetary carrier (12) and hub (25). Remove retaining bolts (11) and remove carrier (12). Detach snap-ring (15) and remove sun gear (16).

To install carrier assembly (12), make sure alignment marks on carrier and hub (25) are aligned, then secure carrier assembly to hub (25). Bolts (11) should

be tightened to a torque of 18 ft.-lbs. Fill assembly with clean oil (specification M2C105-A, MIL-L-2105B SAE90 or API GL-5 SAE90), until sight plug hole overflows. Capacity of carrier assembly is approximately 2.1 pints for 8700 and 9700 models, 1.8 pints for TW-10 and TW-20 models and 3.2 pints for TW-30 models.

47. OVERHAUL. Unscrew slotted nut (17—Fig. 48) and remove planetary ring gear (18) and carrier (19) from hub

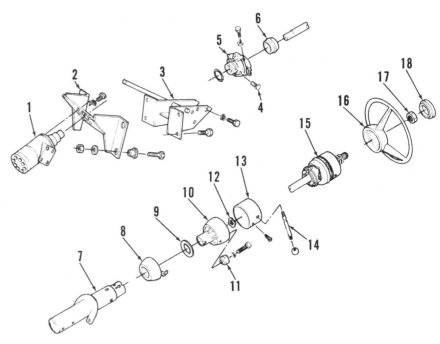

Fig. 42—View of tilting steering column used on Models 8700, 9700, TW10, TW20 and TW30.

1. Motor	6. Seal	10. Flange extension	14. Tilt lever
2. Lower bracket	7. Tube	11. Retainer	15. Tilt mechanism
3. Upper bracket	8. Flange cover	12. Spring washer	16. Steering mechanism
4. Pinch bolt	9. Thrust washer	13. Cover	17. Nut
5. Coupler			18. Cap

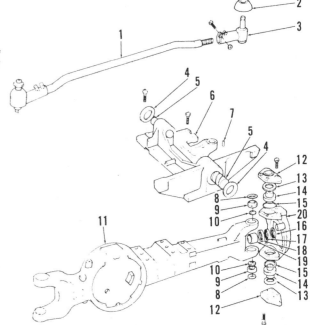

Fig. 47—Exploded view of front axle housing and steering knuckle.

1. Tie rod
2. Cap
3. Tie rod end
4. Shim
5. Pin
6. Pivot support
7. Roll pin
8. Dust cap
9. Bearing
10. Dished cover
11. Axle housing
12. King pin cap
13. Shim
14. King pin
15. "O" ring
16. Retaining ring
17. Ring
18. Shaft seal
19. Bushing
20. Steering knuckle

(25). Remove snap ring (20) from planetary ring gear (18) and remove ring gear carrier (19). Pull hub and bearing assembly (25) carefully off hub carrier (31).

On 8700, 9700, TW-10 and TW-20 models, extract shim (22) from hub carrier. Remove bolts securing hub carrier (31) to steering knuckle and remove carrier. Axle shaft (36) may now be removed from housing. Disassemble planetary carrier (12), inspect components and renew as necessary. Remove oil seal (30), spacer (29—on TW-30 models only), inner bearing (28) cone and dust seal (27) from hub if damaged or worn excessively. If damage or excessive wear is apparent, renew bushing (32) and oil seal (33) using suitable tools. Insert axle shaft (36) into axle housing (11—Fig. 47), then install hub carrier (31—Fig. 48) and secure to steering knuckle (20—Fig. 47) with hex head bolts. Tighten bolts to a torque of 155 ft.-lbs.

NOTE: Two shorter bolts must be installed in two lowest holes to prevent interference between steering knuckle and axle yoke on full lock.

Install bearing (21 and 28—Fig. 48) cups in hub (25) and bearing (28) cone to inboard side of hub (25). On TW-30 models, install washer (29), then on all models install seals (27 and 30). Install hub (25) on hub carrier (31) so carrier contacts hub.

Place ring gear carrier (19) in ring gear (18) and secure with snap ring (20). Position ring gear carrier (19) over splines of hub carrier (31) and push in. Apply grease to slotted nut (17) then install finger tight onto carrier shaft (31).

48. Hub roller bearings (21 and 28—Fig. 48) are preloaded by turning slotted nut (17). Refer to Fig. 49 and measure rolling resistance of hub roller bearings. Wrap a strong cord around wheel stud and hub, then attach a pull scale (S) to cord. Pull scale and note reading as hub rotates. Nut (17) is properly tightened when pulling resistance is between 9-14 lbs. If original bearings are used, reading may be 4.5-7 lbs.

NOTE: If slotted nut (17) locks before proper pulling resistance is obtained on Models 8700, 9700, TW-10 and TW-20, it will be necessary to replace shim (22— Fig. 48). Shim (22) is available in metric sizes from 5.2 to 5.4 mm. On Model TW-30, install a lockwasher and a second slotted nut (17) to serve as a locknut. Bend lockwasher tabs to engage both nuts.

Install sun gear (16) on axle shaft (36) and secure with snap ring (15). To check axle shaft (36) end play, push axle shaft inward. While shaft is being turned by an assistant, turn hub (25) to left and right steering lock positions. With wheel at full lock position, refer to Fig. 50 and measure depth of shaft end (36) from joint face of hub (25). Refer to Fig. 51 and using a similar setup, determine height of detent plug (14—Fig. 48) above joint face of carrier (12—Fig. 51). Subtract height of detent plug (14—Fig. 48) from previously measured depth of shaft end to obtain shaft end play. Drive out detent plug (14) and install shims (13) equal to measured shaft end play so end play will be zero with hub at full steering lock. Shims (13) are available in metric

thicknesses of 0.3, 0.5, 1.0 and 3.0 mm.

49. **R&R AXLE SHAFTS.** To remove axle shafts (36—Fig. 48), refer to paragraph 46 and remove wheel hub and planetary assembly. Axle shaft can now be removed. Inspect bushings (19—Fig. 47 and 32—Fig. 48) and seals (18—Fig. 47 and 33—Fig. 48) and renew if necessary. To install axle shaft, reverse removal procedure. Refer to paragraph 48 to adjust axle shaft end play and hub bearing preload.

STEERING KNUCKLE AND KING PINS

50. **R&R AND OVERHAUL.** Refer to Fig. 47 for an exploded view of steering

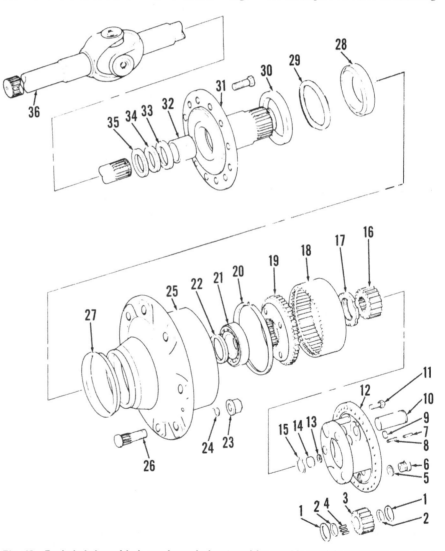

Fig. 48—Exploded view of hub, carrier and planetary drive used in models equipped with front wheel drive. Washer (29) is used in TW-30 models only.

1. Thrust washer	11. Cap screw	20. Snap ring
2. Spacer	12. Planetary carrier	21. Bearing
3. Planetary pinion	13. Spacing washer	22. Shim (all models except
4. Needle bearings	14. Detent plug	TW-30)
5. Sealing ring	15. Snap ring	23. Wheel nut
6. Screw plug	16. Sun gear	24. Spring washer
7. Roll pin	17. Slotted nut	25. Hub
8. Roll pin	18. Ring gear	26. Wheel bolt
9. Sealing cover	19. Ring gear carrier	27. Dust shield
10. Pin		

28. Bearing
29. Washer (TW-30)
30. Seal
31. Hub carrier
32. Bushing
33. Seal
34. Ring
35. Retaining ring
36. Axle shaft

knuckle assembly. Knuckle (20) may be removed without disassembly of final drive components.

Refer to paragraph 49 for removal of axle shaft. Remove upper and lower king pin caps (12) and tie rod (1). If left knuckle assembly is to be serviced, remove steering cylinder from steering arm. Remove shims (13) and set aside for installation in original location. Remove king pins (14) and bearing cones and mark for return to original location. Remove steering knuckle (20) from axle housing. If renewal is necessary, remove bearing (9) cups from axle housing yoke. Dished covers (10) should be renewed if damaged. Examine all components for burrs, scoring and excessive wear and renew as necessary.

51. If removed, install dished covers (10) and bearing (9) cups into axle housing yokes. Install "O" ring (15), dust cap (8) and bearing (9) cone to each king pin (14). Mount steering knuckle on axle housing yoke and install upper king pin (14). Coat originally installed shim (13) with grease and position it over king pin (14). Install king pin cap (12) on steering knuckle (20). Tighten socket head screws to 100 ft.-lbs. and hex head screws to 140 ft.-lbs. Repeat procedure on lower king pin assembly. Install axle shaft and hub carrier and tighten hub carrier cap screws to a torque of 155 ft.-lbs. Check steering torque by attaching Ford tool No. T3119 steering knuckle adaptor and a torque wrench to upper cap (12) just above king pin. Measure torque required to turn steering knuckle. If steering torque is not 13.3-15.5 ft.-lbs., install or remove shims (13) as required to obtain desired torque reading. Shims are available in metric thicknesses from 2.6 to 3.3 mm. Thickness of upper and lower shim packs should be equal, however, it may be necessary to transfer shims between shim packs so drive shaft is centered in axle housing. With steering knuckle in straight ahead position, rotate axle shaft and check for binding. Transfer shims between shim packs (13) to remove any binding.

DIFFERENTIAL AND BEVEL GEARS

52. **REMOVE AND REINSTALL.** Remove front wheels, disconnect drive shaft from differential input flange and drain axle housing oil. Remove cap screws securing hub carriers (31—Fig. 48) to steering knuckles. Support hubs then pull hubs and front axle drive shafts out far enough to disengage drive shaft inner ends from differential side gears.

NOTE: Remove two differential housing cap screws and replace with

Fig. 49—Measure hub bearing pre-load by measuring rolling torque using a spring scale (S). Turn nut (17) to adjust pre-load. See text.

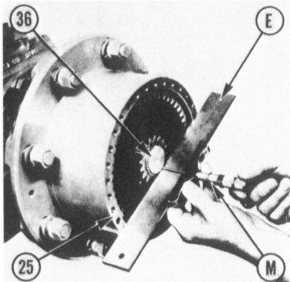

Fig. 50—View showing shaft projection measurement.

E. Straight edge
M. Depth micrometer
25. Hub joint face
36. Axle shaft

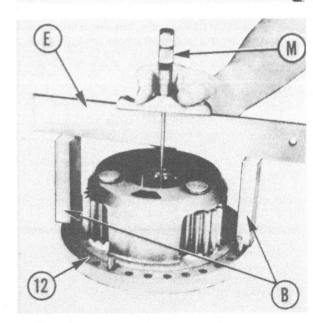

Fig. 51—Measure height of detent plug as outlined in text.

B. Gage blocks
E. Straight edge
M. Depth micrometer
12. Hub carrier

guide studs for ease of differential housing removal and installation.

Carefully remove differential housing assembly using a suitable floor jack or hoist. To reinstall reverse removal procedure. Differential housing to axle housing screws should be tightened to a torque of 61 ft.-lbs. Fill axle housing with API GL-5 SAE 90 lubricant. Axle capacity is 14.8 pints on 8700 and 9700, 16.9 pints on TW-10 and TW-20 or 15.8 pints on TW-30.

53. OVERHAUL (ALL MODELS EXCEPT TW-30). Mount differential housing in a vise, then unscrew bevel pinion nut (15—Fig. 52) and adjusting nuts (6). Mark bearing caps (2) according to position, remove caps and lift out differential.

Remove bearing (7) cones from differential case halves (4 and 9–Fig. 53), using a suitable puller. Remove bevel ring gear retaining cap screws and separate gear (4—Fig. 52) from differential case halves then separate halves.

Remove washer (14—Fig. 52), drive flange (13), and dust seal (12) from pinion shaft. Using a suitable press or hammer, remove bevel pinion gear (3) from differential housing (8). If required, press bearing (5) cone off bevel pinion shaft (3) and remove bearing (5) cup and shim (17) from differential housing. Remove oil seal (11) and drive bearing (10) out from rear of differential carrier.

Inspect all components and renew as necessary. Bevel ring (4) and pinion gear (3) are available as a matched set only. Identical serial numbers will be stamped on outer edge of bevel ring gear (4) and on end of bevel pinion gear (3).

If gears require renewal, the following procedure must be followed to determine pinion bearing shim (17) thickness: All measurements should be metric or conversion will be necessary. Refer to Fig. 54 and install pinion setting mandrel No. T3123 and dummy pinion No. T3123-3 on Models 8700 and 9700, or pinion setting mandrel No. T3131 and dummy pinion No. T3131-1 on Models TW-10 and TW-20. Apply pressure to mandrel when measurements are taken. Height of dummy pinion (a) + gap between mandrel and dummy pinion (b) + half the diameter of the differential bores (c) = dimension (x).

Note that dummy pinion height (a) is marked on side of tool. Dimension (c), is 1.96 inches for Models 8700 and 9700, 1.93 inches for TW-10 and TW-20 models. Measure overall width of assembled pinion bearing (5—Fig. 52) cone and cup, then add bearing width to etched pinion setting number (in millimeters) next to serial number on gear

end of bevel pinion (3). Subtract result from dimension (x—Fig. 54) to obtain required thickness of pinion bearing shim (17—Fig. 52). Shims are available in thicknesses of 0.1 through 0.5 mm. Install shims so tolerance is within plus or minus 0.05 mm of calculated thickness. Install shim (17) and bearing (5) cup in differential housing and press bearing (5) cone on pinion gear shaft until seated against shoulder.

To determine pinion bearing pre-load,

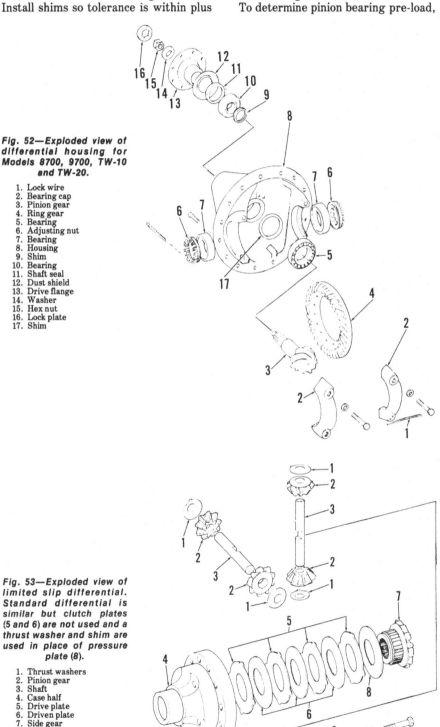

Fig. 52—Exploded view of differential housing for Models 8700, 9700, TW-10 and TW-20.

1. Lock wire
2. Bearing cap
3. Pinion gear
4. Ring gear
5. Bearing
6. Adjusting nut
7. Bearing
8. Housing
9. Shim
10. Bearing
11. Shaft seal
12. Dust shield
13. Drive flange
14. Washer
15. Hex nut
16. Lock plate
17. Shim

Fig. 53—Exploded view of limited slip differential. Standard differential is similar but clutch plates (5 and 6) are not used and a thrust washer and shim are used in place of pressure plate (8).

1. Thrust washers
2. Pinion gear
3. Shaft
4. Case half
5. Drive plate
6. Driven plate
7. Side gear
8. Pressure plate
9. Case half

install bearing (10) cup without shim (9), into small bore at rear of differential carrier (8). Install pinion shaft (3) into differential carrier, then insert adjustable spacer, tool No. 3120 for 8700, 9700 models or No. T3130 for TW-10 and TW-20 models (T—Fig. 55) so it surrounds pinion shaft (3—Fig. 52). Spacer roll pins should protrude as much as possible. Install bearing (10) cone without oil seal (11), then install drive flange (13), washer (14) and retaining nut (15). Hold drive flange and tighten nut in small increments until torque required to turn pinion (3) is 10-20 in.-lbs. Remove pinion shaft (3) from differential housing (8). Carefully remove adjustable spacer. Using a suitable gage, measure overall height of adjustable spacer. This height will be required thickness of shim (9). Shims are available in thicknesses from 5.6 mm to 6.3 mm.

Reinstall pinion shaft (3) into differential housing (8) and place proper shim (9) on pinion shaft with chamfer towards gear end. Press bearing (10) cone into position. Install new oil seal (11) with lip facing inward. Coat threads of retaining

nut (15) with grease, then install drive flange (13), washer (14) and retaining nut (15). While holding flange, tighten retaining nut to a torque of 260 ft.-lbs. Install new cotter key on 8700, 9700 models or new lock plate (16) on TW-10 and TW-20 models.

Install bevel ring gear (4) on differential case half (9—Fig. 53), install bolts and tighten to a torque of 155 ft.-lbs. If equipped with limited slip differential, refer to next paragraph for differential assembly. On models not equipped with limited slip, install new lock plate over differential case to bevel ring gear bolts. Install pinion gears (2) with thrust washers (1) on pinion shafts (3) and install assembly with side gear (7) into differential case half (9). To determine backlash of pinion gears (2), hold three pinion gears and measure backlash of fourth gear. Backlash should be 0.006-0.008 inch. To reduce backlash install a

thicker shim behind side gear. Shims are pinned into the differential case half. When new shims are installed, securing pins should be below surface of shim. Shims must be installed with oil grooves facing side gear. Repeat procedure for other case half. After obtaining correct backlash, assemble differential case halves making sure projections on pinion thrust washers are positioned as shown in Fig. 56. Part numbers are stamped on outer edge of both differential case halves (4 and 9—Fig. 53) and should align when halves are assembled.

To assemble limited slip differential on models so equipped, proceed as follows: Refer to Fig. 53 and assemble pressure plate (8) and clutch plates (5 and 6) on side gear (7) being sure that polished side of pressure plate (8) is towards gear. Carefully insert gear and clutch pack assembly into differential case half (4 or 9) so clutch plate (5) tabs engage

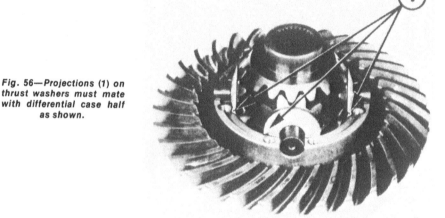

Fig. 56—Projections (1) on thrust washers must mate with differential case half as shown.

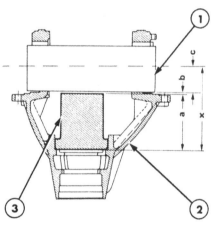

Fig. 54—Diagram of pinion shimming tools and measuring points. Refer to text.

1. Pinion setting mandrel
2. Differential housing
3. Dummy pinion

Fig. 57—Measure clutch pack free play by inserting dial gage (1) through case half (2) to contact clutch plate.

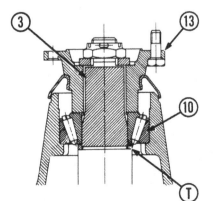

Fig. 55—Refer to text to adjust pinion bearing preload.

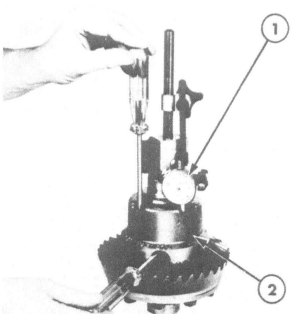

slots in case half. Assemble remaining differential case half, side gear and clutch pack. Install bevel pinions (2), thrust washers (1) and differential shafts (3). Assemble differential halves (4 and 9) and tighten retaining screws to a torque of 85 ft.-lbs. To check free play of each clutch pack, refer to Fig. 57 and mount a dial gage on one end of differential case so the plunger contacts the outer clutch plate. Use two screwdrivers to move clutch plate assembly up and down while noting movement of dial gage needle. Allowable free play is 0.004-0.008 inch. If free play is excessive, side gear pressure plate (8—Fig. 53) must be changed. Pressure plates (8) are available in thicknesses of 2.8, 2.9 and 3.0 mm. If after installation of thickest pressure plate free play remains excessive, then a new set of clutch plates (5 and 6) must be installed. Repeat procedure for opposite side of differential.

On all models, note difference in widths of bearings (7—Fig. 52) then install bearing cones on differential case halves (4 and 9—Fig. 53) so wider bearing is on ring gear case half (9). Install differential assembly into housing (8—Fig. 52). Install bearing caps (2) making sure alignment marks (made during disassembly) are aligned. Screw adjusting nuts (6) hand tight, then tighten bearing cap bolts to a torque of 140 ft.-lbs. Bevel ring gear (4) backlash is reduced by moving gear (4) towards pinion (3) by adjustment of slotted nuts (6). Correct backlash is 0.008-0.011 inch. Backlash should be measured with a dial gage mounted to differential case (8).

To check differential bearing (7) pre-load, mount a dial gage so tip of dial gage contacts back (flat) side of bevel ring gear (4). Move differential back and forth while tightening adjusting nuts (6) until no movement of differential is noted on dial gage. Pre-load bearings (7) by tightening adjusting nut (6) opposite bevel ring gear (4) 1½ to 2½ slots, as necessary to align cotter key with a slot in nut (6). Re-check bevel ring gear (4) backlash as previously described. Repeat steps as necessary to obtain proper bearing pre-load and gear backlash. Measure bevel ring gear (4) runout on back surface of gear. If runout exceeds 0.003 inch then gear is seated improperly on differential case and should be removed and inspected.

Refer to Fig. 58 for proper bevel ring and pinion gear tooth contact pattern. If pinion shimming procedure wasn't performed properly, ideal tooth pattern will not be obtained and shimming procedure will have to be repeated.

Insert new cotter keys to secure adjusting nuts (6—Fig. 52). On models without a limited slip differential, safety wire differential case half securing bolts.

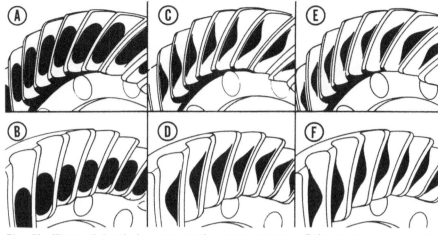

Fig. 58—Views of bevel ring gear tooth contact patterns. Refer to text to determine thickness of shim (17-Fig. 52 or 16-Fig. 59) if appropriate tools are available.

A. Proper tooth contact-coast side pattern
B. Proper tooth contact-drive side pattern
C. Pinion gear requires thicker shim-coast side pattern
D. Pinion gear requires thicker shim-drive side pattern
E. Pinion gear requires thinner shim-coast side pattern
F. Pinion gear requires thinner shim-drive side pattern

To install differential assembly refer to paragraph 52.

54. **OVERHAUL (TW-30 MODELS).** Mount differential housing (7—Fig. 59) in a vise and loosen twelve cap screws that retain bevel ring gear (3) to differential case half (9—Fig. 53). Drive out roll pins that lock adjusting nuts (4—Fig. 59), then unscrew and remove adjusting nuts (4). Remove bearing (5 and 6) cups. Note that differential components must be disassembled in differential housing and cannot be removed as a unit assembly. Remove ring gear retaining screws then reposition housing so pinion shaft is horizontal. Separate differential case halves (4 and 9—Fig. 53) and remove both pinion shafts (3), pinion gears (2) and thrust washers (1). Remove two halves (4 and 9) from housing separately. Separate bevel ring gear (3—Fig. 59) from differential case half. If necessary, remove bearing cones from differential case halves. Remove retaining nut (14), washer (13) and drive flange (12)

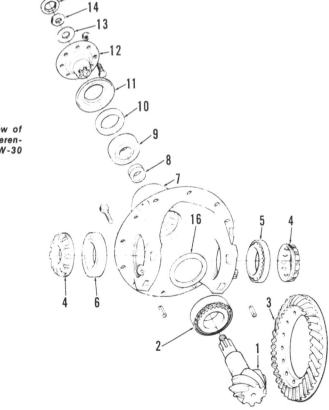

Fig. 59—Exploded view of front wheel drive differential housing for TW-30 models.

1. Bevel pinion gear
2. Bearing
3. Bevel ring gear
4. Adjusting nut
5. Bearing
6. Bearing
7. Housing
8. Sleeve
9. Bearing
10. Shaft seal
11. Dust shield
12. Drive flange
13. Washer
14. Nut
15. Lock plate
16. Shim

from bevel pinion gear (1), then drive or press bevel pinion gear from housing. Remove and discard collapsible sleeve (8). Remove oil seal (10) and bearing (9) cone from rear of housing (7), and if necessary, drive out bearing (2) cup.

Clean and inspect all components and renew as necessary. Bevel ring and pinion gears are available only as a matched set. Identical serial numbers will be stamped on outer edge of bevel ring gear (3) and on end of bevel pinion gear (1). If housing, bearings, differential case or ring and pinion gears have been renewed, then the following shimming procedure must be followed to determine pinion bearing shim (16) thickness: All measurements should be metric or conversion will be necessary. Refer to Fig. 54 and install dummy pinion tool No. T3131 into bore of differential housing, then install pinion setting mandrel, tool No. T3131 into differential bearing bores.

NOTE: Mandrel is smaller at one end, install small end of mandrel tool No. T3131 into bearing bore on bevel ring gear side.

Apply pressure to mandrel as following measurements shown in Fig. 54 are taken: Height of dummy pinion (a) + gap between dummy pinion and mandrel (b) + half the diameter of differential bores (c) = dimension (x).

Dummy pinion height should be 1.93 inches. Measure overall width of assembled pinion bearing (2—Fig. 59) cone and cup, then add bearing width to etched pinion setting number (in millimeters) next to serial number on gear end of bevel pinion (1). Subtract result from dimension (x—Fig. 54) to obtain required thickness of pinion bearing shim (16—Fig. 59). Shims are available in thicknesses of 0.1 through 0.5 mm. Install shims so tolerance is within plus or minus 0.05 mm of calculated thickness. Install shim (16) and bearing (2) cup in housing and press bearing (2) cone on pinion gear shaft until seated against shoulder.

To determine pinion bearing pre-load, install bearing (9) cup into small bore at rear of differential carrier (7) and install pinion gear (1). Install a new collapsible sleeve (8) on bevel pinion shaft, then install bearing (9) cone and oil seal (10). Coat threads of retaining nut (14), install washer (13) and drive flange (12) then install nut (14) finger tight. Before determining pinion (9) bearing pre-load, the resistance of oil seal (10) must be determined. Apply a torque wrench to flange nut (14), turn pinion (1) shaft and record torque required to turn shaft. This result will be oil seal (10) resistance. While holding drive flange

(12) with a suitable tool, tighten retaining nut (14) in small increments until torque required to turn assembly is between 10-20 in.-lbs., after adding resistance of oil seal (10). If desired torque reading is exceeded, then bearing pre-load is excessive and a new sleeve (8) must be installed and bearing pre-load readjusted.

Install bevel ring gear (3) onto differential case half (9—Fig. 53). Install pressure plates (8) on side gears so polished side is towards gear and assemble clutch disc plates (5 and 6) on side gears (7). Install gear and clutch components into differential case halves (4 and 9). Refer to Fig. 53 for proper clutch pack assembly sequence. Mate differential halves (4 and 9), install ring gear retaining screws and tighten to a torque of 155 ft.-lbs. Refer to Fig. 57 and check clutch pack free play as follows: Mount a dial gage on differential case so gage plunger contacts outer clutch plate. Use two screwdrivers to move clutch plate assembly up and down and note movement of gage needle. Allowable free play is 0.004-0.008 inch. Adjust free play by changing thickness of pressure plate (18—Fig. 53). Pressure plates (8) are available in 2.8, 2.9 and 3.0 mm sizes. If after installation of thickest pressure plate (8) free play remains excessive, a new set of clutch plates (5 and 6) must be installed. Repeat instructions for opposite differential case half.

Separate case halves (4 and 9) for installation in differential housing (7—Fig. 59). Note that bearing (5) is wider than bearing (6) and bearing (5) must be installed on ring gear differential case half (9—Fig. 53). While supporting differential housing with pinion gear in a horizontal position, install ring gear with differential case half (9) and clutch assembly. Install opposite differential case half and clutch assembly then while holding case half (4), install differential shafts (3) complete with bevel pinion gears (2) and thrust washers (1). Be sure projections on thrust washers mate with case half as shown in Fig. 56. Mate differential halves and install ring gear screws but do not tighten at this time. Install bearing (5 and 6—Fig. 59) cups and adjusting nuts (4). Tighten bevel ring gear cap screws to a torque of 155 ft.-lbs.

Backlash between ring and pinion gears should be 0.008-0.011 inch and is adjusted by turning adjusting nuts (4—Fig. 59). To check differential bearing (5 and 6) pre-load, mount a dial indicator so tip of gage contacts back (flat) side of bevel ring gear (3). While prying differential side to side, turn adjusting nut (4) opposite bevel ring gear just until no movement of differential assembly is noted on dial gage.

Pre-load bearings (5 and 6) by tightening adjusting nut (4) opposite bevel ring gear 1½-2½ slots, as necessary, to align locking roll pin with a slot in nut (4). Re-check bevel ring gear to pinion gear backlash. With dial gage tip still contacting back of bevel ring gear, check for ring gear runout. If runout exceeds 0.003 inch then gear is seated improperly on differential case and should be removed and inspected.

Refer to Fig. 58 for proper bevel ring and pinion gear tooth contact pattern. If pinion shimming procedure wasn't performed properly, ideal tooth pattern will not be obtained and shimming procedure will have to be repeated.

To install differential assembly, refer to paragraph 52.

FRONT SUPPORT

56. **REMOVE AND REINSTALL.** To remove front support, refer to paragraph 45 and separate front axle assembly from tractor then refer to paragraph 33 for front support removal and installation.

POWER STEERING SYSTEM

57. The power steering system is the same as used on two wheel drive models. Refer to paragraphs 34 through 38 for service.

DRIVE SHAFT

58. **R&R AND OVERHAUL.** To remove drive shaft, remove eight locknuts at transfer box output flange and lower rear of shaft to ground. Remove eight cap screws and locknuts at differential input flange and lower shaft to ground.

Inspect all components for damage or excessive wear and renew as necessary. If U-joints are renewed, drive shaft must be balanced.

To install drive shaft, reverse removal procedure. Tighten all cap screws and locknuts to a torque of 36 ft.-lbs.

TRANSFER BOX

59. **OPERATION.** The transfer box is mounted on the transmission handbrake housing as shown in Fig. 60. The transfer box output shaft is supported at the rear by needle roller bearing (2) located within the hollow shaft of the transmission handbrake while the front is supported by tapered roller bearing (11). The transfer box assembly is actuated by a control rod that passes through the cab floor and connects to a control valve designed to divert oil from the tractor hydraulic system to clutch piston

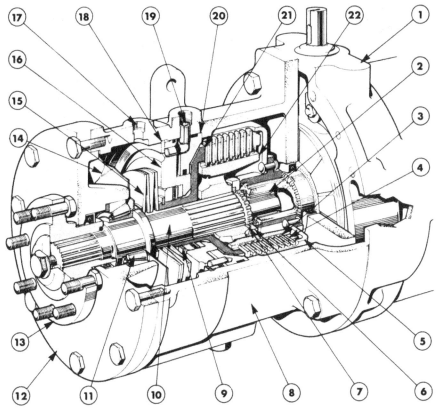

Fig. 60—Sectional view of transfer box assembly.

1. Handbrake housing	7. Split ring	13. Output shaft flange	
2. Roller bearing	8. Transfer box housing	14. Oil deflector	19. Roll pin and clutch oil
3. Driven plates	9. Snap ring	15. Belleville washers	feed port
4. Drive plates	10. Output shaft	16. Clutch piston	20. Thrust ring
5. Clutch drum	11. Roller bearing	17. Clutch oil return port	21. Pressure piece
6. Retainer ring	12. Cover	18. Snap ring	22. Plate carrier

(16). When the rod is pushed down, the control valve allows hydraulic oil to pressurize clutch piston (16) and disengage the clutch. When the rod is pulled up, the oil will return to the sump through an oil pipe leading from the control valve to the front of the transfer box case. In this position, clutch plates (3 and 4) are engaged as pressure of Belleville washers (15) against pressure piece (21) forces clutch plates together. With clutch plates engaged, plate carrier (22) and output shaft (10) rotate and transfer power to front wheel drive.

60. **R&R TRANSMISSION HAND-BRAKE AND TRANSFER BOX.** Jack up left side of tractor to divert transmission oil away from transmission handbrake housing. Remove left rear wheel and drain oil from transfer box. Remove locknuts securing drive shaft rear flange to transfer box output flange. Remove cotter pin, washer and clevis pin from operating rod. Disconnect main oil feed pipe at top of control valve. Disconnect lower end of handbrake cable from control rod. While supporting handbrake/transfer box assembly remove hex head bolts that secure assembly to rear axle center housing. Remove complete unit. To reinstall reverse removal procedure. Refer to paragraph 176 and refill transmission oil as required.

61. **OVERHAUL.** Disconnect clutch oil feed and return lines from transfer box housing ports (17 and 19—Fig. 60). Remove hex head bolts securing control valve retaining plate to housing and remove valve assembly, retaining plate and oil pipes. Remove remaining bolts and separate transfer box from handbrake assembly. Using special tool Ford No. T3122 or equivalent remove hex bolt, lockplate disc and pin (25—Fig. 61) from end of output shaft then remove flange (13). Mark housing and cover for proper alignment during reassembly. Remove cover bolts and transfer box cover (12). It may be necessary to use a hammer and brass drift to aid in

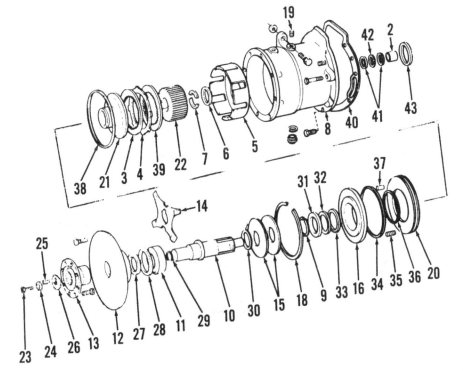

Fig. 61—Exploded view of transfer box.

2. Bearing	24. Lock plate
3. Driven plate	25. Grooved pin
4. Drive plate	26. Disc
5. Clutch drum	27. Oil seal
6. Retainer ring	28. Shim
7. Split rings	29. "O" ring
8. Case	30. Shim
9. Snap ring	31. Thrust washer
10. Output shaft	32. Bearing
11. Bearing	33. Shim
12. Cover	34. Ring
13. Output flange	35. Detent spring
14. Oil deflector	36. Ring
15. Belleville washers	37. Dowel pins
16. Piston	38. Seal ring
18. Snap ring	39. Pressure plate
19. Roll pin	40. Gasket
20. Thrust ring	41. Thrust washers
21. Pressure piece	42. Bearing
22. Plate carrier	43. Oil seal
23. Cap screw	

removal. Remove clutch and output shaft assembly snap ring (18). Drive roll pin (19) into transfer box housing from outside. Drive or press clutch and output shaft assembly out of housing from rear and remove clutch drum (5). Roller bearings in transfer box may be removed if necessary.

Using suitable tools, remove bearing inner race and oil deflector (14) from output shaft. Remove retainer ring (6) from rear of output shaft. Support output shaft and with Ford tool No. 1312 or equivalent, compress Belleville washers (15) and remove split rings (7). Remove output shaft components. Remove snap ring (9) and disassemble piston (16), springs, dowel pins, seals and thrust ring (20). Remove oil seal at rear face of transfer box case. Remove roller bearing if necessary from front of transfer box handbrake shaft. Inspect transfer box components and renew as necessary.

Control valve must be renewed as a unit assembly if defective.

62. When reassembling, coat periphery of oil seals with sealant meeting Ford specification ESA-M4G129-A or equivalent. Coat inner seal lips and "O" rings with petroleum jelly.

Install quad ring (34) on piston (16) and quad ring (36) and seal ring (38) on thrust ring (20). Install two dowel pins (37) and four springs (35) in piston (16) and mate thrust ring (20) with piston. Install piston and thrust ring assembly on pressure piece (21) and assemble shim (33), thrust bearing (32) and small thrust washer (31) on pressure piece. Compress unit and retain with snap ring (9).

Install clutch plate carrier (22) in clutch drum. Coat drive plates (4) with oil, then insert components in following order: pressure plate (39) with flat side up, drive plate (4) and driven plate (3) then continue alternating installation of plates (4 and 3) until remaining plates are installed. If old Belleville washers are to be installed disregard following paragraph. However, if new Belleville washers are necessary use procedure in following paragraph to determine shim thickness.

To determine shim (30) thickness, place clutch assembly on output shaft with clutch drum up. Remove drum without moving plates. Coat split rings (7) with petroleum jelly and install to secure clutch components. Carefully turn assembly so shaft collar (C—Fig. 62) is facing up and measure gap between pressure piece (21) and underside of output shaft collar (C) using a 0.787 inch long gage block (B). Record this measurement. Note dimension of Belleville washer pack as indicated on packing slip then subtract dimension of washer pack from previously measured gap

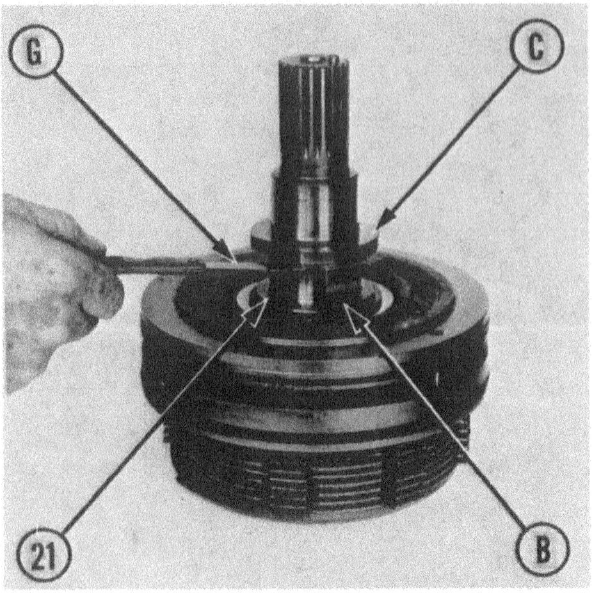

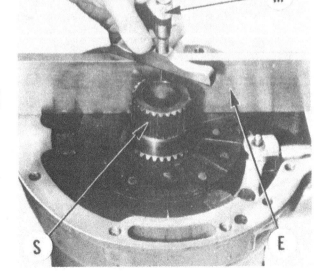

Fig. 62—View showing transfer box clutch shimming procedure.

B. Gage block (0.787-in.)
C. Output shaft collar
G. Feeler gage
21. Pressure piece

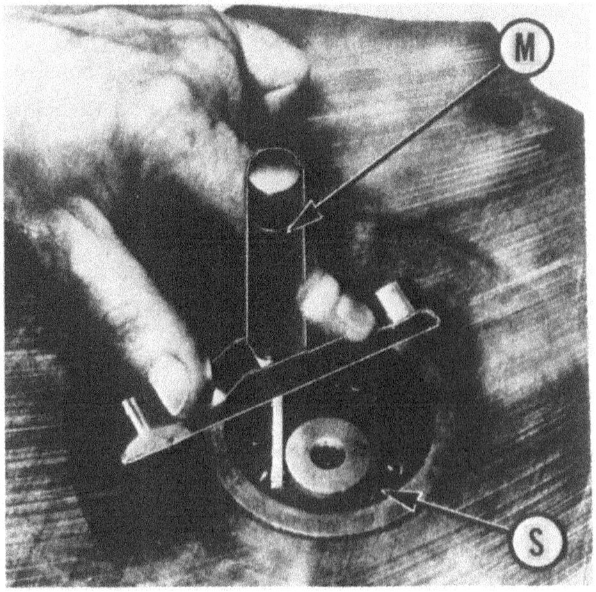

Fig. 63—View checking cover shim thickness.

M. Depth micrometer
S. Output shaft shoulder

Fig. 64—View showing proper method to check brake shaft projection.

E. Straight edge
M. Depth micrometer
S. Brake driven shaft

between shaft collar and pressure piece. Result is thickness of shim (30—Fig. 61). Shim is available in metric thicknesses from 2.0 mm through 5.0 mm in 0.5 mm increments. Separate output shaft from clutch and piston assemblies. Install oil deflector (14) with lip engaging flat on collar then press roller bearing (11) cone on front of shaft until oil deflector (14) and bearing are seated against shaft collar. Install original or new shim (30) then install Belleville washers (15) with concave sides facing each other. Install clutch and piston assemblies. Remove drum (5) without disturbing components, compress clutch and install split rings (7) on output shaft. Cover retaining ring (6) with petroleum jelly and install over split rings. Reinstall clutch drum.

Install seal (1) in housing (8) using a suitable sealant on seal periphery. Coat transfer box housing bores (where piston and thrust ring are located) with petroleum jelly and lower housing (8) over clutch and output shaft components. Align one recess in outer edge of thrust ring (20) with clutch piston oil pressure feed hole (19—Fig. 60) then drive roll pin into feed hole 0.006 inch from outer edge of hole. Roll pin must engage thrust ring recess to prevent turning. Secure clutch piston with snap ring (18). Coat outer edge of oil seal (43) with a suitable sealant and press into transfer box cover with sealing lip towards roller bearing. If front bearing cover outer race was removed install original size shim. Obtain correct output shaft end play by performing following shimming procedure: Protect front cover oil seal from output shaft splines and install "O" ring (29) on output shaft stem. Coat cover (12) face with sealant then install cover on housing using alignment marks made during disassembly. Tighten cap screws to a torque of 32 ft.-lbs. Mount transfer box assembly on blocks face down. Measure distance from rear face of case to output shaft shoulder as shown in Fig. 63. Distance should be between 2.216-2.232 inches. Install correct thickness of shim (28—Fig. 61) to

obtain proper dimension. After shimming cover to proper limits, record measured distance. Install output shaft flange bolts then secure flange (13) to output shaft. Tighten flange retaining screw to 88 ft.-lbs.

63. To determine proper end play between transfer box output shaft and drive shaft in brake housing assembly, remove brake housing gasket and place a straight edge of known height across brake housing face as shown in Fig. 64. Measure distance from face of brake housing to edge of brake driven shaft. Subtract previously measured distance from transfer box rear face to output shaft shoulder (Fig. 63). The result (Dimension C – Fig. 65) is the total thickness of thrust needle bearing (42 – Fig. 61) and two thrust washers (41) needed to obtain desired end play. See Fig. 65 to determine correct bearing/washer thicknesses. Coat thrust needle bearing and thrust washers with petroleum jelly and install on rear of transfer box output shaft. Washer's polished face must face bearing. Install new needle bearing in bore of brake driven shaft if necessary. Apply sealer, Ford specification ESE-M4G-11-A or equivalent to gasket, then install transfer box assembly and control valve plate to brake assembly. Tighten housing cap screws to a torque of 38 ft.-lbs. Tighten control valve retaining plate bolts to a torque of 18 ft.-lbs. Install clutch, oil return lines and transfer case drain plug.

NOTE: Check for proper alignment of drive shaft splines if drive shaft has been separated.

ENGINE AND COMPONENTS

R&R ENGINE WITH CLUTCH

76. To remove engine and clutch assembly, proceed as follows: Remove

hood side panels, air pre-cleaner, exhaust pipe and hood top panel. Drain radiator and disconnect radiator hoses. Disconnect transmission oil cooler lines on units with pressure lubricated transmissions. If engine is to be disassembled, drain oil pan. Disconnect engine oil cooler lines from engine and power steering tubes from steering motor in front support or cylinder on front axle. Unbolt side braces from radiator shell and support tractor under front end of transmission. On wide adjustable front axle models, drive a wooden block between front axle and pedestal on both sides, so that front assembly cannot tip sideways. Attach hoist to radiator shell, remove side plates and unbolt front support from engine.

NOTE: Be careful not to lose shims on front support to oil pan bolts. Roll front end assembly away from engine.

Disconnect battery cables (negative cable first), then disconnect power steering tubes from pump and hydra-motor steering unit. Disconnect air cleaner hose at intake manifold and muffler from exhaust manifold on 8000, 8600, 8700, and TW-10 models. Disconnect turbocharger oil lines and manifold connections on 9000, 9600, 9700, TW-20 and TW-30 models. Plug all openings on turbocharger, oil lines and manifolds. Attach hoist to air cleaner and hood support, unbolt side braces from instrument panel support and lift assembly from tractor. Remove fuel tank on 8700, 9700, TW-10, TW-20 and TW-30 models. Remove turbocharger intercooler on TW-30 models.

Disconnect electrical wiring, Proof-Meter cable, fuel shut-off cable, fuel line and throttle linkage from engine. Remove battery cables from starter, remove starter motor, then remove flywheel access cover. Attach hoist to engine with lifting fixture attached with cylinder head bolts. Unbolt engine from transmission and lift assembly away with hoist.

To reinstall engine, reverse removal

Dimension 'C' (inch)	0.279-0.284	0.285-0.288	0.289-0.292	0.293-0.296	0.297-0.299	0.300-0.303	0.304-0.309	0.310-0.313	0.314-0.317	0.318-0.324	0.324-0.330
Washer (1) thickness (inch)	0.110	0.110	0.118	0.118	0.110	0.124	0.118	0.124	0.118	0.124	0.124
Washer (2) thickness (inch)	0.0927	0.0984	0.0937	0.0984	0.110	0.0984	0.110	0.110	0.118	0.118	0.129
Needle Bearing thickness (inch)	0.0787	0.0787	0.0787	0.0787	0.0787	0.0787	0.0787	0.0787	0.0787	0.0787	0.0787

Fig. 65—Chart showing bearing-washer combinations available to maintain proper clearance between brake driven shaft and shoulder of transfer box output shaft.

procedures and observe the following: Remove transmission input shaft from transmission and place into clutch splines. As engine is moved toward transmission, the input shaft will swallow pto shaft. Tighten the 5/8-inch engine to transmission bolts to a torque of 95-130 ft.-lbs. and ¾-inch bolts to 180-220 ft.-lbs.

NOTE: If oil pan was removed, refer to paragraph 102.

Refer to paragraph 6, 33 or 56 for installation of front support to engine. Bleed the fuel injection system as outlined in paragraph 115 and bleed the power steering system as outlined in paragraph 7 or 34.

Make connections on 9000, 9600, 9700, TW-20 and TW-30 series turbocharger and prime with oil following procedure in paragraph 137.

ENGINE COMPRESSION PRESSURES

77. Engine compression should be checked at cranking speed of 200 rpm with a maximum allowable variation between cylinders of 20 psi.

NOTE: Considerable variation in compression pressures will be noted at speeds under 200 rpm.

Compression pressure should be 300-400 psi on Model TW-10, 275-375 psi on Models TW-20 and TW-30 or 380-480 psi on all other models.

CYLINDER HEAD

78. **REMOVE AND REINSTALL.** To remove cylinder head, drain cooling system and proceed as follows: Remove engine hood side panels, air pre-cleaner, exhaust pipe and hood top panel. Disconnect power steering tubes from Hydramotor unit, power steering pump and steering motor. On 8000, 8600, 8700 and TW-10 models, disconnect air cleaner hose from intake manifold and muffler from exhaust manifold. Remove turbocharger intercooler on TW-30 models. Disconnect and remove turbocharger oil lines and manifold connections on 9000, 9600, 9700, TW-20 and TW-30 models and remove turbocharger. Plug all openings on turbocharger, oil lines and manifolds. Unbolt side braces from radiator shell and instrument panel support casting. Attach hoist and lift air cleaner and hood support assembly from tractor. Remove fuel tank from 8700, 9700, TW-10, TW-20 and TW-30 models. Disconnect injector lines from injectors and pump and remove lines. Cap all exposed fittings to prevent entry of dirt. Clean area around injectors, remove

mounting nuts and if necessary, pry injector gently from head. Remove intake and exhaust manifolds, then remove rocker arm cover, rocker arms assembly and push rods. Disconnect upper radiator hose and temperature gage sender. Unbolt and remove cylinder head assembly.

NOTE: If cylinder head gasket failure has occurred, check cylinder head and block mating surface for flatness. Maximum allowable deviation from flatness is 0.006 inch overall or 0.003 inch in any six inches. If cylinder head is not within flatness specified or is rough, the surface may be machined providing the depth from valve seat inserts to cylinder head surface is not less than 0.117 inch after machining. If the cylinder block is not within flatness specified, it may be machined providing the distance between top of pistons at top dead center and top surface of cylinder block is not less than 0.002 inch after machining. Also, install cylinder head to block without gasket, install rocker arm supports, washer and all head bolts finger tight. Then using feeler gage, measure clearance between underside of bolt heads and cylinder head or rocker arm supports. If clearance is 0.010 inch or more, cut threads of bolt hole deeper in block with a suitable tap.

Models 8700, 9700, TW-10, TW-20 and TW-30 are equipped with 9/16-inch cylinder head bolts while all other models have ½-inch bolts. New cylinder blocks have 9/16-inch bolt holes so a stud kit must be installed to mate an early cylinder head with a new cylinder block. When reassembling, make sure head gasket is properly positioned on two dowel pins.

Tighten ½-inch bolts or nuts to 110 ft.-lbs. using steps of 90 and 100 ft.-lbs. Tighten 9/16-inch bolts to 160 ft.-lbs. using steps of 120 and 140 ft.-lbs. Refer to Fig. 69 for tightening sequence.

NOTE: Torque values given are for lubricated threads; tighten cylinder head bolts only when engine is cold.

Adjust intake valve gap of 0.014-0.016 inch and exhaust valve gap to 0.017-0.019 inch. Complete the reassembly of engine by reversing disassembly procedure. Tighten the intake manifold bolts to a torque of 23-28 ft.-lbs., the exhaust manifold bolts to a torque of 25-30 ft.-lbs. Make connections on 9000, 9600, 9700, TW-20 and TW-30 models turbocharger and prime with oil by following procedure in paragraph 137. With reassembling complete, bleed the diesel fuel system as outlined in paragraph 114A or 115A.

VALVES, GUIDES, STEM SEALS AND SEATS

80. Exhaust valves are equipped with positive type rotators and an "O" ring type seal is used between valve stem and rotator body. Intake valve stems are fitted with umbrella type oil seals on 8000, 8600, 8700 and TW-10 engines. On 9000, 9600, 9700, TW-20 and TW-30 engines, no seals are used on intake valve stems. Both the intake and exhaust valves seat on renewable type valve seat inserts that are a shrink fit in cylinder head. Inserts are available in oversizes of 0.010, 0.020 and 0.030 inch as well as standard size.

Fig. 69—Tighten cylinder head retaining bolts in sequence shown.

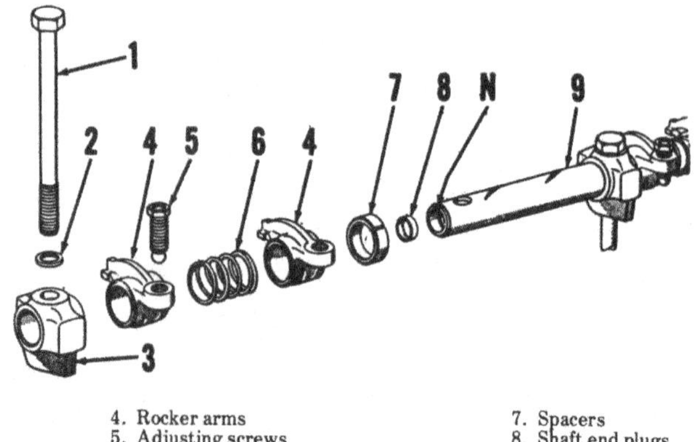

INTAKE SIDE

EXHAUST SIDE

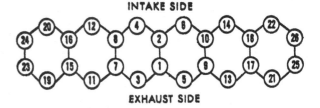

Fig. 70—Exploded view showing rocker arms, spring and rocker arm spacer sequence for No. 1 cylinder; sequence for installation of remaining rocker arms, springs and spacers is identical. Install shaft with notch (N) up and towards front of engine.

N. Notch
1. Cylinder head bolt
2. Flat washer
3. Rocker arm support
4. Rocker arms
5. Adjusting screws
6. Springs
7. Spacers
8. Shaft end plugs
9. Rocker arm shaft

NOTE: Valve seat inserts of 0.010 and 0.020 inch oversize have been installed in some cylinder heads in production. Cylinder heads so fitted are stamped "SO 10/OS" or "SO 20/OS" on exhaust manifold side in line with oversize valve seat. Counterbore in cylinder head for standard exhaust valve seat is 1.597-1.598 inches and for standard intake valve seat is 1.897-1.898 inches.

Refer to Fig. 70A to determine valve face and seat angle. Valve margin is 1/16-inch for intake valves of Models 9600, 9700, TW-20 and TW-30 and 1/32-inch for all other valves.

Valve guides are integral with cylinder head. If wear on guides or valves is excessive, renew valves and ream valve guide bore to nearest oversize if necessary. Desired stem to guide clearance is 0.001-0.0024 inch for intake valves and 0.002-0.0037 inch for exhaust valves. New (standard) stem diameter is 0.3711-0.3718 inch for intake valves and 0.3701-0.3708 inch for exhaust valves. Valves with 0.003, 0.015 and 0.030 inch oversize stems are available as well as reamers for enlarging valve guide bore to 0.003, 0.015 or 0.030 inch oversize.

NOTE: Some production cylinder heads have been fitted with 0.015 inch oversize valve stem. Heads so fitted are stamped "15" or "VO 15/OS" on exhaust side of head opposite the oversize valve stem guide.

VALVE SPRINGS

81. Intake and exhaust valve springs are interchangeable. Valve spring free length should be 2.15 inches. Springs should exert a force of 61 to 69 pounds when compressed to a length of 1.74 inches, and a force of 125-139 pounds when compressed to a length of 1.32 inches. Valve springs should also be checked for squareness by setting spring on flat surface and checking with a square; renew spring if clearance between top end of spring and square is more than 1/16-inch with bottom end of spring against square. Also, renew any spring showing signs of rust or erosion.

VALVE TAPPETS (CAM FOLLOWERS)

83. Valve tappets are semi-mushroom type and can be removed after removing camshaft as outlined in paragraph 91. Tappet diameter is 0.9889-0.9894 inch and bore in cylinder block is 0.990-0.991 inch. Desired tappet to bore clearance is 0.0006-0.0021 inch.

Model	Intake face	Exhaust face	Intake seat	Exhaust seat
8000, 8600, 8700, 9000	44	44	45	45
9600, 9700	29½	44	30	45
TW-10	44½	44½	45	45
TW-20, TW-30	29½	44½	30	45

Fig. 70A—Table listing valve angles.

VALVE CLEARANCE ADJUSTMENT

84. Valve clearance should be adjusted every 600 hours on Models 8000, TW-10, TW-20 and TW-30 and every 300 hours on all others. To adjust valve clearance, rotate engine and adjust intake valves to 0.014-0.016 inch and exhaust valves to 0.017-0.019 inch. Do not adjust valves if engine is over normal operating temperature.

NOTE: Torque required to turn the self-locking adjusting screws in rocker arms should be from 9 to 26 ft.-lbs. if less than 9 ft.-lbs., install a new adjusting screw. If turning torque is still less than 9 ft.-lbs., install a new rocker arm assembly. Refer to paragraph 85.

After valve clearance is properly adjusted, reinstall rocker arm cover using a new gasket and tighten cover retaining cap screws to a torque of 10-15 ft.-lbs.

ROCKER ARMS

85. To remove rocker arms, remove air cleaner on TW-30 or fuel tank on all other models, then remove rocker arm cover. Unscrew the seven cylinder head bolts that retain rocker arm assembly to cylinder head, but do not remove bolts from rocker arm supports. Lift out the rocker arm assembly and head bolts as a unit.

To disassemble, withdraw the cylinder head bolts. Rocker arm to shaft clearance should be 0.002-0.004 inch. Shaft diameter is 1.000-1.001 inches; rocker arm inside diameter is 1.003-1.004 inches. Renew rocker arm if clearance is excessive or if valve contact pad is worn more than 0.002 inch. Torque required to turn valve adjustment screw in rocker arm should be 9 to 26 ft.-lbs.; renew rocker arm and/or screw if torque required to turn screw is less than 9 ft.-lbs.

When reassembling, be sure notch (N—Fig. 70) in end of rocker arm shaft is up and towards front end of engine; this will correctly place the rocker arm oiling holes. Back each rocker arm adjusting screw out two turns, then tighten all retaining bolts evenly until valve springs are compressed and rocker arm supports are snug against the cylinder head; then, tighten all cylinder head bolts as outlined in paragraph 78.

After rocker arm shaft is installed, adjust valve clearance as described in paragraph 84.

R&R TIMING GEAR COVER

86. To remove timing gear cover, first split tractor between front support and engine as outlined in paragraph 76, then proceed as follows:

Drain and remove engine oil pan. Remove cap screw and washers from

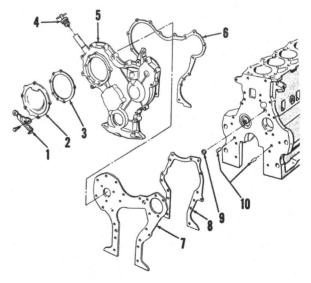

Fig. 71—Exploded view showing fuel injection pump gear cover (2), timing gear cover (5), engine front plate (7), gaskets and related parts.

1. Timing pointer
2. Injection pump gear cover
3. Gasket
4. Oil filler cap
5. Timing gear cover
6. Gasket
7. Front plate
8. Gasket
9. Cup plug
10. Dowel pins

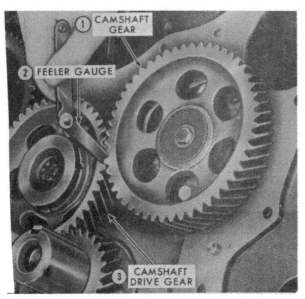

Fig. 72—Checking camshaft drive (idler) gear to camshaft gear backlash with feeler gage.

front end of crankshaft, then remove crankshaft pulley using suitable pullers. The timing gear cover, with fuel injection pump drive gear cover (2—Fig. 71) attached, can then be unbolted and removed from engine.

With timing gear cover removed, the crankshaft front oil seal (5—Fig. 76) and dust seal (6) can be renewed as outlined in paragraph 99.

When reinstalling timing gear cover, tighten cap screws to a torque of 13-18 ft.-lbs. Tighten crankshaft pulley retaining cap screw to a torque of 130-160 ft.-lbs. Install oil pan following procedure outlined in paragraph 102 and refer to paragraph 6, 33 or 56 when reconnecting front support to engine.

TIMING GEARS

87. Before removing any gears in the timing gear train, first remove rocker arms assembly to avoid the possibility of damage to piston or valve train if either the camshaft or crankshaft should be turned independently of the other.

The timing gear train consists of the crankshaft gear, camshaft gear, injection pump drive gear and a camshaft drive gear (idler gear) connecting the other three gears of the train. Refer to

paragraph 129 for information on fuel injection pump drive gear.

Timing gear backlash between crankshaft gear and camshaft drive gear, or between camshaft drive gear and camshaft gear (see Fig. 72) should be 0.001-0.009 inch. Backlash between camshaft drive gear and fuel injection pump drive gear should be 0.001-0.012 inch. If backlash is not within recommended limits, renew the camshaft drive gear, gear shaft and/or any other gears concerned.

88. CAMSHAFT DRIVE GEAR AND SHAFT. To remove, unscrew the self-locking cap screw and remove the camshaft drive gear and shaft (see Fig. 73) from front face of cylinder block. Renew shaft and/or gear if bushing to shaft clearance is excessive, or if bearing surfaces are scored. Shaft oil hole must be free of dirt or foreign material. Inspect gear teeth for wear or score marks; small burrs can be removed with fine carborundum stone.

To reinstall the camshaft drive gear, turn crankshaft so that No. 1 piston is at top dead center on compression stroke and turn camshaft and fuel injection pump drive gear so that timing marks point to center of camshaft drive gear location. Place the camshaft drive gear within the other three gears so that all timing marks are aligned as shown in Fig. 74, then install the shaft (2—Fig. 73) and tighten the self-locking cap screw (1) to a torque of 100-105 ft.-lbs.

89. CAMSHAFT GEAR. To remove the camshaft gear, remove cap screw (1—Fig. 75), lockwasher (2) and washer (3) then, pull gear from shaft. Gear should be a hand push fit on shaft. With gear removed, inspect drive key (10), thrust plate (6) and spacer (7) and renew if damaged in any way.

To reinstall gear, first install spacer, thrust plate and drive key, then install gear, washer, lockwasher and gear retaining cap screw. Tighten the cap screw to a torque of 40-45 ft.-lbs.

90. CRANKSHAFT GEAR. If not removed with timing gear cover and seal assembly, slide the spacer (7—Fig. 76)

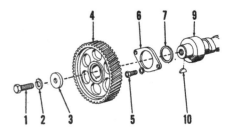

Fig. 75—View showing assembly of camshaft gear to camshaft. Camshaft end play is controlled by plate (6) which fits between shoulders on gear and shaft.

1. Cap screw
2. Lockwasher
3. Flat washer
4. Camshaft gear
5. Cap screw
6. Thrust plate
7. Spacer
9. Camshaft
10. Woodruff key

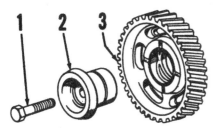

Fig. 73—Camshaft drive (idler) gear and adapter as removed from front end of cylinder block. Bushing in gear is not available separately.

1. Retainer bolt
2. Adapter shaft
3. Gear & bushing assy.

Fig. 74—View showing timing marks aligned on crankshaft gear, camshaft drive gear, camshaft gear and fuel injection pump gear.

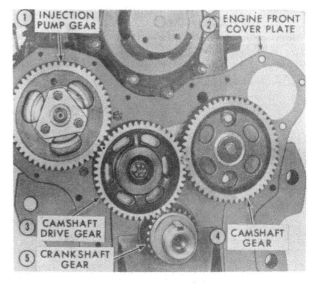

from crankshaft; then, using remover-replacer (Ford tool No. 2134) and insert (Ford tool No. 1237) or equivalent tool, pull gear from crankshaft. Inspect the gear and crankshaft pulley drive key (2) and renew if damaged in any way.

To reinstall gear, first drive the key (2) into crankshaft keyway until fully seated, then install the gear with timing mark outward using remover-replacer tools as used in removal procedure, or with bolt threaded into front end of crankshaft, use a nut, large washer and a sleeve.

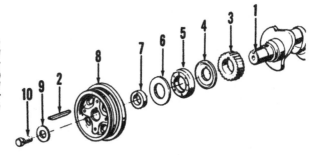

Fig. 76—View showing crankshaft gear, front oil seal and crankshaft pulley installation. Dust seal (6) and oil seal (5) are pressed into timing gear cover from inside and ride on pulley spacer (7).

1. Crankshaft
2. Drive key
3. Crankshaft gear
4. Oil slinger
5. Oil seal
6. Dust seal
7. Pulley spacer
8. Crankshaft pulley
9. Flat washer
10. Cap screw

CAMSHAFT AND BEARINGS

91. To remove camshaft, first remove timing gear cover as outlined in paragraph 86. Turn engine so that the timing marks on crankshaft gear, camshaft gear and fuel injection pump drive gear point to center of idler gear. Remove rocker arm cover, rocker arm assembly and push rods as outlined in paragraph 85. Drive suitable size wood dowels into the hollow valve tappets (cam followers), then lift the tappets up away from camshaft and hold with pincher type clothes pins. On all models except TW-10, TW-20 and TW-30, remove oil filter, plug and oil pump drive gear; refer to Fig. 85.

Desired camshaft end play is 0.001-0.007 inch. If end play is excessive, renew thrust plate (6—Fig. 75) during reassembly. Remove the cap screw (1), lockwasher and flat washer and pull camshaft gear from shaft. Remove the Woodruff key (10), thrust plate and spacer (7). Withdraw camshaft from front end of engine, taking care not to strike camshaft lobes as shaft is removed.

The camshaft is supported in five renewable bearings. Check camshaft and bearings against the following values:

Camshaft journal
diameter . . . . .2.3895-2.3905 in.
Desired journal to
bearing clearance . . . . .0.001-0.003 in.
Camshaft end play . . . . . .0.001-0.007 in.

If excessive bearing wear is indicated, the bearings can be removed and new bearings installed with bearing driver (Ford tool No. 1255) and handle (Ford tool No. 1442) or equivalent tools.

NOTE: It will be necessary to remove engine and remove oil pan, clutch, flywheel, engine rear cover plate and camshaft rear cover plate to remove and install bearings. Pay particular attention that the oil holes in bearings are aligned with the oil passages in cylinder block.

Refer to Fig. 78. Remove camshaft drive (idler) gear, then reinstall gear

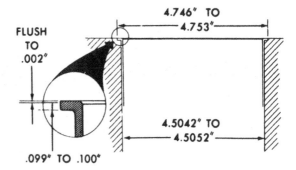

Fig. 77—Diagram showing machining dimensions necessary for installation of service sleeve. Refer to text.

with timing marks aligned as shown in Fig. 74. Refer also to paragraph 88. New bearings are pre-sized and should not require reaming if carefully installed.

Lubricate tappets and camshaft and reinstall camshaft by reversing removal procedure. Rotate camshaft to be sure there is no binding in new bearings before installing drive gear. Tighten thrust plate cap screws to a torque of 12-15 ft.-lbs. and tighten the camshaft gear retaining cap screw to a torque of 40-45 ft.-lbs.

CONNECTING ROD AND PISTON UNITS

92. Connecting rod and piston units are removed from above after removing the cylinder head and oil pan. Be sure to remove top ridge from cylinder bores before attempting to withdraw the assemblies.

Connecting rod and bearing cap are numbered to correspond to their respective cylinder bores. When renewing the connecting rod, be sure to stamp the cylinder number on new rod and cap.

When reassembling, it is important that the identification notch in top face of piston is towards the front end of engine. Assemble connecting rod to piston with cylinder numbers to right side of engine (away from camshaft). Refer to Fig. 79.

Fig. 78—View showing special tool for camshaft bearing installation. Be sure that bearings are installed with oil holes aligned with oil passages in cylinder block.

1. Bearing
2. Handle no. 1442
3. Oil holes
4. Driver no. 1255

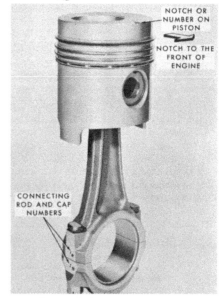

Fig. 79—View showing proper assembly of piston to connecting rod.

When installing connecting rod cap, be sure that bearing liner tangs, and the cylinder identification number, of rod and cap are towards same side of engine. Tighten the connecting rod nuts to a torque of 60-65 ft.-lbs.

PISTON RINGS

93. Pistons are fitted with three compression rings and one oil control ring. The two top compression rings and the oil control ring are chrome plated. All 8000 models and early 9000 models are equipped with a barrel face type top compression ring which must be installed with identification mark up. Later 9000 models and all 8600, 8700, 9600, 9700, TW-10, TW-20 and TW-30 models are equipped with a keystone type top compression ring. Second compression ring is black with chrome edge. Third compression ring is dull gray or black. Both are marked with an "O", which must be installed towards top of piston.

The oil control ring and slotted expander may be installed with either side up. Detailed instructions are packaged with most service ring sets.

Piston ring sets are available in oversizes of 0.020, 0.030 and 0.040 inch as well as standard size. The standard size rings are to be used with standard size pistons and also with 0.004 inch oversize pistons. Refer to the following specifications for checking piston ring fit:

Ring End Gap
Top compression ring ...0.012-0.038 in.
Second and third compression
   rings ....0.010-0.035 in.
Oil control ring ....0.013-0.033 in.

Ring Side Clearance In Groove
Top compression ring ..0.0044-0.0061 in.
Second and third compression
   rings ....0.0039-0.0056 in.
Oil control ring ....0.0024-0.0041 in.

PISTONS AND CYLINDERS

94. Engine is fitted with trunk type aluminum alloy pistons with a continuous skirt. Pistons are tapered in the upper land area. Replacement pistons may have a drill point dimple in their crown, which will necessitate the use of the newer, thicker head gasket, since crown height is higher on the new type pistons.

Cylinder bores in engine block are unsleeved. Where excessive cylinder wear has occurred, cylinders can be rebored to fit next larger oversize piston. Pistons 0.004 inch oversize are available for standard size cylinders which must be honed. Where necessary to rebore cylinders, pistons of 0.020, 0.030 and 0.040 inch oversizes are available. If cylinder taper is 0.005 inch or more, cylinders should be rebored or honed to next larger oversize piston. Refer to paragraph 95 for fitting pistons to cylinder bores.

A cylinder sleeve is available for service installation. Sleeve should be chilled to aid in installation. After sleeve is started in bore, use puller in kit No. 2757 to complete installation. Do not drop chilled sleeve into bore as sleeve lip may be cracked. Cylinder is bored to 4.5042-4.5052 inches. Counterbore cylinder to 4.746-4.753 inches at a depth of 0.099-0.100 inch. Depth of lip of sleeve must be no greater than 0.002 inch as shown in Fig. 77. Standard or 0.004 inch oversize pistons must be used in sleeve. Sleeve must not be overbored.

95. **FITTING PISTONS.** Recommended method for fitting piston is as follows: Before checking piston fit, deglaze cylinder wall using a hone or deglazing tool. Using a micrometer, measure piston diameter at centerline of and at right angle to the piston pin bore. Then, using an inside micrometer, measure cylinder bore diameter of cylinder block crosswise with the block at the smallest point. Subtract the piston diameter from the cylinder bore diameter; the resulting piston to cylinder bore clearance should be within range of 0.008 to 0.009 inch for proper piston fit on pistons No. 1 through No. 5. Piston No. 6 should have 0.002 inch more clearance than the others, with the range being 0.010 to 0.011 inch clearance.

NOTE: After honing or deglazing cylinder bore, wash bore thoroughly with hot water and detergent until a white cloth can be rubbed against cylinder wall without smudging, then rinse with cold water, dry thoroughly and oil to prevent rusting.

PISTON PINS

96. A 1.4997-1.5000 inch diameter floating type piston pin is used on 8000, 8600, 8700 and TW-10 engines. Diameter of floating type piston pin on 9000, 9600, 9700, TW-20 and TW-30 engines is 1.6246-1.6251 inches. Piston pins are retained in the piston pin bosses by snap rings and are available in standard size only. Piston pin should have a clearance of 0.0005-0.0007 inch in connecting rod bushing and a clearance of 0.0003-0.0005 inch in piston bosses. After installing new piston pin bushings in connecting rods, the oil hole in bushing must be drilled as shown in Fig. 80 being sure to drill through bushing on 9000, 9600, 9700, TW-20 and TW-30 models to oil passage in rod. Final size bushing with a spiral expansion reamer to obtain the specified pin to bushing clearance. Bushing inside diameter should then be 1.5003-1.5006 inches on 8000, 8600, 8700 and TW-10 engines and 1.6253-1.6256 inches on 9000, 9600, 9700, TW-20 and TW-30 engines. When assembling, identification notch or number in top face of piston must be to front end of engines and the identification number on rod and cap towards right side of engine (away from camshaft).

CONNECTING RODS AND BEARINGS

97. Connecting rods for all models are similar except 9000, 9600, 9700, TW-20 and TW-30 model connecting rod has an oil passage going from big end of rod to small end and a larger pin bore. Connecting rod bearings are of the nonadjustable, slip-in precision type, renewable from below after removing the oil pan and connecting rod bearing caps.

Crankpin bearing liners may be of two different materials, copper-lead or aluminum-tin alloy. The bearings will have an identification marking as follows:

Copper-lead...................PV or G
Aluminum-tin...............G and AL

NOTE: Copper-lead bearings only are used on 9000, 9600, 9700, TW-20 and TW-30 connecting rods. Bearings for these models have an oil hole which must match oil passage in connecting rod. Standard size bearing liners of each material are available in two different

Fig. 80—Oil hole in connecting rod bushing must be drilled after bushing is installed, but before reaming bushing to size. Hole in top of connecting rod is 0.045-0.050 inch and must not be drilled oversize.

thicknesses and are color coded to indicate thickness as follows:

Copper-lead bearing thickness:
Red . . . . .0.0943-0.0948 in.
Blue . . . . .0.0947-0.0952 in.
Aluminum-tin alloy bearing thickness:
Red . . . . .0.0939-0.0944 in.
Blue . . . . .0.0943-0.0948 in.

In production, connecting rods and crankshaft crankpin journals are color coded to indicate bore and journal diameters as follows:

Connecting rod bore diameter:
Red . . . . .2.9412-2.9416 in.
Blue . . . . .2.9416-2.9420 in.
Crankpin journal diameter:
Red . . . . .2.7500-2.7504 in.
Blue . . . . .2.7496-2.7500 in.

When installing a new crankshaft and the color code marks are visible on connecting rods, the crankpin bearing liners may be fit as follows: If the color code markings on both rod and crankshaft crankpin journal are red, install two red bearing liners; if both color code markings are blue, install two blue coded bearing liners. If color code marks on rod and crankpin do not match (one is red and the other is blue) install one red and one blue bearing liner.

NOTE: Be sure that both bearing liners are of the same material; that is, either both are copper-lead or both are aluminum-tin alloy. If color code mark is not visible on connecting rod or crankshaft, bearing fit should be checked with plastigage for the proper clearance according to bearing material as follows:

Crankpin journal to bearing liner clearance:
Copper-lead
bearings . . . . .0.0017-0.0038 in.
Aluminum-tin
bearings . . . . .0.0025-0.0046 in.

As well as being available in either red-coded or blue-coded standard size, bearing liners are also available in undersizes of 0.002, 0.010, 0.020, 0.030 and 0.040 inch. When installing undersize crankpin bearing liners, the crankpin must be reground to one of the following exact undersizes:

Bearing Undersize	Crankpin Journal Dia.
0.002	2.7476-2.7480 in.
0.010	2.7400-2.7404 in.
0.020	2.7300-2.7304 in.
0.030	2.7200-2.7204 in.
0.040	2.7100-2.7104 in.

NOTE: When regrinding crankpin journals, maintain a 0.12-0.14 inch fillet radius and chamfer oil hole after journal is ground to size.

When reassembling, tighten the connecting rod nuts to a torque of 60-65 ft.-lbs.

CRANKSHAFT AND MAIN BEARINGS

98. Crankshaft is supported in seven main bearings. Crankshaft end thrust is controlled by the flanged main bearing liner which is used on the second main journal from rear. Before removing main bearing caps, check to see that they have an identification number so that they can be installed in same position from which they are removed.

Main bearing liners may be of two different materials, copper-lead or aluminum-tin alloy. The bearings will have an identification marking to indicate bearing material as follows:

Copper-lead . . . . .PV or G
Aluminum-tin alloy . . . . .G and AL

Standard size bearing liners are available in two different thicknesses and are color-coded to indicate thickness as follows:

Red . . . . .0.1245-0.1250 in.
Blue . . . . .0.1249-0.1254 in.

In production, main bearing bores in block and main bearing journals on crankshaft are color coded to indicate bore and journal diameter as follows:

Main bearing bore diameter:
Red . . . . .3.6242-3.6246 in.
Blue . . . . .3.6246-3.6250 in.
Main journal diameter:
Red . . . . .3.3718-3.3723 in.
Blue . . . . .3.3713-3.3718 in.

When installing new crankshaft and the color code marks are visible in crankcase at main bearing bores, new main bearing liners may be fit as follows: If the color code marks on bore and journal are both red, install two red coded bearing liners; if both marks are blue, install two blue coded bearing liners. If color code mark on bore is not the same as color code on journal (one is blue and the other is red), install one red coded bearing liner and one blue coded bearing liner.

NOTE: Be sure both liners used at one journal are of the same material; however, copper-lead bearing liners may be used on one or more journals with aluminum-tin alloy liners on the remain-

ing journals. If color code marks are not visible at the main bearing bores in block, check bearing fit with plastigage and install red or blue, or one red and one blue liner to obtain a bearing journal to liner clearance of 0.0022-0.0045 inch.

As well as being available in either red-coded or blue-coded standard size, new main bearing lines are also available in undersizes of 0.002, 0.010, 0.020, 0.030 and 0.040 inch. When installing undersize main bearing liners, the crankshaft journals must be reground to one of the following exact undersizes:

Bearing Undersize	Main Journal Dia.
0.002	3.3693-3.3698 in.
0.010	3.3618-3.3623 in.
0.020	3.3518-3.3523 in.
0.030	3.3418-3.3423 in.
0.040	3.3318-3.3323 in.

NOTE: When regrinding crankshaft main bearing journals, maintain a fillet radius of 0.12-0.14 inch and chamfer oil holes after journal is ground to size.

When reinstalling main bearing caps, proceed as follows: Be sure the bearing bores and rear main bearing oil seal area are thoroughly clean before installing bearing liners. Be sure tangs on bearing inserts are in the slots provided in cylinder block and bearing caps. Refer to paragraph 100 for rear cap side seals. Tighten bearing cap bolts to a torque of 115-125 ft.-lbs.

CRANKSHAFT OIL SEALS

99. **FRONT OIL SEAL.** Crankshaft front oil seal is mounted in timing gear cover and cover must be removed to renew seal. Timing gear cover removal procedure is outlined in paragraph 86. To renew seal, drive dust seal (6—Fig. 76) and oil seal (5) out towards inside of timing gear cover. Install new dust seal in timing gear cover, then using a seal installer (Ford tool No. 7536 or equivalent), install new oil seal with spring loaded lip towards inside of cover.

The crankshaft front oil seal rides on pulley spacer (7). Carefully inspect spacer for wear at seal contact surface and renew spacer if wear or scoring is evident.

100. **REAR OIL SEAL.** To renew rear oil seal, engine must be removed from tractor and the clutch, flywheel, engine rear plate, oil pan and rear main bearing cap must be removed.

Production crankshafts are "scrolled" to aid in feeding oil back from the rear oil seal. Whenever necessary to renew rear oil seal, either with a used or new service installed crankshaft, it will be necessary to apply a scroll finish to rear oil seal contact surface. Refer to Fig. 81; it is very important that only Grade 240 aluminum oxide cloth be used and that the scroll marks be applied at angle shown. Finish accessible area of crankshaft, then turn shaft a little at a time and finish adjacent area until complete circumference of shaft sealing surface is finished.

NOTE: "Scroll" lines should not be apparent to the naked eye.

After "scrolling" shaft, proceed as follows to install rear main bearing and bearing cap: Clean the mating surfaces of block and bearing cap and apply a light coating of gasket sealing compound to both surfaces. Install new side seals in rear main bearing cap with seals projecting slightly beyond block face of cap (see Fig. 82) and install cap with bearing insert and side seals. Tighten bearing cap bolts to a torque of 115-125 ft.-lbs., then trim bottom ends of side seals so that seals project 1/64-inch. Apply a light coating of gasket cement to split lines of bearing cap and block in seal

bore, taking care not to allow any cement on shaft. Pack grease between the two lips of the new seal and apply light coat of high temperature grease to seal bore in block and bearing cap and to crankshaft seal journal.

Install new crankshaft seal using special installation tool (Ford tool No. 1301) as shown in Fig. 83. Use flywheel retaining cap screws; tighten cap screws evenly until tool bottoms against crankshaft flange, then tighten each cap screw to a torque of 25 ft.-lbs. to be sure that seal is square with crankshaft centerline.

NOTE: If special tool is not available, use a 4-7/8 inch I.D. sleeve and carefully press seal into block until rear face of seal is 0.060 inch inside rear face of block. Then, using dial indicator mounted on rear end of crankshaft check to see that runout of seal does not exceed 0.015 inch.

Apply a liberal amount of penetrating oil to the cap side seals to cause them to swell, then install engine rear plate

using new gasket. Be sure that bottom edge of gasket is parallel to bottom edge of cylinder block and cap. Reinstall flywheel, clutch and oil pan and reinstall engine in tractor.

NOTE: Do not tighten oil pan retaining cap screws until after engine is bolted to transmission; refer to paragraph 102.

FLYWHEEL

101. The flywheel can be removed after splitting tractor between engine and transmission and removing the clutch. Flywheel can be installed in one position only. The flywheel bolts retain the power take-off drive adapter. Before installing flywheel, inspect flywheel to crankshaft cap screws to make sure they are proper type grade 8 screws and in good condition. After applying "Loctite No. 721" or equivalent to threads, tighten screws to a torque of 155-165 ft.-lbs.

Starter ring gear is installed from front face of flywheel; therefore, the flywheel must be removed to renew the ring gear. Heat the gear to be removed with a torch from front side of gear and knock off of flywheel. Heat new gear evenly by applying heat from inside only, so that heat expands out into gear teeth, and only until gear expands enough to slip onto flywheel. Tap gear all the way around to be sure it is properly seated and then quench with water to cool gear rapidly.

NOTE: Be sure to heat gear evenly; if any portion of gear is heated to a temperature higher than 450 degrees F., rapid wear will result. Heat sensing crayons which melt at 400 degrees F. and

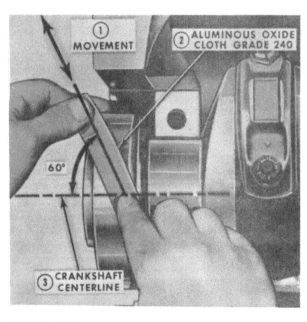

Fig. 81—Whenever necessary to renew crankshaft rear oil seal, crankshaft sealing surface must be "scrolled" using Grade 240 aluminum oxide cloth and at angle shown. Refer to text.

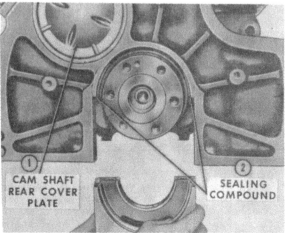

Fig. 82—Installing rear main bearing cap with side seals; seals should protrude slightly above block face of cap as shown. Apply light coat of sealing compound to block and cap mating surfaces.

Fig. 83—Using special tool No. 1301 (2) to install crankshaft rear oil seal. Tighten capscrews (1) to 25 ft.-lbs. If special tool is not used, seal rear face must be checked for runout with dial indicator mounted on rear end of crankshaft; maximum allowable runout is 0.015 inch.

450 degrees F., can be used to avoid overheating, if they are available.

OIL PAN

102. To remove oil pan, proceed as follows: One at a time, replace existing cylinder block-to-front support bolts with bolts 8-inches long. Drain oil pan and remove oil level dipstick. Support tractor with a jack under transmission and a hoist at the front support and radiator assembly. Remove hood side panels, air pre-cleaner, exhaust pipe and hood top panel. Remove two bolts securing oil cooler to steering motor housing. Support air cleaner support assembly and remove bolts which secure support assembly to front hood. Remove bolts securing side frame members to front support. Move front support assembly forward until there is suffi-

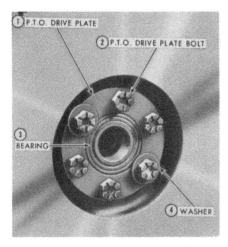

Fig. 84—Flat washers are installed under two of the flywheel and pto drive plate retaining bolts to hold clutch shaft pilot bearing in position.

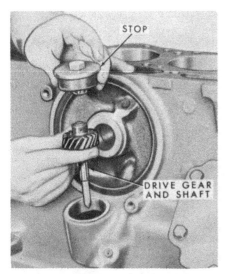

Fig. 85—Removing oil pump drive gear and shaft on all models except TW-10, TW-20 and TW-30. The shaft may remain in oil pump as gear is removed.

cient clearance to remove front oil pan bolts. Do not lose shims. Remove the two transmission to oil pan bolts and support oil pan. Unbolt and lower oil pan from engine.

If a new oil pan is being installed, refer to paragraph 6, 33 or 56 for shimming front support to oil pan. To reinstall oil pan, proceed as follows:

If timing gear cover is removed, reinstall cover before installing oil pan. Be sure all gasket surfaces are clean and cement gasket to pan with thin film of gasket sealer. Lift pan into place with

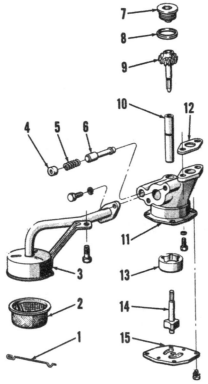

Fig. 86—Exploded view of oil pump assembly, suction (inlet) screen and oil pump drive gear for all models except TW-10, TW-20 and TW-30. Stop (7) and drive gear (9) are removed from above as shown in Fig. 85.

1. Retainer clip	
2. Inlet screen	
3. Suction tube	9. Drive gear
4. Plug	10. Floating shaft
5. Relief valve spring	11. Oil pump body
6. Relief valve plunger	12. Gasket
7. Stop	13. Outer pump rotor
8. Gasket	14. Inner pump rotor
	15. Cover plate

Fig. 87—Checking pump rotor to pump cover clearance and pump outer rotor to body clearance; refer to text for specifications and to Fig. 88 for checking pump rotors.

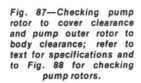

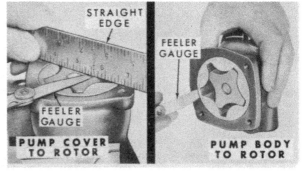

floor jack and reinstall oil pan to engine cap screws finger tight only. Install and tighten, then loosen the two transmission to oil pan cap screws. Then, starting from center of pan, tighten the oil pan to engine cap screws to a torque of 30-35 ft.-lbs. Tighten the transmission to oil pan cap screws to a torque of 180-220 ft.-lbs. Reinstall front end assembly by reversing removal procedure, making sure the shims removed from between oil pan and front on disassembly are reinstalled. Tighten the front support to cylinder block bolts and cap screw to a torque of 180-220 ft.-lbs., then tighten front support to oil pan cap screws to a torque of 270-330 ft.-lbs. Fill and bleed the power steering system as outlined in paragraph 7, 34 or 57.

OIL PUMP AND RELIEF VALVE

All Models Except TW-10, TW-20 And TW-30

NOTE: Later oil pumps have a six-flute inner rotor which is thicker than early four-flute inner rotor. All clearances are the same, however it may be necessary to grind away interfering metal on early oil pan if later pump is fitted.

103. To remove oil pump first remove oil pan as outlined in paragraph 102, then unbolt and remove pump and inlet filter screen from cylinder block. Refer to Fig. 86 for exploded view of oil pump and the oil pump drive gear assembly. The floating drive shaft (10) will usually be removed with the pump. To remove drive gear, remove full flow oil filter assembly, then unscrew plug and remove gear as in Fig. 85.

To disassemble pump, remove clip (1—Fig. 86) and screen (2), then remove screws retaining suction tube (3) and pump cover (15) to pump body (11). Remove the covers and pump rotor set (13 and 14), noting which direction outer rotor was placed in pump body. Thread a self-tapping screw into plug (4) and pull plug from pump body. Remove the spring (5) and oil pressure relief valve (6).

Check the pump for wear as shown in Figs. 87 and 88. Pump cover to rotor clearance (rotor end play) should be 0.001-0.0035 inch; pump body to rotor clearance should be 0.006-0.011 inch; and rotor clearance should be 0.001-0.006 inch when measured as shown in Fig. 88. Renew the rotor set and/or pump body if clearances are excessive. Renew pump cover plate if excessively worn or scored. Relief valve spring should exert a force of 10.7 to 11.9 pounds when compressed to a length of 1.07 inches. Engine oil pressure should be 60-70 psi at 1000 engine rpm.

Assemble the oil pump and reinstall by reversing removal and disassembly procedure. Press a new relief valve plug (4—Fig. 86) into pump body so that plug is flush with body. Tighten the oil pump retaining cap screws to a torque of 23-28 ft.-lbs.

Models TW-10, TW-20 And TW-30

104. To remove oil pump on TW-10, TW-20 and TW-30 models, first remove oil pan as outlined in paragraph 102. Oil pump idler gear (4—Fig. 89) and bearing (3) should be checked for excessive end play. If end play exceeds 0.008 inch, renew bearing. Renew bearing whenever gear is renewed. Remove idler gear (4) using tool No. 7537 or equivalent and remove drive gear (11) by hand or use a suitable press. Remove relief valve plug (23) and spring (24). Clean and inspect all parts for excessive wear or scoring and replace as necessary.

Measure clearances as in Figs. 87 and 88. Rotor clearance is 0.001-0.006 inch. Rotor to pump housing clearance is 0.006-0.011 inch. Rotor end play is 0.001-0.0035 inch. Renew shaft and rotor assembly if not within specifications.

NOTE: Prime pump by filling pump inlet with clean oil while rotating pump

drive gear and shaft.

Install pump assembly by reversing removal procedure. Tighten pump to block bolts to a torque of 23-28 ft.-lbs. Tighten oil pan bolts to a torque of 30-35 ft.-lbs. Normal oil pump pressure is 60-80 psi at 2200 rpm.

DIESEL FUEL SYSTEM

The diesel fuel system consists of three basic components: The fuel filters, injection pump and injection nozzles. When servicing any unit associated with the fuel system, the maintenance of absolute cleanliness is of utmost importance. Of equal importance is the avoidance of nicks or burrs on any of the working parts.

Probably the most important precaution that service pesonnel can impart to owners of diesel powered tractors is to urge them to use an approved fuel that is absolutely clean and free from foreign material. Extra precaution should be taken to make certain that no water enters the fuel storage tanks.

TROUBLESHOOTING

105. If the engine will not start, or does not run properly after starting, refer to the following paragraphs for possible causes of trouble.

106. **FUEL NOT REACHING INJECTION PUMP.** If no fuel will run from line when disconnected from pump, check the following:

Be sure fuel supply valve is open.

Check the filters for being clogged (Including filter screen in fuel supply valve).

Bleed the fuel filters.

Check lines and connectors for damage.

Check for low fuel pump pressure.

107. **FUEL REACHING NOZZLES BUT ENGINE WILL NOT START.** If, when lines are disconnected at fuel noz-

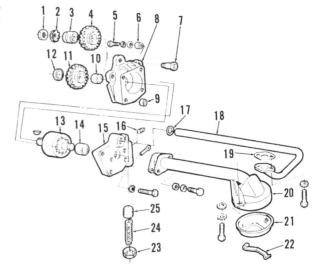

Fig. 89—Exploded view of oil pump used on Models TW-10, TW-20 and TW-30.

1. Idler gear shaft nut
2. Lock washer
3. Idler gear bearing
4. Idler gear
5. Bolt
6. Pin
7. Adaptor shaft
8. Housing
9. Dowel
10. Front bushing
11. Drive gear
12. Plug
13. Rotor and shaft
14. Rear bushing
15. Pump plate
16. Plug
17. "O" ring
18. Outlet tube
19. Gasket
20. Pump inlet
21. Screen
22. Spring
23. Pressure relief plug
24. Relief spring
25. Relief valve

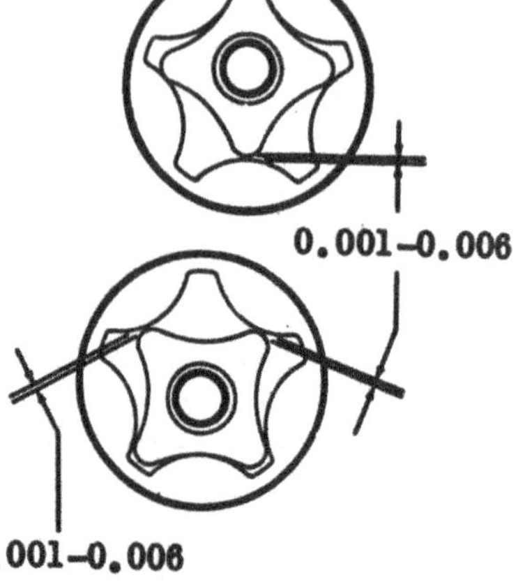

0.001—0.006

0.001—0.006

Fig. 88—Measure inner to outer rotor clearance as shown; renew rotors if clearance exceeds 0.006 inch. Refer also to Fig. 87.

Fig. 89A—Front view of oil pump assembly used on TW-10, TW-20 and TW-30 model tractors.

1. Oil pump support plate
2. Oil pump drive gear
3. Oil pump housing
4. Idler gear
5. Front support bolts
6. Rear support bolts
7. Oil pump and screen
8. Support bolt

zles and engine is cranked, fuel will flow from connections, but engine will not start, check the following:

Check cranking speed.

Check throttle control rod adjustment.

Check pump timing.

Check fuel lines and connections for pressure leakage.

Check engine compression.

108. ENGINE HARD TO START. If the engine is hard to start, check the following:

Check cranking speed.

Injection pump timing.

Bleed the fuel filters.

Check for clogged fuel filters.

Check for water in fuel or improper fuel.

Check for air leaks on suction side of fuel lift pump.

Check engine compression.

Injection pump gallery may be draining back to auxiliary tank if there is no fuel in upper tank.

109. PUMP FAILS TO DELIVER FUEL TO ONE OR MORE INJECTORS. Check for the following:

Air in fuel lines to injectors.

Plunger spring broken.

Scored barrel.

Control rod stuck in off position.

Delivery valve defective.

110. ENGINE STARTS, THEN STOPS. If the engine will start, but then stops, check the following:

Check for clogged or restricted fuel lines or fuel filters.

Check for water in fuel.

Check for restrictions in air intake.

Check engine for overheating.

Check for air leaks in lines on suction side of fuel lift pump.

Check for faulty lift pump.

111. ENGINE SURGES, MISFIRES OR POOR GOVERNOR REGULATION. Make the following checks:

Bleed the system.

Check for clogged filters or lines or restricted fuel lines.

Check for water in fuel.

Check pump timing.

Check injector lines and connections for leakage.

Check for faulty or sticking injector nozzles.

Check for faulty or sticking engine valves.

Check for faulty governor.

112. LOSS OF POWER. If engine does not develop full power or speed, check the following:

Check throttle control rod adjustment.

Check maximum no-load speed adjustment.

Check for clogged or restricted fuel lines or clogged fuel or air filters.

Check for air leaks in suction line of transfer pump.

Check pump timing.

Check engine compression.

Check for improper engine valve gap adjustment or faulty valves.

113. EXCESSIVE BLACK SMOKE AT EXHAUST. If the engine emits excessive black smoke from exhaust, check the following:

Check for restricted air intake such as clogged air cleaner.

Check pump timing.

Check for faulty injectors.

Check engine compression.

FILTERS AND BLEEDING

Models TW-10, TW-20 and TW-30

114. MAINTENANCE. These models are equipped with one filter element and a separate sediment bowl. Drain sediment bowl when water or sediment can be seen in separator. Renew fuel filter every 600 hours. Close fuel shut-off valve before removing filter.

114A. BLEEDING. Open bleed screw on filter head and operate priming lever until bubble free fuel flows from opening. Close filter bleed screw, then open front bleed screw on injection pump and operate lever until fuel flowing from bleed screw is free of bubbles. Closed bleed screw while operating lever. Loosen fuel injector lines at the injectors and crank engine until fuel appears at all injectors, then tighten the

fuel injector line connections and start engine.

All Other Models

115. MAINTENANCE. These models are equipped with two spin-on type filter elements as shown in Fig. 90. Water drain plugs (bottom caps) should be removed after 50 hours of operation and any water in sediment bowls drained. Sediment bowls should be drained more often if excessive condensation is noted. Fuel filter elements should be renewed after 900 hours of operation. Close fuel shut-off valve at tank before removing filters.

115A. BLEEDING. Refer to Fig. 90. Open rear bleed screw on filter head and operate lever on fuel lift pump until fuel flowing from bleed screw is free of bubbles. Close the rear bleed screw and open front bleed screw on filter head and operate primer lever until bubble free fuel flows from opening. Then, close the fuel filter bleed screw, open front bleed screw on fuel injection pump and operate primer lever until fuel flowing from bleed screw is free of bubbles. Close the pump bleed screw while operating primer lever. Loosen the fuel injector lines at the injectors and crank engine until fuel appears at all injectors, then tighten the fuel injector line connections and start engine.

INJECTION NOZZLES

CAUTION: Fuel leaves the injection nozzles with sufficient force to penetrate the skin. When testing keep your person clear of the nozzle spray.

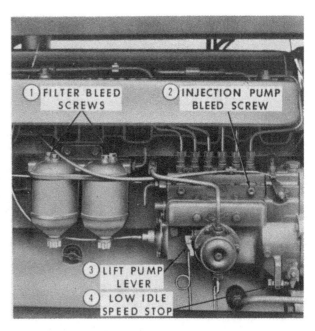

Fig. 90—Operate lever (3) on lift pump to bleed filter and fuel injection pump; refer to text for procedure. Pump on all models except TW-10, TW-20 and TW-30 is fitted with low idle speed stop screw only; high idle speed is adjusted by adjusting throttle linkage; refer to paragraph 130.

① FILTER BLEED SCREWS   ② INJECTION PUMP BLEED SCREW

③ LIFT PUMP LEVER

④ LOW IDLE SPEED STOP

116. TESTING AND LOCATING A FAULTY NOZZLE. If engine does not run properly and a faulty injection nozzle is indicated, such a faulty nozzle can be located as follows: With engine running, loosen the high pressure line fitting on each nozzle holder in turn, thereby allowing fuel to escape at the union rather than enter the injector. As in checking for faulty spark plugs in a spark ignition engine, the faulty unit is the one which, when its line is loosened, least affects the running of the engine.

117. NOZZLE TESTER. A complete job of testing and adjusting the fuel injection nozzle requires use of a special tester such as shown in Fig. 91. The nozzle should be tested for opening pressure, spray pattern, seat leakage and leak back.

Operate the tester until oil flows and then connect injector nozzle to tester. Close the tester valve to shut off pressure to tester gage and operate tester lever to be sure nozzle is in operating condition and not plugged. If oil does not spray from all four spray holes in nozzle, if tester lever is hard to operate or other obvious defects are noted, remove nozzle from tester and service as outlined in paragraph 123. If nozzle operates without undue pressure on tester lever and fuel is sprayed from all four spray holes, proceed with following tests:

118. OPENING PRESSURE. While slowly operating tester lever with valve to tester gage open, note gage pressure at which nozzle spray occurs. This gage pressure should be as follows:

TW-10 . . . . .2600-2850 psi
TW-20, TW-30 . . . . .3000-3250 psi
All other models . . . . . . .2720-2794 psi

If gage pressure is incorrect, remove cap nut and turn adjusting screw (Fig.

92) as required to bring opening pressure within specified limits. If opening pressure is erratic or cannot be properly adjusted, remove nozzle from tester and overhaul nozzle as outlined in paragraph 123. If opening pressure is within limits, check spray pattern as outlined in paragraph 123. If opening pressure is within limits, check spray pattern as outlined in following paragraph.

119. SPRAY PATTERN. Operate the tester lever slowly and observe nozzle spray pattern. All four (4) sprays must be similar and spaced at approximate intervals of 100°, 90°, 70° and 90°. Each spray must be well atomized and should spread to a 3-inch diameter cone at approximately 3/8-inch from nozzle tip. If spray pattern does not meet these conditions, remove nozzle from tester and overhaul nozzle as outlined in paragraph 123. If nozzle spray is satisfactory, proceed with seat leakage test as outlined in following paragraph.

120. SEAT LEAKAGE. Close valve to tester gage and operate tester lever quickly for several strokes. Then, wipe nozzle tip dry with clean blotting paper and open valve to tester gage. On all

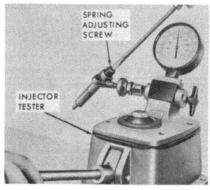

Fig. 92—Adjusting nozzle opening pressure; refer to paragraph 118 for specifications.

models except TW-10, TW-20 and TW-30, push tester lever down slowly to bring gage pressure to 200 psi below nozzle opening pressure and hold this pressure for one minute. On Models TW-10, TW-20 and TW-30, apply pressure to 150 psi below nozzle opening pressure and hold this pressure for six seconds. Apply a piece of clean blotting paper (Fig. 91) to tip of nozzle; the resulting oil blot should not be greater than 1/2-inch in diameter. If nozzle tip drips oil or blot is excessively large, remove nozzle from tester and overhaul nozzle as outlined in paragraph 123. If nozzle seat leakage is not excessive, proceed with nozzle leak back test as outlined in following paragraph.

121. NOZZLE LEAK BACK. Operate tester lever to bring gage pressure to approximately 2300 psi, release lever and note time required for gage pressure to drop from 2200 psi to 1500 psi. Time required should be from 10 to 40 seconds. If time required is less than 5 seconds, nozzle is worn or there are dirt particles between mating surfaces of nozzle and holder. If time required is greater than 40 seconds, needle may be too tight a fit in nozzle bore. Refer to paragraph 123 for disassembly, cleaning and overhaul information.

NOTE: A leaking tester connection, check valve or pressure gage will show up in this test as excessively fast leak back. If, in testing a number of injectors, all show excessively fast leak back, the tester should be suspected as faulty rather than the injectors.

122. REMOVE AND REINSTALL INJECTORS. Before removing injectors, carefully clean all dirt and other foreign material from lines, injectors and

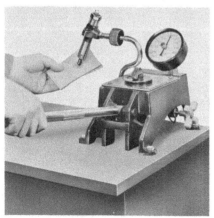

Fig. 91—A fuel injector tester such as the one shown is necessary for checking and adjusting fuel injector assemblies. Nozzle seat leakage check is illustrated; refer to paragraph 120.

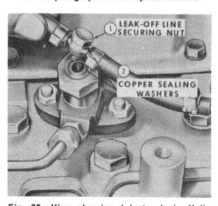

Fig. 93—View showing injector leak-off line disconnected from injector. Note copper sealing washers placed on each side of banjo fitting.

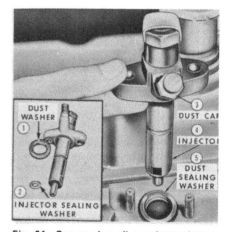

Fig. 94—Cap or plug all openings when removing injector assembly. Dust washer (1) keeps dirt out of injector bore in cylinder head. Be sure seat in bore is clean and install sealing washer (2) when reinstalling injector.

cylinder head area around the injectors. Disconnect the injector leak-off line (Fig. 93) at each injector and at the fuel return line. Disconnect the injector line at the pump and at the injector. Cap all lines and openings. Remove the two retaining nuts and carefully remove the injector from cylinder head (Fig. 94).

Prior to reinstalling injectors, check the injector seats in cylinder head to see that they are clean and free from any carbon deposit. Install a new copper washer in the seat and a new cork dust sealing washer around the body of the injector. Insert the injector in cylinder bore, install retaining washers and nuts and tighten the nuts evenly and alternately to a torque of 10-15 ft.-lbs. Position new leak-off fitting gaskets below and above each banjo fitting and install the banjo fitting bolts to a torque of 5-7 ft.-lbs. Reconnect leak-off line to return line. Check the fuel injector line connections to be sure they are clean and reinstall lines, tightening connections at pump end only. Crank engine until a stream of fuel is pumped out of each line at injector connection, then tighten the connections. Start and run engine to be sure that injector is properly sealed and that injector line and leak-off line connections are not leaking.

123. OVERHAUL INJECTORS. Unless complete and proper equipment is available, do not attempt to overhaul diesel nozzles. Equipment recommended by Ford is Ford tool No. 2021-Poptester, No. 1719-Injector Holding Fixture and No. 1720-Injector Cleaning Kit. Tools may be ordered by using Common Tool Order Form, FTO 8091, available at Ford Tractor Dealerships then forwarding order to nearest district office.

Secure injector holding fixture, No. 1719, in a vise and mount injector assembly in fixture. Never clamp injector body in vise. Remove cap nut and back-off adjusting screw then lift off upper spring disc, injector spring and spindle. Remove the nozzle retaining nut using nozzle nut socket (8136), or equivalent, and remove the nozzle and valve. Nozzles and valves are a lapped fit and must never be interchanged. Place all parts in clean fuel oil or calibrating fluid as they are disassembled. Clean injector assembly exterior as follows: Soften hard carbon deposits formed in the spray holes and on needle tip by soaking in a suitable carbon solvent, then use a soft wire (brass) brush to remove carbon from needle and nozzle exterior. Rinse the nozzle and needle immediately after cleaning to prevent the carbon solvent from corroding the highly finished surfaces. Clean the pressure chamber of the nozzle with a 0.043 inch reamer as shown in Fig. 97. Clean the spray holes with the proper

size wire probe held in a pin vise as shown in Fig. 98. To prevent breakage of wire probe, the wire should protrude from pin vise only far enough to pass through the spray holes. Rotate pin vise

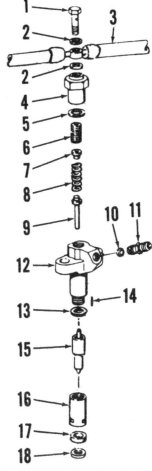

Fig. 96A—Exploded view of early injector. Refer also to Fig. 105.

1. Leak-off bolt	10. Adapter washer
2. Washer	11. Connector
3. Leak-off hose	12. Holder assembly
4. Cap	13. Shim
5. Seal washer	14. Dowel
6. Adjusting screw	15. Nozzle assembly
7. Spring plate	16. Nozzle nut
8. Spring	17. Seal
9. Spindle	18. Sealing washer

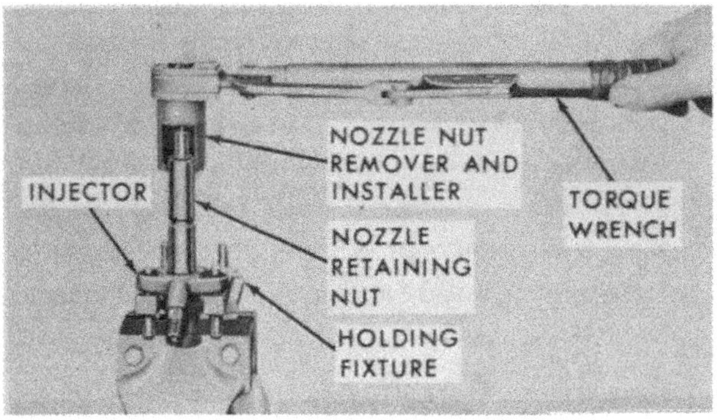

Fig. 95—View showing injector assembly mounted on holding fixture; nozzle retaining nut is being tightened using special wrench and torque wrench.

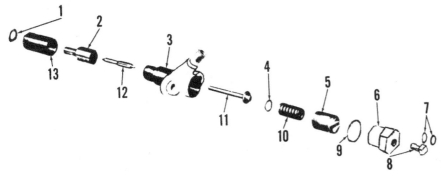

Fig. 96—Exploded view of late model injector.

1. Copper washer	4. Washer	7. Leak off line washers	10. Spring
2. Nozzle body	5. Adjusting screw	8. Leak off line fitting	11. Spindle
3. Nozzle holder	6. Cap nut	9. Washer	12. Needle valve
			13. Nozzle retaining nut

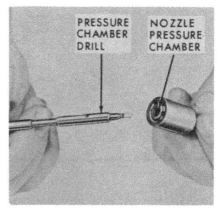

Fig. 97—Cleaning nozzle tip cavity with pressure chamber drill.

without applying undue pressure. Use a 0.012-inch diameter wire probe.

The valve seats in nozzle are cleaned by inserting the valve seat scraper into the nozzle and rotating scraper. Refer to Fig. 99. The annular groove in top of nozzle and the pressure chamber are cleaned by using (rotating) the pressure chamber carbon remover tool as shown in Fig. 100.

When above cleaning is accomplished back flush nozzle and needle by installing reverse flushing adapter (8124) on the nozzle tester and inserting nozzle and needle assembly tip end first into the adapter and secure with knurled nut as shown in Fig. 101. Rotate the needle in nozzle while operating tester lever. After nozzle is back flushed, the seat can be polished by using a small amount of

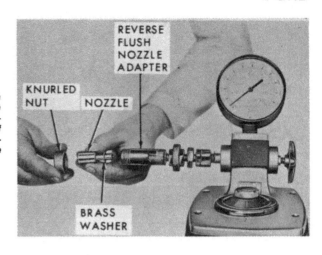

Fig. 101—A back flush attachment is installed on nozzle tester to clean nozzle by reverse flow of fluid; note proper installation of nozzle in the adapter.

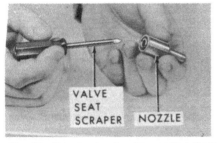

Fig. 98—Cleaning nozzle spray holes with wire probe held in pin vise.

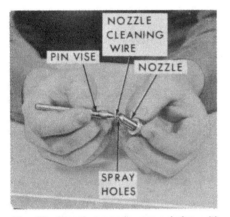

Fig. 99—Use scraper to clean carbon from valve seat in nozzle body.

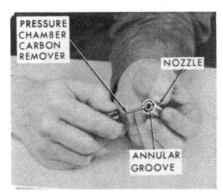

Fig. 100—Pressure chamber carbon remover is used to clean annular groove as well as clean carbon from pressure chamber in nozzle.

tallow on the end of a polishing stick and rotating stick in nozzle as shown in Fig. 102.

If the leak-back test time was greater than 40 seconds (refer to paragraph 121), or if needle is sticking in bore of nozzle, correction can be made by lapping the needle and nozzle assembly. This is accomplished by using a polishing compound (Bacharach No. 66-0655 is suggested) as follows: Place small diameter of nozzle in a chuck of a drill having a maximum speed of less than 450 rpm.

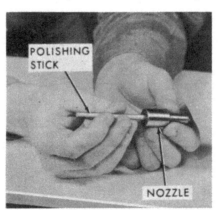

Fig. 102—Polishing needle valve seat with tallow and polishing stick.

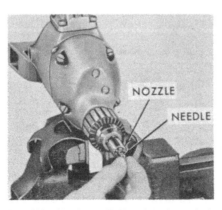

Fig. 103—Chuck small diameter of nozzle in slow speed electric drill to lap needle to nozzle if leak back time is excessive or if needle sticks in nozzle. Hold pin end of needle with vise grip pliers.

Apply a small amount of polishing compound on barrel of needle taking care not to allow any compound on tip or beveled seat portion, and insert needle in rotating nozzle body. Refer to Fig. 103.

NOTE: Do not lap valve for more than five seconds at a time and allow parts to cool between lapping. It is usually necessary to hold upper pin end of needle with vise-grip pliers to keep the needle from turning with the nozzle. Work the needle in and out a few times taking care not to put any pressure against seat, then withdraw the needle, remove nozzle from chuck and thoroughly clean the nozzle and needle assembly using back flush adapter and tester pump.

Prior to reassembly, rinse all parts in clean fuel oil or calibrating fluid and assemble while still wet. Position the nozzle and needle valve on injector body and be sure dowel pins in body are correctly located in nozzle as shown in Fig. 104.

NOTE: Place injector in holding fixture (1719) and tighten nut and new washer with socket (8126).

Install the spindle, spring, upper spring disc and spring adjusting screw. Connect the injector to tester and adjust opening pressure as in paragraph 118. Use a new copper washer and install cap nut. Recheck nozzle opening pressure to

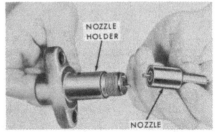

Fig. 104—Be sure dowel pins in nozzle holder enter mating holes in nozzle body.

be sure that installing nut did not change adjustment. Retest injector as outlined in paragraphs 119 through 121; renew nozzle and needle if still faulty. If the injectors are to be stored after overhaul, it is recommended that they be thoroughly flushed with calibrating fluid prior to storage.

FUEL INJECTION PUMP

125. **LUBRICATION.** The Minimec or Simms fuel injection pump is lubricated by oil sump in the pump cambox. After each 300 hours of Simms pump operation, the pump should be drained, the cambox breather cleaned and the cambox refilled to proper level with new, clean engine oil. Periodic oil change is not required on Minimec pump. Use

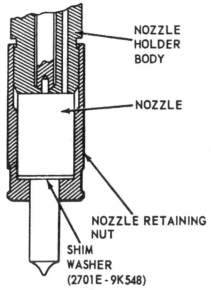

Fig. 105—Cross-sectional view showing shim washer (13—Fig. 96A) installed between nozzle body and nozzle retaining nut.

same weight and type oil as for engine crankcase; refer to Fig. 106 for location of drain plug, oil level plug and filler plug.

Whenever installing a new or rebuilt fuel injection pump, be sure the cambox is filled with engine oil before attempting to start engine. There will be some oil dilution with diesel fuel during engine operation and after engine is stopped, some of the fuel-oil mixture may run from overflow tube.

126. **PUMP TIMING.** To check and adjust pump timing, proceed as follows: Remove Proof-Meter drive cable, cover and gasket from rear end of fuel injection pump. Remove cover plate from front side of engine timing gear cover and remove the flywheel timing hole cover plate from right rear side of engine. Turn engine clockwise until punch marked V-notch in rear end of injection pump camshaft points toward timing mark on pump housing (Fig. 107), then continue to turn engine slowly until the 23° BTDC timing mark on flywheel (Fig. 108) is aligned with arrow at edge of timing hole. The timing marks on fuel injection pump· should then be exactly aligned as shown in Fig. 109; if not, loosen the three cap screws that retain gear to fuel injection pump drive hub and rotate the pump camshaft until marks are aligned. Tighten the cap screws to a torque of 20-25 ft.-lbs. Recheck timing marks and if aligned, reinstall the injection pump timing gear cover and reconnect Proof-Meter drive.

127. **R&R FUEL INJECTION PUMP.** Thoroughly clean the pump, lines and connections and the area around the pump. Proceed as outlined in paragraph 126 to bring the flywheel and pump timing marks into alignment and shut off the fuel. Remove the pump to injector lines, disconnect the fuel inlet and outlet

lines from fuel lift pump and the filter to injection pump line and immediately cap all openings. Disconnect the throttle and fuel shut-off controls. Remove the three cap screws retaining gear to injection pump drive hub and remove the gear clamping plate. Remove the cap screws retaining pump to engine front plate and remove the pump assembly. The pump drive gear will remain in the engine timing gear cover and cannot become out-of-time; however, the engine should not be turned with pump removed. Remove excess leak-off line from pump and plug the opening.

To reinstall the fuel injection pump, reverse the removal procedures and time the pump as outlined in paragraph 126. Tighten the fuel injection pump retaining cap screws to a torque of 26-30 ft.-lbs. and the gear to drive hub retaining cap screws to a torque of 20-25 ft.-lbs. Bleed fuel injection pump as outlined in paragraph 114A or 115A.

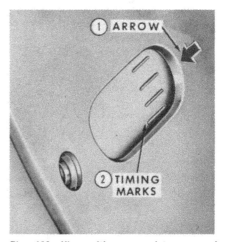

Fig. 108—View with cover plate removed showing timing pointer on engine rear plate and timing marks on flywheel.

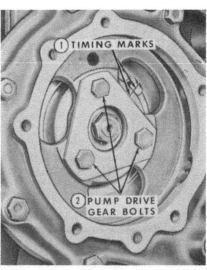

Fig. 109—Pump drive gear cover is removed to show timing marks on drive gear adapter hub and on pump housing.

Fig. 106—View showing filler plug (1), level plug (2) and drain plug (3) for fuel injection pump cambox and governor lubricating oil.

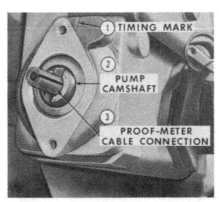

Fig. 107—View of fuel injection pump with Proof-Meter drive cover removed. "V" notch with two punch marks on rear end of pump camshaft should align with timing mark on housing when No. 1 piston is at 23° BTDC on compression stroke.

FUEL LIFT PUMP

128. The fuel lift pump is mounted on the outside of the fuel injection pump and is driven from a cam on the injection pump camshaft. The fuel pump is of the diaphragm type, and the component parts are available for service. Refer to exploded view of the fuel pump assembly in Fig. 110. The inlet and outlet valves (4) are staked into the outer body (3); when renewing valves, insert in body as shown and carefully stake in position. The primer lever retaining pin (12) must be installed with outer end below flush with the machined surface of inner body (7), to prevent damage to diaphragm (5). After inserting pivot pin (10) in inner body, securely stake pin in place.

To test pump, operate the primer lever with lines disconnected and with fingers closing the inlet port; pump should hold vacuum after releasing primer lever. With finger closing outlet port, there should be a well defined surge of pressure when operating primer lever.

INJECTION PUMP DRIVE GEAR

129. To remove the fuel injection pump drive gear, first remove the engine timing gear cover as outlined in paragraph 86. Then, remove the three cap screws, retainer plate and the fuel injection pump drive gear.

Prior to installing gear, turn engine crankshaft so that timing marks on crankshaft gear and camshaft gear point towards center of idler (camshaft drive) gear. Then, remove the self-locking cap screw retaining idler gear to front face of cylinder block, remove the idler gear and reinstall with timing marks aligned with marks on crankshaft gear and camshaft gear. Tighten idler gear cap screw to a torque of 100-105 ft.-lbs.

Place the pump drive gear on adapter hub with timing mark aligned with timing mark on idler gear. Place retainer plate on gear, install socket wrench on nut on front end of pump camshaft and turn pump camshaft until timing mark on adapter is aligned with timing mark (pointer) on pump front plate. Then, turn the pump slowly in a counter-clockwise direction (as viewed from front of engine), if necessary, so that the cap screws can be installed through retainer plate and drive gear into the pump drive adapter. Turn the engine clockwise until No. 1 piston is at the 23 degrees BTDC mark on flywheel, on compression stroke. With pump drive gear retaining cap screws loose, turn the pump camshaft with socket wrench so that pump timing marks are aligned (Fig. 109), then tighten the drive gear retaining cap screws to a torque of 20-25 ft.-lbs.

GOVERNOR AND THROTTLE LINKAGE ADJUSTMENT

130. Idle speed must be adjusted with engine at normal operating temperature. High idle speed for Models 9000, 9600, 9700, TW-20 and TW-30 is 2420-2470 rpm. High idle speed for all other models is 2530-2580 rpm. Low idle speed for all models is 700-800 rpm. To adjust low idle speed on all models, readjust stop screw on fuel injection pump, (4—Fig. 90 or 3—Fig. 112).

To adjust high idle speed on Models TW-10, TW-20 and TW-30, disconnect throttle linkage and turn maximum no load stop screw (1—Fig. 112), until specified idle speed is obtained. To adjust high idle on other models, disconnect throttle rod (POINT "A"—Fig. 111), loosen jam nut (6) and shorten rod to increase speed or lengthen rod to decrease speed. Push hand throttle lever forward against stop and check high idle speed. If not within recommended speed range, loosen rear clip (1) and while holding foot pedal depressed against platform and hand throttle forward against stop, tighten clip. Measure distance between front clip (2) and rear clip; if not within 4.44 to 4.70 inches, loosen front clip, reposition so that distance measures 4½ inches and tighten clip. Recheck high idle speed for both foot throttle and hand throttle operations and readjust if necessary.

On models with hand throttle only, check high idle speed with throttle lever against forward stop. If high idle speed is not within recommended range, disconnect throttle rod (POINT "A"), loosen jam nut (6) and shorten throttle rod to increase speed or lengthen rod to decrease speed. Reconnect throttle rod and tighten jam nut. Recheck high idle speed and readjust if necessary.

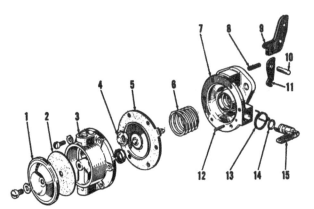

Fig. 110—Exploded view of diaphragm type fuel lift pump.

1. Cover
2. Pulsator diaphragm
3. Outer body
4. Valves
5. Pump diaphragm
6. Diaphragm spring
7. Inner body
8. Cam lever spring
9. Cam lever
10. Pivot pin
11. Diaphragm lever
12. Retaining pin
13. Return spring
14. "O" ring
15. Primer lever

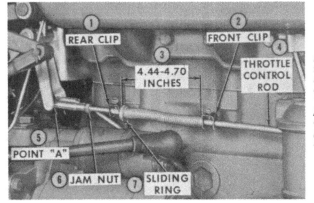

Fig. 111—Adjusting points for throttle linkage are shown for all models except TW-10, TW-20 and TW-30. Linkage is adjusted to obtain desired engine high idle speed; pump is not fitted with high idle speed stop screw.

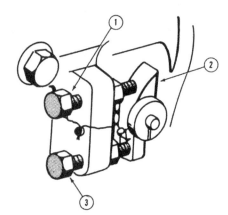

Fig. 112—Pump speed adjustments for Models TW-10, TW-20 and TW-30.

1. Maximum no load stop screw
2. Control stop
3. Low idle speed stop

TURBOCHARGER

134. 9000, 9600, TW-20 and TW-30 models are equipped with a turbocharger. These models will have either a Schwitzer Model 3LD or AiResearch turbocharger. Refer to Figs. 115 or 117 for identification. Major components consist of turbine, turbine housing, compressor, compressor housing and bearing housing. Model TW-30 is equipped with an intercooler as noted in paragraph 142.

TROUBLESHOOTING

135. If turbocharger malfunctions, review following troubleshooting sections to determine cause:

TURBOCHARGER NOISY. If turbocharger is noisy check the following:
   Dirty air cleaner.
   Foreign material or object in compressor to intake manifold duct.
   Foreign object in engine exhaust system.
TURBOCHARGER BINDING OR DRAGGING. If turbocharger binds or drags check the following:
   Damaged compressor wheel.
   Damaged turbine wheel.
   Worn bearings, shaft journals or bearing bores.
   Excessive dirt build up in compressor.
   Excessive carbon build up behind turbine wheel.
SEAL LEAKS AT COMPRESSOR END OF TURBOCHARGER. Check the following:
   Dirty air cleaner.
   Restricted duct between air cleaner and turbocharger.
   Loose compressor-to-intake manifold duct connections.
   Leaks at intake manifold.
   Restricted turbocharger oil drain line.
   Plugged crankcase breather.
   Worn or damaged compressor wheel.
   Excessive engine blow-by.

SEAL LEAKS AT COMPRESSOR END OF TURBOCHARGER. Check the following:
   Excessive pre-oiling.
   Plugged engine crankcase breather.
   Restricted turbocharger oil drain line.
   Worn bearings, bores or shaft journals.
WORN TURBOCHARGER BEARINGS, BORES OR JOURNALS. Check for:
   Inadequate pre-oiling following turbocharger installation or engine lube servicing.
   Contaminated or improper grade of engine oil used in engine.
   Restricted oil feed line.
   Plugged engine oil filter.
   Insufficient oil to turbocharger caused by oil lag.
   Insufficient oil supply caused by oil pump malfunction.

137. **REMOVE AND REINSTALL.** During removal procedure, be sure to cap or plug all manifold, turbocharger and oil tube openings to prevent damage to turbocharger due to foreign matter.

To remove turbocharger, remove hood side panels, exhaust extension, hood top panel, exhaust pipe, exhaust pipe flange and sealing ring. Remove fuel tank on TW-20 models and intercooler on TW-30 models. Disconnect air cleaner and intake manifold tubes from turbocharger. Disconnect oil supply and oil return tubes from turbocharger. Remove oil supply tube adapter. Remove four nuts and lockwashers securing turbocharger to exhaust manifold adapter and remove turbocharger.

To reinstall, place turbocharger on exhaust manifold adapter with a new gasket, use new tabbed lockwashers and tighten nuts to 30-35 ft.-lbs. Install oil supply tube adapter and new gasket and connect oil supply tube. Do not connect oil return tube at this time. Position exhaust seal ring in turbocharger. Install exhaust pipe flange and secure to

exhaust manifold adapter. Connect intake manifold tubes to turbocharger and tighten clamp bolts to 15-20 in.-lbs. Be sure there is no strain on compressor cover from intake manifold tube. If necessary, loosen compressor cover bolts and realign cover with intake manifold. Be sure compressor cover is seated properly and tighten cover bolts to 60 in.-lbs. Tighten air cleaner to turbocharger clamp bolts to 15-20 in.-lbs. Install exhaust pipe and tighten exhaust pipe flange bolts to 20-26 ft.-lbs. and the support bracket nuts to 8-12 ft.-lbs.

To prime turbocharger, proceed as follows: With oil return tube disconnected, place a container under the oil return passage of the turbocharger bearing housing and crank engine with diesel engine stop control out until there is a steady flow of oil from oil return passage of bearing housing. Connect oil return tube to turbocharger, install top hood panel, hood side panels and exhaust extension. Retighten all bolts after several hours of operation.

138. **OVERHAUL (Schwitzer).** Remove turbocharger as outlined in paragraph 137. Before disassembling, mark relative positions of compressor cover, bearing housing and turbine housing to aid in reassembly. Remove compressor cover and note that bolts are of a special design and same type must be used if the bolts are to be renewed. Remove clamp band (14—Fig. 115) and separate core assembly from turbine housing. It may be necessary to tap lightly on turbine housing to dislodge core assembly. The turbine wheel and shaft can be conveniently held by mounting a 5/8-inch 12-point box end wrench in a vise, so that turbine shaft can be inserted into wrench, with compressor wheel up. Remove compressor wheel retaining nut (2) being careful not to apply pressure to turbine or

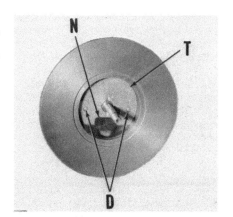

Fig. 116—Thrust plate (T) is installed on dowel pins (D) with bronze face up. Note cut-out (N) in thrust plate for lip of oil deflector.

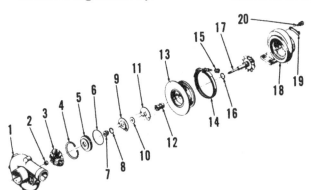

1. Compressor housing
2. Nut
3. Compressor wheel
4. Snap ring
5. Insert
6. "O" ring
7. Flinger sleeve
8. Seal ring
9. Oil deflector
10. Thrust ring
11. Thrust plate
12. Bearing
13. Bearing housing
14. Clamp band
15. Nut
16. Seal ring
17. Turbine wheel & shaft
18. Turbine housing
19. Clamp plate
20. Cap screw

Fig. 115—Exploded view of Schwitzer Model 3LD turbocharger used on 9000, 9600, 9700 tractors.

compressor wheel fins. Remove compressor wheel (3) and withdraw turbine wheel (17) and shaft assembly. Remove bearing (12) and mark bearing so it can be reinstalled in its original position. Remove bearing housing snap ring (4) and using two screwdrivers, lift insert (5) and "O" ring (6) from housing. Remove remainder of components in bearing housing. If sealing rings (8 and 16) are removed, note that they are different in diameter and should be marked to prevent incorrect assembly.

Soak turbocharger components in a suitable solvent and clean using a soft brush, plastic blade or compressed air.

CAUTION: Do not use a wire brush, steel scraper or caustic solution for cleaning as this will damage turbocharger parts.

139. Inspect turbine wheel and compressor wheel blades, which should be renewed if any blades are broken, bent or cracked. If any one blade is bent less than 20°, wheel may be used. If more than one blade is bent discard wheel. Do not attempt to straighten blades.

Inspect all bearing and thrust surfaces for wear, excessive scoring, grooves or heavy scratch marks. Inspect seal rings and grooves, inspect bearing housing for cracks and wear in bearing bore. Inspect oil supply holes in housing and be sure they are clear of obstacles.

To assemble, refer to exploded view in Fig. 115. Install copper-coated seal ring (16) on turbine wheel. Be sure correct sealing ring is installed. Lubricate sealing ring and turbine shaft with engine oil and install in bearing housing, being careful not to damage seal ring recess in bearing housing or bearing bore. Lubricate bearing (12) and install in bearing housing or turbine wheel shaft. If original bearing is being used, it should be reinstalled in its original position. Install thrust plate (11) on dowel pins with bronze face up as shown in Fig. 116. Lubricate and install thrust ring (10—Fig. 115). Install oil deflector (9) on dowel pins so that lip is in towards housing. Install seal ring (8) on flinger sleeve (7) and install flinger sleeve and seal ring in insert (5) so that flat side of sleeve (7) is flush with flat side of insert (5). Lubricate "O" ring and place in groove of insert (5). Install insert (5) in bearing housing with flat side out and press into counterbore until snap ring groove in bearing housing is clear. Be sure flinger sleeve remains in place when installing insert. Install snap ring (4) with beveled side out, and place compressor wheel on turbine wheel shaft. Coat the threads of a new locknut with graphite grease and install locknut on turbine wheel shaft and tighten to 156

in.-lbs. Spin compressor wheel and note any binding or rubbing of components. Assembly must spin freely. Check for turbine wheel shaft end play. Normal end play is 0.006 inch. If end play is excessive, check for abnormal wear on thrust surfaces. If there is no end play, check for carbon build-up on turbine wheel surface.

Reassemble remainder of assembly by reversing disassembly procedure. Tighten clamp band nut to 120 in.-lbs. When installing compressor cover, align cover with reference marks made during disassembly. Tighten compressor cover bolts using a diagonal pattern to 60 in.-lbs. Check that assembly will spin freely and disassemble if any drag is felt. To reinstall turbocharger, refer to paragraph 137.

140. OVERHAUL (AiResearch). Remove turbocharger as outlined in paragraph 137. Clean housing and mark for proper alignment during reassembly. Bend lockplate tabs (3 and 20—Fig. 117), remove bolts (4 and 19), clamps (2 and 21) and carefully separate housing assemblies. Remove locknut (18) while supporting center housing in a vertical position. Avoid bending turbine wheel shaft (5) while removing compressor wheel (17) from turbine wheel assembly. Turn over housing and withdraw turbine wheel assembly with piston ring (6). Bend locktabs on lockplate (23) and remove remaining bolts, lockplate (23) and backplate assembly (16). Do not disassemble backplate assembly. Center housing pins (14) should not be removed unless renewal is required.

Small parts should be inspected for burns, rubbing, scoring and other damage before cleaning, then clean parts in a non-caustic solution using a soft bristle brush, plastic scraper and dry compressed air. To make sure oil squirt hole (1—Fig. 118) isn't plugged, insert a 0.053-0.060 inch diameter wire into oil squirt hole of center housing. Refer to paragraph 139 for component inspection procedure.

The following parts should be renewed whenever turbocharger is overhauled or disassembled: Bolts (4, 19 and 24—Fig. 117), lockplates (3, 20 and 23), sealing ring (12), piston rings (6 and 11), snap rings (8), and bearings (9).

Refer to Fig. 117 to assemble turbocharger. Reassembly procedure is the reverse of disassembly, however the following steps should be taken: Fill piston ring groove with high vacuum silicon grease manufactured by Dow-Corning or equivalent. Install piston ring (6) and shroud (7) on turbine wheel assembly (5), then guide wheel assembly shaft through bearings (9). Slide shaft into center housing (15) as far as it will go. Place serrated end of turbine wheel assembly in a suitable socket or box end wrench clamped in a vise. Install thrust collar (10) on turbine wheel assembly (5) and slide the assembled parts down against center housing (15) so pins (14) engage thrust bearing holes. Install backplate (16) over turbine wheel shaft (5) and guide piston ring (11) into backplate bore. Install screws (24) and lockplate (23) then tighten screws to a torque of 75-90 in.-lbs. Install compressor wheel (17) on turbine wheel shaft.

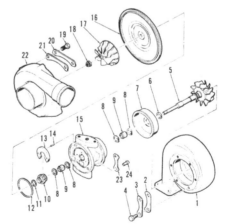

Fig. 117—Exploded view of AiResearch turbocharger.

1. Turbine housing	13. Thrust bearing
2. Clamp	14. Spring pin
3. Lockplate	15. Center housing
4. Bolt	16. Backplate assembly
5. Turbine wheel assembly	17. Compressor wheel
6. Rear piston ring	18. Locknut
7. Wheel shroud	19. Bolt
8. Snap ring	20. Lockplate
9. Bearing	21. Clamp
10. Thrust collar	22. Compressor housing
11. Front piston ring	23. Lockplate
12. Seal ring	24. Bolt

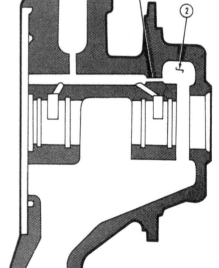

Fig. 118—Sectional view of turbocharger center housing.

1. Oil squirt hole	2. Turbine end oil cavity

After oiling washer face and threads of locknut (18), install locknut on shaft and tighten to a torque of 18 in.-lbs. then turn nut an additional 90 degrees. Apply high temperature sealer to threads of housing screws and attach turbine housing (1) to center housing (15). Tighten screws (4) tight enough to prevent turbine wheel from contacting housing. Install compressor housing (22) on backplate (16) and install clamps (21), lockplates (20) and screws. Tighten screws only enough to prevent housing contact with compressor wheel. If alignment marks are positioned properly, tighten screws to a torque of 100-130 in.-lbs. and bend lockplate tabs.

INTERCOOLER

142. Model TW-30 is equipped with a turbocharger intercooler. Air from the turbocharger is cooled before it enters the intake manifold. This process increases air density and permits increased fuel delivery to the engine, resulting in an increase in horsepower. Main parts of the intercooler are: Tip turbine fan and housing, pre-cleaner and heat exchanger. Components may be removed as necessary for repair after tractor hood and side panels are removed.

143. **OVERHAUL TIP TURBINE FAN.** Mark both housings for proper alignment during reassembly. Remove fan connector (4 – Fig. 119), with check valve flapper (5) from turbine housing (8) then separate valve and connector. Remove gasket (6) and bridge (7) from housing. Remove baffle assembly (22) from inlet housing. Remove inlet housing screws, lockwashers and separate inlet housing (21) from turbine fan assembly (8).

NOTE: It may be necessary to break face of seal by tapping with a brass mallet.

Remove locknut (20) while using suitable tools to keep the shaft from turning. Remove cover (19) and fan wheel (18) from turbine and discharge housing (1).

NOTE: It is not necessary to separate the turbine housing and discharge housing if there is no damage.

To disassemble discharge housing shaft and bearing assembly remove snap ring (9) and grease retainer (10) then remove shaft far enough so "O" ring (14) can be removed. Remove "O" ring, withdraw retainer (11) and shims (12) from shaft. Remove bearing spacer (15). It is not necessary to remove washer shields (16) and spring (17) unless spring is damaged.

Inspect parts before cleaning, then clean all parts in a non-caustic solution using a soft bristle brush, plastic scraper and compressed air.

144. After cleaning, inspect housing (21), wheel (18), and turbine housing (8), for wheel rubbing. Turbine housing should be inspected for nozzle blade erosion (both blade edges must be round).

Inspect shaft (13) for worn, crossed or stripped threads. Bearings and shaft assembly should be separated only if installing new components. Shaft diameter at bearing location should be 0.4725 inch and at wheel location 0.4717 inch. Fan wheel (18) bore diameter shouldn't exceed 0.4724 inch and bearing bores of discharge housing assembly (1) should not exceed 1.2604 inches.

Inspect turbine and cooling air passages for salt build up and flapper valve (5) for dirt. Replace if outer rim is bent more than 0.150 inch. Inspect bearing assemblies (13), snap ring and grease retainer (10) for excessive wear. Renew parts as necessary.

145. Reassembly procedure is the reverse of disassembly, however the following steps should be taken: Install shaft and bearing components after first applying all grease supplied in bearing and shaft kit, Ford part No. D8NN-6N839-AA, to free area between bearings.

Assemble discharge (1) and turbine (8) housings with marks aligned and tighten retaining screws to a torque of 100-120 in.-lbs. Install locknut (20) but do not tighten until shim (12) thickness is determined. With fan, bearings and shaft components assembled, mount a dial gage on mating surface of turbine housing as shown in Fig. 120. Zero dial gage on mating surface (M) then reposition dial gage to read difference in height between housing mating surface and fan wheel at surface (F). Rotate fan wheel and take at least four readings. Average the readings and the result will be thickness of shim (12 – Fig. 119).

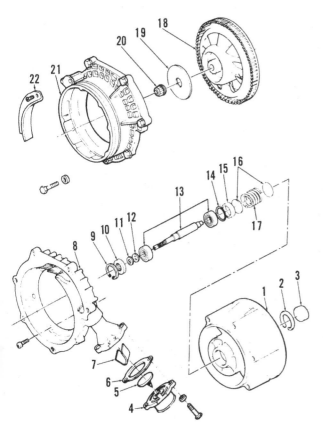

Fig. 119—Exploded view of heat exchanger fan.

1. Fan discharge housing
2. Snap ring
3. Fan housing cup plug
4. Fan connector
5. Check valve flapper
6. Housing gasket
7. Bridge
8. Turbine housing
9. Snap ring
10. Grease retainer
11. Bearing retainer
12. Shim
13. Bearing and shaft
14. "O" ring
15. Bearing spacer
16. Washers
17. Shaft spring
18. Fan wheel
19. Cover
20. Locknut
21. Inlet housing
22. Baffle assembly

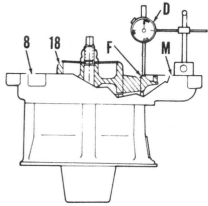

Fig. 120—View showing dial gage placement to determine fan wheel shim size as outlined in text.

D. Dial gage
F. Fan wheel surface
M. Housing mating surface
8. Turbine housing
18. Fan wheel

Shims are available in thicknesses of 0.002, 0.005 and 0.010 inch. With fan wheel properly positioned on shaft, tighten locknut (20) to 140-160 in.-lbs. If wheel drag is noticed when turning by hand, recheck clearance. Replace fan if housing rub can't be eliminated by shimming. Press bridge (7—Fig. 119 and 121) into turbine housing (8) so top of bridge is 0.05-0.06 inch below surface (M—Fig. 121). Reinstall all remaining components.

147. OVERHAUL HEAT EXCHANGER. General maintenance may be performed with unit on engine. Remove tip turbine fan for inspection of fan side of intercooler. Accumulated dirt and oil should be removed. Replace gaskets whenever heat exchanger is removed. Before disassembly, mark top of cooler core for proper alignment during reassembly.

Remove intercooler from intake manifold to prevent dirt from entering intake system. Remove old gaskets, steam clean, rinse with water and blow dry.

NOTE: Do not use petroleum based solvents for cleaning cooler on engine.

Clean any remaining gasket material

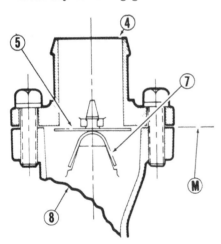

Fig. 121—View showing bridge (7-Fig. 119) installation. Press bridge (7) into turbine housing so top of bridge is 0.05-0.06 inch below surface M. Refer to text and Fig. 119.

from surface of unit. Use Dow-Corning No. 732 or equivalent sealer to secure new gaskets. After gasket installation, let unit stand until sealer has cured. Reinstall intercooler assembly.

COOLING SYSTEM

RADIATOR PRESSURE CAP AND THERMOSTAT

151. A 7 psi radiator pressure cap is used on 8000, 9000, 8600 and 9600 models. All other models use a 13 psi radiator cap. TW-10, TW-20 and TW-30 models are equipped with two thermostats. All other models have one thermostat. Thermostats on Models TW-10, TW-20 and TW-30 are located in a thermostat housing at front of cylinder head, while the thermostat on all other models is located in front of cylinder head. Thermostat opening temperature on TW-10, TW-20 and TW-30 models is 178°F. Thermostat opening temperature on all other models is between 168°F. and 178°F. Thermostat(s) may be removed after draining radiator and removing water outlet connection.

RADIATOR

152. To remove radiator, first drain cooling system and proceed as follows: Evacuate a/c system if so equipped, disconnect lines to condenser then remove grille from radiator shell for access to bolts. Remove engine hood side panels and unbolt hood top panel and side braces from radiator shell. Disconnect radiator hoses, oil cooler and hydraulic lines then unbolt radiator shell from front support. Attach hoist to radiator shell and lift the radiator, a/c condenser (if equipped) and shell unit from tractor. Remove shroud nut and radiator from shell. Reinstall by reversing removal procedure.

WATER PUMP

153. Water pump can be unbolted and removed after draining cooling system,

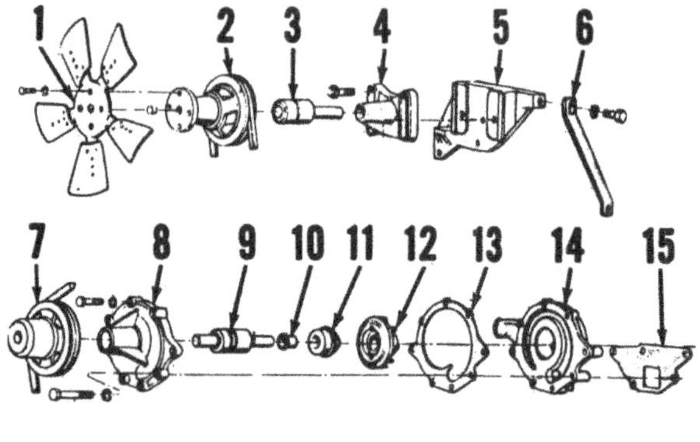

Fig. 122—Exploded view of fan and bracket assembly (top) and water pump assembly (bottom).

1. Fan assembly
2. Fan pulley
3. Shaft & bearing assy.
4. Fan bracket
5. Fan support
6. Brace
7. Water pump pulley
8. Pump housing
9. Shaft & bearing assy.
10. Water slinger
11. Water pump seal
12. Water pump rotor
13. Gasket
14. Pump rear cover
15. Gasket

removing pump and alternator or generator drive belt and disconnecting hose from pump.

Using standard two-bolt puller, remove pulley from shaft. Remove rear

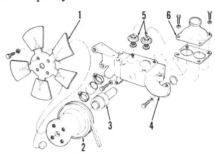

Fig. 122A—Exploded view of thermostat housing and fan drive assembly on Models TW-10, TW-20 and TW-30.

1. Fan
2. Pulley
3. Shaft and bearing assy.

4. Thermostat housing
5. Thermostats
6. Outlet housing

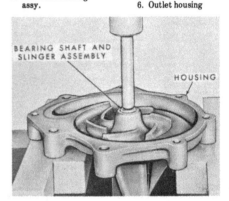

Fig. 123—Press bearing, shaft and slinger assembly from rotor and pump housing.

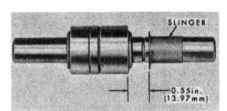

Fig. 124—Press new slinger onto shaft so that distance from edge of bearing outer race to slinger is 0.55 inch.

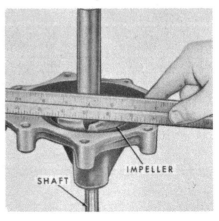

Fig. 125—Press impeller onto shaft with ¾-inch I.D. pipe so that impeller is flush with rear face of housing.

cover (14—Fig. 122) and press the shaft and bearing assembly (9) out towards front of housing as shown in Fig. 123. Drive the seal (11—Fig. 122) out towards rear of housing.

Using a length of 1-5/16-inch I.D. pipe, press new seal into housing. Check to see that flange on water slinger (10) is located 0.55 inch from edge of bearing race as shown in Fig. 124, then press shaft into front of housing until outer bearing race is flush with front end of housing. Using a length of ¾-inch I.D. pipe, press impeller onto shaft as shown in Fig. 125 so that impeller is flush with rear end of housing. Press pulley onto shaft so that center of belt groove in pulley is 2½ inches from rear face of housing as shown in Fig. 125A, for 8000, 9000, 8600 and 9600 models and 3½ inches for all other models. Install rear cover with new gasket and tighten retaining cap screws to a torque of 18-22 ft.-lbs.

Reinstall water pump by reversing removal procedure and tighten retaining cap screws to a torque of 23-28 ft.-lbs. Drive belt is properly adjusted when a force of 25 pounds applied midway between pulleys will deflect belt 5/8 to 7/8 inch.

ENGINE OIL COOLER

155. A water jacketed oil cooler (3—Fig. 126) is attached to top of power steering motor behind the radiator on all models except 8700, 9700, TW-10, TW-20, and TW-30. Oil flows to the cooler from the main oil gallery of engine and returns to the oil pan via right side of cylinder block. On 8000 and 8600 models only, the fitting (6) that connects return tube (2) to cooler contains a relief valve to maintain oil pressure in the engine oil gallery at low engine speeds. This relief valve is not required on 9000 and 9600 models.

Engine oil on Models 8700, 9700, TW-10 and TW-20 is cooled by an oil cooler located in the radiator lower tank.

Engine oil on TW-30 models is cooled by an oil cooler located on the left side of cylinder block. Two spin on oil filters are attached to it.

FAN AND FAN BRACKET

All Models Except TW-10, TW-20 And TW-30

156. To remove fan and fan bracket assembly, unbolt fan from pulley (2—Fig. 122) hub, then unbolt bracket (4) from support (5) and remove the bracket, bearing and pulley as an assembly. Remove fan from radiator shroud.

To renew fan bearing and shaft assembly (3), remove fan guide plug and insert driver through pulley center hole to press bearing and shaft assembly from pulley. Press bearing and shaft assembly from bracket. Reassemble and reinstall by reversing removal and disassembly procedure.

Fan drive belt is properly adjusted when a force of 25 pounds applied midway between pulleys will deflect belt 5/8 to 7/8 inch.

Models TW-10, TW-20 And TW-30

156A. The fan shaft and bearing assembly is carried in the thermostat housing (4—Fig. 122A). which is attached to front of the cylinder head. Refer to Fig. 122A for an exploded view of thermostat housing and fan shaft and bearing. Adjust fan belt tension so belt will deflect approximately 5/8 to 7/8 inch when a force of 25 pounds is applied midway between pulleys.

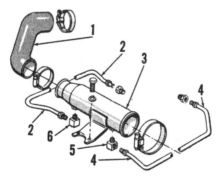

Fig. 126—Engine oil cooler (3) is mounted on steering motor and is connected by lower radiator hose (1) to cooling system. Hose connecting cooler to radiator is not shown. Note that outlet fitting (6) contains a relief valve to maintain engine oil pressure on 8000 and 8600 models only.

1. Cooler to engine hose
2. Oil return line
3. Oil cooler
4. Oil pressure line
5. Inlet fitting
6. Outlet fitting & relief valve assembly

Fig. 125A—View showing water pump pulley position.

1. 3.5 inches—Models
   8700, 9700, TW-10,
   TW-20 and TW-30
   2.5 inches—Models
   8000, 9000, 8600,
   9600

ELECTRICAL SYSTEM

GENERATOR AND REGULATOR

157. A Ford C7NN-10000-C 22-ampere generator and a Ford D0NN-10505-A regulator are used on earlier 8000 models without cab. The generator is a two-pole shunt wound type with type "B" circuit; that is, one field coil terminal is grounded to generator frame and the other field coil terminal is connected to armature terminal through the regulator. Output may be tested as follows: Make sure belt is tight and terminals are tight and clean. Disconnect the two wires from generator and connect a jumper wire across the two terminals. Run engine at idle, connect positive lead of a voltmeter to either generator terminal and the negative lead to the generator frame. As engine speed is increased, voltage should rise rapidly. Do not allow voltage to reach 20 volts. Overhaul generator if voltage will not rise, or is not steady. Specifications are as follows:

C7NN-10000-C Generator
Max. output (hot) at 1350
   Engine rpm and 15 volts . . . . . 22 amps
   Renew brushes if
      shorter than 13/32-inch
   Min. brush spring tension
      with new brushes. . . . . . . .18 ounces
   Field coil current . . . . .2 amps
   Field coil resistance . . . . .6 ohms
Commutator min. diameter . . .1.450 inch
   Max. commutator runout . . .0.002 inch
   Max. armature shaft runout 0.002 inch

If generator output was tested and voltage rise was rapid and steady, the voltage regulator is probably defective. Regulator is a sealed unit and adjustment or disassembly is not recommended.

ALTERNATOR AND REGULATOR

CAUTION: An alternator (A.C. generator) is used to supply charging current on earlier 8000 models with steel cab, later 8000 models without cab and all other models. Due to the fact that certain components of the alternator can be seriously damaged by procedures that would not affect a D.C. generator, the following precautions must be observed.

1. Always be sure that when installing batteries or connecting a booster battery, the negative posts of all batteries are grounded.

2. Never short across any of the alternator or regulator terminals.

3. Never attempt to polarize the alternator.

4. Always disconnect all battery ground straps before removing or replacing any electrical unit.

5. Never operate the alternator on an open circuit; be sure that all leads are properly connected and tightened before starting the engine.

158. A Motorola alternator (Ford part No. C7NN-10300-B 55 amp, D5NN-10300-D 51 amp or D7NN-10300-B 72 amp), and transistorized voltage regulator (Ford part No. C7NN-10316-A, D1NN-10316-A, D7NN-10316-B or D8NN-10316-AA) are used. Service specifications follow:

All Alternators

Max. output at 160°F., 2400 engine rpm and 14.4 volts	*47 amps
Minimum brush length	3/16-inch
Min. brush spring tension with new brush	4 ounces
Field current	1.8-2.4 amps
Field resistance	6 ohms

*Max. output for D7NN-10300-B alternator is 67 amps.

All Regulators
Voltage regulation with 10 ampere load:

Ambient Temperature	Output Terminal Voltage
40°F	14.2-15.0
60°F	14.1-14.9
80°F	13.9-14.7
100°F	13.8-14.6
120°F	13.6-14.4

STARTING MOTOR

159. Engine is equipped with a Ford starting motor and relay assembly. Closing the starter switch energizes the solenoid; movement of the solenoid plunger engages the drive pinion and closes a two-stage switch. If the teeth of the drive pinion butt against teeth on flywheel, only the first stage of the

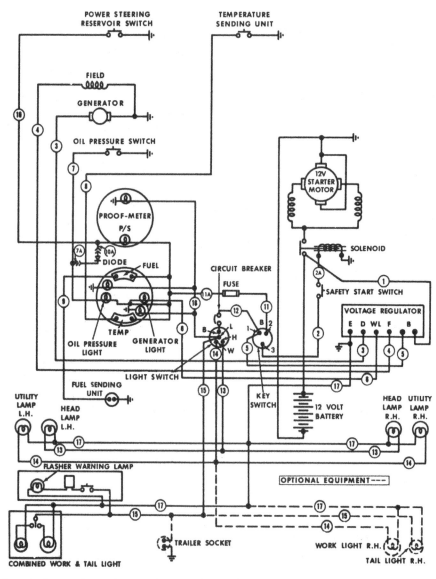

Fig. 128—Wiring diagram for models with generator. Refer to Fig. 130 for models with alternator. Code for regulator connections are as follows: (E) ground, (D) armature, (WL) warning light, (F) field and (B) battery. Color code for wiring is given in legend.

1. Brown	5. Brown-white stripe	10A. Green-blue stripe	14. Red
2A. Red	6. Yellow-black stripe	11A. Black-green stripe	15. Blue-white stripe
3. Yellow-black stripe	7A. White	12. Yellow	16. Blue-red stripe
4. Blue	8. Red-white stripe	13. Green	17. Black
	9. Orange		

switch is closed which will allow current to flow to one field coil. This will provide enough power to turn starter until drive pinion is in position to engage flywheel ring gear teeth; then, full engagement of drive pinion will close second stage of switch energizing all four field coils.

When drive pinion is in engaged position, there should be a clearance of 0.010-0.012 inch between drive pinion and thrust collar. To check clearance, first energize solenoid with 12-volt power source, then check clearance with

feeler gage as shown in Fig. 133. If clearance is not within 0.010 to 0.020 inch, refer to Fig. 134, loosen locknut and turn eccentric pivot pin as required to obtain proper clearance. Then, tighten locknut and recheck clearance.

Service specifications are as follows:

Starting Motor And Relay Assembly
Brush spring tension (min. with
  new brushes) . . . . .42 ounces
Min. brush length 5/16-inch

Commutator min.
  diameter . . . . .1.50 inches
Max. armature shaft end play . 0.025 inch
Max. armature shaft runout . . 0.005 inch
Drive pinion clearance
  (engaged) . . . . .0.015-0.025 inch
No-load test:
  Volts 12
  Amps 117
  Rpm .5500-8000
Loaded test (with warm engine):
  Amps .250-300
  Engine rpm 150-200

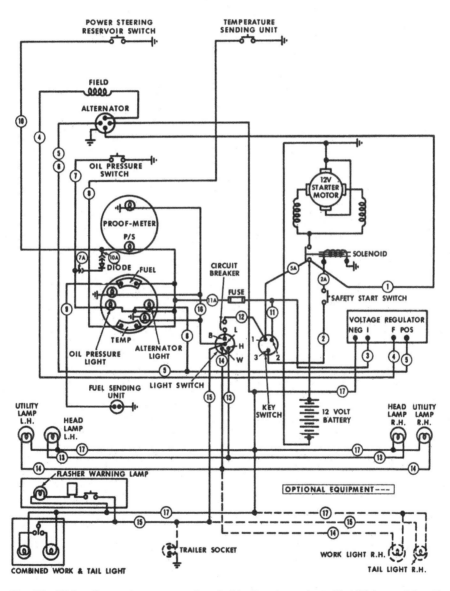

Fig. 130—Wiring diagram for tractor equipped with alternator; refer to Fig. 128 for models with generator. Voltage regulator is pre-wired with plug-in type pigtail. Color code for wiring is given in legend.

1. Brown	5A. Brown-white stripe	9. Orange	13. Green
2A. Red	6. Yellow-black stripe	10A. Green-blue stripe	14. Red
3. Yellow	7A. White	11A. Black-green stripe	15. Blue-white stripe
4. Green	8. Red-white stripe	12. Yellow	16. Blue-red stripe
5. Red			17. Black

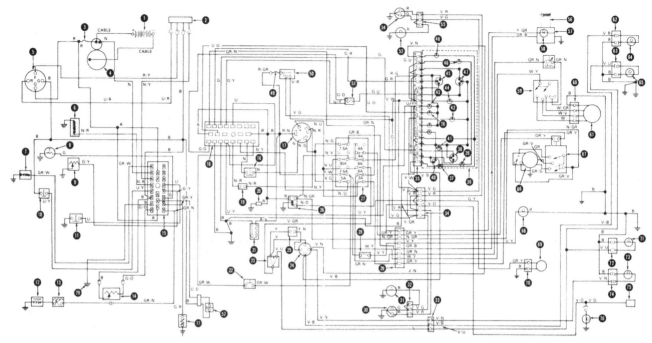

Fig. 132—Wiring schematic for Models 8700, 9700, TW-10, TW-20 and TW-30 equipped with a cab. Note that TW models are equipped with instrument panel gages while 8700-9700 models are equipped with a temperature gage and warning lights. Component (52) is a diode on 8700-9700 models and an oil pressure sender on TW models. Models 8700 and 9700 aren't equipped with components (77, 78 or 79).

B. Black	10. Air cleaner	24. Turn signal switch	37. Turn signal	52. Diode or oil sender
G. Green	restriction switch	25. Flasher unit	indicator, R.H.	53. Work lamp, R.H.
GR. Gray	11. Oil pressure sender	26. Ignition relay	38. Temperature gage	54. Hi-lo beam lamp,
N. Brown	12. Air compressor	27. Fuse block	39. Panel light	R.H.
O. Orange	clutch	28. Dome light switch	40. Alternator warning	55. Work & headlamp
R. Red	13. De-icing switch	29. Cab roof harness	light or gage	connector
U. Blue	14. **Auxiliary fuel level**	connector	41. Panel light	56. Speaker
V. Violet	**sender**	30. Hi-lo beam lamp,	42. Hi-lo beam indicator	57. Radio
W. White	15. Front main harness	L.H.	43. Air cleaner warning	58. Radio dial light
Y. Yellow	connector	31. Work lamp, L.H.	light	59. Blower motor switch
1. Battery	16. Rear main harness	32. Work & headlamp	44. Panel light	60. Blower motor
2. Regulator	connector	connector	45. Oil pressure	connector
3. Fuse link	17. Ignition switch	33. Implement & flasher	warning light or gage	61. Blower motor
4. Starter motor	18. Safety start switch	lights connector,	46. Panel light	62. Flasher connector,
5. Alternator	19. Ether start button	rear	47. Fuel gage	R.H.
6. Ether solenoid	20. Condenser	34. Main extension	48. Turn signal	63. Implement lamp
7. Horn	21. Cigar lighter	harness connector	indicator, L.H.	connector, R.H.
8. **Temperature sender**	22. Horn button	35. Instrument cluster	49. Fuse	64. Flasher light, R.H.
9. **Main fuel level**	23. Implement/hazard	connector	50. Headlight switch	65. Implement lamp,
sender	light switch	36. Instrument cluster	51. Fuel selector switch	R.H.

66. Wiper motor
67. Wiper/washer
switch
68. Dome light
69. Washer pump motor
70. Washer pump
connector
71. Implement lamp,
L.H.
72. Implement lamp
connector, L.H.
73. Flasher light, L.H.
74. Flasher light
connector, L.H.
75. Trailer socket
76. Tail light
77. Trans. oil pressure
switch
78. Trans. oil pressure
warning light
79. Front hood R.H. support

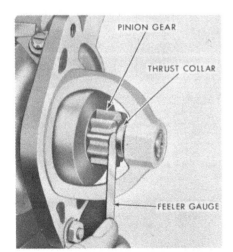

Fig. 133—Measuring starting motor drive pinion clearance. Refer to text for procedure and to Fig. 134 for adjustment of clearance.

Fig. 134—Adjusting drive pinion clearance; refer also to Fig. 133 and to text.

CLUTCH

All except TW-20 and TW-30 models are equipped with a 13-inch diameter single plate clutch assembly; TW-20 and TW-30 models are equipped with a 14-inch diameter clutch. The clutch friction disc on 8000 models is fitted with five cera-metallic pad inserts, while 8600, 9000 and TW-10 models have six cera-metallic pad inserts. 8600 model discs can be used on prior 8000 models. The disc used on 8700, 9600 and 9700 models is equipped with nine pad inserts and can be used on earlier 9000 models. The disc used on TW-20 and TW-30 models has eight cera-metallic pad inserts. Clutch disc is renewable as an assembly only. The clutch cover, thrust plates, pressure springs and adjusting nuts are available separately. There are 12 pressure springs in pressure plate and cover assembly on 8000, 8600 and 8700 models, while 9000, 9600, 9700 and TW-10 models have 16 pressure springs. TW-20 and TW-30 models have 15 pressure springs. The clutch pressure plate and release levers are available as a sub-assembly or in complete clutch cover and pressure plate assembly only.

PEDAL FREE TRAVEL ADJUSTMENT

All Models Except 8700, 9700, TW-10, TW-20 And TW-30

160. Clutch pedal free play is measured at pedal as shown in Fig. 135. Clutch pedal free travel should be 1-3/8-1-5/8 inches. To adjust free travel, disconnect clutch pedal rod from release lever, loosen locknut and turn clevis to lengthen rod until correct free travel is obtained. Left hood panel on some tractors must be removed for access to clevis. On tractors with cabs, remove left step assembly, tool box side panel and door, and cab support bracket in order to adjust clevis.

Models 8700, 9700, TW-10, TW-20 And TW-30

161. Measure clutch pedal free play at pedal as shown in Fig. 136. Clutch pedal free play should be 1-3/8-1-5/8 inches. To adjust free play, loosen locknut (L) and rotate turnbuckle (T) on clutch pedal control rod (R).

NOTE: If the difference between clutch control rod exposed threads exceeds 0.010 inch, clutch control rod is not assembled properly. See Fig. 137.

R&R CLUTCH ASSEMBLY

162. **SPLIT TRACTOR.** To split trac-

tor for clutch removal, proceed as follows: On tricycle models, attach support braces to keep front end assembly from tipping. On wide front axle models, drive wood wedges between front axle and front support. Remove tool box and battery covers and disconnect battery cables from battery, ground cable first. Remove hood side panels, air pre-cleaner, exhaust pipe and hood top panel. Disconnect clutch pedal return spring, then unbolt and remove side plates. Disconnect throttle linkage from bellcrank at instrument panel support. Disconnect electrical wiring, engine stop cable, fuel supply line and Proof-Meter cable at engine. On models with pressure lubricated transmissions disconnect lines to oil cooler. Unbolt and remove starting motor. Disconnect power steering fluid lines from steering motor and unbolt hood side braces from instrument panel (rear hood) support.

On wide front axle or front wheel drive models, support rear end of engine with rolling floor jack and front end of transmission with safety stand. On tricycle models, place the rolling floor jack under front end of transmission. Unbolt engine from transmission and separate tractor.

163. To reconnect tractor, first remove clutch shaft from transmission and insert it into clutch friction disc and pilot bearing, then complete reassembly by reversing split procedures. Tighten the 5/8-inch engine to transmission bolts to a torque of 95-130 ft.-lbs. and the 3/4-inch bolts to a torque of 180-220 ft.-lbs. Bleed the fuel injection system as outlined in paragraph 114A or 115A and bleed power steering system as in paragraph 7 or 34.

164. **R&R CLUTCH.** With tractor split as outlined in paragraph 162, alternately

and evenly loosen the cap screws retaining clutch cover to flywheel to avoid distortion of cover. Remove clutch cover and pressure plate assembly, taking care not to drop the clutch friction disc.

The clutch shaft pilot bearing can now be removed from flywheel by removing the two cap screws and flat washers; refer to Fig. 84.

To reinstall clutch assembly, lightly lubricate splines with a high temperature silicone grease and align clutch disc to pilot bearing with transmission input shaft. Determine which is the larger

Fig. 136—Clutch pedal free play on 8700, 9700, TW-10, TW-20 and TW-30 models should be 1-1/8—1-5/8 inches. Rotate turnbuckle (T) to adjust free play.

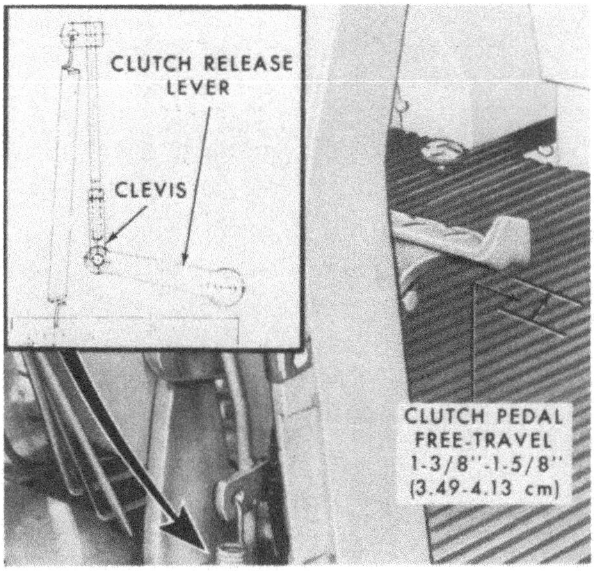

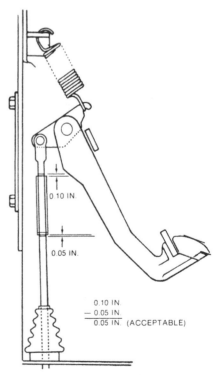

Fig. 137—If control rod and adjusting nut are properly assembled, the difference between exposed threads of upper and lower rods will be 0.010 inch or less. See above example.

Fig. 135—Clutch pedal free travel should be 1-3/8 to 1-5/8 inch when measured at pedal. Adjust by turning clevis on pedal to clutch release lever rod.

CLUTCH RELEASE LEVER

CLEVIS

CLUTCH PEDAL FREE-TRAVEL 1-3/8"-1-5/8" (3.49-4.13 cm)

dowel pin in flywheel so clutch cover can be installed correctly. Alternately and evenly tighten clutch cover retaining cap screws, taking care that the cover seats correctly on the dowel pins in flywheel, until cover contacts flywheel. Then, alternately and evenly tighten cover cap screws to 26-35 ft.-lbs. on TW-10, 20-30 ft.-lbs. on TW-20 and TW-30, or 20-27 ft.-lbs. on all other models. Reconnect engine to transmission as outlined in paragraph 163.

NOTE: If a new clutch disc is installed on 8000, 8600 or 8700 model tractor, place new disc on flywheel to be sure hub rivets on disc do not contact inner diameter of flywheel. If contact exists, flywheel must be removed and machined out to 6.44 inch I.D. to clear hub rivets.

OVERHAUL CLUTCH

165. With clutch cover and pressure plate assembly removed as outlined in paragraph 164, place assembly in press and position a square, flat steel plate over cover so that adjusting nuts and thrust plates are accessible (Fig. 138). Compress the assembly until release levers are free, then unbolt and remove the thrust plates and unscrew the adjusting nuts from the finger yoke studs. Release cover assembly and remove cover and springs from the pressure plate and release lever assembly. Note positions of springs on 8000, 8600 and 8700 models, before removing springs.

Inspect all parts and renew as necessary. Heat discoloration of the clutch disc friction pads is normal; friction pad thickness (new) is 0.463 inch for Models TW-20 and TW-30, or 0.405 inch for all other models. Renew the pressure plate and release lever assembly if levers are damaged or excessively worn, or if pressure plate is cracked, distorted or excessively scored. Check the clutch springs against the following specifications:

Models TW-20 & TW-30
Part number D8NN-7572-AA
Color code. . . . . None
Lbs. pressure at 1.539 in. 220-210

All Other Models
Part number C7NN-7572-C
Color Code. . . . . . . Pink with black stripe
Free length, inches. 2¾
Lbs. pressure at
    1.539 inches 135.5-145.5

When reassembling clutch on 8000, 8600 and 8700 models, refer to Fig. 138 for placement of springs on the pressure plate and note that second boss ("X") in clockwise direction from each release lever is not fitted with a spring. On all

other models, there is an equal number of springs and locating bosses. See Fig. 139 for view of clutch and flywheel used on Models TW-20 and TW-30. Reassemble by reversing disassembly procedure, but on all models except TW-20 and TW-30, tighten thrust plate retaining cap screws finger tight only until the release levers are adjusted. To adjust release levers, mount cover and pressure plate assembly on flywheel with new friction disc or with several equally placed 0.405 inch thick spacers between pressure plate and flywheel. Models TW-20 and TW-30 will require 0.463 inch spacers. Release lever height on Models TW-20 and TW-30 is adjusted by turning adjusting screws at lever ends, while lever height on all other models is adjusted by turning adjusting nuts (Fig.

138). Adjust release lever height above friction surface of flywheel to 1.969-2.031 inches on 8000, 8600 and 8700 models, 2.125-2.187 inches on 9000, 9600, 9700 and TW-10 models and 2.26-2.56 inches on TW-20 and TW-30 models. On all models except TW-20 and TW-30, stake adjusting nuts in position and tighten thrust plate retaining cap screws to a torque of 15 ft.-lbs.

DUAL POWER

Dual Power consists of a planetary gear set and two clutch assemblies enclosed in a housing and mounted in forward compartment of transmission case. Dual Power is hydraulically actuated to provide either direct drive or

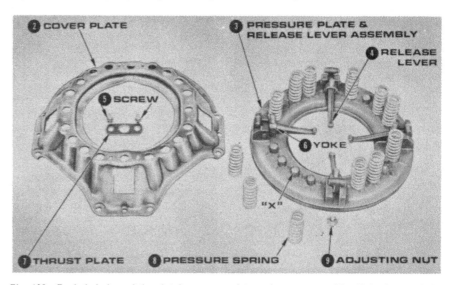

Fig. 138—Exploded view of the clutch pressure plate and cover assembly. Note that on 8000, 8600 and 8700 models, second boss ("X") in a clockwise rotation from each release lever is not fitted with a pressure spring. Pressure plate and release levers are serviced as an assembly only.

Fig. 139—Assembled view of clutch and flywheel used on Models TW-20 and TW-30.

1. Flywheel
2. Clutch disc
3. Clutch cover
4. Cover to flywheel bolts

underdrive to transmission input shaft. Dual Power is controlled by pedals and linkage from operator's position to hydraulic control valve on side of Dual Power housing.

LUBRICATION

167. Dual Power assembly is used on pressure lubricated transmissions. On all models except TW-10, TW-20 and TW-30, lubricating oil is directed from transmission oil cooler to control valve on side of Dual Power housing and into oil passages of housing. Oil exits into transmission by tube connecting Dual Power housing and transmission case and through common passages. Lubrication pressure for Models TW-10, TW-20 and TW-30 is provided by the low pressure oil system. Refer to paragraph 174.

HYDRAULIC OIL PRESSURE

168. Clutch assemblies of Dual Power unit are engaged by hydraulic pressure against their repective pistons. To check hydraulic oil pressure, first operate tractor until oil is at normal operating temperature (120° F.). Stop engine and tee a pressure gage into the ¼-in. pressure line at the hydraulic pump cover and hydraulic system oil filter adapter plate. Start engine and set engine speed at 1000 rpm. Engage direct drive of Dual Power then engage underdrive. On models except TW-10, TW-20 and TW-30, pressure gage should read 150-180 psi with Dual Power in direct drive and 155-185 psi when in underdrive. Pressure gage readings on TW-10 and TW-20 models should be between 185-220 psi. The difference in readings should not exceed 10 psi.

On TW-30 models, check pressure reading with dual power in direct drive and pto engaged. Apply left brake, check pressure then apply right brake and check pressure. Gage reading at 2200 rpm with dual power in direct drive should be 185-235 psi.

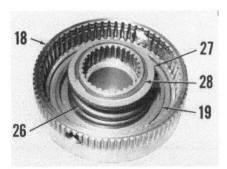

Fig. 140—Special Nuday tool No. 1312 is used to press against spring retainer (27) to reduce spring tension on snap ring (28) so that it may be removed. Refer to Fig. 142 for identification.

If pressure is as stated and trouble exists, the hydraulic system and Dual Power control valve can be considered satisfactory and Dual Power assembly should be inspected.

If pressure is higher than specified in both direct drive and underdrive operation, then remove hydraulic pump adapter plate as outlined in paragraph 233, remove Dual Power regulating valve and inspect valve for sticking during operation. The regulating valve threads into inner side of adapter plate. If unable to free up valve, renew as an assembly. A cross section of valve is shown in Fig. 144A. Valve is also shown in Hydraulic Section (10—Fig. 212). If oil pressure is lower than specified in both direct drive and underdrive operation, inspect hydraulic pump and pto system relief valve. Also check for a weak or broken spring in Dual Power regulating valve, or valve stuck open. If oil pressure is below specified pressure in either direct drive or underdrive, but normal for the other operation, check for a leaking control valve spool or for a leaking gasket between control valve and Dual Power housing. If direct drive operation has low oil pressure, disassemble Dual Power unit as outlined in paragraph 171 and inspect direct drive clutch assembly for leaking seals or a broken or cracked piston. If underdrive operation has low oil pressure, disassemble Dual Power unit and inspect underdrive clutch assembly for leaking seals or broken or cracked piston.

LINKAGE

169. Dual Power pedal or buttons are connected by a control rod to the control valve on the Dual Power housing. Length of control rod is adjusted by disengaging clevis end of rod, backing off locknut and turning clevis. Length of control rod should be adjusted so that both underdrive and direct drive operation is obtained when the corresponding operator's pedal is depressed. Adjust control rod on Models 8700, 9700, TW-10, TW-20 and TW-30 so control pedal heel is 3/8-5/8 inch from floor when in Low range. If control rod length is improperly adjusted, either underdrive or direct drive operation will be absent. If rod is too long, underdrive will not function. If rod is too short, direct drive will not function. If rod adjustment will not cure problem, refer to control valve, paragraph 173, or to Dual Power paragraph 171, for overhaul of defective assembly.

R&R DUAL POWER

170. To remove Dual Power assembly, drain lubricant from transmission and

rear axle center housing (approximately 17 to 23 gallons depending on model; see paragraph 176) and split tractor between engine and transmission as outlined in paragraph 162. Disconnect clevis pin from yoke on Dual Power control pedal linkage. Mark and remove lubrication (upper) and hydraulic pressure (lower) tubes from control valve. Unplug transmission oil switch wires on TW-10, TW-20 and TW-30 models. Oil return tube has been eliminated on these models. Remove clevis pins from clutch release fork and remove clutch fork cross shaft, clutch release bearing and clutch fork. Disconnect and remove control valve control rod.

To remove Dual Power housing from transmission, proceed as follows: Remove housing cover (41—Fig. 143), ring gear (36), planetary gear set (33), shaft (31) and sun gear (29). Remove direct drive clutch assembly (18) and thrust washer (16).

NOTE: Do not allow transmission mainshaft (16—Fig. 160) to move forward when removing Dual Power housing or thrust bearing (38) may dislodge. Fabricate a tool made of pipe which bears against mainshaft and is secured to pto shaft.

Remove five retaining bolts and separate Dual Power housing from transmission case. Lubrication supply tube from housing to transmission case may be removed by pulling straight out.

To reassemble Dual Power unit, reverse disassembly procedure. Install lubricating tube (2—Fig. 143) with new "O" rings on tube and a new gasket on housing before attaching housing to transmission case. Be sure housing outlet is aligned with tube when installing housing. Tighten Dual Power housing-to-transmission case bolts in a diagonal pattern to 65-90 ft.-lbs. on TW-10, TW-20 and TW-30, or to 50-60 ft.-lbs. on all other models. Tighten lubrication inlet tube (upper tube) fitting to 8-10 ft.-lbs. Tighten pressure inlet tube (lower tube) fitting to 50-67 inch-pounds. Tighten housing cover (41) bolts in a

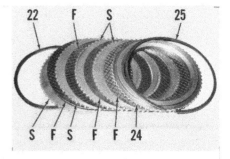

Fig. 141—View of direct drive clutch pack showing order of friction plates (F), steel plates (S), feathering spring (22), plate (24) and snap ring (25).

diagonal pattern to 27-37 ft.-lbs. Before installing control valve actuating rod, check length of rod to be sure there is 7½ inches from center to center of attaching points. Place thrust washer (16) in housing (1) with tab up and toward rear. Hold new gasket (40) to housing with grease and be sure drilled oil passage in cover (41) is up when installed. With installation complete, refer to paragraph 169 for further adjustment if required.

OVERHAUL DUAL POWER

171. To overhaul Dual Power unit, remove unit as outlined in paragraph 170, separate major assemblies and proceed as follows: Disassemble direct drive clutch by removing snap ring (25—Fig. 143) and withdraw plate (24), clutch pack (23) and feather spring (22).

NOTE: To reduce a complexity of Dual Power drive clutch, feathering spring was deleted and piston modified starting in November, 1979.

Note order of plates in clutch pack when removing. Place housing in a press and using Ford tool No. 1312 or equivalent, compress the piston return spring (26—Fig. 140) and remove retaining snap ring (28). Gradually release spring pressure making sure spring retainer (27) does not catch in snap ring groove, and remove retainer and spring. Remove piston (19—Fig. 143) by applying air pressure to hole between second and third sealing rings (17) on clutch housing hub. Carefully clean and inspect all parts. Inspect sealing rings for imperfections. Inspect sealing ring bore for damage. Check splines for cracked, broken or missing teeth. Renew any parts not suitable for further service.

To disassemble underdrive clutch, remove outer snap ring (15—Fig. 143) and clutch pack components (8 through 14). Note position of clutch pack plates. Remove piston (5) and seal rings. Clean and inspect all parts for wear. Check for broken clutch springs. Renew any part not suitable for further service. During assembly, the locating dowel pins (10) must be evenly spaced and installed so there is a dowel pin in every fifth notch.

Use a suitable puller to remove pilot bearing (34) from shaft (31) and separate planetary gear assembly (33) from shaft. Planetary gear assembly (33) is available as a unit only and should be serviced as a unit assembly. Using a suitable puller, remove bearing (38) from ring gear (36) being careful not to damage shims (37). Shims are used to adjust ring gear shaft (36) end play.

172. To reassemble Dual Power unit, lubricate components with transmission oil and proceed as follows: Install underdrive piston and clutch pack in housing being sure not to damage piston seals. Refer to Fig. 143 for sequence of clutch plates. Install locating dowel pins (10) in every fifth notch of housing.

To assemble direct drive clutch, reverse disassembly procedure. Refer to Fig. 141 for view of clutch pack. Be careful not to damage seal rings of piston when installing. Renew main bearing in housing (1—Fig. 143) if worn. Install thrust washer (32) and planetary gear assembly (33) on shaft (31) and press bearing (34) on shaft (31).

To determine correct size of shims (37) to obtain correct end play of 0.004-0.020 inch for ring gear shaft (36), assemble all components except ring gear and cover in housing (1) without attaching housing to transmission. Install ring gear without shim (37) and bearing (38). Place tool No. 1303 on step of ring gear shaft in place of shims and bearing. Make sure all components are properly seated in housing and place cover (41) on housing. At three equi-distant points, measure gap between cover and housing and average the three measurements. If average dimension is 0.046-0.060 inch, no shims are needed. If gap is less than 0.046-0.060 inch, refer to following table for correct shim size. (Dimensions are in inches.)

Average Gap Measurement	Required Shim Thickness
0.001-0.013	0.041-0.049
0.014-0.026	0.030-0.034
0.027-0.032	0.022-0.030
0.033-0.045	0.011-0.015

If tool No. 1303 is not available, assemble Dual Power unit with bearing (38) and original shim (37) on ring gear shaft (36). Install cover (41) being sure components are seated in housing and measure end play of ring gear shaft (36). End play should be 0.004-0.020 inch. Install needed size of shims (37) to obtain correct end play. Shims are available in sizes of 0.013 and 0.032 inch.

To reinstall Dual Power unit on transmission, remove cover, ring gear, shaft and planetary gear assembly, sun gear and direct drive assembly. Dual Power unit may now be installed as outlined in paragraph 170.

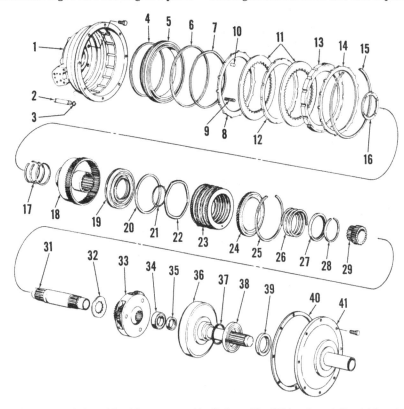

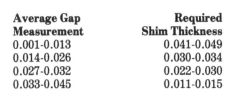

Fig. 143—Exploded view of Dual Power assembly. Refer to Fig. 141 for view of direct drive clutch pack assembly. Note: clutch feathering spring (22) deleted in Nov., 1979.

1. Dual Power housing	11. Clutch plate	22. Clutch feathering spring
2. Lubrication tube	12. Center plate	23. Direct drive clutch assembly
3. "O" ring	13. Pressure plate	24. Plate
4. Piston seal	14. Front plate	25. Snap ring
5. Piston	15. Snap ring	26. Spring
6. Piston seal	16. Thrust washer	27. Spring retainer
7. Clutch feathering spring	17. Seal rings	28. Snap ring
8. Pressure plate	18. Direct drive housing	29. Sun gear
9. Spring	19. Piston	31. Input shaft
10. Locating pin	20. Seal	32. Washer
	21. "O" ring	33. Planetary gear
		34. Bearing
		35. Seal
		36. Ring gear
		37. Shims
		38. Bearing
		39. Seal
		40. Gasket
		41. Cover

CONTROL VALVE

173. The Dual Power control valve directs flow of lubricating oil and oil from the pto hydraulic pump to engage either direct drive clutch pack or underdrive clutch pack.

To remove control valve on Models TW-10, TW-20 and TW-30, disconnect hydraulic line and remove access plate on right side of transmission center housing just forward of transmission shift cover. On all other models, access to control valve is obtained after splitting tractor between engine and transmission as outlined in paragraph 162.

With control valve removed, pull valve spool (2—Fig. 144) to full out

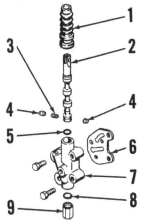

Fig. 144—Exploded view of Dual Power control valve.

1. Boot
2. Spool
3. Spring
4. Detent cups
5. "O" ring
6. Gasket
7. Body
8. "O" ring
9. Sleeve

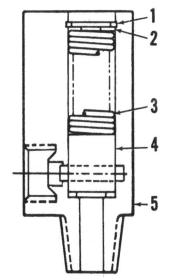

Fig. 144A—Cross-sectional view of Dual Power regulating valve.

1. Snap ring
2. Washer
3. Spring
4. Valve
5. Body

position. Unscrew sleeve (9) from valve body (7) and pull sleeve from end of spool (2) being careful not to lose detent cups (4) and spring (3) in end of spool. Remove valve spool (2) by pulling from bottom of valve body. Remove "O" rings (5) and (8).

Inspect spool (2), body (7) and sleeve (9) for scratches, wear or other damage. Inspect "O" rings, detent cups and spring. To reassemble, reverse disassembly procedure. Tighten sleeve (9) to a torque of 25-30 ft.-lbs. Tighten control valve mounting bolts to a torque of 27-37 ft.-lbs.

LOW PRESSURE OIL SYSTEM

174. All TW-10, TW-20 and TW-30 models are equipped with a low pressure oil system. An eight gpm gerotor hydraulic pump supplies low pressure oil, drawn from the rear axle center housing, to all tractor hydraulic functions except the hydraulic power lift and remote cylinders. See schematic in Fig. 145. The pump also provides low

pressure oil to operate the power assist brakes on the TW-30 models. The pump (71—Fig. 160) is mounted on the transmission rear cover and is driven by the main pto shaft. If the pump is damaged or worn excessively, it must be replaced as a complete unit. Pressure is regulated by a pressure regulating valve mounted on the transmission housing just above the shift cover. The pressure valve used on TW-30 models differs from the valve used on TW-10 and TW-20 models as TW-30 models are equipped with power brakes and the TW-30 pressure valve uses a brake priority valve (2—Fig. 145B) that maintains a minimum pressure of 150 psi to the power assist brake valve. Refer to Fig. 183.

174A. **PRESSURE TESTING (ALL MODELS).** Refer to paragraph 213 for low pressure system pressure testing.

GEROTOR PUMP

175. **R&R AND OVERHAUL GEROTOR PUMP.** Refer to paragraph 191 for removal of gerotor pump. If

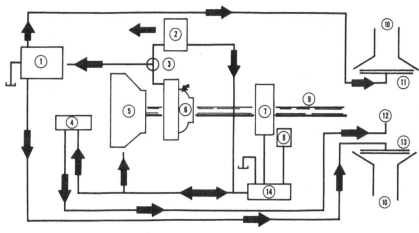

Fig. 145—View showing low pressure hydraulic system for TW-10, TW-20 and TW-30 models.

1. Power brake valve (TW-30 only)
2. Pressure regulating valve
3. Brake priority valve (TW-30 only)
4. Differential lock valve (TW-30)
5. Dual power clutch pack
6. Gerotor pump
7. Pto clutch
8. Pto brake
9. Pto shaft
10. Axle housing
11. Brakes
12. Differential
13. Differential lock valve (TW-10 and TW-20)
14. Pto valve

Fig. 145A—Exploded view of hydraulic system regulating valve used on TW-10 and TW-20 models.

1. Valve springs
2. System pressure valve
3. Housing
4. Plugs and seals
5. Lubrication pressure valve
6. "O" ring seals

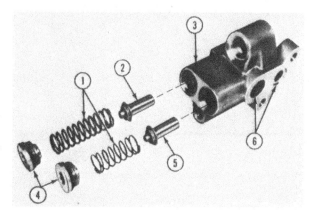

pump is worn or damaged, it must be replaced as a complete unit.

PRESSURE REGULATOR VALVE

175A. **R&R AND OVERHAUL PRESSURE REGULATOR VALVE.** Valve is located above transmission shift cover. To remove valve, first remove oil lines to valve then remove cap screws retaining valve to tractor. Refer to Fig. 145A for exploded view of regulating valve used on TW-10 and TW-20 models and to Fig. 145B for exploded view of valve used on TW-30 models.

When overhauling valve, tag spools and springs for later identification. Note that spool (2—Fig. 145A or 5—Fig. 145B) has an 0.031 inch orifice while remaining valve spool(s) does not have an orifice. The lubrication pressure valve spring is 1.30 inches long with 7½ coils while the system pressure valve spring is 2.09 inches long with 9½ coils. The brake priority valve spring used on

Model TW-30 is 1.37 inches long with 8½ coils.

Inspect components for excessive wear, scoring, burrs and binding in valve bores and renew as necessary. Clean valve housing and components with a suitable solvent and blow dry. Make sure all orifices and passageways are clean. Reassembly procedure is reverse of disassembly procedure. Install new "O" rings in valve, then install valve assembly onto tractor and tighten cap screws to a torque of 20-26 ft.-lbs. Reconnect oil lines to valve.

TRANSMISSION

The gear type dual range transmission provides eight forward and two reverse speeds. Refer to Fig. 146 for view showing all transmission gears except reverse idler. On Models 8700, 9700, TW-10, TW-20 and TW-30 transmission speeds are selected using shift levers

which actuate shift forks on a shift rail. On all other models, transmission speeds are selected by shift rails being moved by a cam wheel (see Fig. 156). An indexing roller holds the cam wheel in selected gear, but does not provide a neutral position. Neutral is provided by holding the range shift collar (109—Fig. 146) disengaged from both the low range (111) and high range (106) gears. A pawl type transmission brake (see Fig. 161) can be engaged with teeth on range shift collar to hold tractor stationary.

LUBRICATION

176. Transmission on later 8000 models and all other models is pressure lubricated and an oil cooler is located in front of engine radiator. The oil reservoir is common with oil reservoir in rear axle center housing. On all models except TW-10, TW-20 and TW-30, oil is pumped from rear axle center housing by the main hydraulic pump to the oil cooler then to the transmission or to the Dual Power housing on models so equipped. TW-10, TW-20 and TW-30 models are equipped with a low pressure oil system which uses a gerotor pump mounted on transmission rear cover and driven off pto mainshaft. Refer to paragraph 174.

Lubricant capacity for transmission and rear axle is as follows: Models except TW-10, TW-20 and TW-30 with Dual Power, 69 quarts, without Dual Power, 64 quarts; Models TW-10 and TW-20 with Dual Power, 79 quarts, without Dual Power, 73 quarts; All TW-30 models require 92 quarts. Recommended lubricant is Ford lubricant No. ESN-M2C53-A or M-2053-B and should be changed every 1200 hours of operation (or once every 12 months).

Transmissions on earlier 8000 models are splash lubricated and have an

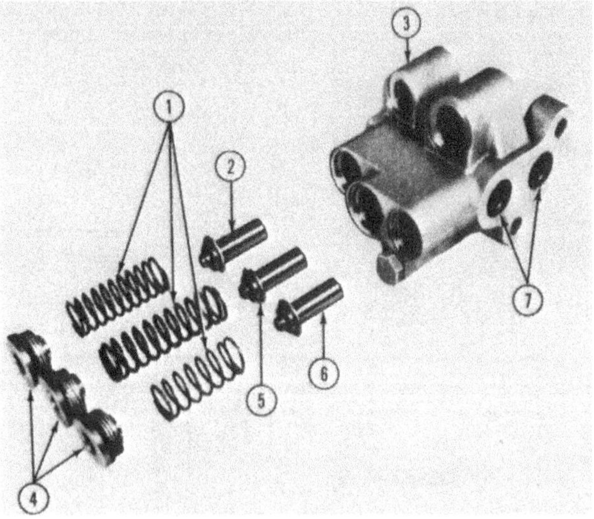

Fig. 145B—Exploded view of low pressure hydraulic system regulating valve used on TW-30 models.

1. Valve springs
2. Brake priority valve
3. Housing
4. Plugs and seals
5. System pressure valve
6. Lubrication pressure valve
7. "O" ring seals

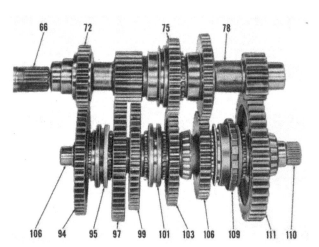

66. Input shaft
72. Main shaft
75. Sliding gear/coupling
78. Secondary countershaft
94. 2nd/6th gear
95. Sliding coupling
97. 1st/5th gear
99. Reverse gear
101. Sliding coupling
103. 3rd/7th gear
106. Main countershaft
109. Sliding coupling
110. Output shaft
111. Output shaft gear

Fig. 146—View showing all transmission gears except reverse idler. Refer also to exploded view in Fig. 160. Sliding coupling (109) incorporates dog teeth for transmission brake pawl. Front end of main countershaft (106) and top shafts and gears are supported in needle roller bearings.

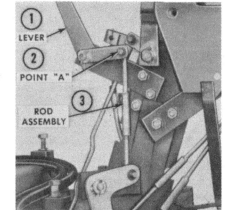

Fig. 147—On 8000—9000 models, adjust length of transmission brake rod assembly (3) so that lever (1) clears ends of slot in console in both "ON" and "OFF" detent positions. Disconnect rod at point "A" (2) to make adjustment.

independent oil reservoir within the transmission case. Splash lubricated transmissions are identified by an external filler plug in the top cover, a vent plug in the side of the case and an oil level plug in the cam cover. Transmission lubricant capacity is 12 quarts. Recommended lubricant is SAE 80-EP gear lube (Ford specification No. M-4864-A or ESN-M2C77-A); lubricant should be changed after each 1200 hours of operation. On splash lubricated transmissions, oil level plug is located in gearshift cover plate; refer to Fig. 154. On early models, a pipe (5—Fig. 157) extended filler opening plug (4) to hole in operator's platform. On later models, filler plug (4A) threads directly into top cover (8) and is accessible after removing center plate from platform. On pressure lubricated transmissions, filler plug for hydraulic system common reservoir is located at top rear center of hydraulic lift cover. Drain plug is located in center rear of transmission housing.

LUBRICATION PRESSURE

177. To check oil pressure of pressure lubricated transmissions, proceed as follows: Operate tractor until hydraulic oil is at normal operating temperature. Stop engine and connect 0-150 psi gage to outlet tube from oil cooler located in front of radiator. Start engine and set speed at 1000 rpm. Oil pressure for all models except TW-10, TW-20 and TW-30 tractors should be 20-35 psi. Pressure for TW-10, TW-20 and TW-30 models should be 130 psi. If pressure is too low cause may be clogged or restricted oil cooler or a weak oil pressure regulating valve spring. Regulating valve (10—Fig. 212) for all models except TW-10, TW-20 and TW-30 threads into back of hydraulic pump adapter plate (39—Fig. 216). Regulating valve for TW-10, TW-20 and TW-30

models is located on the right hand side of the transmission; refer to paragraph 175A for service.

PARKING BRAKE

Models 8000 and 9000 are equipped with a separate parking brake lever, while 8600, 8700, 9600, 9700, TW-10, TW-20 and TW-30 models have the parking brake and High-Low shift lever on the same lever. Models 8000 and 9000 must have the High-Low shift lever in neutral before parking lever can be engaged. Models 8600, 8700, 9600, 9700, TW-10, TW-20 and TW-30 can not be shifted to "PARK" unless High-Low shift lever is in neutral. Refer to appropriate following paragraph for linkage adjustments.

178. **8000-9000 MODELS.** The pawl type parking brake can be engaged only when the tractor is stopped and range (high/low) shift lever is in neutral position. The brake detent ball (15—Fig. 153 or 10—Fig. 157) should engage detent hole in transmission cover in both the "ON" and "OFF" position. Also, shift lever should clear end of slot in console in both detent positions. If necessary to adjust parking brake linkage, remove right fender and refer to Fig. 147. Disconnect rod assembly from lever at "point A" and adjust length of rod so that when disconnected, detent will be fully engaged and lever will not contact

end of slot in both the "ON" and "OFF" positions. Reinstall right fender.

Any necessary overhaul of the parking brake external linkage can be made after removing right fender and center plate from operator's platform. Overhaul procedure is obvious from inspection of unit.

179. **8600-9600 MODELS.** The parking brake lever is also the High-Low shift lever. When tractor is stopped and lever is moved to "PARK" position, parking brake actuating lever (Fig. 148) moves parking pawl (23—Fig. 157) into engagement with lugs on sliding coupler (51—Fig. 160). If parking pawl contacts the top of a lug on sliding collar, spring on parking brake rod (12—Fig. 150) will compress until tractor moves far enough to allow parking pawl to drop into full engagement with sliding collar.

To adjust for internal parking brake engagement, remove right hood panel, floor mat and center platform on operator's platform. One rear wheel (or both) must be jacked up enough to rotate. Rotate rear wheel until parking lever can be moved into "PARK" position without fully engaging parking pawl inside transmission. When proper adjusting position has been reached, the parking brake actuating lever stop arm (Fig. 148) should align with index line on pivot bracket, with spring on parking

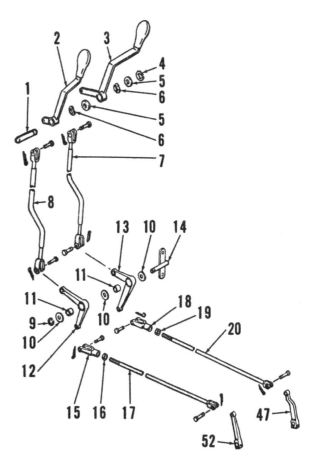

Fig. 149—Exploded view of 8000—9000 transmission shift levers and linkage. Shift lever pivot pin (1) is mounted in power steering Hydramotor support bracket. Control levers are synchronized to detent positions of shift levers (47 and 52) by adjusting length of lower rods (17 and 20).

1. Pivot pin
2. High/low lever
3. Main shift lever
4. Snap ring
5. Flat washer
6. Wave (spring) washers
7. Main shift rod (21.4 in.)
8. High/low shift rod (20.3 in.)
9. Snap ring
10. Washers
11. Bushings
12. High/low bellcrank (4.2 x 4.2 in.)
13. Main shaft bellcrank (3 x 5¾ in.)
14. Bellcrank pivot
15. Yoke, high/low shift rod
16. Jam nut
17. High/low shift rod (28.5 inches)
18. Yoke, main shift rod
19. Jam nut
20. Main shift rod (21.88 in.)
47. Main shift lever
52. High/low shift lever

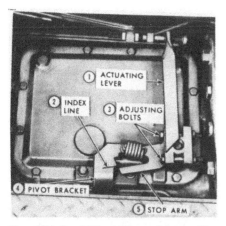

Fig. 148—On 8600—9600 models, parking brake is properly adjusted when stop arm is aligned with index line as shown (with park pawl not engaged). Refer to paragraph 179.

brake control rod compressed. When rear wheel is rotated, parking pawl should drop into full engagement and stop wheel from rotating further. Spring on parking brake control rod MUST be compressed with index line and stop arm aligned, and parking pawl on top of lug on sliding coupler. To adjust to index line, alternately loosen one adjusting bolt on actuating lever then tighten the other until stop arm is aligned with index line on pivot bracket. This adjustment must be made anytime transmission cover is removed.

If necessary to adjust the parking brake rod so that the neutral-park lever (1—Fig. 150) will be centered in the neutral slot of quadrant on instrument panel, loosen the two adjusting nuts (13) and adjust rod (12) longer or shorter as

necessary. When properly adjusted, the spring on end of rod (12) should measure 1⅝-1¾ inches, with park lever in neutral position. Nut (17) can be adjusted to obtain proper length.

On all models, the parking brake detent can be serviced after removing transmission top cover as outlined in paragraph 184. To renew parking brake pawl or the high-low shift coupling, refer to transmission overhaul procedure.

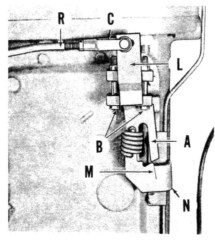

Fig. 151—View of Model 8700, 9700, TW-10, TW-20 and TW-30 parking brake linkage. Refer to text for adjustment.

180. 8700-9700, TW-10, TW-20 and TW-30 MODELS. The High-Low shift lever on Models 8700, 9700, TW-10, TW-20 and TW-30 is also used to engage parking brake in transmission. When lever is moved to park position, park pawl (23—Fig. 157) engages lugs on sliding coupler (51—Fig. 160).

To adjust parking brake linkage, jack up rear of tractor so one or both rear wheels may be rotated. Remove transmission access cover. Move shift lever to park position so that park pawl rests on top of a tooth on sliding coupler. Rotate a rear wheel if necessary to position pawl on coupler. Index mark (M—Fig. 151) should be aligned with inside edge of stop arm (A). If not, alternately loosen then retighten each adjusting bolt (B) until stop arm (A) is aligned with index mark (M). Move shift lever out of park position. Detach control rod (R) from lever (L) and turn clevis (C) to obtain maximum rod length but still be able to attach rod (R) to lever (L).

TRANSMISSION SHIFT LEVERS AND LINKAGE

181. 8000-9000 MODELS. Refer to Fig. 149 for exploded view of transmission shift levers and linkage on 8000 and

Fig. 150—Exploded view of 8600-9600 transmission shift linkage and parking brake linkage. Bushings in levers (1, 3 and 4) are not serviced separately. Control lever (2) must engage either parking brake lever (1) or high-low range lever (3). Parking pawl override spring on end of rod (12) was inside transmission on 8000—9000 models. Items (15 and 16) replace item (1—Fig. 157), on 8000—9000 models. (Item 16 is shown reversed.) Item (15) was not used on some early models.

1. Park lever	9. High/low shift rod
2. Park, high/low shifter	10. High/low arm
3. High/low lever	11. Gear shift arm
4. Gear shift lever	12. Parking brake rod
5. Spring	13. Adjusting nut
6. Snap ring	14. Pivot bracket
7. Washer	15. Brake shaft hub
8. Gear shift rod	16. Actuating lever
	17. Adjusting nut

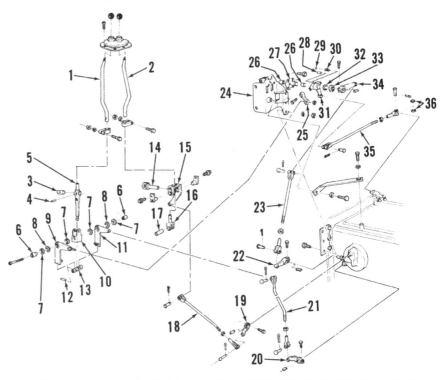

Fig. 152—Exploded view of shift linkage on Models 8700, 9700, TW-10, TW-20 and TW-30.

1. High/low shift lever	10. Yoke	19. Gearshift arm	28. Pin
2. Gearshift lever	11. High/low bellcrank	20. High/low arm	29. Detent ball
3. Pivot pin	12. Pin	21. High/low rod	30. Spring
4. Pin	13. Interlock pin	22. Interlock arm	31. Park brake bellcrank
5. High/low shift finger	14. Pivot shaft	23. Interlock rod	32. Washer
6. Trunnion	15. Gearshift bellcrank	24. Shift bracket	33. Snap ring
7. Washer	16. Yoke	25. Park arm	34. Interlock arm
8. Shim	17. Pin	26. Bushing	35. Park brake rod
9. Park bellcrank	18. Gearshift rod	27. Support	36. Washer

9000 models. The high-low shift lever (2) should be aligned with neutral (N) position and main shift lever should be at 4th-8th gear position when the levers on transmission are aligned with reference marks on transmission housing as shown in Fig. 154. Also, the high-low shift lever should not strike end of slot in instrument panel when in either high (H) or low (L) position; main shift lever should not strike end of slot when in either reverse or 4-8th speed position. If necessary to adjust shift linkage, remove the pins connecting yokes (15 and/or 18—Fig. 149) on lower shift rods to bellcranks (12 and/or 13) and loosen jam nut (16 and/or 19). Shorten or lengthen lower rods as necessary to obtain proper shift action, then secure yoke pins with cotter pins and tighten jam nuts.

If shift linkage cannot be properly adjusted, either the linkage has been improperly assembled or overhaul of linkage is indicated. Wear of pins and pin holes in shift levers, shift rod ends and/or bellcranks, or bearing wear in shift levers and/or bellcranks will cause improper shift action. Bushings (11) in bellcranks are renewable.

When assembling linkage, note the measurements of shift rods given in legend of Fig. 149 and observe the following: Early models were not fitted with wave (spring) washers (6); installing the wave washers on models not so equipped will help eliminate free play of shift arms. The high-low shift bellcrank (12) arms are the same length. On main shift bellcrank (13), connect the 3-inch arm to upper shift rod (7) and the 5¾-inch long arm to lower rod yoke (18). Pivot (14) is installed on instrument panel support with pin side to rear (against support); foot throttle bellcrank pivots in the hollow bracket pin.

182. 8600-9600 MODELS. Refer to Fig. 150 for exploded view of transmission shift levers and linkage for 8600-9600 models. High-low shift lever (2) engages high-low intermediate lever (3) when shift lever is in the high-low slot in instrument panel. When shift lever (2) is moved through neutral and over to the park slot, intermediate lever (1) is engaged. Gear shift lever (4) and the two intermediate levers (1 and 3) are equipped with non-renewable bushings. When adjusting shift levers on linkage, the reference marks on transmission case must NOT be used. Fig. 155 shows dimensions for installing high-low lever arm and gearshift cam lever arm if arms were removed or if they are out of adjustment. After lever arms are positioned correctly, tighten retaining bolts to 11-15 ft.-lbs. torque. Adjust the forward yokes on rods (8, 9 and 12—Fig.

Fig. 153—View of transmission with top cover removed. Brake lever (14) and rod (7) are removed with cover and are shown installed for illustration only. Note that a spacer ring is installed at (S) on splash lubricated transmissions or at (P) on pressure lubricated transmissions. Refer to text. Numbers following legend indicate identical parts in Fig. 157.

1. Index lever spring (27)
2. Index lever (33)
3. Spring (hair) pins (15 & 22)
4. Brake pawl (23)
5. Sleeve (18)
6. Washer (16)
7. Brake rod (14)
8. Index arm retaining nut (30)

9. Brake lever shaft (3)	13. Flat washer
10. Pivot bolt (37)	14. Brake lever (1)
11. Cotter pin	15. Detent ball (10)
12. Mainshaft	16. Reverse idler

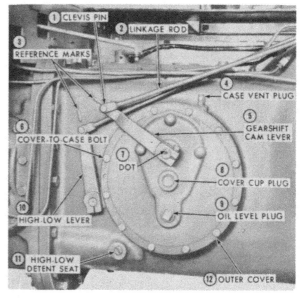

Fig. 154—On 8000-9000 models, reference marks (3) on transmission housing and dot mark (7) on cam drive gear shaft are necessary for proper reassembly of shift cam and shift levers. Note case vent plug; inspect vent to be sure that it is not plugged whenever servicing transmission. Refer to Fig. 155 for 8600-9600 models.

Fig. 155—On 8600-9600 models. DO NOT use reference marks on transmission housing. Adjust levers to dimensions shown, with gearshift cam lever in first gear position and high-low lever in neutral. Refer to Fig. 154 for 8000-9000 models.

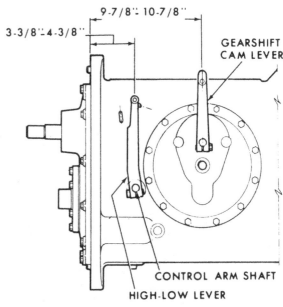

150) so gearshift lever (4) is in 1st-5th position and intermediate levers (1 and 3) are exactly in neutral position in slots in instrument panel.

183. 8700, 9700, TW-10, TW-20 and TW-30 MODELS. Refer to Fig. 152 for exploded view of transmission shift levers and linkage used on Models 8700, 9700, TW-10, TW-20 and TW-30.

To remove shifter assembly, unscrew boot retaining screws, slide boot up shift levers (1 and 2) and detach shift levers. Disconnect shift rods; unbolt and remove shift bracket (24). Remainder of disassembly is evident after inspection of shift assembly. Do not lose shims (8), detent ball (29) or spring (30).

Renew any components causing sloppiness or binding in shift action. Use trial assemblies to locate shifting malfunctions.

Reverse disassembly procedure to assemble shifter components. Install sufficient shims so end play of bellcranks (9 and 11) in shift bracket (24) is 0.000-0.009 inch.

To adjust shift linkage, proceed as follows: Detach shift rod (21—Fig. 152) from high/low shift arm (20) and place shift arm in neutral. Center high/low shift lever (11) over interlock pin (13) then adjust length of shift rod (21) by turning clevis so clevis pin will fit easily through clevis and shift arm (20). Adjust length of interlock control rod (23) by turning clevis so center to center distance between rod ends is 11-3/8 inches. Adjust length of gearshift control rod (18) by turning clevis so center to center distance between rod ends is 13-7/8 inches. Check clearance between gearshift lever (2) and guard with lever in 4th gear. Minimum clearance is 1/2 inch; loosen gearshift lever and rotate lever to obtain clearance. Clearance between high/low shift lever (1) and guard should be at least 1/2 inch while in high range and 1/4 inch when in park position. Loosen lever clamp and rotate lever to obtain desired clearance.

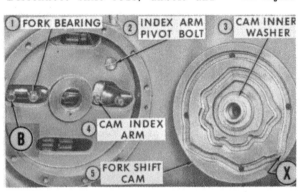

Fig. 156—When installing shift cam, align lobe "X" with fork bearing "B"; refer to text for complete fork and index lever alignment procedure and for cover plate and cam drive gear installation.

TRANSMISSION TOP COVER

184. 8000-9000 MODELS. To remove

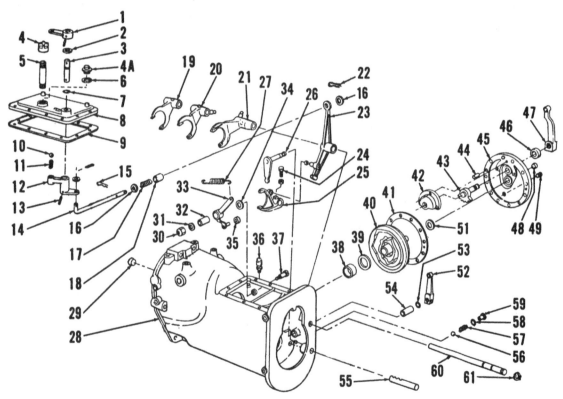

Fig. 157—Exploded view showing transmission housing, top cover and brake linkage, shift cover, shift cam, shift forks and rails. Refer to Fig. 160 for exploded view of gears and shafts with rear cover, bearings and seals and front bearing supports. Index lever (33) acts as detent for shift cam (40); there is no neutral detent position for the shift cam. Three detent notches in high-low shift rail (55) provides transmission neutral position as well as high and low range detent positions.

1. Brake lever	11. Detent spring	21. 4th gear shift fork
2. Flat washer	12. Brake lever	22. Spring (hair) pin
3. Brake lever shaft	13. Roll pin	23. Brake pawl
4. Filler cap (early)	14. Brake rod	24. Set screw
4A. Filler plug (late)	15. Spring (hair) pin	25. High/low shift fork
5. Extension (early)	16. Flat washer	26. High/low shift lever
6. Gasket	17. Spring	27. Index arm spring
7. "O" ring	18. Sleeve	28. Transmission
8. Top cover	19. 1st/2nd gear fork	housing
9. Gasket	20. 3rd/reverse shift	29. Shift rail bore plug
10. Detent ball	fork	

30. Nut	39. Thrust washer
31. Flat washer	40. Shift cam
32. Pivot bushing	41. Gasket
33. Index arm	42. Inner cover
34. Thrust washer	43. Shift cam drive gear
35. Cam roller bearing	44. Dowel pin
36. Transmission vent	45. Shift cam cover
37. Index arm pivot	46. Seal
bolt	47. Main shift lever
38. Bushing	48. Plug
	49. Oil level plug

51. Thrust washer
52. High/low shift lever
53. "O" ring
54. Bushing
55. High/low shift rail
56. Detent ball
57. Detent spring
58. Gasket
59. Detent plug
60. Main shift rail
61. Snap ring

transmission top cover, first remove center plate from operator's platform, then proceed as follows: Disconnect transmission brake linkage rod from lever (1—Fig. 157) on top of transmission cover.

NOTE: Some 8000—9000 models may have been changed over to the later parking brake lever assembly, similar to items 14 and 16 (Fig. 150); Item 15 was not used in the changeover.

Unbolt cover from transmission housing, pivot front end of cover to left only far enough to gain access to right end of brake rod (14—Fig. 157). Pivoting center too far will disengage detent ball (10). Remove the spring (hair) pin (15) and slide washer (16), spring (17) and sleeve (18) towards left end of rod, then reinstall pin (15) in rod. Remove outer pin (22) and flat washer (16), disengage the brake rod from brake pawl (23) and lift cover from transmission.

Drive roll pin from lever (1) and shaft (3) and remove shaft and lever (12) out bottom of cover; take care not to lose detent ball (10) and spring (11). Install new "O" ring (7) in groove on shaft and renew detent ball and/or spring if necessary. Lubricate shaft, "O" ring and bore in cover and install shaft and lower lever (12) with detent spring and ball. Place flat washer (2) on shaft, install lever (1) aligned with detent arm of lever (12) and secure with roll pin.

Reinstall cover with new gasket as follows: Insert left end of rod (14) through pawl and install flat washer and "hairpin" (22). Remove "hairpin" (15), slide spring, sleeve and washer against pawl and reinstall "hairpin" (15) through hole in rod. Swing cover back into position, taking care not to dislodge detent ball, and tighten retaining cap screws to a torque of 12-18 ft.-lbs. Reinstall center plate in operator's platform.

185. 8600-9600 MODELS. To remove transmission top cover, remove operator's platform mat and center platform, then proceed as follows: Shift high-low range shift lever to neutral, slightly loosen the two bolts on pivot bracket (Fig. 148). Shift high-low lever to park position, use a suitable pry bar or adjustable wrench on pivot bracket while front bolt is removed.

CAUTION: Spring on pivot bracket is under tension and may jump out if bracket is removed without holding bracket against spring pressure.

Shift high-low lever to neutral and remove the other bolt on pivot bracket and remove spring and spacer. Remove the two adjusting bolts holding actuating lever to brake lever shaft, remove actuating lever and lay lever aside without disturbing adjustment on spring end of lever. Unbolt cover from transmission housing, pivot front end of cover to the left far enough to remove "hairpin" spring clip (22—Fig. 158), and washer from end of brake rod (14). Cover and brake rod may now be removed. With cover off, drive out roll pin at top of brake lever shaft (3) and remove actuating lever hub from top of shaft. Tap shaft (3) out bottom of cover. "O" ring (7) should be renewed at this time and brake lever (12), brake rod (14) and shaft (3) inspected for wear, as well as shaft bore in transmission cover. Lubricate shaft, bore and "O" ring before reinstalling in cover. Reinstall actuating lever hub on top of shaft (3) and reinstall roll pin.

Install a new cover gasket, insert end of brake rod (14) into parking brake pawl (23) and reinstall washer and "hairpin" (22). Swing cover back into position, install all but the two center bolts on left side of cover and tighten to a torque of 12-18 ft.-lbs. Place the pivot bracket strap spacer over the two remaining bolt holes, so that the offset holes align properly and loosely install actuating lever adjusting bolts onto hub on top of brake shaft (3). Start rear bolt into pivot bracket, swing pivot bracket away from transmission and install spring. Using a suitable pry bar or adjustable wrench, force pivot bracket against spring pressure until front bolt hole aligns with strap spacer and transmission cover, and install front bolt.

NOTE: To avoid excessive spring tension when installing pivot bracket, place high-low shift lever in park position and make sure parking brake pawl (23) falls into full engagement in collar in transmission.

Tighten the two pivot bracket bolts to a torque of 24-30 ft.-lbs. Adjust parking brake linkage as outlined in paragraph 179.

186. 8700, 9700, TW-10, TW-20 AND TW-30 MODELS. To remove transmission top cover, remove operator's platform mat and center platform, then proceed as follows: On TW-30 models, disconnect brake system return to sump

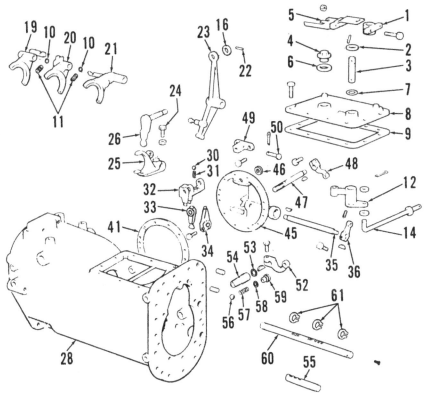

Fig. 158—View showing transmission housing, gearshift and parking brake components.

1. Brake actuator hub	14. Brake rod	28. Transmission housing	48. Gearshift arm
2. Washer	16. Washer	30. Detent ball	49. Support
3. Brake lever shaft	19. 1st/2nd gear fork	31. Spring	50. Clevis pin
4. Filler plug	20. 3rd/reverse shift fork	32. Interlock	52. High/low shift lever
5. Parking brake lever	21. 4th gear shift fork	33. Gearshift finger	53. "O" ring
6. Gasket	22. Pin	34. Interlock finger	54. Bushing
7. "O" ring	23. Brake pawl	35. Shaft	55. High/low shift rail
8. Top cover	24. Set screw	36. Interlock lever	56. Detent ball
9. Gasket	25. High/low shift fork	41. Gasket	57. Spring
10. Detent ball	26. High/low shift lever	45. Shift cover	58. Gasket
11. Spring		46. Seal	59. Detent plug
12. Brake lever		47. Gearshift shaft	60. Main shift rail
			61. Snap rings

hydraulic line and rear axle lubrication line, cap lines and fittings, then on all models, detach parking brake control rod clevis (C—Fig. 151) from lever (L). Use a suitable pry bar or wrench against anchor (N) and remove anchor retaining screws.

CAUTION: Spring on stop arm (A) is under tension and may jump out if anchor is removed without holding against spring pressure.

Unscrew bolts (B) and remove actuator assembly. Unbolt transmission housing cover, pivot front end of cover to the left far enough to remove pin (22—Fig. 158) and washer (16) from brake rod (14). Cover and brake rod may now be removed. With cover off, drive out roll pin at top of brake lever shaft (3) and remove actuating lever hub from top of shaft. Tap shaft (3) out bottom of cover. "O" ring (7) should be renewed at this time and brake lever (12), brake rod (14) and shaft (3) inspected for wear, as well as shaft bore in transmission cover. Lubricate shaft, bore and "O" ring before installing in cover. Reinstall actuating lever hub (1) on top of shaft (3) and reinstall roll pin.

Install a new cover gasket, insert end of brake rod (14) into parking brake pawl (23) and reinstall washer (16) and pin (22). Swing cover back into position, install all but the two center bolts on left side of cover and tighten bolts on 8700-9700 models to a torque of 12-18 ft.-lbs. Tighten cover bolts on TW-10, TW-20 and TW-30 models to a torque of 27-37 ft.-lbs. Install actuating lever assembly onto hub (1). Start rear bolt into anchor (N—Fig. 151), swing anchor away from transmission and install spring. Using a suitable pry bar or adjustable wrench, force anchor against spring pressure until front bolt hole aligns with strap spacer and transmission cover, and install front bolt.

NOTE: To avoid excessive spring tension when installing anchor, make sure parking brake pawl (23—Fig. 158) falls into full engagement in collar in transmission.

Tighten the two anchor bolts to a torque of 24-30 ft.-lbs. On TW-30 models reconnect brake hydraulic line and rear axle lubrication line. Adjust parking brake linkage as outlined in paragraph 180.

R&R TRANSMISSION ASSEMBLY

All Models Except 8700, 9700, TW-10, TW-20 And TW-30

NOTE: The following transmission removal procedure, by deleting removal of rear hood (instrument panel) support and operations outlined in paragraph 162, will also apply to splitting tractor between transmission and rear axle center housing.

187. To remove transmission, drain lubricant (approximately 17 gallons) from transmission and rear axle center housing, split tractor between engine and transmission as outlined in paragraph 162, then proceed as follows:

Disconnect lower shift rods at forward end and the clutch rod at crossshaft end. Remove center plate from operator's platform and disconnect the four brake lines at top of transmission; plug all openings and identify lines for reasssembly. Disconnect wiring at disconnect plug at left side of transmission. Remove tool box, battery and battery box. Unbolt instrument panel (rear hood) support and remove the support assembly from tractor. On models with Dual Power, control pedal assembly and linkage should be removed and disconnected to prevent damage. Disconnect lower shift rods from transmission and brake rod from lever on top of transmission cover, on 8000 and 9000 models. On 8600-9600 models, do not remove park brake rod at end of actuating lever, to prevent disturbing adjustment. Make a reference mark across actuating lever and hub and remove adjusting bolts to parking brake hub. Lay actuating lever aside. To avoid possible breakage of pto clutch locating pin (18–Fig. 197), remove pin from upper left side of rear axle center housing.

Support rear axle center housing, both under front end next to transmission housing and under drawbar to keep rear unit from tilting backward or forward when transmission is removed. Attach hoist to transmission, unbolt transmission from rear axle center housing and remove the transmission assembly or separate tractor between transmission and rear axle center housing.

To facilitate reconnecting transmission to rear axle center housing, remove the hydraulic pump and adapter plate assembly following procedure outlined in paragraph 233. Place new gasket on face of rear axle center housing and be sure that pto clutch hub is fully engaged in the clutch discs. Lift transmission into position with hoist and working through hydraulic pump opening, guide pto input shaft into clutch hub and coupling and onto output shaft of transmission. Tighten the ½-inch cap screws at each side of housing to a torque of 50-65 ft.-lbs. and the 5/8-inch cap screws at top and bottom of housing to a torque of 95-130 ft.-lbs. Install pto clutch locating pin with new "O" ring and

tighten pin to a torque of 70-95 ft.-lbs. Reinstall hydraulic pump as outlined in paragraph 234 or 235, reconnect engine to transmission as in paragraph 163. Reinstall 8600-9600 park brake actuating lever on brake shaft hub on top of transmission cover, with reference mark aligned. If unable to match reference mark to obtain original adjustment of parking brake, adjust as outlined in paragraph 179.

Complete remainder of assembly by reversing disassembly procedure.

Models 8700, 9700, TW-10, TW-20 And TW-30

188. Prior to transmission removal, drain transmission lubricant (approximately 17 gallons on 8700-9700 models, 20 gallons on TW-10 and TW-20 models and 23 gallons on TW-30 models) and split tractor between engine and transmission as outlined in paragraph 162.

Remove screws retaining gearshift boot, raise boot, loosen gearshift lever clamps and detach gearshift levers. On models with Dual Power, control pedal assembly and linkage should be removed and disconnected to prevent damage. Disconnect and cap brake lines. Detach engine clutch linkage. It is not necessary to remove the cab or cabless platform in order to remove the transmission if the sides of cab or platform are properly supported. If cab or platform removal is desired, disconnect wires, hoses, tubing or linkage which will interfere with separation of platform or cab from tractor. On cab equipped models, remove scuff plates in door openings, remove access plates and unscrew front cab mounting bolts being careful not to lose shims. Remove bolts securing front of platform on models not equipped with cab. Unscrew bolts securing rear of cab or platform to rear axle housings and use a suitable hoist to lift cab or platform away from tractor.

Remove brake lines from transmission housing. Detach shift rods and remove gearshift assembly from housing. Remove Dual Power oil tube. Support front of rear axle center housing to prevent tipping. Attach hoist to transmission, unbolt transmission from rear axle center housing and remove transmission.

If transmission or pto shafts will not engage during reassembly, remove hydraulic pump and adapter plate assembly as outlined in paragraph 233. Tighten ½-inch cap screws on each side of housing to 65-90 ft.-lbs. and 5/8-inch cap screws at top and bottom of housing to 140-170 ft.-lbs. Tighten platform retaining bolts to 100-120 ft.-lbs. Tighten cab retaining bolts to 200-220 ft.-lbs.

TRANSMISSION OVERHAUL

All Models

189. **SHIFT CAM, INDEX ARM AND SHIFT COVER (ALL MODELS EXCEPT 8700, 9700, TW-10, TW-20 AND TW-30).** Shift cam, index arm, shift fork rollers and shift cover can be removed without removing transmission from tractor; proceed as follows:

Remove battery cover, disconnect ground cable and remove battery, battery box and center plate from operator's platform. Drain transmission lubricant. Remove transmission top cover as outlined in paragraph 184 or 185. Disconnect lower shift rod from gearshift cam lever; refer to Fig. 154. Unbolt and remove the shift cover (45—Fig. 157), then unbolt inner cover (42), remove gearshift cam lever (47) and withdraw cam drive gear (43). Remove seal (46) from cover.

Remove washer (51) from cam (40), remove cam from transmission and remove inner washer (39). Remove the rollers (35) from index arm (33) and shift forks (19, 20 and 21).

Carefully clean and inspect all parts, renewing any that are excessively worn or damaged. Bushing (38) in transmission housing may be renewed if excessively worn. After installing new bushing, it must be line reamed to an inside diameter of 1.5020-1.5045 inches. The index arm can be removed after removing arm pivot bolt; refer to Figs. 153 and 156.

To reinstall shift cam and cover, proceed as follows: Install index arm and tighten pivot bolt nut to a torque of 48-64 ft.-lbs. Using proper size driver, install new seal in cover with lip of seal inward. Lubricate cam drive gear and

carefully install gear shaft through housing bore and seal. Install inner cover and tighten retaining cap screws to a torque of 20-27 ft.-lbs. Place the two sliding couplings (95 and 101—Fig. 146) in neutral position and engage sliding gear (75) with secondary countershaft. Refer to Fig. 156 and place inner washer on cam and the bearings on shift forks and index arm. Working through top cover opening, move upper end of index arm rearward to align index roller with cam, then install cam with lobe (X) engaging rear shift fork and release index arm when cam is properly seated. Using new gasket, install cover. On 8000-9000 models, be sure to have dot (or stake mark) on outer end of cam drive gear shaft pointing up as shown in Fig. 154. Tighten cover retaining cap screws to a torque of 27-32 ft.-lbs. following a diagonal pattern. Install gearshift cam lever so that it is aligned with reference mark on transmission housing; refer to Fig. 154. On 8600-9600 models, place transmission main countershaft front coupling to the rear and high-low coupling in neutral. Tighten cover retaining cap screws to a torque of 27-32 ft.-lbs. following a diagonal pattern. Refer to Fig. 155 and install levers to dimensions shown. DO NOT use reference marks on outside of housing. Tighten lever clamp bolts to a torque of 11-15 ft.-lbs. Refill transmission and complete remainder of reassembly by reversing disassembly procedure.

190. **SHIFT LEVER, INTERLOCK AND SHIFT COVER (MODELS 8700, 9700, TW-10, TW-20 AND TW-30).** Shift cover with shift lever and interlock can be removed without removing transmission from tractor by using following procedure:

Drain transmission lubricant and remove battery and box. Shift transmission to neutral. Remove oil cooler inlet tube and torque amplifier tube. On TW-30 models, remove Dual Power hydraulic oil pressure line from regulating valve and "tee" fitting. Disconnect lube cooler inlet line from regulating valve, loosen clamps and reposition line. On all models, plug openings to prevent contamination. Detach lower ends of interlock rod (23—Fig. 152) and gearshift rod (18). Be sure transmission is in neutral, remove shift cover screws and remove shift cover.

Remove interlock finger (34—Fig. 158) and key from shaft (35) then withdraw shaft from shift cover. Remove gearshift arm (48) and key from shaft (47). Extract clevis pin (50) from support (49) and pull out shaft (47) just enough to remove shift control finger (33). Do not pull out shaft more than necessary to remove finger as shaft splines may

damage bushing and seal in cover. Remove shift control finger (33) then withdraw shaft (47) from inside of shift cover being careful not to lose detent ball (30) or spring (31). Remove interlock (32) and support (49).

Inspect components for damage or excessive wear which may affect function of gearshift or interlock mechanisms. Inspect shift cover for cracks and distortion.

To assemble gearshift, interlock and shift cover components, reverse disassembly procedure while noting the following: Install shift control finger (33) with chamfer in bore towards shift cover. Finger (33) should travel an equal amount forward and rearward of fingers on interlock (32). If travel is unequal, reposition finger (33) on shaft (47) splines. Tighten interlock finger (34) clamp bolt and gearshift arm clamp bolt to 20-26 ft.-lbs. Tighten shift cover screws to 20-27 ft.-lbs. in a diagonal pattern on 8700-9700 models. Shift cover bolts on Models TW-10, TW-20 and TW-30 should be tightened to a torque of 27-37 ft.-lbs.

191. **OIL PUMP, REAR COVER, REAR OIL SEALS AND HIGH-LOW GEARS.** The low pressure oil pump, transmission rear cover, high-low gears and shift rail can be removed after splitting tractor between transmission and rear axle center housing or with transmission removed as outlined in paragraph 187 or 188.

On TW-10, TW-20 and TW-30 models, remove four screws (3—Fig. 159) securing low pressure oil pump to rear cover (6). Carefully withdraw pto shaft (4) and

Fig. 159—View showing rear of transmission on Models TW-10, TW-20 and TW-30.

1. Jack screw holes
2. Low pressure pump
3. Pump retaining screws
4. Pto shaft
5. Output shaft bearing retainer
6. Rear cover

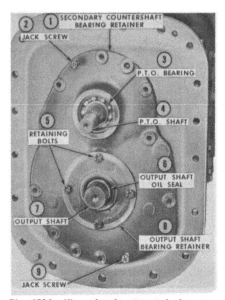

Fig. 159A—View showing transmission rear cover with two retaining cap screws threaded into jack screw holes. Pto shaft is removed with rear cover (secondary countershaft bearing retainer).

oil pump. Remove outer snap ring and drive pto shaft and bearing out of pump housing. Oil pump is available as a unit only and must be serviced as an assembly.

On all models, refer to Fig. 159 or 159A and unbolt then carefully pry output shaft bearing retainer off rear cover to avoid damaging output shaft bearing adjustment shims. Remove the cap screws retaining rear cover to transmission housing and thread two of the cap screws into jack screw locations shown. Evenly tighten the two cap screws to force rear cover from housing, then remove the rear cover and, on all models except TW-10, TW-20 and TW-30, the pto shaft assembly. Remove jack screws from cover. On splash lubricated transmissions pry oil seal (68—Fig. 160) from retainer (70) and on all models, if bearing cup (61) is scored or otherwise damaged, remove the cup. Remove snap ring (66) and bump pto shaft (42) and bearing (65) out rear of cover, then remove pto shaft oil seal (63) (splash lubricated transmissions only). Inspect bearing (45) in rear cover and remove the needle roller bearing if worn or damaged. Withdraw output shaft (52) with gear (58) out rear of transmission; press shaft from gear and rear bearing cone (60) and remove front bearing cone (50) if worn or scored. Shift high-low sliding coupling (51) to rear and remove coupling from shift fork. Withdraw secondary countershaft (41) from rear of transmission, taking care not to drop thrust bearing (38). Inspect needle roller bearing (39) in web of transmission housing; if bearing is excessively worn or damaged, it can be removed and a new bearing installed providing care is taken to remove any dislodged needle rollers from housing when removing old bearing. Inspect bearing cup (49) in rear end of main countershaft (47) and remove if scored or excessively worn. Inspect ball bearing (65) on pto shaft (42); remove snap ring (67) and press bearing from shaft if bearing is worn or rough.

To remove high-low shift fork, remove detent plug (59—Fig. 157), spring (57) and ball (56) and unscrew set screw (24) from fork (25). Remove shift rail (55) out rear of transmission housing, then disengage fork from transmission brake pawl (23) and remove fork from housing.

Excessive wear of gears, bearings and other components in rear section of transmission would indicate that the transmission should be completely disassembled and inspected.

192. To reassemble transmission, proceed as follows: Using suitable drivers, install new needle roller bearing (39—Fig. 160) in transmission web and new

bearing cup (49) in rear end of main countershaft (47). Place high-low shift fork (25—Fig. 157) on pin in transmission brake pawl (23), then insert shift rail (55) and fork retaining set screw (24). Align single indentation in shaft with screw, then tighten screw and locknut to 23-30 ft.-lbs. on Models TW-10, TW-20 and TW-30 and 12-18 ft.-lbs. on all other models. Install detent ball (56), spring (57) and retaining plug (59) with new gasket (58). Place high-low sliding coupling (51—Fig. 160) in fork with brake teeth to rear, then shift coupling forward to engage with dog teeth on main countershaft (47). Refer also to Fig. 161.

NOTE: Transmission gear shift sliding connector and coupling assemblies are furnished as matched sets on TW-10, TW-20 and TW-30 models. When removing countershaft, make sure forward

connector and coupling assembly and rearward connector and coupling assembly each remain together as matched sets. If installing new connectors and couplings, make sure etched timing marks are aligned. During transmission reassembly, position connector and coupling assemblies with chamfered end of connector bores facing rearward for ease of countershaft installation.

NOTE: When installing new coupling (51—Fig. 160) on 8000—9000 models, install spacer C9NN—7N445—A between countershaft gear dog teeth and gear face.

Using grease, stick thrust bearing (38—Fig. 160) to front end of secondary countershaft (41). On pressure lubricated transmission, except TW-10, TW-20 and TW-30, install spacer ring (20) in groove adjacent to sliding coupling splines on secondary countershaft as

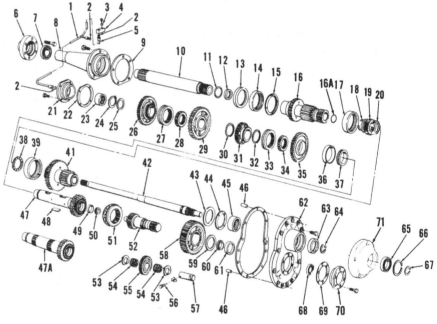

Fig. 160– Exploded view of transmission gears, shafts and bearings with rear cover and bearing supports. Refer to Fig. 157 for exploded view showing transmission housing, top and shift covers, shift forks and rails and transmission brake linkage. Note rear cover differences in Figs. 159 and 159A. On later models except TW-10, TW-20 or TW-30, third gear (35) face is recessed and a thrust washer is fitted. Seals (63 and 68) are not used in pressure lubricated transmissions. Snap ring (64) is not used in splash lubricated transmissions. "O" ring (16A) is installed in mainshaft (16) on models without Dual Power. Oil pump (71) is only used on TW-10, TW-20 and TW-30.

1. Oil inlet tube	18. Spacer	36. Bearing cup	53. Thrust washer
2. "O" ring	19. Sliding gear/ coupling	37. Bearing cone	54. Bearing
3. Spacer	20. Spacer	38. Thrust bearing	55. Reverse idler
4. Manifold	21. Bearing support	39. Needle roller bearing	56. Idler shaft bolt
5. Tube fitting	22. Gasket	41. Secondary counter shaft	57. Idler shaft
6. Pto drive flange	23. Needle roller bearing	42. Pto shaft	58. Output shaft gear
7. Pilot bearing	24. Snap ring	43. Washer	59. Thrust washer
8. Clutch release bearing support	25. Thrust washer	44. Thrust washer	60. Bearing cone
9. Gasket	26. Second gear	45. Needle roller bearing	61. Bearing cup
10. Input shaft	27. Sliding coupling	46. Dowel pin	62. Rear cover
11. Snap ring	28. Connector	47. Main countershaft	63. Oil seal
12. Pto shaft seal	29. First gear	47A. Main countershaft	64. Snap ring
13. Mainshaft seal	30. Thrust washer	48. Key	65. Ball bearng
14. Needle roller bearing	31. Reverse gear	49. Bearing cup	66. Snap ring
15. Thrust washer	32. Thrust washer	50. Bearing cone	67. Snap ring
16. Mainshaft	33. Sliding coupling	51. Sliding coupling	68. Oil seal
16A. "O" ring	34. Connector	52. Output shaft	69. Shim
17. Needle roller bearing	35. Third gear		70. Bearing retainer
			71. Low pressure oil pump

shown at (P—Fig. 153). Note shape of spacer in Fig. 162. Lubricate needle bearing in housing and position secondary countershaft in the needle bearing in housing. Install new bearing cone (50—Fig. 160) on front end of output shaft (52), place gear (58) on shaft with dog teeth forward, then install thrust washer (59) and rear bearing cone (60) on rear end of shaft. Insert output shaft assembly so that splines on front end of shaft enter sliding coupling (51) and bearing cone is seated in cup in main countershaft.

Using suitable drivers, install new needle roller bearing in rear cover and bearing cup in bearing retainer (70). On splash lubricated transmissions, renew seals (63 and 68). These seals are not used on pressure lubricated transmission, since a common reservoir is used. Drive or press new bearing onto pto shaft and secure with snap ring. Lubricate shaft and oil seal if so equipped, in rear cover, then install shaft and bearing assembly, taking care not to damage oil seal. Install snap ring in rear cover at rear side of pto shaft bearing. Position new gasket on dowel pins on rear face of housing. Install steel washer (43) and thrust washer (44) on rear end of secondary countershaft (41) with tab on thrust washer (44) pointing to rear and up. Note that on early 8000 models, steel washer (43) is absent. On all models except TW-10, TW-20 and TW-30, insert pto shaft through secondary countershaft, taking care not to damage oil seal (12) in front end of main shaft (16) as pto shaft is moved forward. Turn shaft as necessary to align splines with drive plate (6) at flywheel. Be sure dowel holes are aligned as cover contacts rear face of housing. Tighten rear cover retaining cap screws alternately and evenly to a torque of 20-27 ft.-lbs.

On Models TW-10, TW-20 and TW-30, install bearing (65) and snap ring (67) on pto shaft (42) then install oil pump (71) on shaft and bearing and snap ring (66). Install rear cover on transmission housing and tighten rear cover retaining screws alternately and evenly to 27-35 ft.-lbs. Insert pto shaft with oil pump into transmission and tighten oil pump retaining screws to 21-29 ft.-lbs.

On all models, install output shaft bearing retainer (70) with sufficient thickness of shims (69) to adjust output shaft bearing preload. With retainer cap screws tightened to the proper torque, measure torque required to turn output shaft with pull scale attached to cord wrapped around shaft. Bearings are correctly adjusted when a pull of 12 pounds or less is required to steadily rotate shaft and there is no end play of shaft in bearings. Remove shim thickness to remove end play or add shim thickness if pull required to turn shaft is greater than 12 pounds. Shims are available in thicknesses of 0.002/0.006 and 0.012/0.014 inch.

194. RENEW INPUT SHAFT OIL SEALS. Oil leakage into clutch compartment can be from damaged or worn oil seal (13—Fig. 160) in clutch release bearing support (8) or pto shaft oil seal (12) in front end of transmission main shaft (16). The following procedure for input shaft oil seal renewal does not apply to Dual Power as seal is not used with Dual Power unit. Refer to paragraph 195 for renewal of pto shaft oil seal on all models including Dual Power models.

Oil seal (13) in release bearing support can be renewed after splitting tractor between engine and transmission housing as outlined in paragraph 162; proceed as follows: On pressure lubricated transmissions, disconnect oil supply tube going to clutch release hub and countershaft bearing retainer by pulling tube straight out. Withdraw input shaft (10), remove pins from clutch fork and remove cross shaft, fork and release bearing, then unbolt and remove clutch release bearing support from transmission. Take care not to lose thrust washer (15) from front end of mainshaft.

Remove and discard the needle roller bearing and oil seal from rear end of release bearing support. Refer to Fig. 163 and install new oil seal with flat side first (lip to rear); drive seal in until it lightly contacts shoulder in support. Install new needle roller bearing by driving on lettered side of bearing (flat end) only until bearing is 1/8-inch below flush with thrust face of support.

To reinstall support, stick thrust washer (15—Fig. 160) on front end of mainshaft using grease and position tab on washer up and forward. Lubricate needle bearing and oil seal, then carefully install support with new gasket (9), placing flat side of gasket and support down, and tighten retaining cap screws to a torque of 20-27 ft.-lbs. on splash lubricated transmissions, 27-37 ft.-lbs. on TW-10, TW-20 and TW-30 models and 30-40 ft.-lbs. on all other models. Install cross shaft, clutch release bearing and fork and insert transmission input shaft into clutch disc and flywheel. Reconnect engine to transmission as outlined in paragraph 163.

195. To renew pto shaft oil seal in transmission mainshaft, first remove Dual Power unit on models so equipped, as outlined in paragraph 170, or remove clutch release bearing support on all other models as outlined in paragraph 194. Remove transmission top cover as in paragraph 184, 185 or 186, then proceed as follows:

Remove cap screw (56—Fig. 160) and locking washer, then slide reverse idler shaft (57) rearward and remove the idler. Remove countershaft bearing

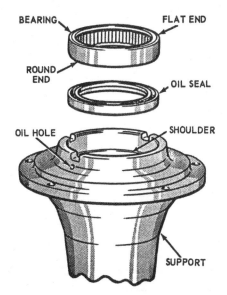

Fig. 163—View showing proper installation of oil seal and needle roller bearing in clutch release bearing support. If bearing is driven in too far on splash lubricated transmissions, it will close off oil hole in support.

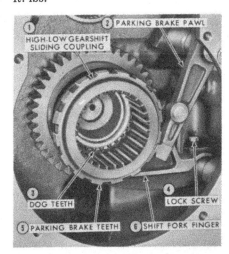

Fig. 161—View of rear end of transmission with output shaft and secondary countershaft removed. Note brake pawl engaged with high/low sliding coupling.

Fig. 162—Note difference in shape of spacer rings used on splash lubricated (S) and pressure lubricated (P) transmissions, except TW-10, TW-20 and TW-30 which are not equipped with spacers (18 or 20-Fig. 160). Refer to paragraphs 192 and 195 and to Fig. 153 for placement.

support (21) from front end of transmission housing to allow countershaft to drop down. Engage sliding gear/coupling (19) with dog teeth on front end of secondary countershaft (41), then remove the mainshaft (16) from front end of transmission. Splash lubricated transmissions have a thrust washer (15) at front of mainshaft. Refer to Fig. 164, remove old seal and install new seal (12) with lip towards bushing as shown.

CAUTION: Take care not to damage bushing as it is not available separately from mainshaft.

NOTE: On splash lubricated transmissions, install spacer ring (18—Fig. 160) on mainshaft between sliding coupling splines and bearing (17) as shown at (S—Fig. 153). Spacer ring chamfer should be towards front. Ring prevents sliding coupling from contacting bearing web and should be installed on all splash lubricated transmissions.

Lubricate bushing, oil seal and pto shaft, then carefully reinstall mainshaft over pto shaft and be sure it engages with sliding gear. Working through input shaft opening in housing, pry countershaft gear up with screwdriver, then install countershaft bearing support, oil screen up, with new gasket and tighten retaining cap screws to a torque of 27-37 ft.-lbs. on TW-10, TW-20 and TW-30 or to 20-27 ft.-lbs. on all other models. On pressure lubricated transmissions, install thrust washers (53—Fig. 160) between reverse idler gear (55) and transmission case walls. Reinstall reverse idler with step in hub to front and shaft with notch to rear and up. Tighten shaft retaining cap screw to a torque of 12-18 ft.-lbs. and secure with locking washer. Reinstall clutch release bearing support or Dual Power assembly and reassemble tractor.

195A. OVERHAUL COMPLETE TRANSMISSION. With disassembly completed as outlined in paragraphs 189 through 195, proceed as follows:

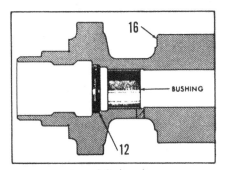

Fig. 164—Cross-section of mainshaft (16) showing proper installation of pto shaft oil seal (12). Bushing is not available separately from mainshaft.

Refer to Fig. 165 and remove snap ring and thrust washer from front end of countershaft. To remove countershaft (47A—Fig. 160) from 8000 models, withdraw countershaft from rear of transmission and remove gears, spacers, couplings and connectors as countershaft is withdrawn.

To remove single key countershaft on TW-10, TW-20 and TW-30 models, remove shift rail and shifter detent balls and springs. Remove 1st-5th, 2nd-6th, the 3rd-7th rev gear, shift fork and parking brake pawl from transmission. The 4th-8th shift fork must be removed following countershaft removal. Reinstall main countershaft bearing retainer (70—Fig. 160) loosely to support front end of countershaft then install in retainer screw holes, substitute bolts or studs that are at least ½ inch longer than original. Extra length studs will allow countershaft to move forward for key (48) removal. Move front sliding gear coupling (27) forward to engage dog teeth of 47 tooth gear (26). Using a long screwdriver, move connector (28) forward so front countershaft key will be accessible. Separate gears until 55 tooth gear (29) is fully rearward. Make sure 47 tooth gear (26) is fully forward contacting front bulkhead. Rear face of 43 tooth gear (35) should remain in contact with countershaft bearing and rear bulkhead. Remove front countershaft key with a magnet. Move rear sliding gear coupler (33) forward to engage dog teeth of 51 tooth gear (31). Move connector forward using a long screwdriver so rear countershaft key will be accessible. Separate gears until 43 tooth gear (35) contacts rear of bulkhead and countershaft bearing. Remove rear countershaft key with magnet. Withdraw countershaft (47) through rear of transmission.

To remove countershaft (47—Fig. 160) on all other models, slide the 55 tooth gear (26) as far forward as possible then slide coupler forward to engage dog teeth of 55 tooth gear. With a long screwdriver, evenly move the connector (28) until retaining keys (48) are removable. Using same procedure, move 49 tooth gear (31) forward and remove gear keys (48) in countershaft by moving connector (34). Withdraw countershaft through rear of transmission. Disengage snap rings on shift rail and on 8700 and 9700 models, turn shift rail so detent balls and springs may be removed through passages in shift forks. Withdraw shift rail from rear of transmission while removing shift forks and brake pawl from shift rail.

Carefully clean and inspect all parts and thoroughly clean transmission housing. Check all bearings and renew any scored, excessively worn or damaged tapered roller bearing cones and cups or

needle roller bearing assemblies.

NOTE: When installing new needle roller bearing assemblies, be sure to select size of driver that will fully contact flat (lettered) face of bearing cage and take care not to drive inner (rounded) end of cage against shoulder in bearing bore.

Inspect all gear teeth for chipping, excessive wear or scoring, and renew any not suitable for further service. Also inspect bearing and thrust surfaces of gears and shafts for scoring or excessive wear.

Prior to reassembling transmission, temporarily assemble gears, thrust spacers, connectors and couplings in proper assembly order on the transmission countershaft, then install front thrust washer and snap ring; refer to Fig. 160 for assembly guide. Check clearance between front thrust washer (25) and snap ring (24); if clearance is not within limits of 0.0073-0.0143 inch, select new thrust washer (25) of thickness to provide proper clearance. Thrust washers are available in thicknesses of 0.091/0.093, 0.101/0.103, 0.111/0.113 and 0.121/0.123 inch. Then, with countershaft and gears disassembled, proceed as follows:

On splash lubricated transmissions, turn transmission housing up on end, clutch housing end down. Place countershaft gears, connectors, couplings and thrust washers in housing in proper order. On later models, third gear (35) face is recessed and a thrust washer is fitted. Insert countershaft down through the gears until tapered roller bearing cone is seated in cup in transmission housing web. It may be necessary to rotate countershaft back and forth to align all splines on components. Carefully turn housing back upright and install previously selected thrust washer and snap ring on front end of the countershaft.

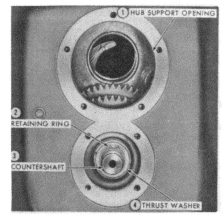

Fig. 165—View showing front end of transmission with mainshaft and main countershaft bearing retainer removed.

To install countershaft assembly in pressure lubricated transmissions, place countershaft gears, connectors, couplings and thrust washers in proper order in transmission case. On later models, third gear (35) face is recessed and a thrust washer is fitted. Insert countershaft in gears until taper roller bearing on shaft is seated in bearing cup in transmission case web. Position countershaft so that keyways are on top and bottom of shaft. With transmission case positioned so that top cover opening is up, install a transmission cover bolt in opposite sides of the cover opening so that the bolts are aligned as close as possible with the rear keyways on countershaft and keyways are accessible. Install a key, beveled edge up, in the rear upper keyway. Move the sliding connector partially over the key and rotate the countershaft until the installed key is on the under side of the shaft. Using a wire of sufficient length, pass the wire under the installed key and tie ends of wire to bolts in top cover opening.

Wire should hold key on shaft as connector is slid off key. Install key with beveled edge up, in upper keyway being careful not to dislodge previously installed key. Slide coupling connector over installed keys. Repeat procedure to install keys in forward keyways of shaft. Install previously selected thrust washer and snap ring on front end of countershaft.

Insert shift rail through rear face of transmission while installing snap rings, brake pawl and shift forks. On Models 8700, 9700, TW-10, TW-20 and TW-30, align detent ball and spring holes in shift rail and shift forks. Install detent balls and springs then rotate shift rail so tapped hole in forward end of shift rail is at three o'clock position. Be sure the shift forks engage the two sliding couplings on countershaft and the shift fork for 4th-8th gear sliding gear/coupling is properly positioned, as shown in Fig. 153. With snap ring at rear end of shaft seated against housing, move the pawl snap rings until they engage proper grooves in rail to hold pawl in position.

Complete remainder of reassembly as outlined in paragraphs 189 through 195. Fill transmission with lubricant designated in paragraph 197. Oil level plug is found on cam cover on splash lubricated transmissions as shown in Fig. 154. Pressure lubricated transmissions have common sump with rear axle. Oil level plug for 9000 models is just forward of left rear axle housing. Oil level dipstick for 8600, 8700, 9600, 9700, TW-10, TW-20 and TW-30 models is at upper left rear axle housing.

TRANSMISSION HANDBRAKE

An auxiliary transmission handbrake is optional on Models 8700, 9700, TW-10, TW-20 and TW-30 and standard on all models equipped with front wheel drive. The transmission handbrake is a parking brake only and should not be used to slow or stop tractor movement.

R&R AND OVERHAUL

196. To remove unit on models equipped with front wheel drive, refer to paragraphs 60 and 61 and remove transmission housing then separate handbrake and transfer box assemblies. On models not equipped with front wheel drive, transmission handbrake components may be serviced with unit mounted on transmission housing. Prior to disassembly, raise left side of tractor to divert transmission oil away from transmission handbrake housing.

To overhaul transmission handbrake, remove rear cover (19—Fig. 167) and brake shaft (15). Heat idler shaft retaining screw (21) to loosen Loctite then remove screw. Withdraw idler shaft (7) while removing idler gear (4), bearings (1) and shim (2). On models so equipped, remove front cover (12). On all models remove brake components (8 through 10).

Inspect brake housing oil seal (22) renew as necessary. Coat outer edge of new seal with Ford sealer ESA-M4G-129-A or equivalent. Inspect and renew the following components as necessary: Brake discs (10), actuators (11) and braking surfaces in housing and rear face of transfer box. Roller bearings, brake shaft assembly (15) and idler (4) should be inspected and renewed as necessary.

NOTE: Lubricate components in reduction gear side of transmission handbrake housing before assembly. Do not lubricate handbrake components.

Install bearing outer races on idler gear (4) if removed. Install inner races. Install idler gear assembly into housing without shaft or shim (2). Measure gap between idler gear bearing and housing. Install shim (2) to reduce end play to zero. Shim (2) is available in thicknesses from 1.00 mm to 1.50 mm in increments of 0.05 mm. Reinstall idler gear and shaft assembly. Apply a suitable locking compound to shaft retaining screw (21).

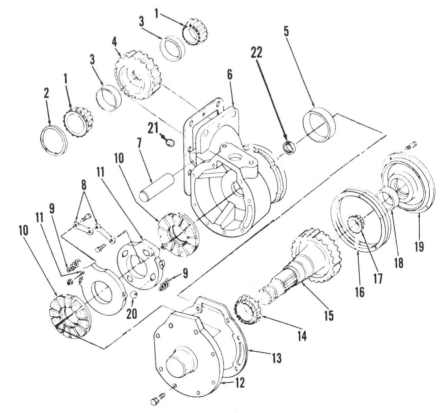

Fig. 167—Exploded view of transmission handbrake.

1. Idler gear bearing	7. Idler shaft	12. Front cover
2. Shim	8. Brake actuating disc link	13. Housing gasket
3. Bearing cup		14. Bearing cone
4. Idler gear	9. Disc return spring	15. Brake shaft
5. Bearing cup	10. Disc	16. Rear cover shims
6. Housing assembly	11. Brake actuating link	17. Bearing cone

18. Bearing cup
19. Rear cover
20. Brake actuating ball
21. Shaft retainer screw
22. Oil seal

Install oil seal (22) and bearing outer race in housing. Press front and rear bearing inner races on brake shaft (15). Wrap tape around splines to avoid damaging oil seal then install brake shaft assembly.

Install rear cover (19) without shims (16) located between cover flange and housing. Rotate cover and apply pressure to seat bearing. Measure gap between cover and housing at three places around circumference. Note average measurement and select proper shim(s) for installation. Refer to following chart to determine thickness and quantity of shims (16). Be sure to measure shim pack to verify actual thickness equals desired thickness.

MEASURING GAP [Inch]	SHIMS [Inch]		
	0.007	0.009	0.012
0.009-0.011	...	1	...
0.012-0.013	1	1	...
0.014-0.015	...	2	...
0.016	1	...	1
0.017-0.018	...	1	1
0.019-0.021	...	...	2
0.022-0.023	...	3	...
0.024-0.025	...	2	2
0.026	1	...	2
0.027-0.029	...	1	2
0.030-0.031	...	...	3
0.032	2	...	2
0.033	...	3	1
0.034	1	1	2
0.035	3	1	1
0.036	...	2	2
0.037	1	...	3
0.038	3	...	2
0.039	...	1	3
0.040	3	3	...
0.041	...	4	1

Install shims (16) and rear cover (19) so dipstick elbow tube on cover is at bottom then tighten cover retaining screws to 32 ft.-lbs. Install remainder of brake assembly by reversing disassembly procedure. Refer to paragraph 176 and refill transmission oil as required.

If tractor is equipped with front wheel drive and brake shaft (15), housing (6), cover (19) or shaft bearings were renewed, then refer to paragraph 63 to adjust end play between brake and transfer box shafts.

DIFFERENTIAL AND BEVEL GEARS

NOTE: Although outward appearance is similar, components may vary in size between models due to differences in power output. Service procedures are the same except as called out in text.

LUBRICATION

197. Early model 8000 tractors are equipped with separate transmission and rear axle lubricating oil sumps. Rear axle lubricating oil capacity for these models is 40 quarts. All other models are equipped with pressure lubricated transmissions having a common sump with the rear axle housing. Rear axle oil capacity for these models equipped with Dual Power, except TW-10, TW-20 and TW-30, is 69 quarts; without Dual Power capacity is 64 quarts. TW-10 and TW-20 models equipped with Dual Power require 79 quarts; models without Dual Power require 73 quarts. TW-30 models require 92 quarts. Recommended rear axle lubricant is Ford specification No. ESN-M2C53-A.

DIFFERENTIAL LOCK

198. The differential lock consists of a hydraulically actuated multiple disc brake located in differential housing. When applied, the brake locks right differential side gear to differential housing, causing both rear wheels to be driven at the same speed, regardless of traction. The fluid pressure for operating the brake is taken from pto hydraulic circuit; refer to paragraphs 212 and 213 for troubleshooting information on the differential lock hydraulic circuit. Service of the differential lock brake components is covered by procedure for servicing the differential; refer to paragraph 199. Refer to paragraph 240 for servicing differential lock valve and to 237, 238 or 238A for pump.

DIFFERENTIAL AND BEVEL RING GEAR

199. **REMOVE AND REINSTALL.** To remove the differential and bevel ring gear assembly, first remove the hydraulic lift cover as outlined in paragraph 245, the pto drive shaft and gears

Fig. 169—Using feeler gage to determine proper shim pack thickness required for adjustment of differential carrier bearings; refer to text.

as in paragraph 217 and both rear axle housing assemblies as in paragraph 204, then proceed as follows:

Disconnect differential lock hydraulic line from top of right differential bearing support (27—Fig. 170). Place a rope sling around differential housing, attach hoist to sling and take weight of differential assembly on hoist. Unbolt and remove the differential bearing supports, taking care not to damage shims (2 and 26) and to keep the shims from under each support separate and identified for reassembly. Lift differential assembly from rear axle center housing.

To reinstall lower differential into rear axle center housing, then reinstall bearing supports with same shims as removed. Tighten support retaining cap screws to a torque of 50-60 ft.-lbs. and reconnect differential lock hydraulic line to right bearing support. Reinstall pto drive shaft as outlined in paragraph 217, the hydraulic lift cover as in paragraph 245 and the rear axle assemblies as in paragraph 204.

NOTE: If differential housing, carrier bearings, bearing supports, rear axle center housing or the bevel ring gear and pinion set have been renewed, it will be necessary to check and adjust carrier bearing preload and bevel gear backlash as in paragraph 200.

200. **ADJUST CARRIER BEARING PRE-LOAD AND BEVEL GEAR BACKLASH.** Carrier bearing pre-load is adjusted by varying shim thickness between bearing supports and rear axle center housing. Main bevel gear backlash is adjusted by transferring shims from under one bearing support to under opposite support. While some backlash must be maintained while adjusting bearing preload, backlash should be rechecked or adjusted after adjusting bearing preload as follows:

Support differential assembly in rear axle center housing with rope sling and hoist. Install left bearing support with same shims removed on disassembly and tighten retaining cap screws to a torque of 50-60 ft.-lbs. Install right bearing support **without** shims and tighten cap screws only enough to remove end play of differential in carrier bearing, but **do not** preload the bearings. Measure gap at several points between right bearing support flange and rear axle center housing using feeler gage as shown in Fig. 169. Loosen and tighten opposite cap screws until measurement is equal at all points. Check to be sure some backlash exists between bevel pinion and ring gear; decrease shim thickness under left bearing support if no backlash is noted, then retighten right bearing support cap screws to just remove end play of differential and obtain measurement between bearing support flange and center housing. Remove right bearing support, then reinstall support with shim pack of total thickness 0.001-0.006 inch less than measured gap. Tighten retaining cap screws to a torque of 50-60 ft.-lbs.

NOTE: Measure shim thickness with micrometer; thickness of individual shims may vary up to 0.004 inch. Shims are available in thicknesses of 0.002/0.006, 0.006/0.010, 0.010/0.014 and 0.018/0.022 inch.

With carrier bearing preload properly adjusted, measure backlash of bevel ring gear to bevel pinion with dial indicator. Make backlash measurement at several points around ring gear to be sure backlash is within limits at all points.

NOTE: Dial indicator plunger should be against outer ends of ring gear teeth and at right angle to radius of ring gear.

Backlash should be within limits of 0.010-0.020 inch; transfer shims from under left bearing support flange to under right bearing support flange if backlash is less than 0.010 inch; or from under right support to under left support if backlash is more than 0.020 inch. Tighten bearing retaining cap screws to a torque of 50-60 ft.-lbs. before rechecking backlash.

201. OVERHAUL DIFFERENTIAL OR RENEW BEVEL RING GEAR. With differential assembly removed as outlined in paragraph 199, remove the cap screws retaining cover (22—Fig. 170) to differential case (6) and ring gear (4). Separate cover from case and remove the differential lock discs (16 and 17), thrust washer (15) and differential side gear (14).

To renew ring gear, support outer edges of gear in bed of press and press differential case out of the gear. To install new gear, support gear on wood blocks (Fig. 171) and insert differential case in gear with bolt holes in case and gear aligned. Thread all retaining cap screws through case into gear as shown, then press case downward into gear. Remove assembly from press and remove the cap screws.

NOTE: When installing new ring gear, refer to paragraph 202 and install mating bevel pinion.

To remove differential pinions and left side gear, drive the roll pin (13—Fig. 170) out of case and pinion shaft (12). Remove the pinion shaft, pinions (10), pinion thrust washers (9), side gear (14) and the thrust washer (8). Bushings (7 and 21) in case and cover may be renewed if excessively worn or scored. Bushings are pre-sized and should not require reaming if carefully installed.

Inspect the differential lock fluid sealing rings (23) on hub of cover (22) and renew if worn or broken. To remove differential lock piston (18) from cover, apply air pressure to hole between sealing ring grooves in hub of cover. Renew the piston seal "O" rings (19 and 20) and inspect cover and piston for any damage. Lubricate piston, rings and bore in cover before installing piston. Check differential lock friction discs (16 and 17) for warpage or excessive wear and inspect friction surface of differential case; renew case and/or discs if

necessary.

To reassemble, lubricate all parts and reverse disassembly procedure. Refer to Fig. 170 for installation of differential lock discs; insert disc (16) with internal splines first. Tighten cover retaining cap screws to a torque of 160-200 ft.-lbs. Renew carrier bearing cones and cups if excessively worn or scored.

MAIN DRIVE BEVEL PINION

202. REMOVE AND REINSTALL. To remove main drive bevel pinion,

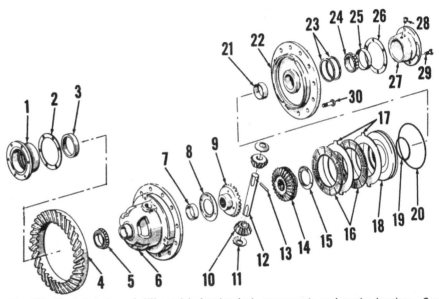

Fig. 170—Exploded view of differential showing lock components and carrier bearings. Cap screws (30) retain cover (22) to differential case (6) and also retain main drive bevel ring gear (4) to differential case; refer to Fig. 171 for installation of ring gear. Fluid for operation of differential lock piston (18) enters right carrier bearing support through fitting (28) and is transmitted to cover (22) through bore in hub of cover. Sealing rings (23) fitted in grooves in cover hub are located at each side of oil passage in bearing support (27) and hub.

1. Left carrier bearing support	9. Left differential side gear	16. Differential lock discs	23. Sealing rings
2. Shims	10. Differential pinions (2)	17. Differential lock plates	24. Bearing cone
3. Bearing cup	11. Thrust washers (2)	18. Differential lock piston	25. Bearing cup
4. Bevel ring gear	12. Pinion shaft	19. Inner "O" ring	26. Shims
5. Bearing cone	13. Roll pin	20. Outer "O" ring	27. Right carrier bearing support
6. Differential case	14. Right side gear & disc hub	21. Bushing	28. Fitting
7. Bushing	15. Thrust washer	22. Cover plate	29. Support cap screws
8. Thrust washer			30. Cover & case to ring gear cap screws

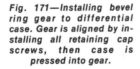

Fig. 171—Installing bevel ring gear to differential case. Gear is aligned by installing all retaining cap screws, then case is pressed into gear.

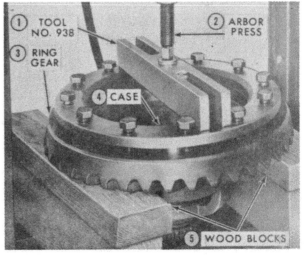

Fig. 172—Remove pto clutch hub for access to pto hydraulic line. Remove locating pin (cap screw) to remove pto clutch and valve assembly.

remove differential as outlined in paragraph 199.

Straighten tabs on washer (3–Fig. 173) and remove the pinion bearing adjusting nuts (2 and 4). Using a soft drift pin, drive pinion rearward until free of front bearing cone (5), then remove pinion with rear bearing cone

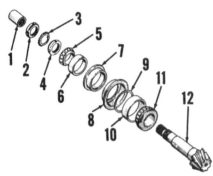

Fig. 173—Exploded view of bevel pinion assembly. Pinion bearing cups (6 and 10) are supported in sleeves (7 and 8) which fit in bores of rear axle center housing. Bearings are adjusted by nuts (2 and 4); refer to text.

1. Drive coupling	7. Bearing sleeve
2. Adjusting nut	8. Bearing sleeve
3. Locking washer	9. Shims
4. Adjusting nut	10. Bearing cup
5. Bearing cone	11. Bearing cone
6. Bearing cup	12. Bevel pinion

Fig. 174—Normal shim (9—Fig. 173) thickness for pinion mesh adjustment is 0.041 inch. If other than this thickness is required, it is indicated by stamping on rear axle center housing in location shown. Refer to text.

(11) from center housing.

Inspect bearing cups (6 and 10) and renew cups if excessively worn or scored. Note that cups are mounted in removable sleeves (7 and 8). Pinion mesh position is controlled by shim spacers (9) between rear sleeve (8) and bearing cup (10); take care not to lose or damage the spacers when removing rear cup. If a new rear axle center housing is being installed, or if spacers are lost or damaged, refer to paragraph 203 for selection of new spacers.

To reinstall bevel pinion, or install new pinion and/or bearings, proceed as follows: Install rear bearing cup and sleeve in housing with same spacers as were removed. Install front bearing cup and sleeve and install rear bearing cone on pinion. Insert pinion and rear bearing cone assembly through rear bearing cup and hold in position with hydraulic jack and two wood blocks as shown in Fig. 175. Then, using a suitable sleeve, drive

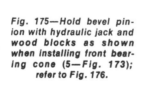

Fig. 175—Hold bevel pinion with hydraulic jack and wood blocks as shown when installing front bearing cone (5—Fig. 173); refer to Fig. 176.

front bearing cone onto pinion until seated in front bearing cup. Install rear adjusting nut (4—Fig. 173) and tighten to obtain proper bearing preload when checked as follows: Wrap a cord around pinion shaft between the front and rear bearings. Attach pull scale to cord and turn pinion steadily with cord and pull scale; bearing adjustment is correct when a steady pull of 10 to 20 pounds is required.

NOTE: If pull required is greater than 20 pounds, it may be necessary to bump front end of pinion shaft after loosening the adjusting nut, then re-tighten nut as required.

With bearings properly adjusted, install locking washer (3) and front adjusting nut (2); tighten front nut securely and recheck bearing adjustment. If adjustment remains within specification, bend tabs of washer into notches of adjusting nuts and install drive coupling (1) on pinion shaft.

To reassemble tractor, reverse disassembly procedure. Do not install pto clutch locating pin until splines of transmission pto shaft and splines of pto clutch are aligned. This will prevent possible shearing of locating pin if splines do not match. Tighten ½-inch housing cap screws to 50-65 ft.-lbs. and 5/8 inch cap screws to 95-130 ft.-lbs. Tighten pto clutch locating pin to 70-95 ft.-lbs.

203. PINION MESH POSITION. Mesh position of the main drive bevel pinion is controlled by shim spacers (9—Fig. 173) located between pinion rear bearing cup (10) and shoulder inside bearing sleeve (8). Thickness of spacers

required is determined by measurement of rear axle center housing and pinion gear shoulder at factory. Normally, required spacer thickness if 0.041 inch; if required thickness is different than 0.041 inch, difference is stamped on rear axle **center housing** in location shown in Fig. 174. In the example shown, required spacer thickness is 0.041+, 0.012, or 0.053 inch. If a negative sign (-) appears in front of number, subtract the value from 0.041 inch.

NOTE: Number in parenthesis marks (.304 inch) below "0.12" marking shown in Fig. 174 is metric (mm) equivalent of 0.012 inch; disregard number in parenthesis when working with decimal inches. Rear axle center housings requiring spacer thickness of 0.041 inch are not marked.

Look at the front of pinion gear just behind splines. If no mark appears on gear, no further spacer adjustment is require. If pinion gear is marked with a figure, such as +0.007 or MD +0.007, that thickness of spacer is to be removed from between bearing cup and gear shoulder. If number on gear is -0.007 inch, that thickness of spacer is to be added to basic (0.041 inch) spacer pack between bearing cup and gear.

Spacers are available in thickness of 0.018/0.022, 0.022/0.026, 0.026/0.030, 0.030/0.034 and 0.033/0.037 inch.

NOTE: As thickness of individual spacer may vary up to 0.004 inch, it will be necessary to select spacers by measuring thickness with micrometer.

In the preceding example, total spacer thickness required was 0.053 inch; this would require a combination such as one 0.033 inch thick spacer and one 0.020 inch thick spacer or any other combination of two spacers that would provide total thickness of 0.053 inch.

FINAL DRIVE AND REAR AXLES

NOTE: Although outward appearance is similar, components may vary in size between models due to difference in power output. Service procedures are the same except as called out in text.

204. **R&R REAR AXLE AND FINAL DRIVE ASSEMBLY.** If removing both assemblies, provide adequate braces for front end of tricycle model, or drive wood wedges between front axle and front support on wide front axle models. Drain lubricant. On Models 8700, 9700, TW-10, TW-20 and TW-30 support rear of platform or cab. Support tractor under rear axle center housing and remove the wheel, hub and tire assemblies from axle shafts. Attach hoist at a point which will balance rear axle housing, then unbolt axle housing from center housing. Move housing outward until clear of dowel pins and the inner axle shaft (planetary drive sun gear). Take care not to allow brake discs or inner axle shaft to fall out as the axle, housing and final drive assembly is removed.

To reinstall, proceed as follows: Insert inner axle shaft into the differential side gear. Remove brake disc guide pins (17—Fig. 177) from rear axle housing and insert them in holes in rear axle center housing. Place the brake

discs on the inner axle shaft and guide pins as shown. Place new gasket on the dowel pins, then lift axle housing assembly into place with hoist, align holes in axle housing with brake disc guide pins and align teeth on planetary gears with sun gear teeth on inner axle shaft. Install the axle housing retaining cap screws and tighten to a torque of 100-120 ft.-lbs. Refill rear axle center housing with proper lubricant; refer to paragraph 197.

205. **REAR AXLE SHAFT AND BEARINGS.** With the rear axle and final drive assembly removed as outlined in paragraph 204, remove axle shaft as follows:

Drill a hole in the flat outer face of axle shaft oil seal (25—Fig. 177) and pry seal out of housing; take care not to damage new seal surface of shaft or seal bore in housing. Remove the lock plate (6) from rear axle cap screw (7), then remove the cap screw, retaining washer (5), shims (4) and planetary carrier (3) with gears. Bump axle shaft out towards outer end of housing and remove bearing cone (32) from inside housing. Remove bearing cups (28 and 31) from housing and outer bearing cone (27) from axle shaft. Inspect oil seal contact ring (26) and install new ring if seal contact surface is scored or rough.

To reassemble, press new oil seal ring tightly against shoulder on axle shaft, then press outer bearing cone tightly

Fig. 176—With pinion supported as shown in Fig. 175, install front bearing cone with proper sized sleeve driver as shown.

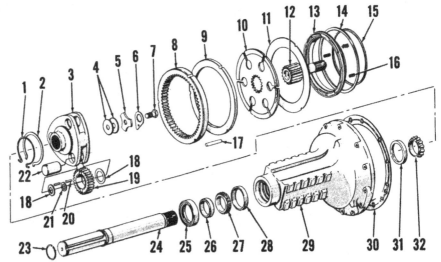

Fig. 177—Exploded view of final drive and rear axle assembly. Hydraulic brake piston (13) is fitted in side of rear axle center housing. Three pins (17) are fitted in mating holes in axle housing (29) and rear axle center housing and hold the planetary ring gear (8), brake outer disc (9) and brake inner disc (11) stationary. Planetary sun gear teeth are machined on outer end of inner shaft (12). Planetary gear carrier (3) is splined to inner end of wheel axle shaft (24). TW-30 models are equipped with four planetary gears and two brake inner discs (11) and two brake friction discs (10) per side.

1. Locking wire	10. Brake friction disc	17. Steel pins	25. Oil seal
2. Retaining ring	11. Brake inner disc	18. Thrust washers	26. Oil seal ring
3. Planetary carrier	12. Axle inner shaft	19. Planet gears	27. Bearing cone
4. Shims	(planetary sun gear)	20. Needle rollers	28. Bearing cup
5. Retainer	13. Brake piston	21. Spacers	29. Axle housing
6. Cap screw lock	14. Sealing ring, large	22. Planet gear shafts	30. Gasket
7. Axle cap screw	15. Sealing ring, small	23. Wheel hub snap ring	31. Bearing cup
8. Planetary ring gear	16. Brake piston springs	24. Axle shaft	32. Bearing cone
9. Brake outer disc			

against seal ring. Install inner and outer bearing cups in axle housing, then insert axle shaft and install inner bearing cone. Install planetary carrier and gear assembly, then install axle retaining cap screw and washer with thickest shim available. Tighten retaining cap screw to a torque of 396-484 ft.-lbs. and measure end play of axle shaft with dial indicator. Remove cap screw, retainer and shim, measure shim with micrometer and subtract measured end play from shim thickness. Select a spacer shim of thickness nearest to result obtained by subtracting end play from shim thickness, then install selected shim, retainer and cap screw. Tighten cap screw to a torque of 396-484 ft.-lbs. and install cap screw lock.

206. OVERHAUL PLANETARY DRIVE. With the axle and final drive assembly removed as outlined in paragraph 204 and the planetary carrier and gear assembly removed as in paragraph 205, proceed as follows:

Cut the locking wire (1—Fig. 177) (refer also to Fig. 178), then pull wire from under planet gear shaft retaining snap ring (2—Fig. 177). The snap ring can then be disengaged from the planet gear shafts by pulling ends of ring together which will allow the shafts to be driven out snap ring side of carrier. There are 54 loose needle rollers (20), two thrust washers (18) and a spacer (21) in each planet gear. Remove snap ring from carrier.

Carefully clean and inspect all parts and renew any that are excessively worn, scored or otherwise not suitable for service. Needle rollers are available in sets of 54 rollers only and should be renewed only as a complete set for each planet gear and shaft.

To reassemble, stick the spacer and the two rows of 27 rollers each in the planet gear using thin layer of heavy grease. Position planet gear in carrier and insert gear shaft from snap ring side of carrier. Push the shafts in flush with bore, then install snap ring, compress the snap ring and align notches in shafts with ring. Push a new locking wire under the ring and secure wire by wrapping ends around snap ring projections as shown in Fig. 178.

Inspect ring gear (8—Fig. 177) and sun gear teeth on inner axle shaft (12) and renew if teeth are chipped or excessively worn. Insert the pins (17) in axle housing to align ring gear when installing gear in housing. If a new ring gear is installed, it may have an identification groove, which should be installed toward center housing.

BRAKES

All Models Except TW-30

207. ADJUSTMENT. The rod connecting trunnion in brake pedal to master cylinder piston must be adjusted in length to provide a pedal free travel of 1/8-inch on 8700, 9700, TW-10 and TW-20 models or 3/16 to ¼-inch on 8000, 9000, 8600 and 9600 models measured at pedal stop pad as shown in Fig. 180. To adjust rod length, loosen jam nut (Fig. 180) and turn rod in or out of trunnion to obtain correct free travel, then tighten jam nut. Right side hood panel must be removed on Models 8000, 8600, 9000 and 9600 for access to linkage.

NOTE: If pedal is spongy, bleed the brakes as outlined in paragraph 208 prior to adjusting. Insufficient free travel will cause brakes to drag.

208. FLUID AND BLEEDING. On some early Model 8000 and 9000 tractors, lubricant from rear axle center housing is utilized as brake hydraulic fluid. On these models, a line connects brake reservoir to hydraulic lift sump return line which supplies master

cylinder with fluid. A second line returns excess fluid to the rear axle center housing. Whenever servicing brake master cylinder, or if cylinder has been drained, refill with proper lubricant (see paragraph 197), then bleed brakes as follows:

With engine running to keep master cylinder supplied with fluid, follow normal hydraulic brake bleeding procedure to remove all air from the system for both master cylinders and brake piston units. Bleeder fitting is located in rear axle center housing at top of rear axle housing on both sides of tractor as shown in Fig. 181.

Later 8000 and 9000 models and 8600, 8700, 9600, 9700, TW-10 and TW-20 models use a separate reservoir for brake fluid. Hydraulic system fluid is NOT USED. The manufacturer recommends that commercial grade brake fluid not be used in these models. Ford part No. IQ-M6C34-A fluid ONLY should be used, to prevent formation of gases in the system, and loss of braking action. If any other fluid has been used, it is recommended that the entire system be thoroughly flushed and approved fluid added.

To bleed these models, follow normal hydraulic brake bleeding procedure to remove all air from system. On models with auxiliary fluid reservoir, make sure that brake fluid is maintained at a visible level while bleeding. On models with dipstick in master cylinder, add fluid while bleeding to maintain adequate fluid level. Refer to Fig. 181 for brake bleeder fitting location.

209. OVERHAUL MASTER CYLINDER. For access to brake master cylinder, remove engine hood right side panel. On 8600, 8700, 9600, 9700, TW-10 and TW-20 models, remove enough of heat shield to allow removal of master cylinder. The master cylinder assembly can then be unbolted and removed from

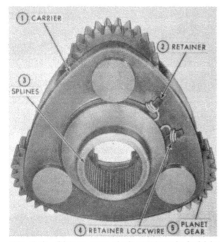

Fig. 178—Planet gear shaft retainer is held in place with lockwire inserted under the retainer and secured by wrapping ends of wire around ends of retainer ring as shown. TW-30 models are equipped with four planetary gears; other models have three gears.

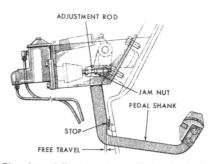

Fig. 180—Adjust master cylinder to brake pedal rods to obtain desired free travel of brake pedal at stop as shown. If free travel is less than specified brakes may drag or seize.

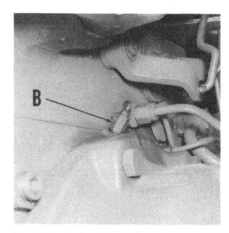

Fig. 181—Brake bleeder fittings (B) are located at each side of rear axle center housing just above axle housing flange.

support after disconnecting the fluid supply and return lines (early models only) and the two pressure lines to wheel pistons.

Refer to exploded view of assembly in Fig. 182 and remove the two rubber boots (not shown), snap (retaining) rings, pistons, piston primary seals and return springs from cylinder bores. Remove reservoir cover and thoroughly clean the cylinder bores and reservoir. Be sure that both the orifice and port in bottom of reservoir are open and clean.

Reservoir and cylinder casting is serviced as a complete master cylinder assembly only. All other parts are available separately or as a master cylinder repair kit which includes parts for servicing both cylinders of unit. Piston secondary seal is integral part of piston.

Lubricate master cylinder parts with recommended rear axle lubricant on early 8000 and 9000 models only. On all other models, lubricate parts with approved Ford brake fluid only and reassemble by reversing disassembly procedure.

TW-30 Models

Power to actuate brakes on TW-30 models is provided by the low pressure oil pump. Refer to paragraph 174. Pressurized oil is routed to power brake valve shown in Fig. 183 then to individual rear brakes.

210. **ADJUSTMENT.** To adjust power brakes on TW-30 models, shut off engine, depress both brake pedals and insert Ford tool No. 7446 (1—Fig. 184) between rocker arm links (2) and power brake valve housing as shown in Fig. 184. Loosen adjusting nuts and locknuts on both pedals then adjust one pedal at a time. Tighten adjusting nut until pedal begins to move. Check corner of gage for movement. If gage can be moved with a heavy drag, adjustment is correct. Gage must be movable without depressing

pedal. Tighten locknut, then repeat procedure for other pedal.

210A. **FLUID AND BLEEDING.** With engine running and hydraulic system oil at normal operating temperature, depress one pedal, loosen corresponding bleed screw in rear axle center housing near top of axle housing (Fig. 181), to expel air from system. Tighten bleed screw while maintaining pedal pressure. Repeat procedure for other brake. Refer to paragraph 176 and replenish transmission oil as required.

210B. **R&R AND OVERHAUL POWER BRAKE VALVE.** To remove power brake from tractor, remove oil tubes (2, 9, 12—Fig. 183), valve mounting bolts (13) and rocker arm retaining pin (10). To overhaul valve, remove check valve (13—Fig. 185) and ball (13A) from housing (10). After removing housing cap screws and cap (23), remove rocker arm assembly (1) by removing retaining rings (8) and retaining pins (7). Withdraw spool assembly (11). Heat valve retainers (3) to break Loctite seal and remove retainers using Ford tool No. 7445. Remove valve plungers (4) then force directional valve needles (6) and "O" rings (5) from housing (10) with a small punch.

Inspect all parts including housing (10) for excessive wear, damage or cracks and renew as necessary. During reassembly apply two or three drops of a suitable locking sealer to threads of retainers (3) making sure no sealer contacts plungers (4) or inside diameters of retainers (3). Tighten retainers to a torque of 10-15 ft.-lbs. Remainder of reassembly procedure is reverse of disassembly procedure. Bleed brakes as outlined in paragraph 210A then adjust brakes as outlined in paragraph 210.

211. **BRAKE PISTONS AND DISCS.** To remove brake pistons and discs on all models, refer to procedure for removing rear axle and final drive

assembly outlined in paragraph 204. The brake discs can then be removed from guide pins and the piston from rear axle center housing. Refer to Fig. 177 for ex-

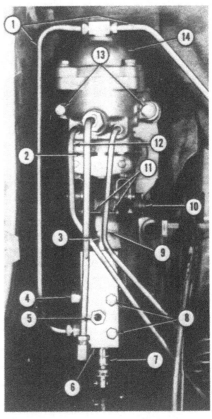

Fig. 183—View of power brake valve and differential lock valve on Model TW-30.

1. Differential lock valve to sump tubes	8. Valve mounting bolts
2. Brake valve oil inlet tube	9. Brake tube to left brake
3. Differential lock valve adjusting bolt	10. Rocker arm pin
4. Valve inlet port	11. Release pins
5. Valve outlet port	12. Brake tube to right brake
6. Differential lock valve	13. Mounting bolts
7. Valve spool	14. Brake valve assembly

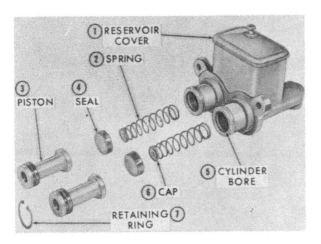

Fig. 182—Exploded view of brake master cylinder assembly. Piston and secondary seal are available as an assembly only. Master cylinder reservoir is kept filled by supply line from hydraulic system and excess oil is returned to rear axle center housing on early 8000—9000 models only.

1. RESERVOIR COVER
2. SPRING
3. PISTON
4. SEAL
5. CYLINDER BORE
6. CAP
7. RETAINING RING

Fig. 184—View of Ford brake adjusting tool No. 7446 (1) installed between rocker arm links (2) and valve body. See text for brake adjustment on Model TW-30.

ploded view showing brake piston and discs. Inspect piston (13) for warpage, scoring or cracks. If piston is serviceable, renew the seal rings (14 and 15). Inspect brake discs (9, 10 and 11) for warping or excessive wear. Remove brake piston springs (16) and discard, since these springs are no longer used.

To reassemble, lubricate piston and sealing ring assembly and install, taking care not to damage the rings. Complete reassembly by following procedure outlined in paragraph 204.

POWER TAKE-OFF

The independent type power take-off is driven by a hydraulically operated multiple disc clutch located in front end of rear axle center housing. On 8000, 8600, 8700, 9700, TW-10, TW-20 and some 9600 models, reduction gears located in rear compartment provide both 540 and 1000 rpm output shaft speeds. On other models, the pto is equipped with reduction gears for 1000 rpm operation only. The output shaft speed on dual speed models is changed by installing either the 540 rpm six-spline output shaft or 1000 rpm output shaft with 21 splines. On 1000 rpm only models a 21-spline 1000 rpm output shaft is used, but a 20 spline heavy-duty 1000 rpm shaft is available. On all models the output shaft supports the reduction gears and acts as a bearing surface.

On all models except TW-10, TW-20 and TW-30, fluid pressure to operate the pto clutch and also operate the differential lock clutch and Dual Power unit, on models so equipped, is provided by a gear type pump that is mounted on rear cover of the hydraulic system pump. Pressure in the pto clutch circuit is limited to 135-165 psi by the regulating valve in pto clutch valve body. Pressure to differential lock clutch is limited to 250-300 psi by the regulating valve sleeve on differential lock valve located in hydraulic system pump adapter plate. Operating pressure for Dual Power units is limited to 150-175 psi by Dual Power regulating valve located on wall of center housing. When both the differential lock and pto clutch are operated at the same time, system pressure is built up to 250-300 psi, but remains limited to 135-165 psi in pto clutch circuit. A relief valve within the pump body limits pump pressure to 450 psi in the event of malfunction of either the differential lock regulating valve or pto regulating valve.

Fluid pressure for operation of pto clutch and differential lock on Models TW-20, TW-20 and TW-30 is provided by the low pressure oil pump discussed in paragraph 174. With engine running at 2200 rpm, operating pressure should be 185-235 psi.

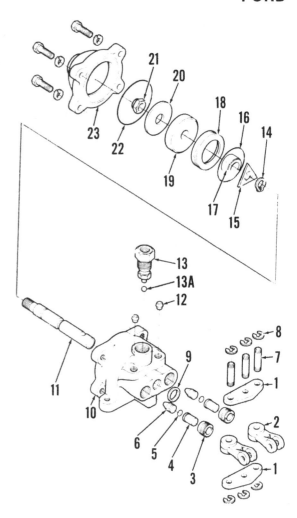

Fig. 185—Exploded view of brake valve used on Model TW-30.

1. Rocker arm
2. Control rod link
3. Valve retainer
4. Valve plunger
5. "O" ring
6. Valve needle
7. Retaining pin
8. Retaining ring
9. Seal
10. Housing
11. Valve spool
12. Seat
13. Check valve
13A. Ball
14. Snap ring
15. Spacer
16. "O" ring
17. Spacer sleeve
18. Actuating piston
19. Seal retainer
20. Washer
21. Lock nut
22. "O" ring
23. Valve housing cap

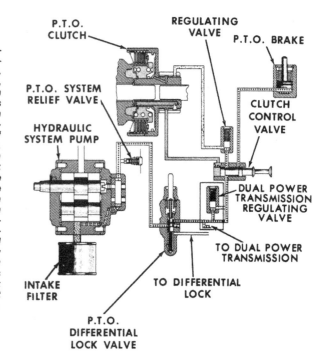

Fig. 186—Schematic diagram of the pto/differential lock hydraulic system. (All models except TW-10, TW-20 and TW-30). When differential lock valve is operated, line to pto clutch control valve is momentarily closed to prevent loss of pressure in clutch circuit, then valve reopens when pressure in differential lock circuit is built up. Pressure is directed to pto brake cylinder to hold clutch housing from turning when pto control valve is in disengaged position. Hydraulic pressure to Dual Power unit is controlled by Dual Power regulating valve. Hydraulic line is connected between lock valve on models not equipped with Dual Power.

When pto clutch valve is in "STOP" position, fluid pressure is directed to pto brake cylinder which applies a disc type brake on TW-10, TW-20 and TW-30 models or band type brake on all other models. The brake stops the pto clutch housing from turning due to fluid drag on the clutch discs. Brake is released by spring pressure whenever clutch is applied or engine is stopped.

CAUTION: As reduction gears are supported by the pto output shaft, do not operate tractor without either the 540 or 1000 rpm output shaft in place. Severe damage may be caused if driven gears are rotated under engine power if not supported by output shaft.

TROUBLESHOOTING

212. OPERATIONAL CHECKS. When troubleshooting pto malfunctions, refer to the following:

A. PTO CLUTCH WILL NOT ENGAGE. Trouble could be caused by:
1. Failure of hydraulic pump.
2. Control lever broken, disconnected or improperly adjusted.
3. Pto actuating arm in center housing disengaged from control valve spool.
4. Extremely cold oil in rear axle center housing.
5. Pto/differential lock relief valve ball or spring missing in hydraulic pump (all but TW models).

B. PTO STOPS UNDER LOAD. Trouble could be caused by:
1. Pto system pressure too low due to worn pump or leaking pump relief valve.
2. Pto clutch plates worn or damaged.

C. PTO WON'T STOP. Trouble could be caused by:
1. Extremely cold oil in rear axle center housing.

2. Pto control linkage disconnected or improperly adjusted.
3. Pto actuating arm in center housing disengaged from control valve spool.
4. Pto clutch plates warped or damaged.
5. Excessive wear of pto clutch brake.
6. Cut, broken or missing "O" ring on pto brake piston.

D. PTO SYSTEM PRESSURE TOO HIGH. Trouble could be caused by:
1. Extremely cold oil in rear axle center housing.
2. Low pressure hydraulic system control valve stuck closed or spring too stiff.

E. PTO SYSTEM PRESSURE TOO LOW, (WARNING LIGHT ILLUMINATES). Trouble could be caused by:
1. Low pressure hydraulic operating and or lube oil system pressure control valve spool(s) stuck open or spring(s) broken or weak.
2. Brake priority valve spool stuck closed or spring too stiff (Model TW-30 only).
3. Low pressure hydraulic system oil pump mounting bolts loose or rotors worn.
4. Low pressure hydraulic system oil pump or oil pick up tube seal damaged.
5. Hydraulic line leading to pto control valve leaking or damaged.
6. Cut or broken o-rings on pto clutch piston.
7. Pto regulating valve stuck open.
8. Worn or broken pto seal rings and/or pto support.
9. Leak in differential lock or Dual Power system.

213. PTO/DIFFERENTIAL LOCK SYSTEM PRESSURE. On all models except TW-10, TW-20 and TW-30, before checking pto/differential lock system pressure, operate tractor until oil in rear axle center housing is at

normal operating temperature. Stop engine and connect a 0-500 psi hydraulic gage at upper ¼-inch pipe plug opening in pump adapter plate as shown in Fig. 188. On models equipped with Dual Power, connect gage to tee fitting in hydraulic pump cover for Dual Power line. Start engine and move control lever back and forth between "start" and "stop" positions several times to purge air from system. With engine running at 700 rpm, check pressure gage reading with clutch engaged and disengaged. Pressure readings should be within the range of 135-165 psi. System pressure of near 165 psi is desirable for positive clutch engagement. A variation of more than 3 psi will indicate internal leakage in pto clutch circuit. While observing pressure gage, depress differential lock pedal. Pressure should drop momentarily then return to same reading. If pressure reading is lower with differential lock engaged, there is an internal leak in the differential lock circuit. Differential lock circuit pressure can be checked by connecting a pressure gage to lower ¼-inch pipe plug opening as indicated in Fig. 188. With engine running at 700 rpm and differential lock engaged, pressure reading should be 250-300 psi.

To check fluid pressure for pto and differential lock on TW-10 and TW-20 models, first operate tractor until oil in rear axle housing is at normal temperature. Stop engine and connect a 0-500 psi hydraulic gage to low oil pressure system tee fitting (3–Fig. 189) on pump adapter plate. Pressure readings on models not equipped with Dual Power

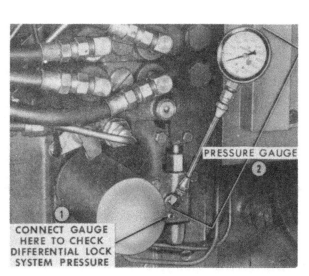

Fig. 188—Pressure gage installed to check pto clutch circuit pressure on all models except TW-10, TW-20 and TW-30. On Dual Power models, connect gage to Dual Power line tee in same location. Install gage in lower pipe plug hole as indicated to check differential lock system pressure. Refer to text for test procedure.

Fig. 189—To check pto and differential lock valve fluid pressure, connect a 0-500 psi gage to low oil system tee fitting (F) and refer to text. Model TW-30 is shown, however Models TW-10 and TW-20 are similar.

should be 185-220 psi with engine running at idle speed of 1000 rpm. On Dual Power models, pressure should be 185-220 psi at engine speed of 1000 rpm under each of the following conditions: Dual Power in underdrive, Dual Power in direct drive. With Dual Power in direct drive, engage the pto, with Dual Power in direct drive, engage the differential lock.

To check pto, differential lock, power brake and Dual Power system pressure on TW-30 models, first operate tractor until oil in rear axle housing is at normal operating temperature. Stop engine and connect a 0-500 psi hydraulic gage to low oil pressure system tee fitting (3–Fig. 189) on pump adapter plate. Pressure readings should be 185-235 psi under each of the following conditions with the engine idle speed set at 2200 rpm: with Dual Power in direct drive first apply left wheel brake then apply right wheel brake; apply both brakes at same time. With Dual Power in direct drive, engage pto then engage differential lock.

CONTROL LINKAGE ADJUSTMENT

215. To check adjustment of the pto clutch control cable on all models except TW-10, TW-20 and TW-30, refer to Fig. 190 and proceed as follows: Move pto control lever to full rearward "STOP" position; the actuating arm should snap to "OFF" position and the cable yoke pin be centered in slot of actuating arm. Then, move control lever fully forward to "START" position; the actuating arm should snap downward on 8700 and 9700 models or upward on all other models to "ON" position and the cable yoke pin should again be centered in slot in actuating lever. If the pin contacts end of slot in actuating lever in

either of the two positions, or the actuating lever will not "snap" into either the "ON" or "OFF" position, readjust control linkage cable as follows:

Disconnect lower end of cable from actuating arm and loosen bottom clip. Move control lever to full rearward "STOP" position and snap actuating arm upward on 8700 and 9700 models or downward on all other models to "OFF" position, then reconnect cable with pin centered in slot of actuating arm and tighten bottom clip with cable in this position. Recheck adjustment as previously described and if correct adjustment is not yet obtained, proceed as follows:

Unbolt and remove console from tractor fender.

NOTE: It will be necessary to remove all control lever handles so that console can be lifted off control linkage.

Loosen top cable clip and position control lever to dimension shown in Fig. 190. Position top clip so that it is 1/16-inch from cable seal on 8700 and 9700 or butts against cable seal on other models, tighten clip and reinstall console. Readjust lower cable stop, if necessary, as previously described.

To check and adjust pto clutch control rod on TW-10, TW-20 and TW-30 models, refer to Fig. 191 and with pto lever (3) in off position, adjust turnbuckle (T) so pto handle is 3¼ inches from front of slot (1).

R&R AND OVERHAUL

216. **PTO/DIFFERENTIAL LOCK HYDRAULIC PUMP.** Fluid pressure for pto and differential lock on Models TW-10, TW-20 and TW-30 is provided by the low oil pressure system pump as

noted in paragraph 174. Refer to paragraph 191 for pump removal.

Fluid pressure for pto and differential lock on all other models is provided by the hydraulic lift system pump. Refer to paragraphs 233 and 236 for pump removal and overhaul.

217. **PTO UPPER (CLUTCH OUTPUT) SHAFT AND REDUCTION GEARS.** To remove the pto clutch output shaft and drive gears, first remove hydraulic lift top link, where required, then refer to Fig. 192 and proceed as follows:

Unbolt the drive gear bearing retainer (top link bracket) (58) from rear axle center housing. Withdraw the retainer, along with drive gear (54), bearings (53 and 55) and shaft (33) as a unit. Remove the shaft from drive gear, taking care not to lose any shims (52) that may be located between gear and snap ring on rear end of shaft. Remove bearing retainer from rear bearing (55). Inspect gear teeth for chips and/or excessive wear.

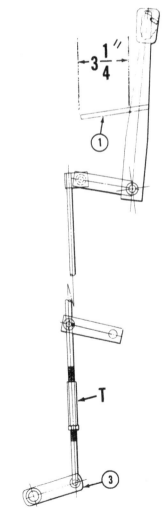

Fig. 191—Rotate turnbuckle (T) on Models TW-10, TW-20 and TW-30 to adjust pto control rod as outlined in text.

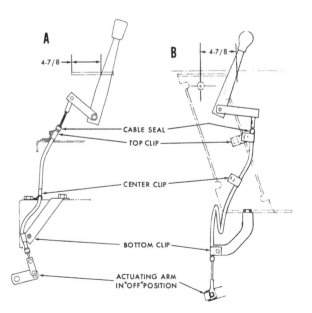

Fig. 190—View showing adjustment points for pto control linkage. Models 8700 and 9700 are shown in View "A" while 8000—9000—8600—9600 models are shown in View "B". Refer to text for procedure. See Fig. 191 for pto adjustment on Models TW-10, TW-20 and TW-30.

NOTE: Teeth on driven gears (61 and 62) may be inspected through drive gear opening in rear axle center housing; turn output shaft to rotate the gears.

Check bearings on drive gear and renew bearings if rough or worn; bearings may be removed using wedge type attachment with two-leg puller and step plate.

On dual speed models there are two driven reduction gears while there is only one driven reduction gear on 1000 rpm only models. To remove driven gear(s), first remove swinging drawbar and drain pto reduction gear housing (rear plug). Remove the cover from bottom of rear axle center housing. Remove pto output shaft after removing safety cap (72). Unbolt and remove output shaft bearing retainer (68), then support driven gear(s) (61 and 62) and remove bearing (64) and sleeve (63) assembly. The driven gear(s) can then be removed from bottom of rear axle center housing. Inspect needle bearing (59) in blind hole in rear axle center housing and remove if worn or damaged. Rear bearing (64) can be pressed from

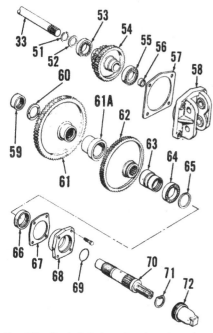

Fig. 192—Exploded view showing pto reduction gears, clutch output shaft and pto output shaft. Refer to Fig. 193 for view of both the 540 and 1000 rpm output shafts used on dual speed models. On 1000 rpm models only, drive gear (54) is a single gear and spacer (61A) is used in place of 540 rpm driven gear (61).

33. Clutch output shaft	61A. Spacer
51. Snap ring	62. 1000 rpm driven gear
52. Shims	63. Bearing sleeve
53. Front bearing	64. Ball bearing
54. Drive gear	65. Snap ring
55. Rear bearing	66. Oil seal
56. Gear bore plug	67. Gasket
57. Gasket	68. Bearing retainer
58. Bearing retainer	69. "O" ring
59. Needle roller bearing	70. Output shaft
60. Thrust washer (3)	71. Snap ring
61. 540 rpm driven gear	72. Safety cap

sleeve (63) after removing snap ring (65).

To reassemble, proceed as follows: Use 1-7/8-inch O.D. sleeve to drive output shaft front bearing (59) into housing, lubricate bearing and stick thrust washer (60) into place using heavy grease. Press rear bearing (64) onto sleeve (63) and secure with snap ring. On dual speed models, pilot the 1000 rpm driven gear (62) onto hub of the 540 rpm gear (61) with a thrust washer (60) between them and position the gears in housing. On 1000 rpm only models, install a thrust washer in spacer (61A) and install spacer and driven gear (62) in housing so that concave side of gear is next to spacer. Insert a thrust washer between sleeve (63) and gear(s) and install sleeve (63) and rear bearing (64). Install new oil seal (66) in retainer (68), lubricate seal and install retainer with new gasket. Install appropriate output shaft (Fig. 193) and secure with snap ring (71—Fig. 192).

Install bearings (53 and 55) on drive gear (54), making sure that sealed side of front bearing (53) faces forward (away from gear). Be sure that cup plug (56) in rear bore of gear is tight and does not have a vent hole; renew plug if loose or if it is early type having a vent hole. If original drive shaft (33) and drive gear (54) are being reinstalled, place shims (52) on rear end of shaft as removed on disassembly and insert shaft in gear. Position retainer on rear drive gear bearing with new gasket (57), then install assembly in rear axle center housing and tighten retaining cap screws to a torque of 68-92 ft.-lbs. Fill pto gear box through fill plug opening on left rear of differential housing with approximately 4 quarts for 8000, 9000, 8600, and 9600 models or 6½ quarts for 8700 and 9700 models. On Models TW-10, TW-20 and TW-30, replenish transmission oil as outlined in paragraph 176. Recommended lubricant for all models is Ford ESN-M2C53-A.

If shaft and/or drive gears have been renewed, drive shaft end play should be checked and correct shim (52) thickness installed on assembly as outlined in following paragraph 218.

218. CLUTCH SHAFT END PLAY. End play of shaft (33—Fig. 192, 197 or 198) should be 0.001-0.029 inch and is controlled by shims (52—Fig. 192) placed between rear snap ring (51) and shoulder on pto drive gear (54) hub. If the hydraulic lift cover is removed, end play can be checked by measuring gap between snap ring and gear hub with a feeler gage and, if not within limits of 0.001-0.029 inch, the assembly should be removed as outlined in paragraph 217 and the correct thickness of shims (52)

installed to provide proper shaft end play. Make sure rear bearing retainer cap screws are tightened to a torque of 68-92 ft.-lbs. when making end play check.

If a new shaft (33) is being installed and the lift cover is not removed, check end play as follows: With the bearing retainer, drive gears and shaft removed as outlined in paragraph 217, install the assembly without a gasket and with total shim (52) thickness of approximately 0.100 inch. Install retainer cap screws and tighten snugly until snap ring on front end of shaft contacts pto clutch housing and there is no end play of shaft, clutch housing or hub, but do not force the retainer in with cap screws. Loosen and tighten alternate cap screws until gap is equal all around retainer, then measure resulting gap with feeler gage. Record the measurement and remove

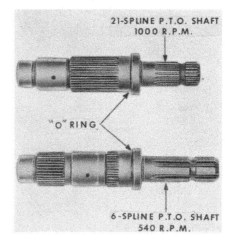

Fig. 193—Views showing 1000 rpm output shaft (top) and 540 rpm output shaft (bottom) used on dual speed models. Output shafts (1000 rpm) for single speed models are similar. Note location for sealing "O" ring (69—Fig. 192).

Fig. 194—View with hydraulic system pump removed showing pto clutch and drive hub. Clutch and hub can be removed out pump adapter plate opening after removing clutch output shaft reduction drive gears.

retainer, drive gear and shaft assembly. Withdraw shaft from drive gear and measure test shim pack thickness with micrometer. Subtract measured gap from test shim pack thickness; then, select shims of total thickness of 0.001 to 0.029 inch **less** than the resulting value of reassembly. For example, if test shim pack thickness was 0.098 inch and gap between retainer and housing was 0.043 inch, resulting value would be 0.055 inch. Desired shim pack thickness would be 0.001 to 0.029 inch less than this value (0.055) or 0.026 to 0.054 inch. Reinstall

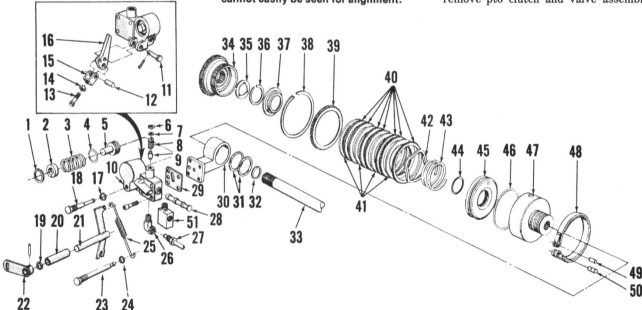

Fig. 195—After removing clutch assembly, lift off valve assembly with valve support and brake band.

shaft gear and retainer assembly with new gasket and selected shim pack thickness as outlined in paragraph 217 and recheck end play.

219. PTO OUTPUT SHAFT. The power take-off output shaft (70—Fig. 192) can be withdrawn after removing cap (72) and retaining snap ring (71).

When installing output shaft, be sure the sealing "O" ring (69) is in proper groove as shown in Fig. 193 and that bearing surfaces of shaft are not scored or burred.

220. R&R PTO CLUTCH AND VALVE ASSEMBLY (TW-10, TW-20 AND TW-30). To remove pto clutch and valve assembly, separate tractor between transmission and rear axle center housing as outlined in paragraph 188. To renew pto clutch or valve assembly, proceed as follows:

NOTE: Removal of hydraulic lift cover is optional. Pto clutch and valve assembly can be removed and installed with lift cover in place, however, installation is difficult because control valve spool (81—Fig. 198) and fork cannot easily be seen for alignment.

Remove hydraulic lift cover, if desired, from center housing, then remove pto drive hub (34—Fig. 198) and thrust washer (35). Disconnect pto lube line (L—Fig. 198A) from fitting on inside left wall of center housing, then disconnect high pressure line (H). Remove shoulder bolt (B—Fig. 200), over center spring (S), locating pin (82) and stop pin (84) from pto control linkage. Turn control valve lever so shift fork will be released from control valve spool (81—Fig. 198).

On TW-10 and TW-20 models, slide pto valve assembly forward and remove through center housing.

On TW-30 models, remove pto upper shaft and gears while supporting pto assembly, then slide pto assembly past hydraulic pump gear (G—Fig. 198A) and out center housing opening.

To reinstall pto clutch assembly, reverse removal procedure. Finger tighten locating pin (82—Fig. 200), connect pto lube line (L—Fig. 198A) and high pressure line (H), then tighten locating pin (82—Fig. 200) to a torque of 70-95 ft.-lbs. On TW-30 models, tighten pto upper shaft and gears bearing retainer to a torque of 68-92 ft.-lbs.

221. R&R PTO CLUTCH AND VALVE ASSEMBLY. (All Other Models). On 8000 models, it is possible to remove pto clutch and valve assembly

Fig. 197—Exploded view of clutch assembly, clutch brake mechanism, control valve assembly and actuating linkage on all models except TW-10, TW-20 and TW-30. Pin (18) is inserted through left side of rear axle center housing and pilots in hole in clutch valve housing (10) to keep the clutch assembly from rotating. Pin (23), also inserted from outer left side of center housing, acts as stop for control arm (21) and also as attaching point for the over-center spring (25). Inset shows view of valve housing from clutch side and the pto brake lever and related parts. On transmissions with Dual Power, regulator valve (51) is used in place of elbow (26). Refer to Fig. 198 for exploded view of pto assembly on Models TW-10, TW-20 and TW-30.

1. Snap ring	10. Valve housing	19. "O" ring	28. Control valve spool	37. Spring retainer
2. Piston guide	11. Lever pin	20. Bushing	29. Gasket	38. Snap ring
3. Piston return spring	12. Clevis pin	21. Actuating arm	30. Valve support	39. Pressure plate
4. "O" ring	13. Adjusting screw	22. Actuating lever	31. Sealing rings	40. External spline
5. Pto brake piston	14. Locknut	23. Stop pin	32. Snap ring	discs
6. Snap ring	15. Clevis	24. "O" ring	33. Clutch output shaft	41. Internal spline discs
7. Washer	16. Brake lever	25. Over-center spring	34. Clutch drive hub	42. Feathering spring
8. Valve spring	17. "O" ring	26. Elbow fitting	35. Thrust washer	43. Piston return spring
9. Regulating valve	18. Locating pin	27. Pressure line	36. Snap ring	44. "O" ring, inner

45. Piston	
46. "O" ring, outer	
47. Clutch housing	
48. Brake band	
49. Band to arm pin	
50. Band to clevis pin	
51. Dual Power	
regulator valve	

through opening in rear axle center housing after the hydraulic pump and adapter plate have been removed. On all other models, it is necessary to separate tractor to remove pto clutch and valve assembly due to interference with other parts. To remove pto clutch, proceed as follows:

On 8000 models, remove the adapter plate and hydraulic pump assembly from right side of rear axle center housing as outlined in paragraph 233. On all other models, separate tractor between transmission and rear axle center housing as outlined in paragraph 187 or 188. Remove pto upper (clutch output) shaft as indicated in paragraph 217. Remove hydraulic pump intake tube, filter screen and disconnect pto pressure line from inside of rear axle center housing. On models with Dual Power, remove Dual Power regulating valve from wall of center housing. On all except 8000 models, remove locating pin (18—Fig. 197), and remove pto clutch assembly.

On 8000 models, remove locating pin (18), slide clutch assembly rearward off input shaft and remove pto clutch assembly through pump adapter plate opening. Care should be taken not to lose thrust washer located between clutch and drive hub if they are separated during removal.

To reinstall pto clutch on 8000 models place clutch drive hub on transmission pto shaft and using heavy grease, stick thrust washer to clutch housing hub so that prongs on thrust washer enter holes in hub of housing (47—Fig. 197). Position clutch and valve assembly on rear of drive hub and rotate hub back and forth to align splines of hub with splines in pto clutch discs. With clutch fully forward, insert locating pin with new "O" ring in left side of rear axle center housing, making sure pin enters hole in pto valve housing, and tighten pin finger tight. Reconnect pto pressure line, reinstall pto upper shaft as in paragraph 217 and tighten locating pin to a torque of 70-95 ft.-lbs. Reinstall hydraulic pump as outlined in paragraph 234 or 235.

To reinstall pto clutch on all except 8000 models, install pto output shaft and drive gear as outlined in paragraph 217. Install thrust washer (35—Fig. 197) using heavy grease so that prongs of thrust washer enter holes in hub of housing (47). Install clutch assembly and control valve on pto output shaft and insert locating pin (18) finger tight to hold assembly in place. Connect control valve hydraulic line and install Dual Power regulating valve, on models so equipped, so that hydraulic connection is facing rearward. Connect Dual Power hydraulic line. Connect rear axle housing and transmission as outlined in para-

graph 187 or 188. Tighten locating pin (18) to 70-95 ft.-lbs.

222. OVERHAUL PTO CLUTCH AND VALVE ASSEMBLY (TW MODELS). After removing pto clutch assembly as outlined in paragraph 220, refer to Fig. 198 and remove brake pad bolts, outer pad (55), spacers (50) and inner pad (54). Remove clutch pack from support and bearing assembly (30). Remove clutch rotor (52) then unbolt and remove valve assembly (80) from support and bearing assembly. Remove brake piston (57) with compressed air. Remove snap ring (38), pressure plate (39), clutch discs (40 & 41) and feathering spring (42) from housing (47). Using Ford tool No. 1312, compress piston return spring (43) as shown in Fig. 199 and remove snap ring (36). Use compressed air to unseat piston (45–Fig. 198). Inspect and renew any components that have excessive wear or damage.

Refer to Fig. 198 for view of control valve. Overhaul is evident after inspection of valve. If regulating valve (77—Fig. 198) is worn or damaged, it must be replaced as a complete unit.

NOTE: Lubricate all parts with hydraulic fluid/differential lubricant before reassembly.

Install a new "O" ring (58) on piston (57), then install feathering spring (42), clutch discs (40 & 41) and pressure plate (39) into housing (47). Reinstall snap ring (38) after compressing clutch spring as

illustrated in Fig. 199.

Assemble brake rotor (52) on clutch housing (47), then attach clutch pack to support assembly (30). Install brake pads (53 & 55) and brake pad bolts. Tighten brake pad bolts to a torque of 27-37 ft.-lbs. Attach pto valve assembly (80) to support (30) and tighten bolts to a torque of 15-21 ft.-lbs.

223. OVERHAUL PTO CLUTCH AND VALVE ASSEMBLY (All Other Models). With pto clutch and valve assembly removed as outlined in paragraph 221, place unit on bench, front face down, and remove the brake band, valve support and valve assembly as a unit from the clutch housing as shown in Fig. 195. Withdraw control valve spool from valve body and remove the pin (11—Fig. 197) retaining band and lever (16) to valve body. Place valve body in a press and using suitable sleeve, compress guide (2) and spring (3) into valve body, then remove snap ring (1). Release spring pressure and remove the guide, spring and piston (5). Remove snap ring (6), flat washer (7), spring (8) and pto regulating valve (9) from bore in valve body.

Remove snap ring (38) from clutch housing (47), then remove the pressure plate (39), clutch discs (40 and 41) and feathering spring (42) from housing. Remove the cast iron sealing rings (31) from rear hub of housing. Refer to Fig. 199, place housing in a press and using

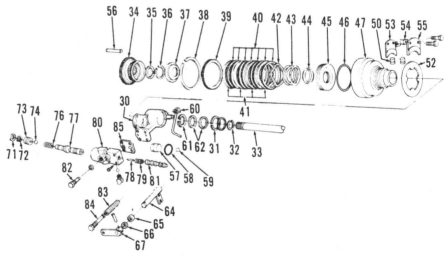

Fig. 198—Exploded view of pto clutch, valve, shaft and related components used on Models TW-10, TW-20 and TW-30. Refer to Fig. 200 for view of shift mechanism.

30. Support	43. Piston return spring	59. Ball (5/16 in.)	74. Valve seal
31. Seal	44. "O" ring, inner	60. Elbow	76. Return spring
32. Snap ring	45. Piston	61. Snap ring	77. Pressure regulator
33. Shaft	46. "O" ring, outer	62. Support bearing	valve
34. Hub	47. Clutch housing	63. Ball	78. Push rod
35. Thrust washer	50. Spacers	64. Shift fork	79. Valve spring
36. Snap ring	52. Brake rotor	65. Bushing	80. Valve assembly
37. Spring retainer	53. Brake pad	66. Seal	81. Control valve
38. Snap ring	54. Brake pad spring	67. Shift lever	82. Locating pin
39. Pressure plate	55. Brake pad	71. Snap ring	83. Spring
40. External spline discs	56. Brake pin	72. "O" ring	84. Stop pin
41. Internal spline discs	57. Piston	73. Valve guide	85. Gasket
42. Feathering spring	58. Seal		

Fig. 198A—View of rear axle center housing on Models TW-10, TW-20 and TW-30. Refer to paragraph 220 for pto clutch removal.

C. Coupler
G. Hydraulic pump gear
H. High pressure line
L. Lube line
V. Pto valve

Ford tool No. 1312 or equivalent, compress the piston return springs (43—Fig. 197) and remove retaining snap ring (36). Gradually release spring pressure and remove the spring retainer (37) and spring (43). Use air pressure through clutch piston port located between the seal ring grooves on housing rear hub to remove piston (45) from housing.

Carefully clean and inspect all parts and remove the "O" rings (44 and 46) from clutch piston and the "O" ring (4) from brake piston. Renew all excessively worn, scored or otherwise damaged

Fig. 199—Using Ford tool No. 1312 to compress piston return spring (43—Fig. 197 or 198), so snap ring (36) can be removed.

36. Retaining ring   T. Tool No. 1312
I. Pto clutch housing   P. Arbor press

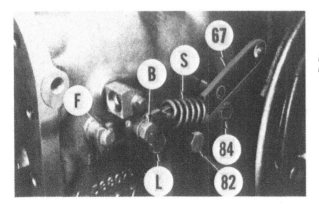

Fig. 200—View of shift mechanism for TW-10, TW-20 and TW-30 models.

B. Shoulder bolt
L. Locating bolt
S. Over center spring
67. Shift lever
82. Locating pin
84. Stop pin

parts not suitable for further service. Lubricate all parts with hydraulic fluid/differential lubricant and using new "O" rings (4, 44 and 46) and cast iron rings (31), reassemble by reversing disassembly procedure. After placing the brake band, valve support and valve assembly back on the pto clutch housing, loosen locknut (14), tighten band adjusting screw (13) to a torque of 9-11 in.-lbs., back screw off 1½ turns, then tighten locknut.

HYDRAULIC LIFT SYSTEM

The hydraulic lift system incorporates automatic draft control, automatic position control and pump flow (rate of lift) control. Provision is also made for installation of optional remote cylinder control valves. Tractor is also available with a hydraulic system for remote cylinder operation only. Fluid for the system is common with differential and final drive lubricant. Refer to paragraph 197 for fluid type and quantity. Hydraulic power is supplied by a gear type pump that is mounted in right side of rear axle center housing. Pump is driven by gear machined on pto clutch input hub. The system is protected by a wire mesh screen filter on intake side of pump on all

models except all 8700, 9700, TW-10, TW-20, TW-30 and late 8600 and 9600, which have a renewable intake filter. The pressure (outlet) side of system is protected by a renewable filter on pump adapter plate on all models.

HYDRAULIC FLUID

226. Fluid used in transmission and rear axle center housing and hydraulic system must be compatible with the wet type disc brakes and differential lock clutch. Refer to paragraph 176 for recommended fluid type and for transmission and rear axle capacity. Rear axle center housing should be drained and refilled with new lubricant after each 1200 hours of service or yearly, whichever occurs first. Before draining, be sure that 3-point lift arms are lowered and any remote cylinders are retracted. Maintain fluid level at oil level plug opening in left side of rear axle center housing at front side of left axle housing flange on 8000-9000 models. All other models have a dipstick located by the left side of the lift cover. Check fluid with the tractor level, with 3-point hitch lift arms in raised position and any remote cylinders extended.

HYDRAULIC SYSTEM FILTERS

227. The external throw-away type filter on pump adapter cover at right side of rear axle center housing should be renewed after each 300 hours of service. If tractor is equipped with external throwaway type intake filter, filter should be renewed after each 300 hours of service. External intake filter is located just below pump adapter plate on 8700, 9700, TW-10, TW-20 and TW-30 models or on left side of center housing on all other models.

On all models without external intake filter, the wire screen filter on pump intake tube should be cleaned each time the hydraulic system is being serviced and is accessible after removing hydraulic pump, lift cover or splitting tractor between transmission and rear axle center housing.

TROUBLESHOOTING

228. When troubleshooting problems are encountered with the hydraulic lift system, refer to the following malfunctions and their possible causes:

A. FAILURE TO LIFT UNDER ALL CONDITIONS. Could be caused by:
1. Low oil level in rear axle center housing.
2. Flow control valve stuck in open position.
3. Faulty hydraulic pump.

4. Control linkage damaged or disconnected.
5. Leak in pump suction (intake) pipe.
6. Sticking exhaust control valve (34—Fig. 219).
7. Sticking or improperly installed exhaust pressure valve (36—Fig. 219).

B. FAILURE TO LIFT UNDER LOAD. Could be caused by:
1. Faulty hydraulic pump.
2. System pressure relief valve opening at too low a pressure or faulty relief valve.
3. Safety relief valve faulty.
4. Exhaust pressure valve (36—Fig. 219) sticking.
5. Clogged pump inlet screen.

C. OCCASIONALLY FAILS TO LIFT NOT DUE TO OVERLOADING. Could be caused by:
1. Flow control valve spool sticking occasionally.
2. Exhaust pressure valve (36—Fig. 219) sticking occasionally or valve installed backwards.
3. Lift control valve improperly adjusted.

D. EXCESSIVE LIFT CORRECTIONS (WILL NOT HOLD LOAD AFTER STOPPING ENGINE). Could be caused by:
1. Drop valve poppet or poppet ball leaking.
2. Lift piston seal leaking.
3. "O" ring seals between control valve body and lift cylinder leaking.
4. Safety valve leaking.

E. LIFTS TOO SLOWLY. Could be caused by:
1. Flow control valve stuck in slow lift position.
2. Flow control valve scored or not properly fit in valve bore.
3. Faulty pump.

F. OVER CORRECTS IN DRAFT CONTROL. Could be caused by:
1. Flow control valve adjusting knob not properly set.
2. Lift control valve not properly adjusted.
3. Binding draft control linkage or lower lift hanger binding in bushings.
4. Flow control override linkage binding and not releasing variable flow valve from fast flow position.

G. SYSTEM DOES NOT RESPOND TO CHANGES IN DRAFT OF IMPLEMENT. Could be caused by:
1. Draft control linkage out of adjustment.

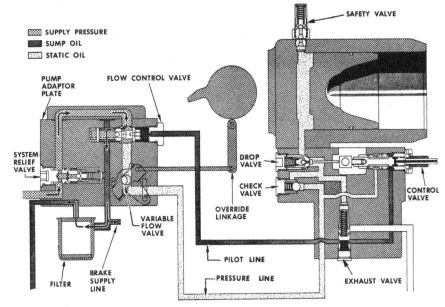

SUPPLY PRESSURE
SUMP OIL
STATIC OIL

Fig. 203—Schematic diagram of hydraulic system with control valve in neutral position. Flow control valve returns oil to sump through filter and brake supply line and maintains static pressure in pressure line with the control valve in neutral or lowering position. When the control valve is placed in raising position, pressure is directed to pilot line, closing the flow control valve, thus directing fluid flow to pressure line and lift cylinder. When control valve is placed in lowering position, drop valve is forced open allowing fluid in cylinder to return to sump past the exhaust valve. Pressure in pilot line holds exhaust valve closed when line is pressurized with control valve in raising position. Variable flow valve restricts flow to lift cylinder pressure line, causing flow control valve to direct part of pump flow to sump depending upon position of variable flow valve. Thus, rate of lift can be varied by varying position of variable flow valve with control (F—Fig. 209) on console. When control valve lever is placed in full raise position, override linkage returns variable flow valve to full flow position.

Fig. 204—Selector lever (S—Fig. 209) controls position of selector shaft which moves rear end of actuating rod attached to control valve lever to position control (top view) or draft control (bottom view) position in position control arm. Also, detent positions between full position control and full draft control provide varying degrees of draft control. Refer to exploded view of detent mechanism in Fig. 223.

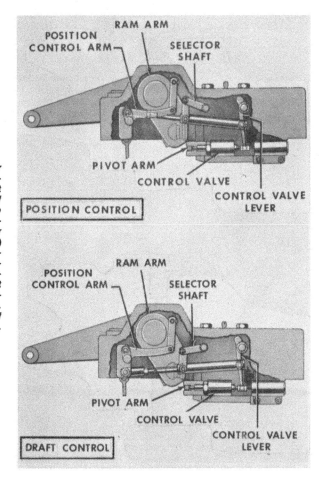

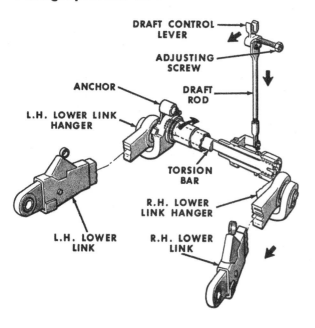

Fig. 205—Cut-away view showing draft control torsion bar and control linkage. Draft on lower links rotates lift link hangers against tension of torsion bar, thus rotating draft control lever. When selector lever (Fig. 204) is in draft control position, control valve position is controlled by draft on lower links.

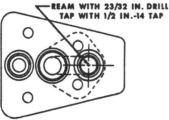

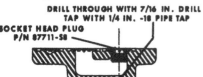

Fig. 206—On models not equipped with remote control valve, manifold plate must be modified as shown above to permit checking hydraulic system relief pressure. Refer also to Fig. 207.

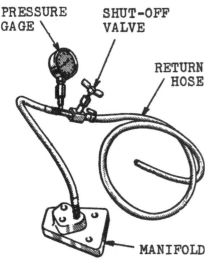

Fig. 207—With manifold on top of pump adapter plate modified as shown in Fig. 206, connect pressure gage and shut-off valve hose to manifold and insert return hose through filler plug opening in lift cover.

2. Linkage disconnected; roller disengaged from draft sensing yoke.
3. Lower link hanger stop not properly adjusted.
4. Reaction too slow due to flow control valve in slow flow position.

HYDRAULIC PRESSURE CHECK

229. Hydraulic system relief pressure should be 2500-2600 psi for TW-10, TW-20 and TW-30 models and 2450-2550 for all other models. Pressure can be adjusted by adding or removing shims (31—Fig. 216 or 3—Fig. 217), between relief valve body cap and relief valve spring.

Shims are available in thicknesses of 0.010 and 0.026 inch. Maximum allowable shim pack thickness is 0.080 inch. Adding one 0.010 inch thick shim will increase pressure approximately 42 psi on TW-10, TW-20 and TW-30 models and approximately 100 psi on all other models. If adding shims does not increase relief pressure or system pressure remains below desired pressure with maximum thickness (0.080

Fig. 208—Eccentric stop on left side of rear axle center housing is adjusted to provide a gap of 0.33-0.34 inch between stop and left hand lift link hanger. Hanger stop on torsion bar hanger is non-adjustable.

inch) shims installed, a worn or faulty pump should be suspected.

If equipped with remote control valves, connect pressure test gage at remote quick disconnect coupling. If not equipped with remote control valves, modify the manifold plate on top of pump adapter plate as shown in Fig. 206 and connect pressure test gage as shown in Fig. 207. Insert return hose in rear axle center housing filler plug opening, open the shut-off valve and start engine. With engine running at 700 rpm, close shut-off valve only long enough to observe pressure reading. After checking system pressure with modified manifold installed, either install new manifold or remove internal ½-inch pipe plug and close external opening with ¼-inch pipe plug.

ADJUSTMENTS

NOTE: The adjustments outlined in following paragraph 230 through 233 are those that can be made externally; internal adjustments are outlined in reassembly procedure in paragraphs outlining overhaul of the system components. Inability to make system function properly by external adjustment will indicate need of system overhaul and/or internal adjustment.

230. **LOWER LINK HANGER STOP.** Refer to Fig. 208; gap between left hand lower link hanger and eccentric stop should be 0.33-0.34 inch (⅓-inch). To adjust gap, loosen cap screw and turn eccentric, then retighten cap screw to a torque of 50-60 ft.-lbs.

231. **CONTROL LEVER LINKAGE ADJUSTMENT.** The lift control lever (L—Fig. 209) should move the control lever (6—Fig. 210) on lift cover through full range of travel without the lever (L—Fig. 209) contacting either end of slot in console. If the control lever can be moved against end of slot, adjust link (1—Fig. 210) as follows: Disconnect lower end of link (1) from lever (6) and

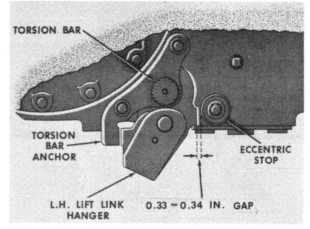

move end of lever down until it pushes override link (5) as far as it will go. Move lift control lever (L—Fig. 209) rearward so that it is 1/8 to 1/4-inch from rear end of slot in console. Then, turn lower end of link (1—Fig. 210) to shorten or lengthen link as necessary so that it can be reconnected to lever (6) without moving either the control lever or lever (6).

After adjusting and reconnecting link (1), check to be sure that control lever can be moved through full range of travel without contacting either end of slot in console.

232. POSITION CONTROL ADJUSTMENT. Place selector lever (S—Fig. 209) in position control (fully rearward). With control lever adjusted as outlined in paragraph 231, start engine and move control lever (L) fully forward; the lift linkage should then move to fully lowered position. Slowly move control lever rearward; lift linkage should start to raise when lever is 1/2 to 1 inch away from front end of slot in console. If lift arms do not go to fully lowered position with control lever fully forward, or if lift arms do not start to raise until after lever is more than 1 inch away from front end of slot in console, adjust position control pivot as follows:

Position the control lever so that it is 1/2 to 1 inch from front end of slot in console. Refer to Fig. 211 and loosen locknut (N1) on position control adjusting pivot (P) while holding pivot with Allen wrench. Slowly turn position control pivot until lift arms just start to raise without moving control lever, then tighten locknut to a torque of 70-80 ft.-lbs. Recheck position control by moving control lever fully forward, allow lift arms to reach fully lowered position, then move lever slowly rearward until

lift arms start to raise. Readjust position control pivot (P) if control lever position at time lift arms start to raise is not as stated.

232A. DRAFT CONTROL ADJUSTMENT. After adjusting control lever linkage as in paragraph 231 and position control pivot as in paragraph 232, check draft control adjustment as follows: Place selector lever (S—Fig. 209) in draft control position (fully forward). With engine running, move control lever (L) forward until lift arms are lowered, then slowly move lever towards rear of slot in console. When the rear side of lever is 1/2 to 1 inch away from rear end of slot, the lift arms should start to raise. If the lift arms do not start to raise with control lever in this position, refer to Fig. 211 and loosen locknut (N2) on draft control pivot (D) while holding pivot with Allen wrench. With the control lever 1/2 to 1 inch away from end of slot in console, slowly turn draft control pivot until lift arms just start to raise, then tighten locknut to a torque of 120-140 ft.-lbs. Recheck draft control adjustment and readjust pivot if necessary.

233. R&R PUMP AND ADAPTER PLATE. To remove hydraulic pump and adapter plate from right side of rear axle center housing, proceed as follows: Drain oil from rear axle center housing.

Remove access panel from right fender on early models and thoroughly clean adapter plate and surrounding area. Disconnect control rod or cable from flow control valve arm at adapter and pull up out of way.

If equipped with remote control valves, remove remote control valve access panel then back out manifold and remote control valve attaching bolts until they are free of pump adapter plate. Raise remote control valve stack slightly to disengage from pump adapter plate. Wire valve stack to adjacent tractor components as required to hold in place above pump cover. Remove inlet manifold (26—Fig. 212).

Refer to Fig. 212. Remove the hydraulic pressure tube and the pilot pressure tube connecting adapter plate to lift cover. Disconnect the flow control override spring and link from flow control valve inner arm. Remove cotter pin connecting differential lock pedal to valve and pull pedal up out of way.

On TW-10 and TW-20 models, disconnect pto/differential lock valve pressure line from pump adapter plate and at tee fitting on top right of transmission. It may be necessary to remove transmission access panel on cab floor. On TW-30 models, disconnect pto and differential lock pressure line from pump adapter plate and valve assembly. Disconnect pto pressure line at rgulating valve.

On early models which use hydraulic oil instead of brake fluid in brakes, disconnect brake fluid lines from adapter plate and push lines downward out of way. If necessary to renew hydraulic system filter, unscrew and discard filter at this time. Remove cap screws around outside edge of adapter plate and pry plate and pump assembly from rear axle center housing.

234. If the pump intake tube and internal intake filter have not been removed, carefully reinstall pump and

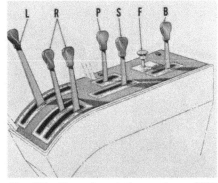

Fig. 209—View of control console on 8000, 8600, 9000 and 9600 models. Models 8700, 9700, TW-10, TW-20 and TW-30 are similar but positions of pto and selector control levers are reversed. Transmission brake lever (B) is not used on 8600, 8700, 9600, 9700, TW-10, TW-20 or TW-30.

B. Transmission brake lever	R. Remote control levers
F. Flow control knob	S. Selector (draft/ position) control lever
L. Lift control knob	
P. Pto control lever	

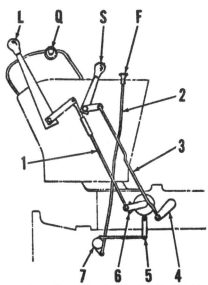

Fig. 210—View showing hydraulic lift lever to lift cover linkage. Lift control link (1) is adjustable. Quadrant stop (Q) is adjusted so that lever may be returned to desired setting after raising implement.

F. Flow control knob	3. Selector link
L. Lift control lever	4. Selector shaft arm
Q. Quadrant stop	5. Override linkage
S. Selector lever	6. Lift control arm
1. Control lever link	7. Variable flow valve arm
2. Flow control link	

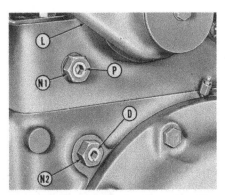

Fig. 211—View showing location of position control eccentric pin (P) and draft control eccentric pin (D) at right rear corner of lift cover and rear axle center housing. Lift shaft arm is (L).

adapter plate (use new gasket between plate and center housing) by reversing removal procedure and observing the following: Take care to be sure that pump opening fits over the intake pipe and install, but do not tighten, adapter plate retaining cap screws. Reconnect brake lines (if so equipped) and reinstall hydraulic pressure tube and pilot pressure tube. Tighten pilot pressure tube fittings to a torque of 20-25 ft.-lbs. Tighten pressure line to lift cover hollow bolt to a torque of 20-25 ft.-lbs. and tighten pressure tube to adapter plate nut to a torque of 50-55 ft.-lbs. Then, tighten adapter plate cap screws to a torque of 30-40 ft.-lbs. Reinstall the remote control valves (if so equipped) using new "O" rings between the valves, adapter plate and manifold and tighten retaining cap screws to a torque of 20-25 ft.-lbs. Reconnect all linkage and reinstall access plate in right fender. Install new external oil filter.

235. If the pump intake pipe and intake filter were removed, or pipe shifted in retainer bracket, reinstall pump and adapter plate separately as follows: Be sure pipe bracket cap screws are tight and separate pump from adapter plate. Position the pump on intake pipe and stick pump to adapter plate "O" rings in counterbores with heavy grease. Install adapter plate with new gasket, taking care not to dislodge the "O" rings, and install adapter plate to pump cap screws. Loosely install the adapter plate cap screws, then tighten pump retaining cap screws to a torque of 20-25 ft.-lbs. Complete remainder of reinstallation procedure as outlined in paragraph 234.

236. **OVERHAUL PUMP.** With pump and adapter plate removed as outlined in paragraph 233, remove pump retaining cap screws and separate pump from adapter plate. Refer to following sections.

237. **TW-10 AND TW-20.** Refer to Fig. 213 and remove cotter pin and nut securing pump drive gear then remove drive gear using suitable puller. Remove woodruff key from shaft. Remove 16 bolts and lock washers securing front and rear cover to body. Remove rear cover by lightly tapping pump shaft with a rubber mallet. After rear cover is removed, lightly tap gear shafts with mallet to remove front cover. Keep components in order as removed. Remove pressure loading rings (5) and seal rings (4) from front and rear covers (7 and 12).

Carefully clean and inspect all parts. Minor burrs and score marks can be removed using "O" grade emery paper and kerosene. Check gear track wear in pump body. Maximum permissible gear track wear is 0.0025 inch. Maximum runout across gear face to tooth edge is 0.001 inch and width of gears must be within 0.002 inch of each other. Gear journals must be within 0.0005 inch of each other. If wear or scoring is extensive, it is usually more economical to renew pump rather than attempt repairs.

Using new parts as necessary, reassemble pump using new seal and sealing rings by reversing disassembly procedure and observing the following:

Lightly lubricate each component with a high temperature grease to protect pump from heat damage during initial start up. Place front and rear bearings (11) over drive/driven gears (10) making sure recess in each bearing is against gear faces and that relief grooves will be on pump outlet side. Install gears and bearings in pump body (3). Pack shaft seal cavity with high temperature grease. Install front cover (7), then tighten front cover bolts to a torque of 65-70 ft.-lbs. Install woodruff key, drive gear and retaining nut. Tighten retaining nut to a torque of 40-55 ft.-lbs. Install pump as outlined in paragraph 235.

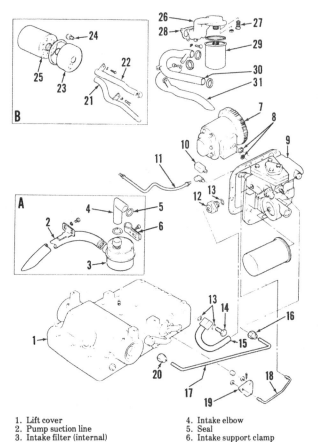

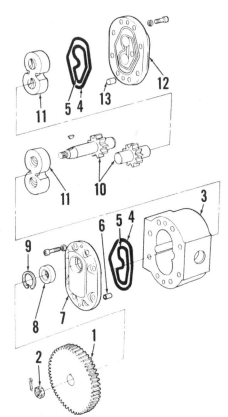

Fig. 212—View showing connections between lift cover and pump adapter plate. Pump (7) mounts on adapter plate (9). Flare nut (20) connects pilot line to pilot pressure sleeve (18—Fig. 221) and hollow bolt (14) connects system pressure tube (15) to pressure sleeve (10—Fig. 221). Components (26 through 31) are used on Models 8700, 9700, TW-10, TW-20 and TW-30; items in insets (A or B) are used on other models.

7. Pump assembly
8. Seal rings
9. Pump adapter plate
10. Dual Power valve
11. Pressure line
12. Pressure line fitting
13. "O" rings
14. Hollow bolt
15. Pressure line
16. Pilot line fitting
17. Pilot line
18. Override link
19. Override arm
20. Pilot line fitting
21. Pickup tube
22. Pump inlet
23. Manifold
24. Valve
25. External inlet filter
26. Manifold
27. Valve
28. Gasket
29. External inlet filter
30. Pump inlet
31. Pickup tube

Fig. 213—Exploded view of hydraulic pump assembly used on TW-10 and TW-20 models.

1. Pump drive gear
2. Slotted nut
3. Pump body
4. Outer seal
5. Pressure seal
6. Hollow dowel
7. Front cover
8. Seal
9. Snap ring
10. Gear set
11. Bearing assembly
12. Rear cover
13. Solid dowel

1. Lift cover
2. Pump suction line
3. Intake filter (internal)
4. Intake elbow
5. Seal
6. Intake support clamp

238. TW-30. Refer to Fig. 214 and remove cotter pin and nut securing pump drive gear; remove drive gear (5) using a suitable puller. Remove woodruff key from shaft. Remove five bolt and lock washers holding pump together while noting location of shorter bolt. Remove rear cover (11) by lightly tapping pump shaft with a rubber mallet. After removing rear cover, lightly tap gear shafts with mallet to remove front cover (4). Keep components in order as removed. Minor burrs and score marks can be removed using "O" grade emery paper and kerosene. Check gear track wear in pump body. If excessively worn, replace complete pump. Maximum allowable runout across gear face-to-tooth edge should not

exceed 0.00035 inch. Face width of paired gears must be within 0.0005 inch of each other and journal size on each side of individual gear must be within 0.0005 inch of each other. Inspect remaining components and renew as necessary. Lightly lubricate components with high temperature grease to protect pump from heat damage during initial start up.

Install a black pressure pre-load seal (6) in front and rear covers (4 and 11). Install a blue pressure load seal (7) on top of black pressure pre-load seals. Install ring seal (10) in front and rear covers (4 and 11). Install wear plates (8) in covers with bronze side facing out. Note that diamond shaped holes in wear plates are placed over larger holes in

front and rear covers. Wear plates fit inside diameters of ring seals (10). Install drive/driven gear (2) in front cover (4) then install pump body (9) over gears making sure dowels enter dowel holes in cover. Install five bolts and washers then tighten to a torque of 35 ft.-lbs. Install woodruff key and pump drive gear. Install retaining nut and tighten to a torque of 40-55 ft.-lbs.; install new cotter key. Install pump as outlined in paragraph 235.

238A. ALL OTHER MODELS. Refer to exploded view of pump assembly in Fig. 215 and proceed as follows:

Straighten tab washer (2), remove nut (1) and washer, pull gear (3) from pump shaft and remove Woodruff key (13). Remove pto pump body (33) from pump rear cover (23) and remove bearing (29) and gears (30 and 31). Unbolt and remove rear cover (23) and remove drive coupling (16). Unbolt and remove front cover (8) and push the bearings (12 and 17) with gears (14 and 15) from pump body (11) as a unit. Remove snap ring (4) and seal (5) from front cover; discard the seal, sealing rings and "O" rings. Remove pto pump pressure relief valve plug (18), washer (19), spring (20) and ball (21).

Separate the bearings and pump gears, keeping the parts in relative position. Carefully clean and inspect all parts. Minor burrs and score marks can

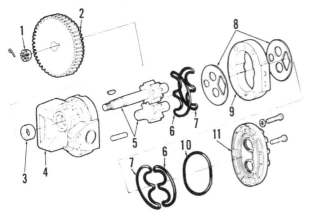

Fig. 214—Exploded view of hydraulic pump used on Model TW-30.

1. Slotted nut
2. Drive/driven gear
3. Shaft seal
4. Front cover
5. Gear set
6. Preload seal (black)
7. Pressure load seal (blue)
8. Wear plate
9. Pump body
10. Ring seal
11. Rear cover

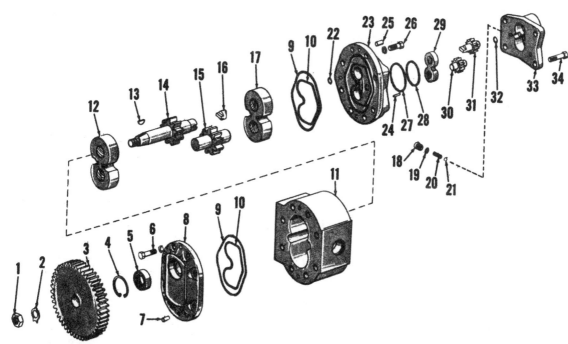

Fig. 215—Exploded view of hydraulic pump assembly for all models except TW-10, TW-20 and TW-30. Pto pump gear (31) is driven by hydraulic system pump gear (14) via coupling (16). Only one bearing (29) is used on pto pump section; body (33) serves as rear bearing.

1. Nut	7. Dowel pins	13. Woodruff key	19. Washer	25. Dowel pin	30. Pto pump idler gear
2. Locking washer	8. Front cover	14. Driven gear	20. Relief valve spring	26. Cap screws	31. Pto pump driven
3. Drive gear	9. Outer sealing ring	15. Idler gear	21. Pto relief valve ball	27. Outer sealing ring	gear
4. Snap ring	10. Inner sealing ring	16. Drive coupling	22. "O" ring	28. Inner sealing ring	32. "O" ring
5. Oil seal	11. Pump body	17. Rear bearing	23. Rear cover	29. Front pto pump	33. Pto pump body
6. Cap screws	12. Front bearing	18. Plug	24. Dowel pin	bearing	34. Cap screws

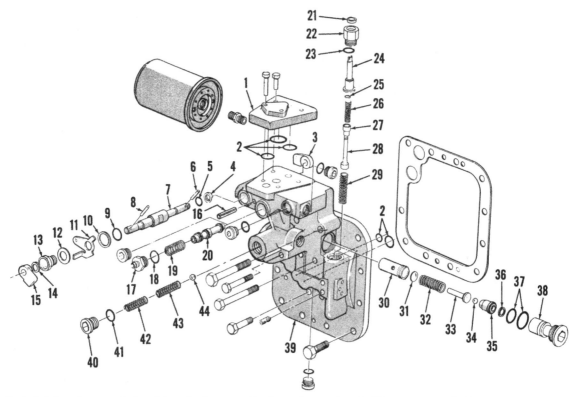

Fig. 216—Exploded view of pump adapter plate and valve assembly. On models equipped with remote control valves, valve units are mounted between adapter plate and manifold (1). Adapter plate (39) is serviced as complete plate and valve assembly only; all other parts are available separately. Flow control valve spool (20) is select fit to bore size in adapter plate. See Fig. 217 for exploded view of pressure relief system used on Models TW-10, TW-20 and TW-30.

1. Manifold	8. Roll pin	16. Roll pin
2. "O" rings	9. "O" ring	17. Connector
3. Variable flow valve	10. Friction washer	18. "O" ring
arm	11. Valve lever	19. Flow control valve
4. Back-up washer	12. Spring washer	spring
5. "O" ring	13. Gland	20. Flow control valve
6. Roll pin	14. Shims (as required)	21. Seal ring
7. Variable flow valve	15. Valve arm	22. Guide

23. "O" ring	29. Spring	37. "O" rings
24. Valve rod	30. Valve cap	38. Relief valve body
25. Snap ring	31. Shims	39. Pump adapter plate
26. Spring	32. Relief valve spring	40. Retainer
27. Pressure regulating	33. Spring guide	41. "O" ring
valve (pto)	34. Relief valve ball	42. Spring
28. Differential lock	35. Valve seat	43. Ball guide
valve	36. Seal ring	44. Check ball

be removed using "O" grade emery paper and kerosene. Check gear wear track in pump body, especially at intake side of pump. Maximum permissible gear track wear is 0.0025 inch. Maximum runout across gear face to tooth edge is 0.001 inch and width of gears must be within 0.002 inch of each other. Gear journals must be within 0.001 inch of each other on either side of each gear. If wear or scoring is extensive, it is usually more economical to renew the pump than attempt repairs.

Using new parts as necessary, reassemble pump using new seal and sealing rings by reversing disassembly procedure and observing the following: Install new seal in front cover with lip of seal towards inside of pump. Assemble the pump gears and bearings with relief groove side of bearings towards gears and with "V" notched side of bearings to intake side of pump. Tighten cover and pto pump body retaining cap screws to a torque of 45-50 ft.-lbs. and the gear retaining nut to a torque of 35-40 ft.-lbs. Bend tab of retaining washer against flats on nut. If pump intake pipe and filter were not moved from original position, assemble pump to adapter plate us-

ing new "O" rings and tighten cap screws to a torque of 20-25 ft.-lbs.

239. PUMP ADAPTER PLATE OVERALL. Refer to exploded view of the pump adapter plate and hydraulic valves in Fig. 216 or 217 and to appropriate following paragraph:

240. DIFFERENTIAL LOCK VALVE. To remove differential lock valve assembly from pump adapter plate, unscrew the valve rod guide (22–Fig. 216). Remove valve rod (24) and guide, then lift out the valve spool (28), regulating valve (27) and spring (26) assembly and withdraw valve return spring (29). Remove snap ring (25) from upper end of valve (28) to remove spring and regulating valve. Remove rod (24) from guide and remove the seal and "O" ring (21 and 23). Carefully clean and inspect all parts including valve bore in pump adapter plate. Renew any worn or damaged parts and reassemble as follows: Install new seal (21) in guide and place new "O" ring on guide. Lubricate guide and rod (24) and insert rod through guide. Install regulating valve (27),

small end first, on valve spool (28), then install spring (26) and snap ring (25). Insert spring (29) in valve bore, then install valve assembly and the actuating rod and guide assembly. Tighten guide securely.

241. PRESSURE RELIEF VALVE TW-10, TW-20 And TW-30. To remove hydraulic system relief valve on TW-10, TW-20 and TW-30 models, first remove differential lock valve as outlined in paragraph 240. The relief valve can then be unscrewed and removed from pump adapter plate.

To disassemble relief valve, refer to Fig. 217 and remove snap ring (1) and cap (2); take care not to lose any shims

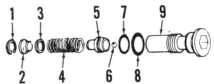

Fig. 217—Exploded view of hydraulic pressure relief valve used on TW-10, TW-20 and TW-30 models.

1. Snap ring	
2. Valve cap	
3. Shim	6. Ball
4. Valve spring	7. "O" ring
5. Ball retainer	8. "O" ring
	9. Relief valve body

(3) or valve ball (6). Carefully clean and inspect all parts. If valve body (9) is damaged beyond further use, a new relief valve assembly will have to be installed. Compare relief valve spring (4) with new spring. If any parts are replaced, system relief pressure should be checked and adjusted as outlined in paragraph 229.

Assemble valve using new "O" rings (7 and 8) and same shims (3) as removed unless necessary to adjust relief pressure. Lubricate "O" rings then reinstall valve in pump adapter plate. Tighten valve securely.

242. **All Other Models.** To remove the hydraulic system pressure relief valve assembly first remove the differential lock valve as outlined in paragraph 240. The relief valve can then be unscrewed and removed from pump adapter plate.

To disassemble relief valve, unscrew cap (30—Fig. 216) from valve body (38); take care not to lose any shims (31) or the valve ball (34). Carefully clean and inspect all parts. If the cap or valve body are damaged beyond further use, a new relief valve assembly must be installed; all other parts are serviced separately. Compare relief valve spring with new spring. If any parts are renewed, system relief pressure should be checked and adjusted as outlined in paragraph 229.

Assemble valve using new seal ring (36) and same shims (31) as removed unless necessary to adjust relief pressure. With valve assembled, install new "O" rings (37) on valve body, lubricate the "O" rings and reinstall valve in pump adapter plate. Tighten valve securely.

243. **FLOW CONTROL VALVE.** The flow control valve (20—Fig. 216) can be removed from bore in pump adapter plate after removing connector fitting (17) and spring (19). If valve is stuck in bore, it may be removed by threading a puller screw into internal threads in outer end of valve spool.

The flow control valve is a selective fit in bore in pump adapter plate and should have a clearance of 0.0005-0.0011 inch in bore. The valves and pump adapter plate are color coded for size identification; refer to the following chart:

VALVE BORE IDENTIFICATION
Color Code	Valve Bore I.D. [in.]
Blue/white	0.9376-0.9379
White	0.9379-0.9382
Blue	0.9382-0.9385
Yellow	0.9385-0.9388
Green	0.9388-0.9391

VALVE SPOOL IDENTIFICATION
Color Code	Valve Spool O.D. [in.]
Blue/white	0.9368-0.9371
White	0.9371-0.9374
Blue	0.9374-0.9377
Yellow	0.9377-0.9380
Green	0.9380-0.9383

Lubricate valve spool and insert in bore with hollow end first (tapped end out), then install spring and connector fitting with new "O" ring (18). Tighten connector fitting securely.

244. **VARIABLE FLOW VALVE.** To remove variable flow valve (7—Fig. 216), proceed as follows: Drive the roll pins (6 and 8) from each end of valve and remove the arms (3 and 15). Remove shims (14) if so equipped. Unscrew gland (13) and remove friction washer (10), lever (11) and spring washer (12); discard "O" ring (9). Push valve out towards outside of adapter plate and remove the two "O" rings (5) and back-up rings (4) from valve. Clean and inspect all parts and renew any excessively worn or damaged parts. Reassemble using new "O" rings and back-up rings as follows:

Install a new "O" ring (5) on each end of valve, then install back-up ring at outside of each "O" ring. Lubricate valve and bore and install valve with flow control notch up and long end of valve towards inside of adapter plate. Place the spring washer (12), lever (11) and friction washer (10) on gland and install new "O" ring (9) on gland shoulder. Install the gland over valve shaft and into adapter plate; tighten gland securely. Reinstall shims (14), if so equipped, unless arm (15), gland (13) and/or valve (7) have been renewed. If a new arm, gland and/or valve have been installed, install arm (15) without shims and check end play of valve. If end play is more than 0.012 inch, remove arm and install shims as necessary. Shims are available in thicknesses of 0.012 and 0.024 inch. Be sure valve does not bind after reinstalling arm.

LIFT COVER, CYLINDER AND CONTROL VALVE

245. **R&R LIFT COVER, CYLINDER AND VALVE ASSEMBLY.** Because of construction differences, refer to the appropriate following paragraphs for removal notes. Paragraphs 246 and following describe procedure for 8000, 8600, 9000 and 9600 models; paragraphs following 247 describe removal for 8700, 9700, TW-10, TW-20 and TW-30 models.

246. Proceed as follows to remove the lift cover (with cylinder and valve attached) from 8000, 8600, 9000 and 9600 models: Remove battery cover and disconnect battery ground cable. Unbolt and remove seat assembly from operator platform. Unbolt and remove both

fenders. Loosen set screws in console control lever handle and remove the handles. Unbolt console and lift off from over the control levers. Shut off fuel supply valve, disconnect fuel line at valve and disconnect fuel gage sender wire at tank. Drain fuel tank to lighten as necessary, then unbolt and lift tank from mounting plate. Unbolt and remove fuel tank mounting plate and cover plate from top of lift cover. Unbolt remote control hose bracket from rear end of lift cover and allow remote hoses to hang over right axle housing. Disconnect control linkage from remote control valves, lift cover and flow control valve. Remove the hydraulic system pressure tube and the pilot pressure tube connecting lift cover to hydraulic pump adapter plate. Disconnect pto control cable at lower end and unbolt control cable bracket. Unbolt control lever support and lift off the support and control levers with linkage and pto cable attached. Disconnect rear (trailer) light socket and disconnect lift rods from cross-shaft lift arms. Bolt lifting fixture or plate to the lift cover, then unbolt lift cover from rear axle center housing. Using a hoist, raise the lift cover, cylinder and valve assembly from rear axle center housing. Set assembly on work bench with front end down and support unit to keep it from tipping while performing service work on the cylinder, control valve and/or linkage.

To reinstall lift cover assembly, place new gasket on the lift cover, then with hoist attached so that lift cover is level, lower the unit onto rear axle center housing so that roller on draft control linkage enters draft control lever yoke and dowels in lift cover enter mating holes in rear axle center housing. Reinstall remote control hose bracket, control lever support with levers and linkage and the pto cable bracket. Install remaining lift cover retaining cap screws and tighten all cap screws to a torque of 95-130 ft.-lbs. If lift cylinder was removed from cover, tighten the four lift cylinder attaching cap screws to a torque of 200-240 ft.-lbs. at this time. Complete balance of reassembly by reversing removal procedure and observing the following: Tighten pilot pressure tube nuts to a torque of 20-25 ft.-lbs. Tighten the hollow bolt retaining system pressure line to lift cover to a torque of 20-25 ft.-lbs. and the nut at pump adapter plate end of pressure line to a torque of 50-55 ft.-lbs.

247. To remove the lift cover from 8700, 9700, TW-10, TW-20 and TW-30 models, first remove the cab or platform as follows: Remove screws retaining gearshift boot, raise boot, loosen gearshift lever clamps and detach gear-

shift levers. On models equipped with Dual Power, control pedal assembly and linkage should be removed to prevent damage. Disconnect and cap brake lines. Disconnect engine clutch linkage. Disconnect wires, hoses, tubing or linkage which will interfere with cab or platform removal. On cab equipped models, remove scuff plates in door openings, remove access plates and unscrew front cab mounting bolts being careful not to lose shims. Remove bolts securing front of platform on models so equipped. Unscrew bolts securing rear of cab or platform to rear axle housing and using a suitable hoist lift cab or platform away from tractor.

Disconnect tubes from remote valves and remove quick coupler bracket and tubes from lift cover. Remove pto control rod, lever and bracket assembly from lift cover on TW-10, TW-20 and TW-30 models. Detach lift rods from lift arms. Disconnect pressure line (15—Fig. 212), pilot pressure line (17) and override link (18). Bolt lifting fixture or plate to the lift cover, then unbolt lift cover, cylinder and valve assembly from rear axle center housing. Set assembly on work bench with front end down and support unit to prevent tipping while performing service work.

To reinstall lift cover assembly, place a new gasket on lift cover then with hoist attached so that lift cover is level, lower the unit onto rear axle center housing so that roller on draft control linkage enters draft control lever yoke and dowels in lift cover enter mating holes in rear axle center housing. Tighten lift cover retaining bolts to a torque of 95-130 ft.-lbs. on 8700-9700 models and to 140-170 ft.-lbs. on TW-10, TW-20 and TW-30 models. Complete remainder of reassembly by reversing disassembly procedure. If lift cylinder was removed from cover, tighten lift cylinder attaching cap screws to a torque of 200-240 ft.-lbs. on 8700-9700 models and to 250-310 ft.-lbs. on TW-10, TW-20 and TW-30 models. Tighten pilot pressure tube nuts to a torque of 20-25 ft.-lbs. Tighten hollow bolt retaining system pressure line to lift cover to a torque of 20-25 ft.-lbs. and the nut at pump adapter plate end of pressure line to a torque of 50-55 ft.-lbs. Tighten platform retaining bolts to a torque of 100-120 ft.-lbs. or cab retaining bolts to a torque of 200-240 ft.-lbs.

250. OVERHAUL CONTROL VALVE. With the lift cover removed as outlined in paragraph 246 or 247, remove the control valve assembly as follows: Remove the pilot pressure tube connecting sleeve (18—Fig. 221) and pressure sleeve (10). Remove the four cap screws (33) retaining lift cylinder to cover and remove the cylinder and control valve as an assembly. Unbolt control valve block from the lift cylinder. To disassemble control valve unit, refer to exploded view in Fig. 219 and proceed as follows:

Unscrew plug (17) and remove poppet spring (19), ball seat (20), ball (21), drop valve poppet (22) and poppet actuator (23). Unscrew plug (31) and remove spring type valve ball guide (29), check valve spring (28) and check valve ball (27). Unscrew control valve bushing (43) and remove bushing along with control valve (44), valve keeper (40), spring (45) and spring seat (46). Withdraw valve and keeper from bushing. Remove snap ring (47), spring seat and spring from bushing. Unscrew plug (39) and remove exhaust valve pressure spring (37), exhaust pressure valve (36), exhaust valve control spring (35) and exhaust control valve (34). Carefully clean and inspect all parts and renew any that are scored, excessively worn or damaged. If valve body (24) is worn or damaged beyond further use, a complete new valve assembly must be installed; all other parts are available separately. If necessary to renew control valve, insert a 1/8-inch diameter rod through hole in valve (44) and unscrew keeper (40). If keeper threads into new valve freely, either renew keeper or treat threads with Grade A "Loctite" on reassembly.

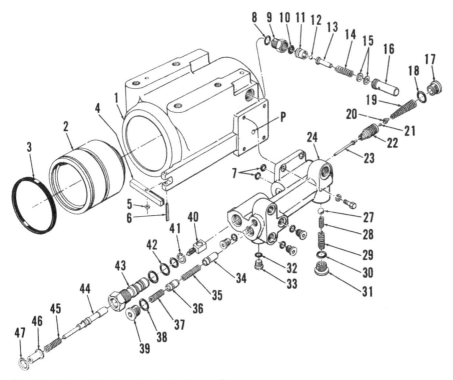

Fig. 219—Exploded view showing hydraulic lift cylinder and control valve assemblies. Cylinder and control valve are removed as a unit, then control valve body can be unbolted from cylinder. When in fully raised position, piston (2) skirt contacts pivot arm (4) moving control valve (44) to neutral position. Safety valve (items 8 through 16) limits pressure in lift cylinder to 2650-2750 psi, thus preventing damage due to shock loads.

P. Pressure port	13. Spring guide	28. Check valve spring	spring
1. Lift cylinder	14. Safety valve spring	29. Spring type guide	38. "O" ring
2. Piston	15. Shims	30. "O" ring	39. Plug
3. Piston ring	16. Valve cap	31. Plug	40. Control valve
4. Shut-off pivot arm	17. Plug	32. "O" rings	keeper
5. Washer	18. "O" ring	33. Passage plugs	41. Back-up ring
6. Roll pin	19. Drop poppet spring	34. Exhaust control	42. "O" rings
7. "O" rings	20. Spring & ball seat	valve	43. Control valve
8. "O" ring	21. Drop poppet ball	35. Exhaust valve	bushing
9. Safety valve body	22. Drop poppet valve	spring	44. Control valve spool
10. Seal ring	23. Actuator rod	36. Exhaust pressure	45. Control valve spring
11. Valve seat	24. Control valve body	valve	46. Spring seat
12. Valve ball	27. Check valve ball	37. Exhaust pressure	47. Snap ring

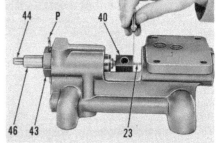

Fig. 220—Checking adjustment of valve keeper (40). Position control valve spool (44) so that a 1/8-inch diameter pin or drill bit (P) can be inserted through spool and bushing (43), then measure gap between keeper and actuator rod (23) for drop poppet. Control valve spring seat is (46).

Reassemble valve unit by reversing disassembly procedure and using all new "O" rings. Tightening torques are as follows: (All values are in ft.-lbs.)

Control valve bushing..........120-150
Drop poppet plug 80-100
Exhaust valve plug 60-70
Check ball plug 80-100

After reassembly, adjust control valve keeper as follows: Insert a 1/8-inch diameter pin or drill bit through hole in control valve bushing and valve as shown in Fig. 220. Then measure gap between keeper (40) and drop poppet actuator (23) with feeler gage. Turn keeper into or out of control valve to obtain a gap of 0.012-0.016 inch between keeper and actuator. When proper gap is obtained, remove pin or drill bit.

Assemble valve block to cylinder using new "O" rings and tighten cap screws to a torque of 20-25 ft.-lbs.

251. OVERHAUL SAFETY VALVE. With lift cylinder removed as outlined in paragraph 246 or 247, unscrew the safety valve body (9—Fig. 219) from cylinder and remove valve assembly from lift cylinder. With appropriate adapters, connect valve assembly to pressure test pump. Valve should pop open at 2650-2750 psi, and should not leak at below pop-off pressure. If valve leaks or pop-off pressure is not correct, disassemble valve by unscrewing cap (16) from body. Inspect valve ball and seat and compare free length of spring with that of new spring. Assemble valve using new seal (10) and other parts as necessary and recheck for leakage and pop-off pressure. Vary shims (15) as required to obtain 2650-2750 psi pop-off pressure. Shims are available in thicknesses of 0.010-0.014 and 0.022-0.026 inch. Reinstall valve assembly using new "O" ring and tighten securely.

252. LIFT CYLINDER AND PISTON. With lift cylinder removed as outlined in paragraph 246 or 247, remove the control valve assembly (see paragraph 250) and the control valve arm pivot roll pin (6—Fig. 219), arm (4) and washer (5). Using low air pressure applied at fluid passage (P) in cylinder, force piston from cylinder.

CAUTION: A sudden blast of air pressure will eject the piston at a dangerous speed.

Remove piston ring (3) and inspect piston and cylinder bore for scoring. Renew cylinder and/or piston if not suitable for further service. Install new piston ring, lubricate piston, seal and

cylinder with hydraulic fluid and install piston, closed end first, into cylinder.

253. LIFT COVER, LIFT SHAFT AND RELATED LINKAGE. With lift cover removed as outlined in paragraph 246 or 247 and with cylinder and control valve removed from cover, refer to exploded views in Figs. 221 and 222, then proceed as follows:

254. CONTROL LINKAGE. Remove snap ring (31—Fig. 221) from eccentric pin (29), snap ring (4) from control lever shaft (5) and clevis pin (68—Fig. 222) from selector shaft (24), then remove the spring loaded actuator rod with control lever (65), position control arm (30) and draft/position control guide (51) assembly from lift cover. If necessary to disassemble spring loaded actuator rod, loosen locknut (N) and unscrew hex rod (58) from yoke (57). Remove internal snap ring (59) from sleeve (63) and withdraw hex rod with spring (61) and bushings (60). Remove snap ring (62) from rod and remove the spring and bushings. To disassemble draft/position guide assembly, refer to Fig. 223 and drive out retaining pins (49) to remove draft roller (53) and pin (54) and the position control lever pin (52). Remove pins (50) to remove detent spring

mounting pins (45), spring (46) and spacer (47). Drive out pin (55) and remove yoke pin (56) and yoke (57) from guide (51).

To remove control lever shaft (5—Fig. 221), remove nut (14), lever (13), Woodruff key (12) and friction disc (11), then withdraw shaft from inside of cover and remove the two flat washers (6) and spring (7) from shaft.

To remove selector lever shaft (24), remove snap ring (17), lever (16) and Woodruff key (25), then withdraw shaft from inside cover.

The position control eccentric pin (29) can be removed after removing nut (27) and lockwasher (28) from outer end of pin and driving the retaining pin (32) from cover.

Reassemble linkage and reinstall in cover by reversing disassembly procedure and using Figs. 221, 222 and 223 as a guide. Length (L—Fig. 224) of the spring loaded actuator rod must be adjusted to 13-49/64 to 13-51/64 inches from center to center of pin holes as shown in Fig. 224. To adjust length, loosen locknut (N) and turn hex rod (58) in or out of yoke (57), then tighten locknut.

255. LIFT SHAFT, LIFT CYLINDER ARM AND BEARINGS. With

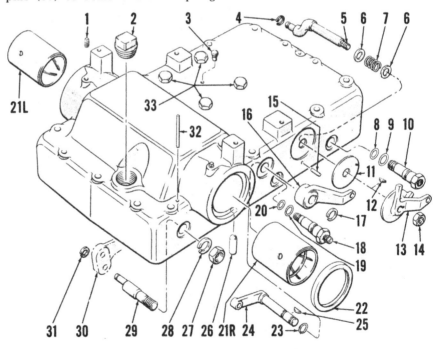

Fig. 221—Exploded view of hydraulic lift cover. Pressure sleeve (10) and pilot sleeve (18) thread into control valve body (24—Fig. 219). Control lever friction disc (11) keeps lever in desired position; spring (7) is adjusted by tightening nut (14). Cap screws (33) retain lift cylinder to cover.

1. Pipe plugs	10. Pressure sleeve	18. Pilot sleeve	26. Dowel pins
2. Filler plug	11. Friction disc	19. "O" ring	27. Locknut
3. Breather	12. Woodruff key	20. "O" ring	28. Lockwasher
4. Snap ring	13. Control lever	21. Lift shaft bushings	29. Position control
5. Control lever shaft	14. Friction adjusting	(L, left; R, right)	eccentric
6. Washers	nut	22. Lift shaft seal (right)	30. Position control arm
7. Friction disc spring	15. Override lever pin	23. "O" ring	31. Snap ring
8. "O" ring	16. Selector lever	24. Selector shaft & arm	32. Retaining pin
9. "O" ring	17. Snap ring	25. Woodruff key	33. Cap screws

control linkage removed as outlined in paragraph 254, remove cap screws (40—Fig. 222) from each end of lift shaft (36) and remove the lift arms (39). Inside diameter of left bushing (21L—Fig. 221) is larger than inside diameter of right bushing; therefore, shaft must be removed out left side of cover. Disengage snap rings (37—Fig. 222) at each side of lift cylinder arm (38) and bump shaft to left side of cover until the lift cylinder arm and snap rings can be removed from shaft, then withdraw shaft from cover.

The bushings (21L and 21R—Fig. 221) and shaft seals (22) can be removed from cover after removing lift shaft. Install new bushings with suitable driver, then install new seals with lips of seals inward.

NOTE: It is not necessary to remove lift shaft to renew seals; remove seals by drilling small hole in seal case, thread a metal screw welded to adapter in hole and pull seal with slide hammer.

When reinstalling lift shaft, note that index marks on shaft must be aligned with index mark on lift cylinder arm and index mark on each lift arm. Lift arms are interchangeable from right to left. Tighten cap screw (40—Fig. 222) in each end of lift shaft to a torque of 65-85 ft.-lbs.

256. LIFT COVER. To renew lift cover, remove the linkage and lift shaft from old cover as outlined in paragraph

254 and 255, then install the linkage and lift shaft in new cover. New cover is fitted with lift shaft bushings, right seal and dowel pins (26—Fig. 221).

LOWER LINK HANGERS TORSION BAR AND DRAFT CONTROL LINKAGE

257. TORSION BAR. To remove torsion bar (12—Fig. 225), proceed as follows: Remove cap screw (15) and lockwasher from right end of torsion bar. Unbolt left lower link pin and remove pin and lift link. Unbolt and remove torsion bar anchor (11), then withdraw bar from lift link hanger (3). If bar is broken or

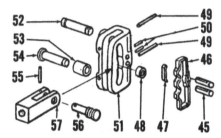

Fig. 223—Exploded view of draft/position control guide and detent mechanism. Head of pin (56) detents in notches formed in spring (46). Spring is mounted in split pins (45) which are retained in guide (51) by roll pins (50). Roller (53) engages yoke on draft control lever (18—Fig. 226).

45. Split pins	52. Position control pin
46. Detent spring	53. Roller
47. Spring spacer	54. Draft control pin
48. Roller for pin (56)	55. Roll pin
49. Roll pins	56. Yoke pin
50. Roll pins	57. Actuator rod yoke
51. Guide	

seized, remove snap ring (14) and flat washer (13) from right end of lift hanger (3), then use suitable drift to drive bar out to left side.

Install torsion bar by reversing removal procedure. Tighten anchor retaining cap screws to a torque of 50-60 ft.-lbs. and cap screw (15) in right end of bar to a torque of 20-25 ft.-lbs.

258. LOWER LINK HANGERS. To remove right lower link hanger (9—Fig. 225), remove right link pin and link from hanger, then remove retaining snap ring (10) and remove right hanger from right end of left hanger (3).

To remove left lower link hanger, first remove lift cover as outlined in paragraph 246 or 247 and remove torsion bar as in paragraph 257, then proceed as follows: Remove snap ring (10) and pull right hanger off right end of left hanger. Loosen locknut (N) and remove set screw (S) from draft control arm (7). Pull left lower link hanger out far enough to remove draft control arm and Woodruff key (4), then remove hanger from rear

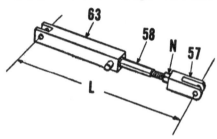

Fig. 224—Length (L) of assembled actuator rod must be adjusted to 13-49/64 to 13-51/64 inches. To adjust, loosen locknut (N) and turn hex rod (58) in yoke (57). Hollow actuator sleeve is (63).

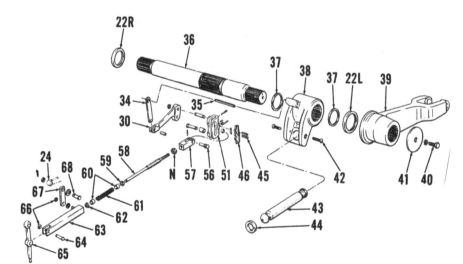

Fig. 222—Exploded view showing lift shaft and control linkage as removed from lift cover. Refer also to Fig. 223 for exploded view of selector detent mechanism and to Fig. 224 for adjustment of control valve actuator link. Bumper ring (44) on piston rod (43) prevents rod from striking cylinder if lift arms are raised with piston in forward (lowered) position.

N. Locknut	38. Lift cylinder arm	45. Detent spring pins	61. Spring
22. Lift shaft seals (L, left; R, right)	39. Lift arms	46. Detent spring	62. External snap ring
24. Selector shaft arm	40. Cap screws	51. Guide	63. Actuator sleeve
30. Position control arm	41. Retaining washers	56. Pin	64. Pin
34. Position control link	42. Round head groove pins	57. Yoke	65. Control valve lever
35. Spring pin	43. Lift piston rod	58. Hex rod	66. Snap rings
36. Lift shaft	44. Bumper ring	59. Internal snap ring	67. Selector arm link
37. Snap rings		60. Bushings	68. Clevis pin

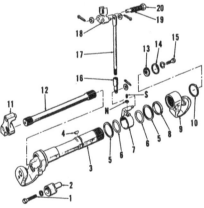

Fig. 225—Exploded view showing torsion bar, lower link hangers and draft control linkage. Refer to Fig. 226 for adjustment of draft control link (16 and 17), and to Fig. 208 for adjustment of hanger stop (2).

1. Cap screw	11. Torsion bar anchor
2. Hanger stop eccentric	12. Torsion bar
3. Left lower link hanger	13. Flat washer
4. Woodruff key	14. Snap ring
5. Washers	15. Cap screw
6. "O" rings	16. Link end
7. Draft control arm	17. Draft control link
8. Spacer	18. Draft control lever
9. Right lower link hanger	19. Eccentric pin
10. Snap ring	20. Locknut

axle center housing. Reverse removal procedure to install hanger.

259. DRAFT CONTROL LINKAGE.
To service draft control linkage shown in Fig. 225, lift cover and cylinder assembly must first be removed as outlined in paragraph 246 or 247. The eccentric pin (19), lever assembly (18) and link (16 and 17) can then be removed. To remove draft control arm (7), follow procedure outlined in paragraph 258 for removal of left lower link hanger (3).

When draft control linkage is reassembled and prior to reinstalling lift cover, the link (16 and 17) must be adjusted for proper length as follows: Turn eccentric pin (19) so that eccentric is in "up" position. Measure distance (D—Fig. 226) between center of dowel pin hole (H) in rear axle center housing to center of yoke in lever (18). If distance (D) does not measure 5-5/8 inches, back out eccentric pin until lever can be removed from pin and turn link rod (17) into link end (16) if measurement is less than 5-5/8 inches, or out of link end if measurement exceeds 5-5/8 inches.

REMOTE CONTROL VALVES

Remote control valves for operation of remote cylinders may be installed between manifold (1—Fig. 227) and top face of hydraulic system pump adapter plate. The valve spools have detents for lift, lowering and float positions. When in either lift or lowering detent position, a pop-off valve releases the spool detent before pressure in remote cylinder reaches hydraulic system relief pressure. The valve must be moved manually from float detent position. Flow of early

remote valves was determined by system volume. Later model valves have a flow control valve knob which may be used to vary the volume of flow to a remote cylinder.

260. CHECK AND ADJUST REMOTE CONTROL VALVE DETENT PRESSURE. Early type without flow control valve. With pressure gage installed at quick coupler remote cylin-

der connection, start engine and adjust engine speed to 700 rpm, then pressurize remote line in which gage is installed by moving control lever to raising or lowering position. Lever should return to neutral position as gage pressure reaches 2000-2300 psi. Check remaining valve (if equipped with two valves) by moving gage to proper hose quick coupler connection.

If control lever will not return to neutral within specified pressure reading, adjust detent pop-off pressure as follows: Using a short screwdriver, pry rubber plugs (1—Fig. 228) from valve end caps (3) and turn screws (12) in to increase, or out to decrease pressure.

261. CHECK AND ADJUST REMOTE CONTROL VALVE DETENT PRESSURE. Late type with flow control valve. With pressure gage installed at quick coupler remote cylinder connection, start engine and adjust engine speed to 1000 rpm on 8700 and 9700, 1500 rpm on TW-10, TW-20 and TW-30, or 700 rpm on all other models. Pressurize remote line in which gage is installed by moving control lever to raising or lowering position. Lever should return to neutral position as gage pressure reaches 2000-2300 psi on 8000, 9000, 8600 and 9600 models and 1900-2200 psi on all other models. If equipped with more than one remote valve, move gage to proper hose quick coupler connection and repeat test.

If control lever will not return to neutral within specified pressure reading, adjust detent pop-off pressure as follows: Remove end cap (27—Fig. 229). Unscrew detent guide (20) from valve spool (31) and remove detent assembly (15 through 26) as a unit. Using an Allen wrench, adjust screw (28) in to increase pop-off pressure. Reassemble detent unit into valve and retest.

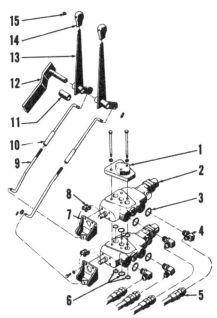

Fig. 227—Remote control valves (2) are mounted between manifold (1) and pump adapter plate top surface. Lever position is adjusted by shortening or lengthening links (9 and 10) to align levers with detent positions of valve. Early valve without adjustable flow control is shown.

1. Manifold
2. Remote control valves
3. "O" rings
4. Remote hose fittings
5. Remote hose
6. "O" rings
7. Bellcrank brackets
8. Connecting links
9. Lower control rod
10. Upper control rod
11. Spacer (single valve)
12. Lever bracket
13. Control levers
14. Lever knobs
15. Set screws

Fig. 228—Exploded view of remote control valve without adjustable flow control valve. Detent pop-off adjusting screw (12) threads into end of valve spool (23). When valve (14) pops-off, pressure lifts detent plunger (18) allowing valve to return to neutral position.

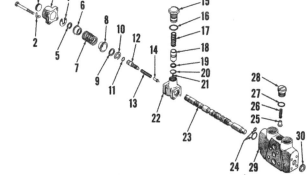

1. Rubber plug
2. Cap screw locks
3. End cap
4. Spool stop
5. Snap ring
6. Stop collar
7. Centering spring
8. Stop collar
9. Teflon back-up ring
10. Quad seal ring
11. "O" ring
12. Adjusting screw
13. Spring
14. Poppet valve
15. Plug
16. "O" ring
17. Detent spring
18. Detent plunger
19. "O" ring
20. Back-up ring
21. Detent ball
22. Detent housing
23. Valve spool
24. "O" rings
25. Check valve poppet
26. Spring
27. "O" ring
28. Plug
29. Valve body
30. Quad seal ring

Fig. 226—With eccentric pin (19) in "up" position, adjust link (17) by turning link into or out of link end (16) so that distance (D) between center of dowel hole (H) in rear axle center housing to center of draft control lever yoke is 5-5/8 inches. Control arm is (7).

262. ADJUST REMOTE CONTROL VALVE LEVER LINKAGE. The links (9 and 10—Fig. 227) may be shortened or lengthened as necessary to align lever with markings on console when remote control valve is in raise, lower or float detent position. Disconnect lower rod (9) from bellcrank on remote valve bracket (7) and turn rod into or out of upper rod (10) as necessary for proper lever alignment.

263. R&R REMOTE CONTROL VALVES. To remove the remote control valve, first remove access panel from right fender, then proceed as follows: Disconnect the remote hoses (5—Fig. 227) from valve fittings (4) and remove the fittings if desired. Plug all openings and disconnect control lever rods (9) from bellcranks on brackets (7). Back out the cap screws retaining manifold (1) and valves (2) to pump adapter plate and slide the valves and manifold unit outward. Separate the manifold and remote valves and cover openings in pump adapter plate.

Reinstall valves by reversing removal procedure and using all new "O" rings. Tighten the cap screws retaining manifold and valves to pump adapter plate to a torque of 20-25 ft.-lbs. Tighten the fitting jam nuts to a torque of 45-55 ft.-lbs.

264. OVERHAUL REMOTE VALVE UNIT. Remove linkage bracket (7—Fig. 227) then refer to Fig. 228 or Fig. 229 for guide to disassembly. Count turns required to remove adjusting screw (12—Fig. 228 or 28—Fig. 229), so setting may be duplicated during assembly.

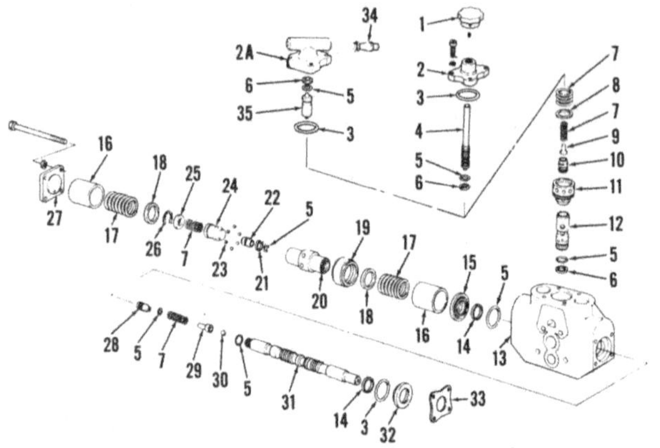

CAB

265. REMOVE AND REINSTALL. To remove and reinstall cab on all models so equipped, remove hood, grill panels, grill, precleaner, muffler and

Fig. 229—Exploded view of late remote control valve with adjustable flow control valve. Detent pop-off adjusting screw (28) threads into valve spool (31). Components (2A, 34 and 35) are used on 8700, 9700, TW-10, TW-20 and TW-30 models.

1. Knob
2. & 2A. Flow control cap
3. Seal
4. Adjuster valve
5. "O" ring
6. Back-up washer
7. Spring
8. Washer
9. Relief poppet
10. Flow valve piston
11. Flow valve sleeve
12. Flow valve guide
13. Valve body
14. Quad seal ring
15. Plate
16. Sleeve
17. Centering spring
18. Stop
19. Detent ring
20. Detent guide
21. Detent seal
22. Piston end
23. Detent ball (7)
24. Detent piston
25. Spacer
26. Snap ring
27. End cap
28. Detent adjusting screw
29. Valve seat
30. Adjusting valve seat
31. Valve spool
32. Valve plate
33. Retainer
34. Adjuster
35. Poppet

hood. On TW-30 models also remove cowl panel next to cab, air intake and turbo grill. Disconnect following components on all models: negative battery cable, a-c receiver/dehydrator, if so equipped, pto cable at axle housing, hydraulic lift control rod, system selector rod, remote control rods, flow control rod and remote cylinder flow control cables. Unhook differential lock link pedal linkage. Unplug main electrical harness and windshield washer harness. Note that washer wire is grey with yellow stripe on tractors with C prefixed serial numbers while tractors with A or B numbers have a green wire with black stripe.

Disconnect brake lines at master cylinder, throttle cable, tachometer cable and fuel shut off cable at fuel injection pump. Let cables hang from front of cab. Close heater valves and remove hoses. Mark and remove power steering hoses, brake lines and clutch linkage. Remove roof retaining bolts and

attach Ford spreader bar tool No. 2420 or similar lifting device to cab. Remove gear shift levers, Dual Power foot pedal and control rod. Remove scuff plates then lift floor mat and remove mounting bolt access plates and front mounting bolts.

NOTE: Front mounts may have shims between cab rails and insulators. Identify shims for proper reinstallation.

Remove any remaining components that will interfere with cab removal then remove rear mount bolts and raise cab using suitable hoist making sure cab doesn't bind.

To install cab reverse removal procedure. Torque front mounting bolts to 180-220 ft.-lbs. then tighten rear bolts to a torque of 200-240 ft.-lbs.

NOTE: Shims are placed on top of front mount insulators. Reconnect and adjust linkages and hoses as necessary. Purge air from power steering system.

General Torque Recommendations

Use the following torque *recommendations* as a guideline when a specification for a particular fastener is not available. In many cases manufacturers do not provide torque specifications, especially on older models.

Consider fastener condition carefully when referring to either a recommendation or a specification. If fastener reuse is appropriate, select the minimum value to account for fastener stretch. Softer fasteners or those securing softer materials, such as aluminum or cast iron, typically require less torque. In addition, lubricated or unusually long fasteners typically require less torque.

Determine fastener strength by referring to the grade mark on the bolt head. The higher the grade is, the stronger the fastener.

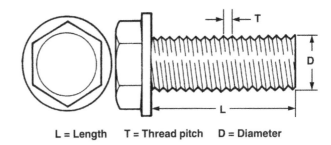

L = Length   T = Thread pitch   D = Diameter

Determine fastener size by measuring the thread diameter (D), fastener length (L) and thread pitch (T).

Size and Pitch	SAE grade 1 or 2 bolts	SAE grade 5 bolts	SAE grade 8 bolts
1/4—20	4-6 ft.-lbs.	6-10 ft.-lbs.	9-14 ft.-lbs.
1/4—28	5-7 ft.-lbs.	7-11 ft.-lbs.	10-16 ft.-lbs.
5/16—18	6-12 ft.-lbs.	8-19 ft.-lbs.	15-29 ft.-lbs.
5/16—24	9-13 ft.-lbs.	15-21 ft.-lbs.	19-33 ft.-lbs.
3/8—16	12-20 ft.-lbs.	19-33 ft.-lbs.	28-47 ft.-lbs.
3/8—24	17-25 ft.-lbs.	26-37 ft.-lbs.	36-53 ft.-lbs.
7/16—14	22-32 ft.-lbs.	31-54 ft.-lbs.	51-78 ft.-lbs.
7/16—20	27-36 ft.-lbs.	40-60 ft.-lbs.	58-84 ft.-lbs.
1/2—13	34-47 ft.-lbs.	56-78 ft.-lbs.	80-119 ft.-lbs.
1/2—20	41-59 ft.-lbs.	64-87 ft.-lbs.	89-129 ft.-lbs.
9/16—12	53-69 ft.-lbs.	69-114 ft.-lbs.	102-169 ft.-lbs.
9/16—18	60-79 ft.-lbs.	78-127 ft.-lbs.	115-185 ft.-lbs.
5/8—11	74-96 ft.-lbs.	112-154 ft.-lbs.	156-230 ft.-lbs.
5/8—18	82-110 ft.-lbs.	127-175 ft.-lbs.	178-287 ft.-lbs.
3/4—10	105-155 ft.-lbs.	165-257 ft.-lbs.	263-380 ft.-lbs.
3/4—16	130-180 ft.-lbs.	196-317 ft.-lbs.	309-448 ft.-lbs.
7/8—9	165-206 ft.-lbs.	290-382 ft.-lbs.	426-600 ft.-lbs.
7/8—14	185-230 ft.-lbs.	342-451 ft.-lbs.	492-665 ft.-lbs.
1—8	225-310 ft.-lbs.	441-587 ft.-lbs.	650-879 ft.-lbs.
1—14	252-345 ft.-lbs.	508-675 ft.-lbs.	742-1032 ft.-lbs.
1 1/8—7	330-480 ft.-lbs.	609-794 ft.-lbs.	860-1430 ft.-lbs.

NOTES